Terrestrial Ecosystem Research Infrastructures

Challenges and Opportunities

Terrestrial Ecosystem Research Infrastructures

Challenges and Opportunities

Edited by

Abad Chabbi and Henry W. Loescher

CRC Press
Taylor & Francis Group
Boca Raton London New York

CRC Press is an imprint of the
Taylor & Francis Group, an **informa** business

CRC Press
Taylor & Francis Group
6000 Broken Sound Parkway NW, Suite 300
Boca Raton, FL 33487-2742

First issued in paperback 2019

ISBN-13: 978-1-4987-5131-5 (hbk)
ISBN-13: 978-0-367-87576-3 (pbk)

Library of Congress Cataloging-in-Publication Data

Names: Chabbi, Abad, editor. | Loescher, Henry W., editor.
Title: Terrestrial ecosystem research infrastructures : challenges and opportunities / edited by Abad Chabbi and Henry W. Loescher.
Description: Boca Raton, FL : CRC Press, 2017. | Includes bibliographical references and index.
Identifiers: LCCN 2016028964 | ISBN 9781498751315 (hbk : alk. paper)
Subjects: LCSH: Biotic communities–Research. | Ecology–Simulation methods. | Ecological monitoring–Methodology.
Classification: LCC QH541.2 .T447 2017 | DDC 577.8/2–dc23 LC record available at https://lccn.loc.gov/2016028964

Visit the Taylor & Francis Web site at
http://www.taylorandfrancis.com

and the CRC Press Web site at
http://www.crcpress.com

Contents

Preface: Goals and Structure of This Book

The future of our planet faces rapid increases in population and, along with it, changes in global consumption patterns and the increased burden on our natural resources and the ecosystem services that they provide. As a result, we are seeing a shortage of these resources, such as energy sources and raw materials, agronomic capability and food production, and the loss or degradation of our soils and water supplies. At the same time, other resources more typically associated with the public commons are also undergoing long-term chronic changes, for example, increases in nitrogen oxides, carbon dioxide, and other pollutants in the atmosphere; climate change; and the loss of biodiversity globally. New discoveries in the natural world are being made almost daily, but we barely understand how it works without the added anthropogenic effects. Yet, we do not know the effects of these chronic, long-term drivers on how ecosystems function and their respective feedbacks to our consumptive demands—and perhaps more importantly, on the ecosystems that provide our food, habitats, fresh water, natural resources, conservation areas, and quality of life. Hence, there is an increased need for environmental scientists to work with other stakeholders to advance our ecological understanding and to provide informed decision-making tools in light of this changing world.

To understand how anthropogenic change affects ecosystems and their economies and the services they provide is a societal and scientific imperative. That we do not know how ecosystems (and their components) will respond to future anthropogenic changes begs the question of how best to approach this challenge. It is difficult (if not impossible) for more traditional research based on 1–5-year funding cycles to provide insights into the long-term, rapidly changing ecological process that can occur slowly over a long time period, or can happen stochastically or nonlinearly in both time and space. In other words, trying to use a magnifying glass to view a whole ecosystem. It is not impossible, but improbable. Rather, scientists have identified the need for long-term, consistent, multidimensional datasets and approaches (experimentation) that can serve to provide new insights in ecological processes that otherwise cannot be achieved by more traditional short-term funding cycles. Environmental research infrastructures (RIs) fulfill this need, but it takes the commitment of researchers, political will, agencies, ministries, and the like to realize such vision. While more and more of these environmental RIs are emerging globally, we wish to point out that they are not meant to supplant traditional research activities, but to augment and advance our science in ways that otherwise would not be possible.

There is no question that environmental RIs are costly. Governmental (tax payer) resources and the efforts from those involved are substantial. Hence, combining these large investments with the large societal and scientific imperatives, environmental RIs must be coordinated, and target specific needs. Environmental RIs cannot be everything to everyone. As such, we advocate that RIs must meet an overarching philosophy for ecological forecasting through (1) the iterative integration of theory, observation, experimentation, and modeling (this notion is discussed further in Chapters 1 and 2), and (2) developing new understandings using data among environmental RIs continentally and globally. To realize this latter point, the mitigation of the impacts of anthropogenic change requires advance warning and prognostic capability of environmental drivers, processes, and their interactions in order to develop the decision-making space to reduce the vulnerability of ecosystem goods and services.

Take, for example, our ability to assess the vulnerability of food production at economically relevant scales requires process-level data on natural and managed species assemblages (community ecology), plant ecophysiology and agronomic techniques (ecosystem experimentation), sustainability (ecosystem science), nutrient cycling and soil physics (microbial microbiology, pedology), water resources (ecohydrology), and social scientists and modelers. The ability to quantify the changes in the drivers of change also has to be included: land use change (landscape ecology), chemical climate (atmospheric chemistry), climate and meteorology (atmospheric science, bioclimatology, micrometeorology), and invasive species (biodiversity). Being able to leverage data and scientific expertise from multiple environmental RIs can bring value-added and new insights, as well as optimizing costs. Moreover, many, if not most of the, scientific discoveries made by other RIs in the fields of geodesy, physics, seismology, astronomy, etc., are those not originally planned for by the RI design, but from utilizing these large investments in new and novel ways.

The goal of this book is to bring together the latest state of the science on environmental RIs, scientific frontiers, the programmatic challenges RIs face in realizing their planned scientific scope, and new perspectives they bring to bear. From the environmental RI vantage point, we have organized the book into five sections: the rationale and challenges to address global change; new generation of controlled environments, field and modeling platforms; new tools and approaches to explore ecosystem processes; the challenges in data management and access, turning information into knowledge; and, finally new developments and perspectives in large-scale RIs.

Each chapter brings fresh ideas to this new, emerging scientific frontier of environmental RIs. This book presents academic viewpoints on the challenges for the design and implementation of sound RIs, which might be valuable to both academics and practitioners wishing to deepen their knowledge in the field of logistics operations and management. We offer a formal and heartfelt thank-you to all the authors for providing such collaborative insights, and for putting up with us during the editorial phase in producing this book.

Our goal for this book is that it would have broad application to a wide readership, such that it can (1) inform the nuanced need for environmental RIs and how different frontier science disciplines have addressed this challenge; (2) stimulate scientific thought and facilitate the needed cross-pollination of ideas; (3) provide insight into community-driven processes to further develop the integration and optimization of experimental, analytical, and modeling platforms; and (4) create the context to innovatively bring together researchers, academicians, the private sector, and decision-makers to advance our knowledge on one of (if not the) most critical issues facing our sustainable future.

We did not wish to present a description of environmental RIs, i.e., a typical cookbook approach toward overall RI design. That type of information can be accessed on webpages and other planning documents. Rather, we wished to present the new perspectives, challenges, and developments in crafting the frontier science and the respective environmental RIs. But this would not happen without a global vision that involves international collaboration, and the need to pool our knowledge together and share lessons learned in crafting large-scale research infrastructures (and the very nature of the research challenges being addressed). We recognize that global research challenges emanate from the interconnectedness of worldwide problems, issues, and implications. It is our hope and commitment to work together with all environmental RIs in advancing a common global ecological understanding for societal benefit.

The culture of shared ideas and how to use and operate these larger environmental RIs is developing concurrently with the efforts to build them. In the private sector, culture overrides strategy, but few acknowledge its role in establishing environmental RIs. Cultural legacies are a big challenge in our ability to realize large-scale environmental RIs. Unfortunately, we did not address this challenge in this book but wanted to say that the scientific imperative to advance our environmental literacy for societal benefit is far too great for past cultural legacies to impede progress, particularly at a time when impactful and collaborative engagement are most needed.

We greatly appreciate the superb editorial work and patience of Irma Shagla Britton and her staff. We extend a heartfelt thank-you to Drs. (alphabetically) Daniele Bailo, Pieter De Frenne, Isabelle Basile-Doelsch, Don Flaten, Alan Franzluebbers, Rick Hooper, Michel Loreau, Russ Monson, Stuart Phinn, Lou Pitelka, Lionel Ranjard, Cornelia Rumpel, Gregory Starr, and Anneke Zuiderwijk-van Eijk for their voluntary and editorial services that helped the authors to greatly improve the quality of the manuscripts and the focus of the book. This book would not have been written without decades of collegial interactions and community engagement with our peers, students and mentors, and our forward-thinking community that has advanced the concepts of environmental RIs for science and society.

Dr. habil. Abad Chabbi
Dr. Henry W. Loescher

Editors

Dr. habil. Abad Chabbi, director of research at the French National Institute for Agricultural Research (INRA), is a plant ecologist and soil biogeochemist. He worked at the Louisiana State University, Baton Rouge, Louisiana; the Faculty of Environmental Science in Cottbus, Germany; the University of Pierre & Marie Curie in Paris, France; and the INRA, where he has been leading the National Observatory for Environmental Research-Agro-Ecosystems, Biogeochemical Cycles and Biodiversity (www.soere-acbb.com) since 2009. His current research centers on understanding the link between soil carbon sequestration, nutrient availability and stoichiometry in the plant–soil system, biodiversity, and the influence that land use management and climate change may have on these dynamics.

During his career, Dr. Chabbi coordinated a number of international multidisciplinary projects, chaired and organized numerous international symposiums, and edited books and several special issues of leading international journals. He has presented a number of keynote lectures and seminars around the world. Currently, Dr. Chabbi is a member/expert of the Joint FAO/IAEA Division of Nuclear Techniques in Food and Agriculture, the Chinese Academy of Science, the Czech Science Foundation, the German Research Science Foundation (DFG), the Hercules Foundation (Hercules Stichting), Belgium; the Biotechnology and Biological Sciences Research Council (BBSRC), UK; the Natural Environment Research Council (NERC), UK; and the European Commission (DG Research & Innovation and DG Agriculture and Rural Development). Dr. Chabbi is the coordinator of two strategic European FP7 projects, "ExpeER" (www.expeeronline.eu) and "ANAEE" (www.anaee.com), and is deeply involved in ENVRIPlus, a cluster H2020 project. Since April 2013, he has also been leading the C2 component of the Ecosystems Task at Group on Earth Observations (GEO).

Dr. Henry W. Loescher's career has been at the nexus of science, engineering, and project development. Formally educated as an ecosystem scientist, he received his PhD and MSc degrees from the University of Florida (forest ecology) and undergraduate degrees from the State University of New York (SUNY) and Vermont State College (environmental science, agronomy, and applied science). His active research projects include determining the biotic and abiotic controls on ecosystem-level carbon, water, and energy balance across a range of spatial and temporal scales, and challenging ecological theory.

Dr. Loescher is the director of strategic projects in the National Ecological Observatory Network (NEON), a first-of-its-kind, continental-scale major research facility that integrates science, engineering, and project management. Dr. Loescher has led multiple NEON Project Science Teams and engineering efforts (instrumentation, mobile deployment, aquatic science, etc.), and he is currently pioneering domestic and international efforts to link large-scale environmental observatories and their "Big Data" to address current and future environmental problems facing society today. As such, he sits on numerous large-project, international advisory boards. Prior to his tenure at NEON, he was at Oregon State University, administrating the DOE AmeriFlux QA/QC Program.

Contributors

Nate Abramson
Biosphere 2
University of Arizona
Tucson, Arizona

Hendrik Albrecht
Plant Sciences (IBG-2)
Forschungszentrum Jülich
Jülich, Germany

Ari Asmi
Department of Physics
University of Helsinki
Helsinki, Finland

Lander Baeten
Faculty of Bioscience Engineering
Department of Forest and Water Management
Ghent University
Gontrode, Belgium

Greg A. Barron-Gafford
Biosphere 2
and
School of Geography and Development
University of Arizona
Tucson, Arizona

Marine Biget
National Center for Scientific Research
Université de Rennes 1
Rennes, France

Heye Bogena
Agrosphere Institute (IBG-3)
Forschungszentrum Jülich GmbH
Jülich, Germany

Magdalena Brus
Department of Physics
University of Helsinki
Helsinki, Finland

Aaron Bugaj
Biosphere 2
University of Arizona
Tucson, Arizona

Andreas Burkart
Plant Sciences (IBG-2)
Forschungszentrum Jülich
Jülich, Germany

M. Pilar Cendrero-Mateo
Plant Sciences (IBG-2)
Forschungszentrum Jülich
Jülich, Germany

Reinhart Ceulemans
Department of Biology
University of Antwerp
Wilrijk, Belgium

Abad Chabbi
Institut National de la Recherche Agronomique
Lusignan, France
and
Ecosys
Thiverval-Grignon, France

Yin Chen
EGI Foundation
Amsterdam, the Netherlands

Jon Chorover
Department of Soil, Water and Environmental Science
University of Arizona
Tucson, Arizona

Timothy F. Clancy
Terrestrial Ecosystem Research Network
The University of Queensland
Brisbane, Queensland, Australia

Scott L. Collins
Department of Biology
University of New Mexico
Albuquerque, New Mexico

Hans J. De Boeck
Department of Biology
University of Antwerp
Wilrijk, Belgium

Katerina Dontsova
Biosphere 2
University of Arizona
Tucson, Arizona

Alexis Dufresne
National Center for Scientific Research
Université de Rennes 1
Rennes, France

Benjamin. H. Ellert
Agriculture and Agri-Food Canada
Lethbridge, Alberta, Canada

Simone Gatzke
Plant Sciences (IBG-2)
Forschungszentrum Jülich
Jülich, Germany

Bert Gielen
Department of Biology
University of Antwerp
Wilrijk, Belgium

Alex Hardisty
School of Computer Science and Informatics
Cardiff University
Cardiff, United Kingdom

Joost van Haren
Biosphere 2
and
Honors College
University of Arizona
Tucson, Arizona

Andy Hector
Department of Plant Sciences
University of Oxford
Oxford, United Kingdom

Harrie-Jan Hendricks Franssen
Agrosphere Institute (IBG-3)
Forschungszentrum Jülich GmbH
Jülich, Germany

David Hudnut
Private Consultant
Lafayette, Colorado

Edward A. Hunt
Biosphere 2
University of Arizona
Tucson, Arizona

Travis E. Huxman
Center for Environmental Biology
University of California, Irvine
Irvine, California

Benedikt Janssen
Plant Sciences (IBG-2)
Forschungszentrum Jülich
Jülich, Germany

H. Henry Janzen
Agriculture and Agri-Food Canada
Lethbridge, Alberta, Canada

Keith Jeffery
British Geological Society
Natural Environmental Research Council
Edinburgh, Scotland, United Kingdom

Jiang Jiang
Laboratory of Soil and Water Conservation and Ecological Restoration
Nanjing Forestry University
Nanjing, Jiangsu, People's Republic of China

Lifen Jiang
Department of Microbiology and Plant Biology
University of Oklahoma
Norman, Oklahoma

Armin Jordan
Max Planck Institute for Biogeochemistry
Jena, Germany

Arnon Karnelli
Remote Sensing Laboratory
Ben-Gurion University of the Negev
Sede Boker Campus, Israel

Beat Keller
Plant Sciences (IBG-2)
Forschungszentrum Jülich
Jülich, Germany

Eugene F. Kelly
Soil and Crop Sciences
Colorado State University
Fort Collins, Colorado
and
National Ecological Observatory Network
Battelle
Boulder, Colorado

Niklas Körber
Plant Sciences (IBG-2)
Forschungszentrum Jülich
Jülich, Germany

Alan K. Knapp
Department of Biology
Colorado State University
Fort Collins, Colorado

Thorsten Kraska
Faculty of Agriculture
University of Bonn
Rheinbach, Germany

Russ Lea
EX-ite LLC
North Bend, Washington

Jinquan Li
Plant Sciences (IBG-2)
Forschungszentrum Jülich
Jülich, Germany

Junyi Liang
Department of Microbiology and Plant Biology
University of Oklahoma
Norman, Oklahoma

Henry W. Loescher
National Ecological Observatory Network
Battelle
and
Institute of Alpine and Arctic Research
University of Colorado
Boulder, Colorado

Denis Loustau
Institut National de la Recherche Agronomique
Bordeaux, France

Andrew J. Lowe
Centre for Conservation Science and Technology
and
Terrestrial Ecosystem Research Network
University of Adelaide
and
Department of Environment and Natural Resources
Science Resource Centre
Adelaide, South Australia, Australia

Yiqi Luo
Department of Microbiology and Plant Biology
University of Oklahoma
Norman, Oklahoma

Paul Martin
System and Network Engineering
University of Amsterdam
Amsterdam, the Netherlands

Shizue Matsubara
Plant Sciences (IBG-2)
Forschungszentrum Jülich
Jülich, Germany

Antonio A. Meira-Neto
Department of Hydrology and Water Resources
University of Arizona
Tucson, Arizona

Sophie Michon-Coudouel
National Center for Scientific Research
Université de Rennes 1
Rennes, France

Carsten Montzka
Agrosphere Institute (IBG-3)
Forschungszentrum Jülich GmbH
Jülich, Germany

Carsten W. Mueller
Technische Universität München
Munich, Germany

Onno Muller
Plant Sciences (IBG-2)
Forschungszentrum Jülich
Jülich, Germany

Mark Müller-Linow
Plant Sciences (IBG-2)
Forschungszentrum Jülich
Jülich, Germany

Stefano Nativi
Institute of Atmospheric Pollution Research
National Research Council of Italy
Sesto Fiorentino, Italy

Julia W. Neilson
Department of Soil, Water and Environmental Science
University of Arizona
Tucson, Arizona

Ivan Nijs
Department of Biology
University of Antwerp
Wilrijk, Belgium

Charles A. Nock
Faculty of Biology
and
Faculty of Environment and Natural Resources
University of Freiburg
Freiburg, Germany

Maarten Op de Beeck
Department of Biology
University of Antwerp
Wilrijk, Belgium

Luke A. Pangle
Department of Geosciences
Georgia State University
Atlanta, Georgia

Dario Papale
Department for Innovation in Biological Agro-Food and Forest Systems
University of Tuscia
and
EuroMediterranean Center on Climate Change
Viterbo, Italy

Roland Pieruschka
Plant Sciences (IBG-2)
Forschungszentrum Jülich
Jülich, Germany

Francisco Pinto
Plant Sciences (IBG-2)
Forschungszentrum Jülich
Jülich, Germany

William T. Pockman
Department of Biology
University of New Mexico
Albuquerque, New Mexico

Antonello Provenzale
Institute of Geosciences and Georesources
National Research Council of Italy
Pisa, Italy

Achim Quaiser
National Center for Scientific Research
Université de Rennes 1
Rennes, France

Uwe Rascher
Plant Sciences (IBG-2)
Forschungszentrum Jülich
Jülich, Germany

Laurent Remusat
Institute of Mineralogy, Materials Physics and Cosmochemistry
National Center for Scientific Research
and
Museum National d'Histoire Naturelle
Paris, France

Pablo Rischbeck
Plant Sciences (IBG-2)
Forschungszentrum Jülich
Jülich, Germany

Cornelia Rumpel
CNRS (ECOSYS and IEES)
Thiverval-Grignon, France

Michael Scherer-Lorenzen
Faculty of Biology
University of Freiburg
Freiburg, Germany

Anke Schickling
Plant Sciences (IBG-2)
Forschungszentrum Jülich
Jülich, Germany

Ulrich Schurr
Plant Sciences (IBG-2)
Forschungszentrum Jülich
Jülich, Germany

Aditi Sengupta
Biosphere 2
University of Arizona
Tucson, Arizona

Michael Sibayan
Biosphere 2
University of Arizona
Tucson, Arizona

Melinda D. Smith
Department of Biology
Colorado State University
Fort Collins, Colorado

Anita K. Smyth
TERN Eco-informatics
University of Adelaide
Adelaide, South Australia, Australia

Sanna Sorvari
Finnish Meteorological Institute
Helsinki, Finland

Alison Specht
School of Geography, Environmental Management and Planning
University of Queensland
St. Lucia, Australia
and
Centre for the Synthesis and Analysis of Biodiversity
Foundation for Research on Biodiversity
Aix-en-Provence, France

Angelina Steier
Plant Sciences (IBG-2)
Forschungszentrum Jülich
Jülich, Germany

Markus Stocker
MARUM
University of Bremen
Bremen, Germany

Nicole Thurgate
Australian Centre for Evolutionary Biology and Biodiversity
and
Terrestrial Ecosystem Research Network
University of Adelaide
and
Department of Environment and Natural Resources
Science Resource Centre
Adelaide, South Australia, Australia

Peter A. Troch
Biosphere 2
and
Department of Hydrology and Water Resources
University of Arizona
Tucson, Arizona

David J. Turner
TERN Eco-informatics
University of Adelaide
Adelaide, South Australia, Australia

Mari R. Tye
National Center for Atmospheric Research
Boulder, Colorado

Philippe Vandenkoornhuyse
National Center for Scientific Research
Université de Rennes 1
Rennes, France

Harry Vereecken
Agrosphere Institute (IBG-3)
Forschungszentrum Jülich
Jülich, Germany

Kris Verheyen
Faculty of Bioscience Engineering
Ghent University
Gontrode, Belgium

Till H. M. Volkmann
University of Arizona
Tucson, Arizona

Craig M. Walker
TERN Eco-informatics
University of Adelaide
Adelaide, South Australia, Australia

Michelle Watt
Plant Sciences (IBG-2)
Forschungszentrum Jülich
Jülich, Germany

Wolfgang W. Weisser
Department of Ecology and Ecosystem Management
Technische Universität München
Freising-Weihenstephan, Germany

Kevin R. Wilcox
Department of Microbiology and Plant Biology
University of Oklahoma
Norman, Oklahoma

Xubin Zeng
Department of Atmospheric Sciences
University of Arizona
Tucson, Arizona

Zhiming Zhao
System and Network Engineering
University of Amsterdam
Amsterdam, the Netherlands

Reviewers and Support

Daniele Bailo
Istituto Nazionale di Geofisica e Vulcanologia
EPOS
Roma, Italy

Isabelle Basile-Doelsch
Institut Universitaire de France
Aix-Marseille University
Marseille, France

Pieter De Frenne
Department of Forest and Water Management
and
Department of Plant Production
University of Gent
Melle, Belgium

Don Flaten
Department of Soil Science
University of Manitoba
Winnipeg, Manitoba, Canada

Alan J. Franzluebbers
Agricultural Research Service
U.S. Department of Agriculture
Raleigh, North Carolina

Leslie Goldman
National Ecological Observatory Network
Battelle
Boulder, Colorado

Rick Hooper
Consortium of Universities for the Advancement of Hydrologic Science, Inc.
Medford, Massachusetts

Ivan A. Janssens
Department of Biology
University of Antwerp
Wilrijk, Belgium

Michel Loreau
Station d'Ecologie Expérimentale du CNRS
Moulis, France

Russell Keith Monson
Department of Ecology and Evolutionary Biology
and
Laboratory of Tree Ring Research
University of Arizona
Tucson, Arizona

Stuart Phinn
School of Geography, Planning and Environmental Management
University of Queensland
Brisbane, Queensland, Australia

Lou Pitelka (retired)
Appalachian Environmental Laboratory
University of Maryland
Frostburg, Maryland

Lionel Ranjard
ECOLDUR
Institut National de la Recherche Agronomique
Dijon, France

Gregory Starr
Department of Biological Sciences
University of Alabama
Tuscaloosa, Alabama

Anneke Zuiderwijk-van Eijk
Engineering Systems and Services
Technical University of Delft
Delft, the Netherlands

Section I

Ecosystem Research Infrastructures

The Need to Address Global Change and Associated Challenges

1

Integrated Experimental Research Infrastructures: A Paradigm Shift to Face an Uncertain World and Innovate for Societal Benefit

Abad Chabbi, Henry W. Loescher, Mari R. Tye, and David Hudnut

CONTENTS

1.1 Introduction

The sustainability of our managed and natural environments is critical for the future of humankind and to ensure long-term quality of life. We also recognize that the environment and our economics are intrinsically linked (PCAST 2011, Holdren et al. 2014). The services provided by these ecosystems (e.g., food and fiber, fuel, biodiversity, biogeochemistry, water and energy, air quality) (MEA 2005, Heinz 2006, FAO 2011, FAO et al. 2012), however, are under threat due to dramatic changes in climate, chemical climate, land use and management, invasive species and species loss, and other human activities (Vitousek et al. 1997, NRC 2001, 2007). Understanding the responses of the biosphere to the anthropogenic drivers of environmental change is both an intellectual grand challenge and a practical necessity. Enhancements or disruptions of these services by human-caused environmental change could alter the fundamental trajectory of society over large parts of the world.

A wide range of biotic and physical processes link the biosphere to the geosphere, hydrosphere, and atmosphere. Despite this link, our understanding of the biosphere does not match our increasingly sophisticated understanding of Earth's physical and biogeochemical dynamics at the regional, continental, and global scales. Because many of these responses and feedbacks are large-scale processes, they need to be studied in an integrated way with standardized, coherent measurements for long periods of time (Doney and Schimel 2007). To date, the existing monitoring programs that collect data to meet regulatory, monitoring, and natural resource management objectives are not designed to address climate change and other new, complex, environmental challenges, that is, "Existing monitoring networks, while useful for many purposes, are not optimized for detecting the impacts of climate change on ecosystems" (Backlun et al. 2008) and "Fragmented federal investment in monitoring ecological change weakens national priorities" (PCAST 2011). While this fragmentation has been recognized, it is partially manifested by the need to maintain agencies and government directives and how resources (funding support) are managed (Vargas et al. 2016). But also, the societal and scientific imperatives to have a more integrated, synergistic, and cross-network approach have fostered the programmatic challenge of how best bring these networks and funding directives closer together.

It is important to note that there are different programmatic structures and rationales/missions for environmental research infrastructures (RIs). Some are principle investigator, bottom-up organizations with minimal governance, others that are more top-down with large institutional and organizational structures, and many in between (see Loescher et al., Chapter 2 for more discussion). RIs can also be characterized by their organizational structure and function: Ecological Observatory Networks (EONs) (centrally organized, primarily top-down institutions), Long Term Research Networks (LTRNs) (primarily Principle Investigator driven), and Coordinated and Distributed Experiments and Observations Networks (CDEOs) (see Peters et al. 2014 for more discussion). It is on this latter classification that we focus the discussion in the chapter. While we adopt the RI classification schema outlined in Peters et al. (2014), we also recognize that new RIs are emerging globally and this is an open area of discussion. Furthermore, each have their own nuances on how they meet their goals and mission, how they interact among RIs and stakeholders, and their funding and operational timelines.

Generally, RIs typically observe either the cause (e.g., climate, air pollution, or land cover change) or the ecological processes that are affected by the drivers, their feedbacks, and interactions (e.g., phenology of plants, changes in plant and animal population dynamics and distribution ranges, changes in biogeochemical cycles, feedbacks to the soil–plant–atmosphere water continuum), that is, one aspect or the other in the cause and effect paradigm. Prior to this decade, environmental RIs rarely provided long-term, consistent integrated observations that include the causes (drivers of change), affected processes (response variables), and their interactions. Only in the recent decade have we seen the advent of continental-scale RIs that embody the cause and effect paradigm in a planned and consistent manner. Examples include The National Ecological Observatory Network (NEON, www.neonscience.org, Chapter 2) and the Terrestrial Ecosystem Network (TERN, www.tern.org.au, Chapters 13 and 16) that embodies EON and CDEO functions in their organizational structures.

Broadly speaking, environmental sciences (ecology, ecosystem, remote sensing, regulatory, etc.) and CDEOs can be classified into two different and complementary experimental approaches:

1. *A noninvasive comparative design*, which may consist of observations and monitoring of different ecosystem types across an expected environmental gradient (cause) or range of responses (process-level information). This experimental approach may also adopt a substitution of time for space in classical chronosequence design (Cowles 1899, Warming 1985). This approach provides a powerful understanding of a range of conditions and their behavior in present time and provides important baseline understanding (e.g., what is the range of conditions today).
2. *An experimental manipulation design*, which may consist of the alteration of relevant forcing or process-level variables within or among ecosystem types. The range of manipulation is often chosen to represent future conditions, because the nature of disturbance is changing from more discrete, stochastic natural forms (e.g., storms, insect outbreaks, fire, flooding) to more chronic forms (e.g., increasing temperature and nitrogen deposition, CO_2 fertilization, population growth, changes in long-term precipitation patterns, changes in plant traits). This approach is particularly illustrative because we expect ecosystem processes will behave differently with chronic, long-term disturbance (Smith et al. 2009). Hence, this approach can provide important data on the unexpected nonlinearities in ecosystem behavior with chronic disturbance, for example, tipping points, process-level sensitivity, species migration, and community changes.

These two approaches are both necessary to advance our understanding of future world and if coordinated can provide unique insights into the causes, effects, and system feedbacks of anthropogenic change.

Used in combination, these two types of experimental design can provide inferences that can span large time and space scales, and is the core rationale for many experimental RIs. Also, when used in tandem, data from these two approaches can inform mechanistic, statistical, and predictive modeling frameworks to further be applied in natural resource management, societal planning, food security, and the like. Both approaches provide rich context to inform our gaps in knowledge, identify which new observations and models are needed, and to challenge theory (cf. Chapters 2 and 7).

There are good examples of both these experimental approaches in environmental RIs. The gradient approach (#1) has been used effectively to assess species richness, turnover and biodiversity (Soininen 2010), and edaphic controls on phosphorus and nitrogen in novel soils (Vitousek and Fields 2001, Vitousek 2004). Likewise, networks focused on large-scale manipulations have also advanced our understanding, for example, manipulating precipitation as a forcing variable on ecosystem responses (PRECIPNET, Weltzin et al. 2003); tree species' interactions with nitrogen deposition (NitroEurope, Sutton et al. 2007); and examining ecosystem-level carbon-use and water-use efficiencies and other biogeochemical interactions with forced Free-Air Carbon Dioxide Enrichment (FACE, Ainsworth and Long 2005). Due to the state of the science at the time of these studies and the pragmatism of funding requirements, many of these experimental studies were reductionist in their approach. Moreover, these studies have been typically restricted geographically (primarily to the United States or Europe), and their experimental designs or data collection methods across sites were not consistent. The associated biases in this restricted/reductionist approach were also apparent in global meta-analyses and did not reflect important patterns within specific biomes (Fraser et al., 2012). Indeed, Ehrlich (1997) also noted the limited capability of this type of reductionist study and called for a broader, more integrated approach among time and space scales and among ecosystems to provide new ecological insights on how these systems behave.

Only since the last decade has there been a widespread recognition of the need to integrate experiments with observations, models, theory, and across ecosystem, biomes and continents (Marshall et al. 2008, Peters et al. 2008, Robertson 2008, Schimel and Keller 2015). Moreover, such an integrated approach can also foster better predictive capabilities by iteratively challenging theory, observations, experiments, and models (and associated analytics) to advance scientific discovery and provide actionable results for societal benefit (see Section 1.2, and also discussed in Jiang et al. 2016, Loescher et al. 2016, Chapters 2 and 6) (Figure 1.1).

Implementing this integrated and novel approach adopts

- The cause and effect paradigm
- The ability to scale results in time and space and across taxa (from genes to biomes)

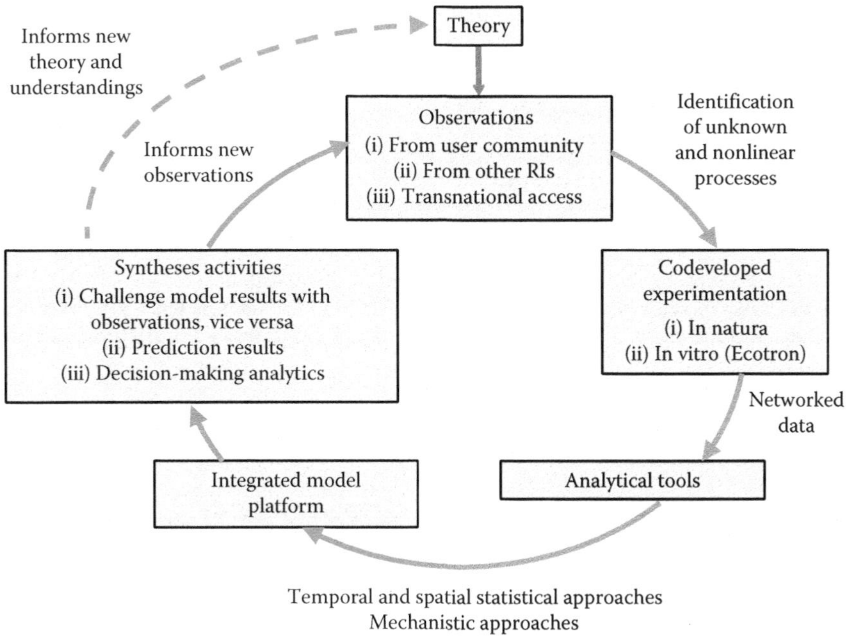

FIGURE 1.1
Conceptual diagram depicting the *overarching philosophy of this project* (and how the spiral management approach will be applied to the science), where (i) theory informs the types of hypotheses and questions to apply, and in turn the observations made, (ii) from observations, unknown and nonlinear behaviors are identified which need to be elucidated by (iii) the codevelopment/execution of experiments, from there, (iv) analytical and statistical data tools are used to reduce the amount of data into a synthetic understanding and further inform (v) the model platform, and (vi) synthesis activities, which, in turn, inform new theory, new observations, and new understandings.

- Specific anthropogenic causes (drivers) of change: changing climate, species loss, invasive species, and land-use change
- Specific ecological processes, feedbacks, and interactions of biodiversity, biogeochemistry, ecohydrology (water), infectious disease, and the transfer of mass and energy (environmental physics)
- A philosophical framework to enhance ecological predictive skill (Section 1.2)
- Ecosystem concepts of stability, resilience, adaptability, and transformability (Table 2.2)

The impact and understandings derived from this integrated approach has yet to be fully realized. With new emergent environmental RIs, we are only beginning to ask ecological and societal relevant questions that span across different continents (Vargas et al. 2016). A common example is how

El Niño, Northern Pacific, and other atmospheric–oceanic oscillations telecommunicate across large regions of the earth and control climate and, in turn, affect ecological processes. Cross-continental spatial and temporal scale synchrony in one ecological process can cascade/feedback to other processes and economies. For example, earlier and/or false spring leaf-out across continents (i.e., the United States) has implications for water usage, earlier midsummer droughts, (migratory) animal habitats, and agronomic economies (Allstadt et al. 2015). In another example, we know the genesis of synoptic drought conditions differs in Australia, China, Southwestern United States, and Europe, but the controls on this genesis are not well understood. This ability to ask questions across different continents is particularly salient given that ecological and societal imperatives transcend geopolitical barriers.

1.2 Predictive Ecology as a Core Premise for an Integrated Distributed Experimental Infrastructure

In order to build a sustainable future society, we must better understand the drivers of environmental change, the ecosystem processes they affect, and their interactions and feedbacks. We do this both for basic research (i.e., scientific discovery) and also to build a twenty-first-century bioeconomy (i.e., applied understanding and actionable results).

Key to this effort is developing our ability to forecast ecology or to develop the prognostic decision space (in engineering terms) that, in turn, provides the information to better preserve, conserve, improve, adapt, and mitigate our ecosystem services, for example, securing our food and water supply. Put another way, we need to understand and forecast how ecosystems will respond to current and future changes, including the ability to inform new management approaches and determine (potential) environmental tipping points, that is, management targets.

Without sufficient understanding of the drivers and mechanisms underlying these sensitive interdependencies and nonlinear behavior between ecosystems and the anthropogenic environment, scientists will be unable to assess the impacts, control the risks, provide decision-making tools, or potentially reap the benefits of anticipated large changes in ecosystem structure and function. Key benefits will include improved protection and management of biodiversity, increased production and sustainability of human-exploited ecosystems, greenhouse gas mitigation, and climate adaptation.

The goal is not to advocate an experimental RI for the sole purpose of experimentation alone. Rather, the core philosophy of experimentation is to elucidate as yet unknown processes and nonlinear behavior in ecosystem functions (e.g., tipping points, stochastic behavior, reorganization of species and ecosystem functions). This cannot be done in isolation from other scientific endeavors. We need key types of data and approaches that require theory, core observations, experiments, analytical capabilities, and modeling framework (Figure 1.1). To enhance predictive skill, we must start with theory that informs the type and signal:noise needed for new observation needs. This, in turn, informs the type of experiments needed to elucidate unknown (but suspected) behavior that, in turn, informs the analytical tools and modeling activities before recommencing the cycle to inform new understandings, theory, and potentially what new observations and experiments need to be performed (Figure 1.1). This cycle is revisited over and over again for specific questions, each time gaining new insights and forecast precision. This integrative philosophy is used very successfully to enhance the predictive skill in weather forecasting and epidemiology (Loescher et al. 2016; Chapter 2), and we advocate the same ecological forecasting philosophy toward building an environmental experiment RI. While this approach is still novel in ecology, it is becoming more broadly accepted worldwide, and other RIs are adopting the same approach (i.e., AnaEE, NEON, TERN).

1.3 Prototyping Integration of Experimental Platforms

Experimentation is by its very nature at the forefront of our knowledge, constantly testing the bounds of our understanding. Establishing infrastructure is typically executed for facilities, both physical and information (see Section 1.4), that are considered "baseline" or considered as a community standard so that the widest possible group of stakeholders can utilize its function—and not constantly be changing to accommodate the latest thinking. Here, the underlying problem is establishing an infrastructure that can be used for ecosystem/ecological experimentation where it has to (1) provide physical and information facilities that have to be standardized in approach and (2) provide dynamic capability to support the current frontier of ecological thought. At the same time, it cannot be everything to everyone.

There are many different kinds of experimentation, and the previous text assists in constraining the scientific scope to estimating nonlinear behavior of ecosystem processes due to (mostly) anthropogenic causes. But still, there have been in situ long-term experimentation on grassland diversity and productivity as those found at Cedar Creek, MN, United States (e.g., Tilman et al. 2001, 2006), and the use of large animal enclosures to challenge theories governing plant biodiversity (Carson et al. 2014, Carson and Schnitzer 2003)—to ex situ short-term experimentation with the monolith extract of whole ecosystems to manipulate drought conditions (Arnone et al. 2008, Malone et al. 2013), and everything in between. The challenge remains: how to best provide both the physical and information facilities for an environmental RI that meet the broadest possible scientific and societal needs? Put another way, experimental infrastructure, instrumentation, ecosystems studied, timescale of experiments, and research focus broadly vary. This large heterogeneity in approach brings great opportunities for advancing knowledge, but is also a large programmatic challenge in constraining scientific scope, optimizing construction and operational budgets, flow of information to respond/utilize the experimental results, and in engagement of stakeholder communities.

There are environmental RIs that provide a single experiment across continents, such as NutNet (ecosystem nutrient addition across the globe, www.nutnet.umn.edu/) and Drought Net (precipitation manipulation, wp.natsci.colostate.edu/droughtnet/). On the European Strategic Forum on Research Infrastructures (ESFRI) roadmap is the Analysis and Experimentation on Ecosystems (AnaEE, www.AnaEE.com): a European-wide effort to develop an integrated open-access experimental user facility—and is faced with the challenge of providing the broadest possible scientific scope while constraining construction and operational budgets. It is a classic and difficult cost–benefit activity pertaining to system engineering.

In addition to the challenge of balancing scope and budget, the ESFRI approach requires buy-in from EU member countries, which provides political and funding challenges. This approach of knitting EU country interests together is part of the strategy, so that only part of the RI construction costs comes from the European Commission in the form of structural funds. The societal needs from partner countries provide the rationale to support the remaining resources needed to fully construct and operate an ESFRI RI, that is, AnaEE. This approach also justifies the in-country support of existing in-country facilities and can contribute to the overall experimental RI. The scope of these in-country contributing facilities has to be overlaid and meshed with the overall scope of the RI, providing additional challenges to define and refine overall scope.

In the case of AnaEE, small-scale end-to-end prototype efforts (from physical infrastructure to information that can be used) are called for to better (1) inform overall scope; (2) inform the needed construction activities to inform and optimize operational needs; (3) develop a dynamic framework for the ongoing integration and optimization of experimental, analytical, and modeling platforms through community-driven processes; (4) inform a sustainable business model (operational model) that can innovatively bring together researchers, academicians, private sector, and planners and decision-makers; and (5) build trust and understanding among stakeholder communities.

FIGURE 1.2
The need for robust lessons learned from prototypes is not a linear process, rather a cyclic workflow that is an ongoing dynamic process, and each aspect is revisited often to assure innovation, cross talk among participants occur, and to optimize efficiencies. This cyclic workflow is a high-level management criteria and maps to all aspects of prototype development.

These prototype efforts also adopt a spiral workflow (Figure 1.2), in that it is management's responsibility to continually iterate design and implementation choices based on lessons learned. In this way, our ability to truly (1) implement a philosophy that integrates theory, observations, experiments, models, and analytics as a truly innovative, transformative, and novel approach and (2) inform the programmatic and technical constraints, schedule and budget, and risks needed to operate key physical experimental infrastructure and improve its scientific capability to tackle grand challenge questions across a range of environmental conditions and ensure the efficiency of data flow. The lessons-learned approach sharpens and defines scope and informs business plans (operational plan). Spiral workflows have been used successfully in many corporate workplaces, but is novel when building RIs and in the environmental sciences.

For example, in the AnaEE, Preparatory Phase has identified in situ (i.e., *in natura*) and ex situ (i.e., in vitro or Ecotron) experimental platforms among several partner countries. Prototype workflows facilitate the links among scientific requirements, experimental design, and experimental data from these platforms to modeling solutions that can run near- real-time simulations that can manage the means to control the experimental protocols and validate flow of information from site(s) to the final product. These prototype efforts also assist in understanding the management and resources needed to integrate such experimental platforms at larger scales, but the challenge still remains in understanding which platforms provide the greatest scientific impact, which platforms will be supported by in-country resources, and how they will be knitted together to address science and societal questions at continental scale. Science and societal questions, in this case, could be considered two sides of the same entity. Their mutual integration in this work drives and is driven by the spiral workflow concept.

1.4 Integrated e-Infrastructure for Efficient Data-Centric Knowledge

Everyone recognizes the need to model our understanding of the system behavior that is elucidated by experiments (Figure 1.1). This activity is key in developing skill in our prognostic capability and providing decision-making tools. Researchers and decision-makers require actionable knowledge, for example, Dilling and Lemos (2011) and Briley et al. (2015). But getting raw data into information and actionable knowledge is not easy and is an inherent design of the RI (also see Stocker 2016; Chapter 15).

1.4.1 Acquisition and Dataflow

A distributed model to collect experimental data from the field or ex situ chambers across countries to be used centrally is a difficult proposition. If such a design is to be constructed from scratch with the same acquisition systems, the data structures, formats, and metadata formats, they can be more easily managed, but the challenge still remains on how to collect the data from the field to a centralized location for QA/QC and further processing. Typically, sending data from the field requires a number of strategies, in part based on the size of the data, and includes some combination of fiber communications, Internet, sneakernet, and satellite phone. Each method has its own nuances and costs to manage. If, as in the case with AnaEE, each distributed experimental location has its own data acquisition system (e.g., datalogging and on-site storage system), then additional complexity arises to develop data and metadata standards that facilitate the transport, storage, analytics, use, and accessibility of the data. There is no simple solution, and this is an active area of research.

Now that we have embraced the era of "Big Data" and given that the experimental data resides at a central repository, researchers will likely ask questions beyond the scope of the experimental data alone. In other words, being able to merge experimental data with other spatially explicit data, economic data, and other drivers of change, modeling environments, other analytical approaches, etc., to increase its sphere of inference. In order to do so, other data will have to be ingested by the researchers from other sources. The challenge then lies in keeping track of the myriads of large data sets and associated software and processing configurations that are used and in making them reproducible—a core premise of the scientific method. The usage of software packages to do this is not necessarily new to computing scientists but is very new to environmental science, and their use is becoming more widespread. Such software packages are Kepler (kepler-project.org), Taverna (www.taverna.org.uk/), BioVal (www.bioval.com), and other workflows, and more advanced functionalities are being developed and tailored specifically for these environmental science purposes. As with Section 1.2, these workflows are being prototyped, applied, and tested across a suite of RIs, and we expect their widespread application to become more routine in the coming years.

1.4.2 Informatics

Many lump much into the discussion about informatics. But there are a few key questions that informatics address for environmental data: (1) What is needed to use and reuse data? (2) How is data archived? (3) How is attribution, providence, and persistence of the data managed? In addition to informatics, there are other data-related functions that are needed by RIs that are best suited for computer scientists, that is, computing and archival systems architecture that facilitates the ease in accessing and moving data from one repository to another, and maintaining data integrity (when using compression software). There are very few standards for informatics, for example, spatial reference standard held by Open Geospatial Consortium (www.opengeospatial.org). Currently, there are numerous, cross-pollinated community forums where these standards and best community approaches are being derived, for example, EarthCube (earthcube.org), Research Data Alliance (rd-alliance.org), Earth Science Information Partnerships (www.esipfed.org), EUdat (https://www.eudat.eu), and CoopEUS (www.neonscience.org/content/coopeus) to name a few. These community forums are advancing informatics in earth and environmental sciences but with no critical mass in a single organization.

There are two notable areas of discussion about informatics. First, when making data from one source interoperable with data from other sources (Table 1.1), the data nomenclature (controlled vocabularies) and the systematic logic that links the nomenclature systematically (semantic ontologies) are still developing. There are some controlled vocabularies and ontologies that are currently being used (e.g., Darwin Core, Ecological Metadata language), but by necessity researchers are quickly moving past these structures and the refresh rate of structures like Darwin Core is slow to catch-up. Hence, new systems are being developed and fostered by the research community. Another reason they are developing is that at the frontier of RI science, new types of data are being generated which are difficult to classify. For example, what to call and classify a sample of ground-up mosquitoes used to generically determine the presence or absence of zika virus?

Second, the collection, quality control, and curation of data are a scholarly activity that is an integral part of advancing our science and deserves to be recognized, that is, attribution. Authorship of data and its provenance (including where it came from and what it means, i.e., metadata) and documenting it as a citable reference can be done digitally via Persistent Object Identifiers (POIs), with Digital Object Identifiers (DOIs) being the most common form. DOIs are managed by International DOI Foundation (www.dio.org), and can be costly to maintain. Moreover, when an end user

TABLE 1.1

Rapid Pace of Large-Scale Environmental Global Changes Underscores the Value of Accessible Long-Term Data Sets

Interoperability Framework	
1. Aligning science questions and hypotheses, requirements, mission statements	• Mapping questions to "what must be done" • Defines joint science scope/knowledge gaps • Define interfaces among respective infrastructures
2. Traceability of measurements	• Use of recognized standards • Traceability to recognized standards, or first principles • Known and managed signal:noise • Managing QA/QC • Uncertainty budgets
3. Algorithms/procedures	• What is the algorithm or procedural process to create a data product? • Provides "consistent and compatible" data • Managed through intercomparisons • What are their relative uncertainties?
4. Informatics	• Standards—Data/metadata formats • Persistent identifiers/open-source • Discovery tools/portals • Ontologies, semantics, and controlled vocabularies

Note: The degree to which research infrastructures are truly interoperable is the degree to which these four elements are adopted by collaborative facilities; This table highlights the interoperability framework to make scientific data useful across research platforms (interoperability may look different for other research areas, for example, education and outreach). Real tasks can be applied to each of these focus areas. This is an active area of research. Signal:noise and uncertainty estimates must be part of this framework, because uncertainties need to be known a priori for data to be used in any Bayesian or data assimilation approach, that is, prognostic capability and ecological forecasting.

downloads a time series, it is a discrete block of time, but the original time series is more often than not being amended and the time series continues in the data repository. POIs also identify a discrete time series that can be reproduced. So the challenge becomes when to apply a POI?, for the entire, continually augmented time series, or an individual POI for each discrete time series that is downloaded. The latter option can become quite costly if an RI expects a lot of use from their data. As such, many environmental RIs are adopting a hybrid approach of developing an in-house POI system for a time series data set that is being continually augmented and a unique identifier (subset) for each download that assures the reproducibility of the data set.

1.4.3 Data Transformation, Analytics, and Modeling

Clearly, the knowledge and new understanding (Stocker 2016; Chapter 15) are derived from transforming raw data into higher-order data products, whether they be statistical analyses, mechanistic models, or used in Bayesian data assimilation for ecological forecasting (Jiang et al. 2016; Chapter 6). Basic transformations of conversion of raw units to calibrated quantities, statistical averaging, and spatial interpolations are commonplace in environmental RI data toolboxes. But based on funding pragmatism alone, Environmental RIs cannot provide all the analytic tools required (or expected) by all researchers. On the other hand, there are many researcher-initiated analytical tools that are being made freely available via the web, for example, R packages, Java and Python codes, and Github development. In parallel, there are also newly developed software(s) that can "wrap" research-level code to make them operate more robustly in the context of larger cyber-infrastructure architectures and make them more available to a broader user base, for example, D4Science. It is this frontier that seems to hold promise of "Big" RI data having more utility while also balancing fiscal constraints.

Models that explain our current understanding come and go with time and new understandings. Many models and model intercomparisons have been made and have provided new understandings, for example, Huntzinger et al. (2016). However, advancing ecological understanding is often hampered by the fragmentation of modeling resources in terms of scope, data requirements, architectures, and modeling approaches. Excellent and well-tested models are available, but these use diverse data formats and software platforms, often conflating data definitions, I/O procedures, and algorithms. RI software architecture requires up-to-date documentation, run-time testing such as tests of pre- and post-conditions, and unit tests; but these methods have rarely been applied when developing ecosystem models. Ecosystem science must therefore facilitate the development of efficient, parsimonious solutions based on articulation of tested modules. Many of these modeling solutions are closed systems, difficult to maintain even by their developers, that is, extend current modeling solutions by attempting to control and centralize development leads slow progress of modeling capabilities and little responsiveness to upcoming societal demands. An alternative is to develop frameworks, which capitalize on existing process modules, build new modeling solutions, and provide functionalities to perform simulations. Environmental RIs are also faced with what/which models to support. There are several initiatives are under development, and there is a strong case to make them interoperable in order to increase their respective capacities. Part of these initiatives are the development of (1) a common model architecture such that submodel routines or modules can be easily altered for a better comparative understanding the contribution of specific model components toward our overall understanding of system behavior.

Similar to the "wrappers" mentioned earlier, "adaptors" are built to allow individual model components or modules to be added or removed from the overall model within the workflow software, (2) standard to report model uncertainty is still nascent, and (3) a community model (or community configured model) that acts as a baseline with which other model behavior can be compared and contrasted to.

1.5 Innovation Capacity, New Market Opportunities, and Strengthen Competitiveness through Distributed Experimental Infrastructure

Generally, environmental RIs represent an untapped resource for innovation worldwide. The need to better integrate large-scale "Big Data" science into the private sector is articulated as a "must do" by stakeholders, governments, and the public (Pulwarty and Maia 2015). Yet opportunities to do so are limited and successes even less common. Often, success occurs by happenstance (see Figure 1.3a) rather than targeted consideration of the joint interests of public and private entities. Here, we change the current (low-success) paradigm of scientist-centric and initiated innovation by developing stakeholder-based needs first, then engaging scientist skillsets (Figure 1.3b). This model has been used for other public/private enterprises and knowledge transfer (e.g., medical industry, cancer research, agrochemical), but has not yet been applied for environmental/ecosystem science. Yet to advance the state of the science, more research is needed to ensure codevelopment of such integration, including lessons learned from previously successful and unsuccessful public/private collaborations; scalability in size, scope, and diversity of partners; how to plan for the extensibility of public/private integrations; how to overcome legal, political, cultural, and institutional barriers; and codevelopment of dynamic business models that can accommodate change.

The innovative merits of experimentation RI are as deep as they are broad. But the utility of integrating environmental RI with the private sector has yet to be fully realized. Academicians, modelers, data scientists, etc., recognize the importance of "exporting" their science to private industries but lack a venue to do so that is fair, equitable, and objective. Further, the opportunities to advertise advances in the science and engage the attention of private industries are limited. That is, it is not clear to either party *how* experimental environmental RI can be used to benefit and meet the needs of the private sector. To date, the need for this integrative imperative has been acknowledged, but the vision and opportunity to integrate science, academia, and the private sector have not occurred within the experimental environmental realm. Such integration could benefit traditional sectors such

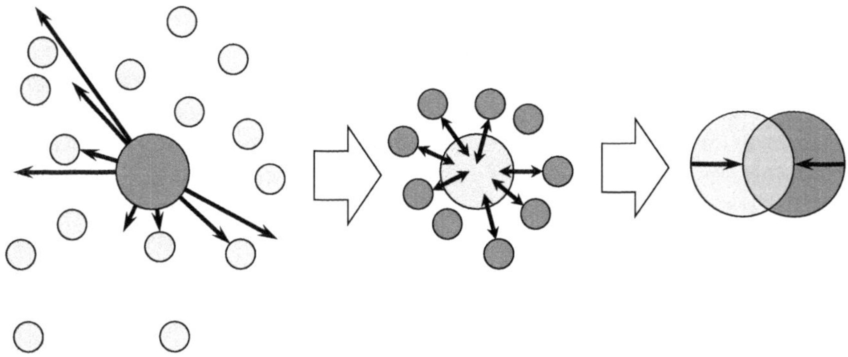

FIGURE 1.3
Conceptual diagram of how the dynamic to engage public/private partnerships will change from the current paradigm. Academicians and researchers already often work closely (shown in dark gray), while the private sector (e.g., planners, government agencies, small-to-medium-size enterprises, and decision-makers) does not have formal established interactions (in light gray). (a) Depicts the current paradigm where much of the interaction stems from academia outward to engage private interests (represented by arrows). Efforts vary in scale and focus, with correspondence to private interest often unplanned and serendipitous. (b) Are current efforts to shift the current paradigm by researching and codeveloping the strategic dynamic that emanates from the private sector, bringing academic partners closer, and specifically targeting core functions and products desired by the private sector and deliverable by academia. We note that (c) represents a desirable sustainable model from longer-term public/private partnerships, which includes formalized interconnectivity, and joint collaborations occur within an overlapping and trusted structure.

as the agronomic economy, rural and urban planners, high-impact weather mitigation natural resource managers, environment regulation, and supporting innovative technologies. In a time of limited public funding, the factors contributing to managing these sectors are just a subset of the plethora of competing needs being weighed by planners who need to prioritize the use of scarce resources (funding and otherwise). While decision-makers may be aware of environmental/ecosystem science advances and likely benefits of inclusion in their policies or projects, they often lack the information to be able to quantify those benefits and make informed choices.

Emergent economies, for example, risk and resilience management, federated data services ("Big Data," Future Earth), and food security, also provide novel areas for innovative research and collaboration. For instance, traditional flood-risk mitigation policies tend to favor solutions that require lower capital investments. While this approach is more affordable initially, the resultant solutions—for example, resistive barriers, such as levees—often prove to be not only unsustainable but also exacerbate the consequences to society and ecosystems when failure does occur (Adger et al. 2012, Tye et al. 2015). Enhancing decision-makers' access to integrated knowledge of changing hazards, exposure, and vulnerability as well as the benefits of ecosystem services to regional biodiversity, improved habitat protection and conservation, and potential mental and physical health benefits is an obvious step toward enhanced resilience. We define resilience as systems that recover to a stable state after a disruptive event that led to systematic failure (Tye 2015). Thus, a resilient system balances ecological, economic, and societal factors and facilitates a "graceful failure" and subsequent recovery.

There are countless position papers that call for the integration of public/private innovation partnerships for policy and decision-making (e.g., Dilling and Lemos 2011, The Royal Society 2014, Tye et al. 2015). And while this is an imperative, we choose to take smaller, more intentional steps toward building this new economy. As such, we have identified there are a few early adopters who seem to be natural partners in the initiation and propagation of these types of public/private partnerships.

Insurance and reinsurance companies are increasingly exposed to economic impacts from weather and climate extremes (Munich Re 2015). There are a variety of reasons for these increases, including societal changes and climate variability and change. Increasingly urbanized societies in vulnerable locations and an associated loss of resiliency have contributed substantially to the trends (Donner and Rodríguez 2008, Flood and Cahoon 2011, Sweet and Marra 2014), and this will continue. Population increases also bring stresses on food and water supplies, which are more likely to collapse under climate extremes such as droughts (Vörösmarty et al. 2000, Rosegrant and Cline 2003), and the built nature of cities amplifies climate variations such as heat waves and intense rain events (Coutts et al. 2007, Rosenzweig et al. 2011).

To date, the analyses used by re/insurance companies to evaluate their exposure rely on the integration of theory–model–observations to advance prognostic capability and evaluation of exposure. Underlying assumptions are based on recent observations, current climate, and other known drivers of changes in ecological processes. These assumptions are not expected to hold true for future conditions given chronic and long-term changes in the environment (e.g., nitrogen deposition, increases in population, temperature, and CO_2) and increases in the frequency and severity of weather and climatic events. The only way to glean insight into the impacts from, and likely future evolution of, these changes is through Experimental RI and integration of results with climate model output. In addition, changing the status quo from one of *risk response* to *risk mitigation* requires an integrated approach that balances the expertise of financial, social, environmental, and academic partners. Hence, codesigned and codeveloped experimental data are of benefit to re/insurance institutions to reduce portfolio exposure and advance socio-ecological–socioeconomic resilience through integrated land-use planning.

1. *Agronomy and Agro-Business*: Food security is becoming more paramount with each passing day (Whitacre et al. 2010, FAO 2011, FAO et al. 2012, Chavez et al. 2015). Traditional agronomic experimental

designs to test crop yield, water-use efficiency, genetics, and management techniques are a natural fit with the experimental approaches found within the RI. Traditional process–based models to manage natural resources, for example, crops, land erosion, and water use, also rely on theoretical and observation-based models to generate estimates of risk. However, the assumptions guiding these models do not always match reality and have limited opportunity to evaluate feedbacks between the agronomic and simulated output. With a focus on a different factor, the need for integrated planning to manage evolving responses to anthropogenic changes is similar to that of the re/insurance industry.

2. *Sensor/instrumentation companies* have a history of working with academicians to advance innovation, and hence are natural partners. There is an identified and immediate need to automate the linkage between observations and data flow activities (e.g., flow of data from sensor to quality assurance and quality control algorithms) to an ensemble of high-level data products and modeling activities. This integrated workflow for environmental RIs (novel sensor design, sensor-to-final data product) is a frontier and a large opportunity for industry partnerships. But there is also a longer-term vision of likely new discoveries and future types of sensing technologies required to manage and balance natural resources, societal well-being, and science. Partnerships with Experimental RI potentially offer explicit climate change scenarios, introduction of (invasive) species, broader comparative capability, diverse modeling platforms, effect data flow, common standards, informatics, and the like, to enhance the production of results and competitive advantage.

We have also identified legal barriers to developing public/private partnerships. For instance, intellectual property rights (IPRs) and data sovereignty issues arise when working across geopolitical borders. Legal frameworks have been developed separately by the public and private communities, respectively, to manage some of these issues. There are active community forums where these discussions and frameworks are developed, for example, RDA, DataOne, OGC, Earth System Integration Partnership, and others. However, there are only a few examples where these frameworks have been codeveloped, for example, National Center for Atmospheric Research's Engineering for Extreme Climate Partnerships (ECEP, Tye et al. 2015) and some relevant University Technical and Innovation parks. The novelty lies with public/private partnerships where competitive advantage and periods of data propriety become negotiable. On one hand, public funding mandates open access to the data. On the other hand, private enterprises wish to maintain a competitive advantage, which implies data propriety. A proven compromise entails "value-added" analytics applied to the openly available data. But the community using the research infrastructures determine the shared goals, culture, and common vision for the multidisciplinary data integration. But the advantage is really only competitive for a few years at a time, for example, 2–3 years, after which the analytics can also become public. This approach has been adopted within ECEP, whereby collaborative research is carried out between academia and the private entity and a moratorium placed on data sharing until after journal articles have been published. The result is a net win for both research and industry, facilitating scientific advances and short-term competitive advantage, and demonstrating that the private entity is at the forefront of technology. Hence, to maintain the corporate advantage and to maintain at the competitive forefront, development of new analytics becomes a long-term, sustainable partnership. This becomes a manageable task where research activities can be codeveloped and roadmapped a priori and also implies the need for new business models for both public and private enterprises to take advantage of such partnerships (Figure 1.4).

This is not to say that there are not both institutional and cultural barriers that still exist. Public/private partnerships are still nascent in the environmental sciences and require a cultural shift for implementation. Building

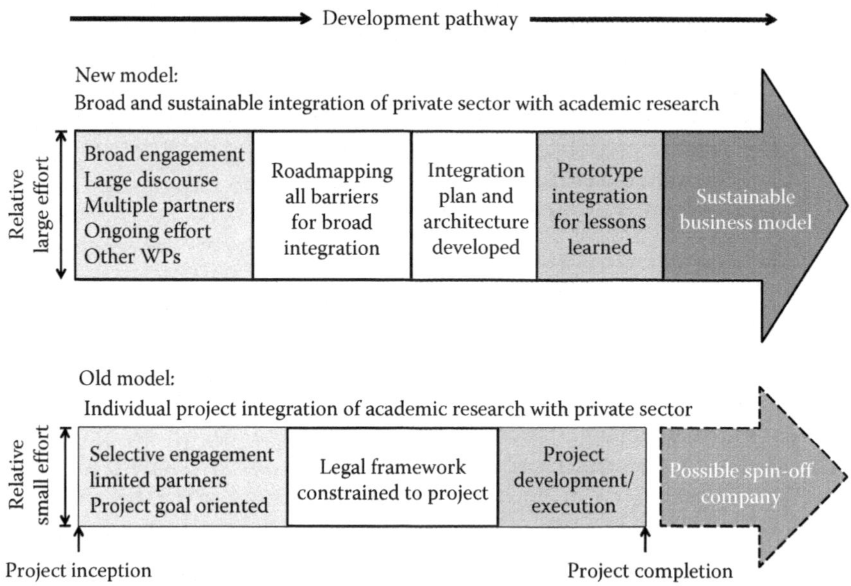

FIGURE 1.4
Here, we change the current (low-success) paradigm of scientist-centric and initiated innovation by developing stakeholder-based needs first, then engaging scientist skillsets (i.e., Figure 1.3). Moreover, we identify barriers to implementation, required architecture, and prototype integration into a sustainable development pathway. This model has been used for other public/private enterprises and knowledge transfer (e.g., medical industry, cancer research, agrochemical), but has not yet been applied for environmental/ecosystem science.

cultural capital determines and disseminates the shared goals and common vision for the multidisciplinary data integration. It is about changing discipline cultures and working toward community building, which generates trust, sharing and providing data, and constructing bridges between experts in different fields. Multidisciplinary community building is a long-term effort as cultures, languages, and approaches are quite different among public and private enterprises, as well as across disciplines. Developing a shared culture does not happen overnight and requires the codevelopment of joint long-term goals and the activities to foster a cultural change, for example, trainings, joint strategic planned efforts, and building new cohorts of stakeholders and early career users. However, the end goal is one of a collaborative community that can work independently, yet balance the needs and expertise of other sectors.

1.6 Future Directions in Working Internationally

Each country designs its stimulus (research) infrastructure to address the needs of its own citizens and better societal benefit. Yet the increasing interconnectedness of the ecology of the planet and our economies loudly calls for better integration of our ecological understanding and the overarching fabric of our global society. We, the environmental RIs, need to demonstrate the need for global integration and not just meet national priorities. The societal discourse and awareness of this global interconnectedness is building, for example, 2015 Paris Climate Talks (www.green-alliance.org.uk/paris2015.php). We are not only in the age of "Big Data," but also how are we going to use "Big Data" to better a global society.

Data from environmental RI play a role in advancing this imperative. The voluntary organization, Group on Earth Observation (GEO, www.earthobservations.org) calls for governmental and "a future wherein decisions and actions for the benefit of humankind are informed by coordinated, comprehensive and sustained Earth observations and information" consists of government and other nongovernment organizations. Top-down efforts like GEO help bring "Big Data" together to address this imperative. It may also seem like environmental RIs are top-down efforts, which is partially true in how they are constructed and funded. Their success, however, will be based on how well their data are used by—and their facilities can adapt to—the changing needs of the stakeholder and user communities. Their success will not solely be judged on how well they meet large-scale government agreements.

Grassroots efforts of environmental RI managers and global thought-leaders recognize the yet untapped scientific potential of virtually linking these infrastructures (Tim Clancy, personal communication). Coordination among international RIs provides added value to the previously made investments worldwide (e.g., physical designs, instrumentation, and human resources, lessons learned) and also leverages the shared scientific and intellectual capital that has already gone into the several RIs, namely, AnaEE, NEON, TERN, Integrated Carbon Observing System (www.icos-ri.eu), LTER (www.lternet.edu), Chinese Ecosystem Research Network (www.cern.ac.cn/0index/index.asp), International LTER (www.ilternet.edu), and many other key in-country national environmental facilities. The bottom-up coordination among these organizations currently revolves around adopting the integrated philosophy of ecological forecasting (Figure 1.1), the interoperability framework (Table 1.1), and building the stakeholder communities to use these environmental RIs as "new instruments."

Each of the environmental RIs is in a different stage of their own development and culture. Working together brings about much more of each of their individual strengths rather than shortcomings. Each RI is unique, with their own unique set of programmatic problems to solve. Lessons learned from one RI is often applicable to another. Some solutions transfer to other RIs, others in part, and others not at all. That said, there are also many countries that cannot afford large-scale environmental RIs, but their inclusion in the global dialog and participation is also paramount to success. Again, efforts at the RI level need to be inclusive and engage less fortunate researchers and countries to broaden participation and our global understanding of the drivers and processes of environmental change. Lastly, the open and collegial international communications and collaborations need more than ever to be reinforced, which will allow us to have ongoing lessons learned. The point is that building RIs is a frontier science, and we are already learning from each other of the potential pitfalls, risk management, and how to best accelerate development and user engagement.

Acknowledgments

This work was supported and benefited from the European Commission through the FP7 projects ExpeER (Experimentation in Ecosystem Research, Grant Agreement Number 262,060) and AnaEE (Analysis and Experimentation in Ecosystems, Grant Agreement Number 312690). The authors also acknowledge the National Science Foundation (NSF) for ongoing support. The National Ecological Observatory Network (NEON) is a project sponsored by the NSF and managed under cooperative support agreement (EF1029808) by Battelle. Any opinions, findings, and conclusions or recommendations expressed in this chapter are those of the authors and do not necessarily reflect the views of our sponsors. This chapter would not have been written without decades of collegial interactions and community engagement with our peers and mentors.

References

Adger, W. N., J. Barnett, K. Brown, N. Marshall, K. O'Brien, 2012. Cultural dimensions of climate change impacts and adaptation. *Nature Climate Change*, 3, 112–117. doi:10.1038/nclimate1666.

Ainsworth, E., S. Long, 2005. What have we learned from 15 years of free-air CO_2 enrichment (FACE)? *New Phytologist*, 165(2), 351–371. doi:10.1111/j.1469-8137.2004.01224.x.

Allstadt, A. J., S. J. Vavrus, P. J. Heglund, A. M. Thogmartin, W. E. Pidgeon, V. C. Radeloff, 2015. Spring plant phenology and false springs in the conterminous US during the 21st century. *Environmental Research Letters*, 10, 104008. doi: 10.1088/1748-9326/10/10/104008.

Arnone J. A. III, P. S. J. Verburg, D. W. Johnson et al., 2008. Prolonged suppression of ecosystem carbon dioxide up take after an anomalously warm year. *Nature*, 455, 383–386.

Backlund, P., A. Janetos, D. S. Schimel et al., 2008. The effects of climate change on agriculture, land resources, water resources, and biodiversity in the United States (SAP 4.3). A Report by the U.S. Climate Change Science Program and the Subcommittee on Global Change Research. US Department of Agriculture, Washington, DC, 362pp.

Briley, L., D. Brown, S. E. Kalafatis, 2015. Overcoming barriers during the co-production of climate information for decision-making. *Climate Risk Management*, 9, 41–49. doi:10.1016/j.crm.2015.04.004.

Carson, W. P., A. A. Royo, C. J. Peterson, 2014. A pox on our land: A case study of chronic deer overbrowsing throughout the Allegheny National Forest region in Pennsylvania. In: *The Herbaceous Layer in Forests of Eastern North America*, 2nd edn, Eds. F. S. Gilliam, M. R. Roberts. Oxford University Press, New York, pp. 400–411.

Carson, W. P., S. A. Schnitzer, 2003. Deep rainforest ecology [Review of: Nouragues: dynamics and plant-animal interactions in a Neotropical rainforest. F Bongers, P Charles-Dominique, P-M Forget and M Théry, editors, 2002]. *Ecology*, 84, 1340–1341.

Chavez, E., G. Conway, M. Ghil, M. Sadler, 2015. An end-to-end assessment of extreme weather impacts on food security. *Nature and Climate Change*, 5, 997–1002. doi:10.1038/nclimate2747.

Coutts, A. M., J. Beringer, N. J. Tapper, 2007. Impact of increasing urban density on local climate: Spatial and temporal variations in the surface energy balance in Melbourne, Australia. *Journal of Applied Meteorology and Climatology*, 46, 477–493. doi:10.1175/JAM2462.1.

Cowles, H.C., 1899. The ecological relations of the vegetation on the sand dunes of Lake Michigan. *Botanical Gazette*, 27, 95–117, 167–202, 281–308, 361–391.

Dilling, L., M. C. Lemos, 2011. Creating usable science: Opportunities and constraints for climate knowledge use and their implications for science policy. *Global Environmental Change*, 21, 680–689. doi:10.1016/j.gloenvcha.2010.11.006.

Doney, S. C., D. S. Schimel, 2007. Carbon and climate system coupling on timescales from precambrian to the anthropocene. *Annual Review of Environment and Resources*, 32, 31–66.

Donner, W., H. Rodríguez, 2008. Population composition, migration, and inequality: The influence of demographic changes on disaster risk and vulnerability. *Social Forces*, 87, 1089–1114.

Ehrlich, P., 1997. A world of wounds: Ecologists and the human dimension. In *Excellence in Ecology Book Series 8*. Ed. O. Kinee, Ecology Institute/Wiley-VCH Verlag GmbH, Oldendorf/Luhe, Germany, 210pp. doi:10.1002/iroh19970820414.

FAO, 2011. *The State of the World's Land and Water Resources for Food and Agriculture (SOLAW)—Managing Systems at Risk*. Food and Agriculture Organization of the United Nations, Rome and Earthscan, London, U.K.

Flood, J. F., L. B. Cahoon, 2011. Risks to coastal wastewater collection systems from sea-level rise and climate change. *Journal of Coastal Research*, 27(4), 652–660.

Food and Agriculture Organization (FAO), WFP and IFAD. 2012. *The State of Food Insecurity in the World 2012: Economic Growth Is Necessary but Not Sufficient to Accelerate Reduction of Hunger and Malnutrition*. FAO, Rome, Italy.

Fraser, L. H., H. Henry, C. N. Carlyle et al., 2012. Coordinated distributed experiments: An emerging tool for testing Global hypotheses in ecology and environmental Science. Department of Forestry & Natural Resources Faculty Publications, Paper 14. doi:10.1890/110279.

Heinz Foundation, 2006. Filling the gaps: Priority data needs and key management challenges for national reporting on ecosystem condition. http://www.heinzctr.org/ecosystem, accessed February, 2016.

Holdren, J., T. Dickenson, G. Paulson et al., 2014. *National Plan for Earth Observations*. National Science and Technology Council. Executive Office of the President, Washington, DC, pp. 71.

Huntzinger, D.N., C.R. Schwalm, Y. Wei et al., 2016. NACP MsTMIP: Global 0.5-deg Terrestrial Biosphere Model Outputs (version 1) in Standard Format. Data set. Oak Ridge National Laboratory Distributed Active Archive Center, Oak Ridge, TN. http://daac.ornl.gov. doi:10.3334/ORNLDAAC/1225, (in press).

Jiang, L., J. Jiang, J. Liang, K. R. Wilcox, S. L. Collins, A. K. Knapp, W. T. Pockman, M. D. Smith, Y. Luo, 2016. Frontiers of ecosystem modeling and large scale experiments. In: *Terrestrial Ecosystem Research Infrastructures: Challenges, New Developments and Perspectives*, Eds. A. Chabbi, H.W. Loescher. CRC Press/Taylor & Francis Group, Boca Raton, FL.

Loescher, H. W., E. Kelly, R. Lea, 2016. National ecological observatory network: Beginnings, programmatic and scientific challenges, and ecological forecasting. In: *Terrestrial Ecosystem Research Infrastructures: Challenges, New Developments and Perspectives*, Eds. A. Chabbi, H.W. Loescher. CRC Press/Taylor & Francis Group, Boca Raton, FL.

Malone, S., G. Starr, C. L. Staudhammer, M. G. Ryan, 2013. Effects of simulated drought on the greenhouse carbon balance of Everglades short-hydroperiod marsh ecosystems. *Global Change Biology*, 19(8), 2511–2523. doi:10.1111/gcb.12211.

Marshall, J. D., Blair, J. M., D. P. C. Peters, G. Okin, A. Rango, M. Williams, 2008. Predicting and understanding ecosystem responses to climate change at continental scales. *Frontiers in Ecology and the Environment*, 6(5), 273–280. doi: 10.1890/070165.

Millennium Ecosystem Assessment (MEA), 2005. *Ecosystems and Human Well-Being: Synthesis*. Island Press, Washington, DC, 137pp.

Munich, R., 2015. NatCatSERVICE Loss events worldwide 1980–2014., Munich, Germany.

National Research Council (NRC), 2001. *Grand Challenges in Environmental Sciences*: The National Academies Press, Washington, DC, 106pp.

NRC, 2007. *Understanding Multiple Environmental Stresses: Report of a Workshop*. The National Academies Press, Washington, DC, 154pp.

Peters D. P. C., P. M. Groffman, K. J. Nadelhoffer, N. B. Grimm, S. L. Collins, W. K. Michener, M. A. Huston. 2008. Living in an increasingly connected world: a framework for continental-scale environmental science. *Frontiers in Ecology and the Environment*, 6(5), 229–237. doi:10.1890/070098.

Peters, D. P. C., H. W. Loescher, M. SanClements, K. M. Havstad, 2014. Taking the pulse of a continent: Role of observatories and long-term research networks to fill critical knowledge gaps. *Ecosphere*, 5(3), Article 29, 1–23. doi:10.1890/ES13-00295.1.

Presidents Council of Advisors on Science and Technology (PCAST), 2011. Sustaining environmental capital: Protecting society and the economy. Report to the President. Office of Science Technology And Policy, Executive Branch, U.S. Government. www.whitehouse.gov/ostp/pcast.

Pulwarty R. S., R. Maia, 2015. Adaptation challenges in complex rivers around the world: The guadiana and the Colorado basins. *Water Resources Management*, 29, 273–293.

Robertson, G. P., 2008. Long-term ecological research: Re-inventing network science. *Frontiers in Ecology and the Environment*, 6(5), 281. doi:10.1890/1540-9295(2008)6[281:LERRNS]2.0CO;2.

Rosegrant, M. W., S. A. Cline, 2003: Global food security: Challenges and policies. *Science*, 302, 1917–1919. doi:10.1126/science.1092958.

Rosenzweig, C., W. D. Solecki, S. A. Hammer, S. Mehrotra (eds.), 2011. *Climate Change and Cities: First Assessment Report of the Urban Climate Change Research Network*. Cambridge University Press, Cambridge, U.K.

Schimel, D., M. Keller, 2015. Big questions, big science: Meeting the challenges. *Oecologia*, 177(4), 925–34. doi:10.1007/s00442-015-3236-3.

Smith, M. D., A. K. Knapp, S. L. Collins, 2009. A framework for assessing ecosystem dynamics in response to chronic resource alterations induced by global change. *Ecology*, 30, 3279–3289.

Soininen, J., 2010. Species turnover along abiotic and biotic gradients: Patterns in space equal patterns in time? *BioScience*, 60(6), 433–439. doi:10.1525/bio.2010.60.6.

Stocker, M., 2016. Advancing the software systems of environmental knowledge infrastructures. In: *Terrestrial Ecosystem Research Infrastructures: Challenges, New Developments and Perspectives*, Eds. A. Chabbi, H. W. Loescher. CRC Press/ Taylor & Francis Group, Boca Raton, FL.

Sutton, M. A., E. Nemitz, J. W. Erisman et al., 2007. Challenges in quantifying biosphere–atmosphere exchange of nitrogen species. *Environmental Pollution*, 150, 125–139. doi:10.1016/j.envpol.2007.04.014.

Sweet, W. V., J. J. Marra, 2014. State of nuisance tidal flooding. NOAA Report. NOAA news, Earth System Research Laboratory, Boulder, CO.

The Royal Society, 2014. *Resilience to Extreme Weather.* The Royal Society London, U.K.

Tilman, D., P. B. Reich, J. M. Knops, 2006. Biodiversity and ecosystem stability in a decade-long grassland experiment. *Nature,* 441, 629–632. doi:10.1038/nature04742.

Tilman, D., P. B. Reich, J. Knops, D. A. Wedin, T. Mielke, C. Lehman, 2001. Diversity and productivity in a long-term grassland experiment. *Sciences,* 294(5543), 843–845. Paper 151. hp://digitalcommons.unl.edu/bioscifacpub/151.

Tye, M. R., 2015. Understanding the risks from extreme rainfall. *Proceedings of the ICE—Forensic Engineering,* 168, 71–80. doi:10.1680/feng.14.00002.

Tye, M. R, G. J. Holland, J. M. Done, 2015. Rethinking failure: Time for closer engineer-scientist collaborations on design. *Proceedings of the ICE: Forensic Engineering,* 168(2), 49–57. doi:10.1680/feng.14.00004.

Vargas, R., D. Alcaraz-Segura, R. Birdsey et al., 2016. Policy-relevant research and interoperability for carbon cycle science in developing countries: Mexico use-case. *Carbon Management* (in press).

Vargas, R., H. W. Loescher, B. Bond-Lamberty et al., 2015. Building global ecological understanding. *Frontiers in Ecology and the Environment* (submitted).

Vitousek, P. M., 2004. *Nutrient Cycling and Limitation: Hawai'i as a Model System.* Princeton University Press, Princeton, NJ, 232pp.

Vitousek, P. M., C. B. Field, 2001. Input–output balances and nitrogen limitation in terrestrial ecosystems. In: *Global Biogeochemical Cycles in the Climate System,* Eds. E.D. Schulze, S.P. Harrison, M. Heimann, E.A. Holland, J. Lloyd, I.C. Prentice, D. Schimel. Academic Press, San Diego, CA, pp. 217–235.

Vitousek, P. M., H. A. Mooney, J. Lubchenco, J. M. Melillo, 1997. Human domination of Earth's ecosystems. *Science,* 277, 494–499.

Vörösmarty, C. J., P. Green, J. Salisbury, R. B. Lammers, 2000. Global water resources: Vulnerability from climate change and population growth. *Science,* 289, 284–288. doi:10.1126/science.289.5477.284.

Warming, E., 1985. *Plantesamfund: Grunträk af den Okologiska Plantegeografi.* Philipsen, Copenhagen, Denmark.

Weltzin, J. F., M. E. Loik, S. Schwinning et al., 2003. Assessing the response of terrestrial ecosystems to potential changes in precipitation. *BioScience,* 53 (10), 941–952. doi:10.1641/0006-3568(2003)053[0941:ATROTE]2.0.CO;2.

Whitacre, P. T., A. P. Fagen, J. L. Husbands, F.E. Sharples (Eds.), 2010. *Planning Committee on Achieving Research Synergies for Food/Energy/Environment Challenges: Implementing the New Biology: Decadal Challenges Linking Food, Energy, and the Environment: Summary of a Workshop, June 3–4, 2010.* National Research Council, Washington, DC, 42pp.

2

National Ecological Observatory Network: Beginnings, Programmatic and Scientific Challenges, and Ecological Forecasting

Henry W. Loescher, Eugene F. Kelly, and Russ Lea

CONTENTS

2.1 Scientific and Programmatic Beginnings

In the United States and elsewhere, the development of large-scale scientific research infrastructure (RI) comes about from needs of the scientific community, as well as imperatives expressed in society and at high levels of government planning. In the 1990s, leaders in ecology began to express the need to address ecological theory at larger time and space domains and the need for large experiments that can elucidate unknown, nonlinear, or stochastic ecological behavior at a scale that could not be typically achieved within normal competitive grant cycles. Data were lacking to address these larger ecology "macrosystem" concepts (Soranno and Schimel 2014), and so the U.S. National Science Foundation (NSF) provided support to explore these notions of large-scale ecology. The idea of a large-scale ecological facility was novel and had never been done before, and the process to develop such a facility of this magnitude was a new challenge for the ecological community.

In fact, the NSF has a history of funding large-scale facilities and infrastructure projects, most notably telescopes, ocean-going vessels, and large geodetic arrays that are designed for multiple user groups and able to address a wide range of scientific questions. This type of support (from the Major Research Equipment and Facilities Construction, MREFC) is quite rare with only a handful of such projects funded every decade. While prominent members of the ecological community articulated the need for a large-scale facility, no one quite knew what it would look like. Some may say that NSF's initial design and development for a large-scale ecological facility was bold, visionary, and risky, because they too recognized its importance, yet did not know what "it" was. And so, NSF funded numerous planning workshops and white papers to clarify the vision of the science that an ecological facility could address as well as to begin defining its overall science scope in response to grand challenge questions. This process of scoping remains to this day very rich in scientific creativity (NSF 2002a–c), and it energized and engendered broad sections of the ecological community (Peters et al. 2008, Robertson 2008, Schimel and Keller 2015).

The deep well of ecological thinking that was brought to bear, in part, from the Long Term Ecological Research network (LTER), which is also funded by NSF, as well as other federally supported activities, for example, U.S. Department of Agriculture (USDA) Agricultural Research Service's Long Term Agroecosystem Research (LTAR), Department of Energy AmeriFlux, U.S. National Oceanic Atmospheric Administration's (NOAA) U.S. Climate Reference Network, and U.S. Geological Survey Earth Resources Observation System (EROS). NSF support for LTER provides a coordinating office, centralized data management, and core operational support at each of the ~23 sites distributed across the United States (with the exception of Palmer Station and McMurdo Dry Valleys of Antarctica and Moorea in Tahiti). And while there may be some science cohesivity and common general themes of research at particular LTER sites, the actual hypothesis-driven research is conducted by principle investigators (PIs) through federal peer-reviewed programs along typical 3-year funding cycles. Governance of LTER can be described as being layered with more complexity since its inception in 1980, and with a hub-and-spoke model, where a Network Science Council (the hub) that is very responsive to the changing needs of the science community, the PIs and site management (the spokes) (www.lternet.edu/node/140/). This structure fosters strong scientific creativity and adaptability (Table 2.1), and in the eye of the science community, scientific creativity trumps all.

When the ecological community was confronted with building a large ecological infrastructure, naturally everyone initially gravitated toward existing organizational models, for example, LTER, DOE AmeriFlux.

TABLE 2.1

Comparison of Programmatic Attributes between a Scientist's Approach (Often Networks) and a System Engineering Approach (Often Large-Scale RIs)

Scientist's Approach		System Engineering Approach	
Programmatic Attributes	**Pro/Con**	**Pro/Con**	**Programmatic Attributes**
Strong in scientific creativity	Pro	Con	Constrains scientific creativity to deliver the RI
Comfort-level for scientists and bottom-up approaches	Pro	Pro	New roles for scientists internal and external to the RI
Complexity becomes an open-ended problem	Con	Pro	Complexity is inherently planned for
Governance is often complex and not extensible	Con	Pro	Governance is inherently planned for, albeit complex
Changing scope, budget, and risk	Pro/Con	Pro	Clearly defines scope, budget, schedule, and risks
Difficult planning for program sponsors	Con	Pro	Develops planning horizons for program sponsors
Problematic for long-term sustainability	Con	Pro	Fosters long-term sustainability and operational models
Does not provide a unique solution for infrastructure design	Pro/Con	Pro/Con	Does not provide a unique solution for infrastructure design

Numerous vision and scoping studies were commissioned (e.g., NSF 2000a–c, 2002a–e, AIBS 2004a–f, and see www.neoninc.org/about/history). The vision and planning activities that NSF supported created a hyper-democratic community where (generally speaking) everyone involved thought that their designs, their concepts, and their areas of study were going to be used in the national facility design. Decadal planning for LTER envisioned more of a national infrastructure like that found in the National Ecological Observatory Network (NEON, more below) than how it is manifested today (LTER 1989). So on the one hand, the large interest in macrosystem-scale ecology demonstrated the scientific imperative for a national ecological facility. On the other hand, having this effort being perceived as everything to everyone created other problems. Expectations of bottom-up governance and with everyone's opinion being considered created difficulties in forming a management structure and defining the scientific scope (rf. Table 2.1). In attempts to include everyone's concept with the design, unrealistic early budgets of the NEON exceeded $1.7 billion for construction alone.

At this point, it is important to reinforce that prior to these efforts, building a large-scale ecological facility was unchartered territory and "near-death" experiences were commonplace in building other large-scale science facilities, such as telescopes, ocean research vessels, and particle accelerators

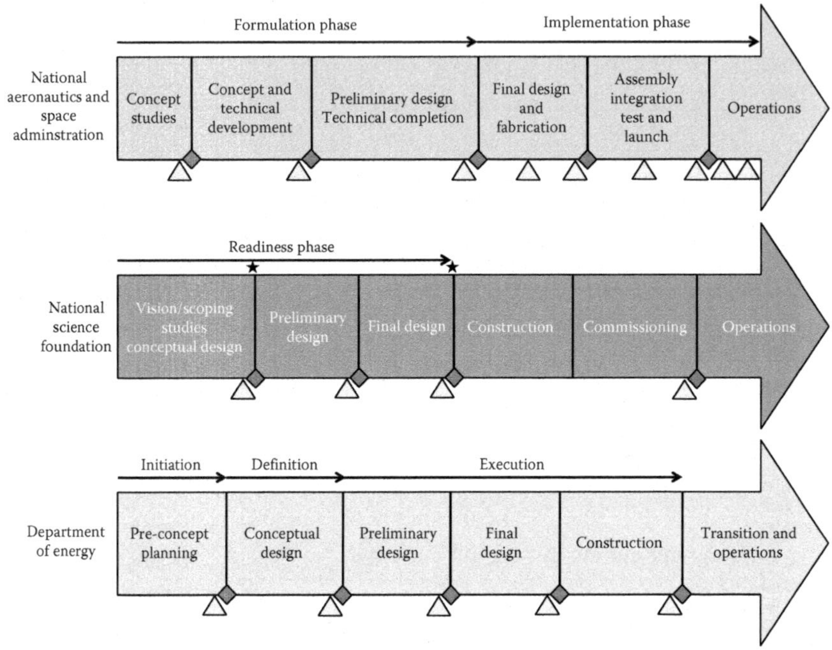

FIGURE 2.1
Conceptual diagram depicting the similarity of project development for large-scale science facilities by the National Aeronautics and Space Administration (NASA), the National Science Foundation (NSF), and the Department of Energy (DOE). Triangles are formal reviews; diamonds signify a post-review decision, often "go/no-go." Each review and decision has specific purpose and name. Decisions transition the project development from one stage to another. Mid-process reviews are often capabilities- or readiness-type reviews.

(A. Beasley, pers comm). There is a common and very rigorous project development cycle used for building large-scale science facilities in federal agencies, for example, Figure 2.1, and other intergovernmental efforts, for example, the Large Hadron Collider. Developing the management ability and the means to bring the scientific community along in how to balance scientific creativity while also establishing a constrained ("baselined") infrastructure is necessary for the success of the facility. This is not an easy task, because it inherently involves changes in culture that can appear to challenge personal philosophies of how the best science is done. Indeed, we see this same issue play out in the development of many other ecological observatories, such as Australia's Terrestrial Ecosystem Research Network, the European Union (EU) Integrated Carbon Observatory System, and the EU Analysis of Experimentation on Ecosystems (AnaEE). In NEON's case, the persistence of key visionary members of the science community was also instrumental in bringing along (small) cultural shifts in approach and maintaining the momentum of project execution.

2.2 Societal Imperatives

Over the past decades, there has been increasing awareness by the public, scientists, and political decision-makers of the need for scientifically valid information to address the rapid pace of global environmental changes. Natural, managed, and socioeconomic systems are subject to complex interacting stresses that play out over extended periods of time and space (NRC 2007, USGCRP 2013, Becknell et al. 2015). Some are rapid and visible, like extreme precipitation, temperature, wind, and wildfire events, while others are subtle and play out over decades, like chronic nitrogen inputs changing ecosystem composition (McDonnell et al. 2014). These societal issues highlight the need for long-term data sets for understanding the context of scientific observations and for forecasting future environmental conditions or cause and effect interactions. Attempts to collate current long-term data sets have been largely accomplished through ad hoc integration of data from existing observation programs that were designed for other purposes (e.g., hypothesis testing), and that also fare poorly when judged against the rubric of being able to broadly address societal problems (Heinz 2006, 2008, Schimel et al. 2008). This lack of integrated, consistent, long-term data to address these societal questions has also been recognized by others (NRC 2011a,b, IOM 2013).

In 2001, the U.S. National Academy of Science (NAS) recognized the need to define the challenges that face society and science in addressing our changing environment (NRC 2001). This study brought together top ecological thinkers of the day and coalesced the societal imperatives as well as the scientific gaps in our understanding into one planning document (NRC 2001). They identified seven Grand Challenge areas of research and cast them into a cause (drivers/controls) and effect (processes/function) paradigm, where the drivers of change are climate change, land-use change, and invasive species and the effects play out in the ecological processes of biodiversity, biogeochemistry, ecohydrology, and infectious disease. This report served as the first "Decadal Survey" for the environmental sciences, and never before had such a robust (top-down) ecological planning document been developed. Recently, these Grand Challenges were reexamined, and it was determined they are as relevant today as they were in 2001 (Loescher et al. 2016). Hence, Grand Challenge approach has become one of the stalwarts in planning large-scale environmental programs and projects, and decades later, is still being used and validated today.

In 2003, the U.S. NAS conducted a second seminal report that contributed toward the evolution of the National Ecological Observatory Network (NEON). Here, the NAS (through the National Research Council [NRC]) advocated for observations and experiments to be made consistently across the North American continent and creating the needed long-term, decadal-scale ecological data sets. While the NAS report did not specifically define science scope, it did contribute toward the discourse of what could be observed at continental scales. Programmatically, this report also raised NEON's mantel to be more broadly considered by the National Science Board (NSB). The NSB has several functions key to NEON's development it (1) aligns the policies of NSF within the framework of national policies set forth by the President and the Congress, (2) identifies issues that are critical to NSF's future, and (3) has NSF budgetary oversight. This is mentioned here because the process to acquire support for such a large-scale endeavor has bearing on how the science is manifested into the final designs and realized in operations.

As part of NEON planning effort in 2006, prominent members of the ecological community again came together and crafted NEON's Integrated Science and Education Plan (ISEP). This linked the ideas of Grand Challenge areas articulated by the 2001 NRC report with the need for a continental-scale observatory outlined in the 2003 NRC report, to formulate NEON Grand Challenge questions:

- How will ecosystems and their components respond to changes in natural- and human-induced forcings such as climate, land use, and invasive species across a range of spatial and temporal scales? And what is the pace and pattern of the responses?
- How do the internal responses and feedbacks of biogeochemistry, biodiversity, ecohydrology, and biotic structure and function interact with changes in climate, land use, and invasive species? And how do these feedbacks vary with ecological context and spatial and temporal scales?

Within these questions, we see the cause and effect paradigm expressed and the need to scale across space and time. These concepts have been slightly refined in Schimel et al. (2011), but the essence of the ISEP still remains in NEON's designs today.

The national relevance and timeliness of NEON's design continued to resonate with other societal imperatives over the course of its development. PCAST (2011) recognized that these changes in ecosystems threaten to erode the nation's (U.S.) environmental capital, resulting in disruptions of these services that would likely alter the fundamental trajectory of society and quality of life manifested across the United States and over large parts of the world (PCAST 2011, Schimel et al. 2011, NRC 2011a). Moreover, the "Fragmented federal investment in monitoring ecological change weakens national priorities" (PCAST 2011). The National Plan for Civil Earth Observations (Holdren et al. 2014) outlined specific Federal actions to "... address the threats to both the environmental and the economic aspects of well-being that derive from the accelerating degradation of the environmental capital." In brief, they call for (1) the continuity of sustained observations for earth system research, which includes the establishment and maintenance of programs to ensure data continuity for high-impact sustained research observations; (2) continued investment in experimental observations, which includes continuing to invest in research and development, incorporating technological advances to improve observations; (3) planned improvements to sustain observation networks and surveys for all observation categories, that is, proceeding with planned improvements to sustain observation systems; and (4) the continuity of, and improvements to, a rigorous assessment and prioritization process. This plan includes a national-level process to prioritize sustained observations for both research and public services and for experimental observations, and includes a process for external advisory input and strategic balance.

These challenges remain today, and from the top-down, the NSB aligns these national priorities with the NSFs policies. NEON as presently designed meets all these national priorities. In addition, NEON expands its scientific footprint through federal agency partnerships and leverages the federal investments in NEON and other agency programs to enhance its overall impact, and continues to be responsive to the changing environmental societal imperatives.

2.3 Hypotheses to Requirements

Often asked when discussing someone's science is "what is your question?" or "what are your hypotheses?" Building a large-scale scientific facility poses a slightly different scenario, where the facility has to be able to address a myriad of hypotheses relevant now, and in the future and at multiple scales. Lacking a crystal ball, the programmatic task then lies with how do we best design a facility that can broadly accommodate the ecological community and still be able to address specific research questions now and as they change in the future. A second (and related) programmatic challenge is how to determine the scientific scope of the facility to such an exacting degree as to be able to estimate (fix) its construction and operational costs, schedule, staffing and staffing needs, institutional support functions, and manage risk. These challenges are very foreign to an ecologist.

System engineering provides unique tools to distill questions and hypotheses into requirements. In turn, having requirements defines what needs to be built to meet the scientific needs and also the fidelity needed to constrain cost, schedule, and risk. Take, for example, the null hypothesis: ecosystem-scale litterfall rates are not controlled by the community of overstory tree species. An investigator may address this hypothesis by estimating the tree diversity by area and describe life histories, place a number of litterfall traps in randomly distributed plots, collect monthly sample, and analyze accordingly, etc. A system engineering approach may include determining a priori what is the expected spatial and temporal signal/noise ratio of the phenomena to be tested, in this case, litterfall and overstory tree diversity, estimate the scale of statistical independence, and craft the science requirements something like (Shishko 1995, Kossiakoff et al. 2011):

- Litterfall traps shall be 1 m × 1 m horizontally square and have a perimeter edge along the top that is 7 mm in width.
- Litterfall traps shall be mounted 0.5 m above mean ground level.
- Litterfall traps shall collect leaves and biotic debris, and not hold water.
- Litterfall traps shall have a minimum of 5 L capacity.
- Litterfall traps shall be placed at least one statistically independent scale length from the next trap.
- The number of litterfall traps shall have the capability of collecting 80% of the spatial mean on a biweekly basis (and so on).

Requirements are crafted so that they can be tested and verified with one testable feature per requirement. Requirements are then reviewed by independent, outside scientists to assure they meet community expectations and needs, and to assure the data are measured consistently over the life of the facility. A control process is put into place to version, test, and approve any changes, as it becomes onerous if a project continually changes its requirements. While this may seem tedious, it also provides the context to communicate the science to non-ecologist staff within NEON, such as site engineers deploying the field experiment, or cyber-infrastructure engineers coding the analytics. Developing requirements for a large science facility only works for scientific approaches that are broadly accepted by the community, or considered best community practices, like measuring tree diameter at breast height, that is, 1.37 m (4.5 ft) above ground. It does not work for approaches that may still be considered as experimental, for example, using a specific algorithm to use to estimate columnar CO_2 through the atmospheric boundary layer that will likely change in a few years. Whether a hypothesis is tested by an individual investigator or distilled into requirements and then tested, in either case, there is no single unique solution in how the experimental design is finally expressed and executed (Shishko 1995, Kossiakoff et al. 2011).

A key challenge in designing NEON was to provide a facility that brings scientific relevance to the largest possible group of ecologists today and in the future. So for a project the size of NEON, system engineering does not end at capturing the science requirements. The process started with the Grand Challenge questions that are meant to be provocative, embody scientific and societal imperative(s), and be quite open ended (see Figure 2.2). From there, the scientific community was asked what would be their hypotheses to address the Grand Challenge questions. Having thousands of hypotheses in hand, NEON staff scientists then asked what are the data products

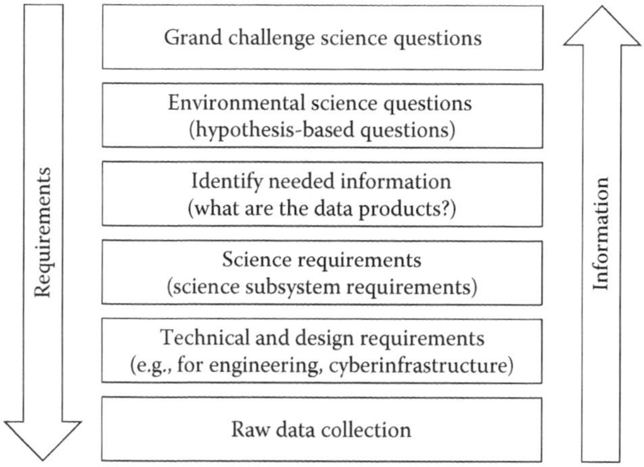

FIGURE 2.2
A simplified, scientist's view of the system engineering approach to distill Grand Challenges and hypotheses into tiered requirements and, conversely, how the flow of information can be used to address the Grand Challenge questions.

needed to test these focused research questions. From there, science requirements were drafted, which in turn could be used to capture technical and design requirements. The ability to assess the requirements at a very fine level of (system engineering) granularity also allows us to verify the data type, quality level, and quantity—the flow of information at the point of data capture, through the facility, in order to address the grand challenge questions (Figure 2.2). The progression to capture requirements and assess the type, quality, and flow of information was not a static process. NEON staff iteratively assesses each requirement and how it is manifested into the overall design, over and over, and often with many internal reviews and external advice. Once the requirements became fixed, the project became baselined and subsequently budgeted, scheduled, risks managed, and transitioned to operations. The capture of requirements is oversimplified here to illustrate the scientist's perspective. But ultimately, availability of this tsunami of data is the payoff to scientists.

Other key system engineering activities are the change control processes that manage these dynamic requirements, and the testing, verification, and commissioning processes that transition construction activities into operational. Many large-scale science facility projects attribute their success to being able to establish and maintain a creative tension between the science and engineering teams. For example, a scientist's choice for a specific sensor that meets the science requirements for accuracy and precision may place undue operational burden to manage that accuracy and precision across the range of environmental conditions found across the North American continent. In another example, there are "best community practices" for capturing small mammals in specific ecosystems, but extending the same protocols for small mammal capture across the continent remains challenging. Moreover, developing adaptive field sampling protocols to meet and manage expected mark-recapture signal:noise ratios also remains challenging. System engineering tools that distill the requirements of both the scientists and engineers craft balance, optimize and standardize the install, design, maintenance, operation, data capture of a particular ecological quantity while still meeting the scientists requirements. System engineering often provides the context to maintain this creative tension between scientists and engineers (Figure 2.3). But of course, one of the roles of leadership is also to ensure that all of the Project Teams are balanced and complementary in responsibilities and authority (and this also applies to Project Sponsors) (Jain and Triandis 1997, Hughes et al. 2014). This is difficult to manage through the course of the project given that the combination of scope, budget, schedule, risk, people management, and politics is inherently a nonlinear, multivariate problem, with some solutions from one project do not necessarily translate to another (A. Beasley, pers. comm.). That said, there are also different philosophies on how best to apply system engineering principles, and how and when to engage scientists. The National Aeronautics and Space Administration (NASA) model to build

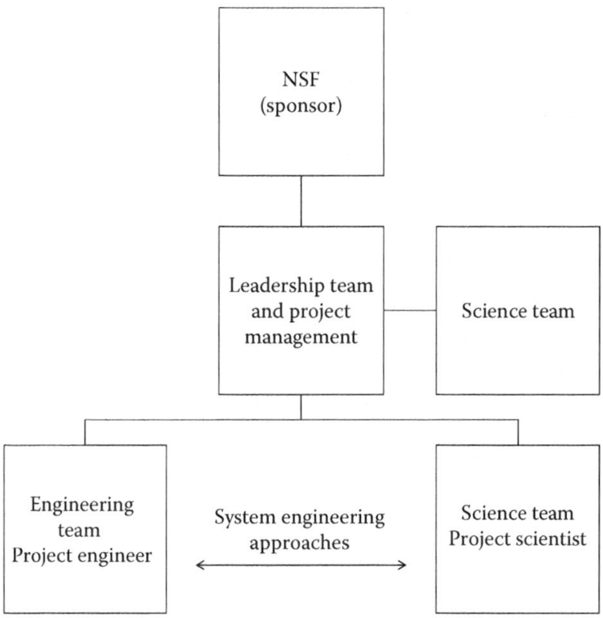

FIGURE 2.3
Simplified project management model to demonstrate the need for creative tension between science and engineering efforts, and the need to balance responsibilities and authority among the associated management strata (and this also applies to project sponsors). (This figure follows similar logic found in Schimel and Keller, 2015, Figure 3 therein.)

satellites involves bringing scientists together initially to capture the science requirements and again during the testing and verification stages. Requirements are often only captured when a design or scope decision needs to be made. This differs from the NSF or Department of Defense (DOD) model where requirements are determined at a very fine degree of fidelity before any construction is initiated. The consequence of this approach is that a longer time is needed than originally planned for, leading to cost overruns (in the case of DOD), or descoping science (in the case of NSF). In Europe, clearly strong system engineering principles were applied to many large-scale science facilities, that is, the Large Hadron Collider. Yet interestingly, the EU FP7 project ENVironmental Research Infrastructures (ENVRI) has taken a slightly different approach given the large plethora of environmental research projects that have been strategically initiated. A common reference model (roadmap process) is used to define cyber- and informatics functional structures that are common among all EU environmental research infrastructures (Chen et al. 2013, 2014, Chen and Hardisty 2014). This helps planning and prototyping activities, but also operational requirements and creates a structure to foster interoperability of data among these facilities. ENVRI Science and Technical reference models are forthcoming. In all cases and for all research infrastructures, managing scientific scope and scope creep is difficult, is an ongoing activity, and requires vigilance and effective programmatic structure, for example, system engineering approaches.

Building new environmental research facilities is still new programmatically and quite a foreign concept for some scientific communities, for example, what do the data mean for my science and how do I apply my science to build something that has never been done before. Each infrastructure is faced with a unique suite of programmatic issues and scientific challenges (otherwise they would be commonplace and formulaic to build). And at the same time, each type of system engineering approach has its strengths, shortcomings, and benefits that have to be weighed against a rigorous assessment of need and feasibility when constructing an infrastructure.

2.4 Unrealized Benefits of Requirements

Extracting requirements from hypotheses was originally based out of need. Requirement capture is a very robust tool to define scope, budget, risk, schedule, and transition to operations (in order to secure funding support) and is a means to communicate what is needed by the scientists to other project teams and nonscience experts, for example, engineer, permitting, or operational staff. Because in the case of ecology and NEON, this activity had never been done before, it is an untapped resource waiting to be exploited.

Requirements give us the basis to communicate the exact design externally as well. This is not to say that they can ever replace hypotheses, but it provides accurate details of exactly what is being constructed and delivered to the community—a consistent, continental, physical infrastructure and long-term, multiscaled data sets and data products that serve as a context for research and education (rf. www.neoninc.org/sites/default/files/basic-page-files/2015-03-23.SCA_.Report.pdf).

Two general categories of users have expressed interest in exploiting NEON requirements. One camp wishes to extract specific requirements to have a NEON-like infrastructure of their own, and a second group, typically other networks and observatories, wishes to link physical and informational infrastructures in an interoperable way to expand both NEONs and their capabilities. The users of first group are, in essence, developing NEON satellite sites, where they may extract protocols, whole or in part, such that they can collect data in the same way as NEON. Others wish to use NEON infrastructure designs in their entirety to augment the NEON's overall capability. For example, to utilize NEON's mosquito sample protocols at higher spatial density in areas where West Nile Virus prevalence is expected to increase, or to construct and operate a fully instrumented tower in an urban environment.

The second category of users is typically other networks, observatories, and infrastructures that wish to become interoperable with NEON. Interoperability is an emerging concept that can mean many different things to different people. Here, we broadly define it as all the efforts needed to enhance the use and transfer of data by removing the technical, scientific, cultural, and geopolitical barriers. Requirements provide us with the specific language to make this possible to a fine degree of fidelity. It is not meant to be prescriptive, rather a dynamic community-driven approach, in which many emerging groups use requirements as a backdrop to steer the forum for discourse (www.coopeus.com, www.esipfed.org, rd-alliance.org).

Interoperability, as defined for scientific utility, has four focus areas previously identified (Chen et al. 2008, Peters et al. 2014). First, why are the data collected in the first place? Identifying the questions, mission statements, hypotheses, or requirements creates the basis to understand the constraints, synergies, and gaps in the data to be shared. Second, is the epistemological question of how to trace the quantities measured to known international standards, first principles, or best community practices? As discussed earlier, it is important to develop uncertainty budgets. Third, how the algorithmic procedures are used to calculate a specific quantity may differ among research groups, networks, observatories, etc., for example, productivity measures? It is fine if one group measures net ecosystem exchange of carbon one way, and another group measures it differently. The important criterion is for the relative uncertainties across algorithmic approaches to be known and estimated, so that they can be used in comparative analyses and predictive Bayesian approaches (Reckhow 2003, Johnson and Omland 2004). This issue also applies to models or protocols at the high level of abstraction such as how to design a national forest inventory, a monitoring network, or a monitoring reporting and verification (MRV) system. And lastly, the issue is broadly defined as informatics. This includes the obvious, such as data and metadata formats and fields, for example, ISO 19115 compliant (www.iso.org/iso/). But it also includes issues of intellectual property rights, citation and attribution, persistent identifiers, open data policies and sovereignty, data portals and discovery tools, and controlled vocabularies. For example, there are conceptual barriers that include syntactic and semantic differences in information to be exchanged; one person's definition of litterfall may be quite different from another's, or how should one label a mashed up archival sample of mosquitos for DNA analyses? In the case of data sovereignty, this includes not only an individual country's rights to its data, but also the organizational barriers within a country that define the country's agency or ministry responsibility and authority over the data, and their mandates to share with other organization structures. This issue is evident in developing countries where the organizational responsibilities are usually not clearly defined, or managed, often hindering collaboration among actors. The degree to which data are truly interoperable is the degree to which these four elements are adopted by collaborative activities and facilities, which would not be possible without explicit requirements.

2.5 The Need and Philosophy of Ecological Forecasting

NEON's science strategy is complex and described in Schimel et al. (2011). NEON staff have taken the Grand Challenge questions, and through requirements capture, they have identified discipline-specific science teams to address specific ecological properties and processes to be measured. A scaling strategy from the site to the continent also was incorporated. At the site level, the science teams are: terrestrial instrumentation; terrestrial organismal sampling; mobile deployment (measurement) platforms; and aquatic instrumentation and organismal sampling (www.neoninc.org/science-design/spatiotemporal-design). NEON has 84 sites distributed across the United States from the north slope of Alaska to Puerto Rico, from Hawaii to New England. They are distributed according to our understanding of how to assess the trends in regional ecological properties and processes, and how the Grand Challenges manifest regionally. Site-based NEON ecology is centered on the ecosystem scale. Our scaling strategy includes aerial flyover of all our sites with an airborne observation platform (remote sensing aircraft) that includes visible imaging hyper-spectrometer and downward facing waveform and discrete LIDAR. These data can then be combined with satellite imagery or other federal data sets to scale ecological quantities from the site to region to continent. All the data are free and available to anyone through a data portal (data.neoninc.org/home), and data product catalogs can be found at data.neoninc.org/data-product-catalog.

Even though the capability of this observatory is profound, as with other continental-scale environmental observatories, for example, Australia's Terrestrial Ecosystem Research Network (TERN, www.tern.org.au), China's China Environmental Research Network, DOE AmeriFlux (ameriflux.lbl.gov), and others mentioned in this book, the scientific and societal imperatives to predict future environmental conditions and ecological functions remain (rf. Katz and Murphy 2005). Reenforcing this notion, Smith and Zeder (2013) suggest that the expectation for science to provide future guidance in the face of growing human population living with limited natural resources is a defining attribute in the Anthropocene. Interpreted another way, these observatories provide information, but that still has to be converted into knowledge and new understandings by the user community. As such, NEON has also been charged by the NSB to "... enable an ecological understanding and forecasting... by providing infrastructure." There are several challenges to meet this objective.

2.5.1 First Challenge

NEON had to define what is meant by "ecological forecasting." Here, we place the cause and effect paradigm into two problem statements (rf. Table 2.2):

1. What is the most likely future state of an ecological system?
2. To provide an applied context; what are possible future ecological outcomes if a given decision is made today?

TABLE 2.2

Ecological Attributes That Help Us Conceptualize How the Internal Ecosystem Structure and Function May Respond to Perturbations and/or Disturbances

Ecological Attributes	Brief Definition	Reference
Resilience	The capacity of an (eco)system to experience "perturbations" or "disturbance" while retaining essentially the same function, structure, feedbacks, and identity	Holling (1973)
Adaptability	The capacity of the (eco)system to manage its internal resilience, this can include all the internal functions, structure, etc., or some in part	Berkes et al. (2003)
Transformability	The capacity to move an (eco)system from existing system cannot maintain itself due to a "perturbation" to a new fundamental (eco) system state	Chapin et al. (2009)

2.5.2 Second Challenge

Ecosystems are continually subject to changing conditions (abiotic environment) and also subject to natural and man-made disturbances (Becknell et al. 2015). These disturbances traditionally have been viewed as episodic, regional, and punctuated and stochastic in time (Dale et al. 2001), for example, ice, snow, and wind storms, fire (made-man and natural), drought, hot spells, earthquakes, volcanic activity, and clear-cutting of forests. Such disturbances affect the ecosystem structure, functions, and process rates (Smith et al. 2009), such as nutrient availability, succession, and broken physical structure. Chapin et al. (2009, Box 1.2) elegantly describe the susceptibility or resistance to change that ecosystems may experience in the presence of disturbance, contributing to the ideas of resilience and adaptability (rf. Table 2.2). But in today's global environment with continual and increase in inputs of reactive nitrogen, and increases in temperature, CO_2, CH_4, atmospheric aerosols, invasive species, land-use change, are a chronic sources of disturbance, the presence or rate of which have not been experienced by ecosystems (Smith et al. 2009). Under these conditions, ecologists do not know how plastic ecosystem functions are (adaptability, Table 2.2), will they transition to different states (transformability, Table 2.2), can redistribute to more favorable environments (Iverson and MacKenzie 2013, Zolkos et al. 2015), or how/if natural selection plays a role (Quintero and Wiens 2012, Thorpe et al. 2015). What ecologists know about ecosystem functions today may not be (if not likely) how they will behave in the future. Being able to understand a posteriori how and why ecosystem states change in response to a given event falls short in being able to deliver on addressing our societal imperatives (NRC 2003). Hence, these theoretical underpinnings provide a context for how these systems may interact, but there is a lack of data to test new theories (Collins et al. 2012), and to provide the context to quantitatively predict future ecological states as an operational function of an observatory.

2.5.3 Third Challenge

How can a prognostic capability be designed when the future temporal and spatial signal/noise ratio of ecological processes is expected to change? Moreover, there are many processes, which express nonlinear and stochastic behavior. For example, temperature tipping points (Wall 2007, Laurance et al. 2011), susceptibility to droughts, and stochastic changes in functions from chronic nutrient inputs, as in the case of the Kuparuk River, AK (Slavik et al. 2004, Benstead et al. 2005). We can only design according to what we know about process-level temporal and spatial signal/noise ratio today. Even though no one has designed a large-scale ecologic observatory with a mandate for forecast ecology before, it is necessary, however, to embody

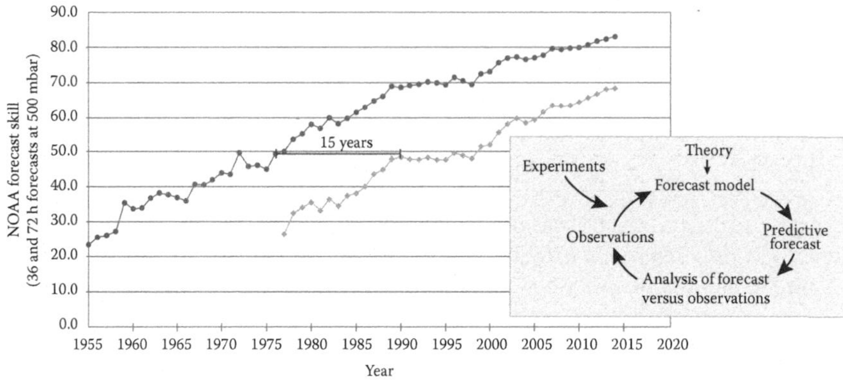

FIGURE 2.4
U.S. National Oceanic Atmospheric Administration's (NOAA) forecast skill from 1955 to date. Forecast skill is for the atmospheric 500 mbar pressure isopleths that control the continental-scale synoptic weather patterns more than 36 hour (dark gray trace) and 72 hour (light gray trace) periods. Note the 15-year lag in the level of forecast skill between 36 and 72 hours, while the rate of improvement (slope) is essentially the same. Data redrawn from www.nco.ncep.noaa.gov/sib/verification/s1_scores/s1_scores.pdf. Inset is the iterative concept of theory, forecast, analysis of observations, and experiments, to advance forecast skill.

a philosophy where new understanding can be accommodated into the designs, measurement suite, and forecast ability.

The philosophy that NEON embodies to forecast ecology mirrors the NOAA's strategy for improving weather forecasts, that is, for the 500 mbar (that controls the continental-scale synoptic weather patterns) more than 36 and 72 hour forecasts (Figure 2.4). Accuracy in NOAA weather forecasts improved almost linearly from 1955 to date, from 20% to more than 80% today. There was no large increase in skill with the deployment of >3000 new weather stations in the 1960s, no increase in skill with weather satellites in the 1970s, and no increase with new supercomputers (particularly in the 1980s and 1990s). Instead, the increase in forecast skill was achieved by starting with theory, constructing theory-informed forecast models, and challenging the model forecasts with observations and improvements in models that, in turn, inform new types of observations and/or capabilities to challenge theory again. It is this iterative approach that enhanced the NOAA forecast skill, and this philosophy is used by NEON to approach how we can forecast ecology.

The ability to incorporate nonlinear behavior, such as tipping points, is still missing within NOAA's forecasting approach. This is the core rationale to include the role of large-scale ecological experiments into continental-scale research infrastructure, such as NEON and AnaEE, to be able to elucidate unknown processes and nonlinear responses. Bringing it all together, the role of an observatory to forecast ecology is to provide (1) estimates of ecosystem state(s), (2) estimates of key ecological state variables and parameters,

(3) experiments to elucidate unknown processes and nonlinear responses, (4) observations collected consistently and systematically over time and space to challenge iterative forecasts, and (5) the ability to augment the infrastructure to improve iterative forecasts. Bringing this philosophy to bear is a new paradigm for ecological research.

2.5.4 Fourth Challenge

The field of continental-scale or macrosystem ecology is still quite nascent (Heffernan et al. 2014). Current ecological theories have to be applied to the new macrosystem context, as well as the very real opportunity for new macrosystem theories to emerge. Clearly, addressing this challenge is a key to inform the iterative forecasting philosophy outlined in Challenge 3. Panarchy may be a leading macrosystem theory (Gunderson and Holling 2002), but deriving testable hypotheses remains a challenge. Many important studies tackle the complex concepts of resiliency, adaptability, and transformability (rf. Table 2.2). But hypotheses and results are all too often specific to an ecological subdiscipline, time period or space domain, and/or use case with little ability to apply these findings to other systems or find new system behavior a posteriori with little prognostic capability. Conversely, some prognostic models exist but lack the long-term, consistent, multiscaled data sets required to distinguish among alternate hypotheses or to assess performance among models (Schimel et al. 2011). Lastly, the current overuse of correlative statistics will not be sufficient when more mechanistic understandings are needed for prognostic applications (rf. Collins et al. 2010).

Again, NEON and other research facilities are charged to "enable an ecological forecasting," which means providing the data needed to forecast the future state of an ecosystem and to have the ability to augment the infrastructure to accommodate the future needs of ecological science. Ecological forecasting will not work without the close, collaborative working relationship among scientists, NEON (observatory) scientists, and sponsors. Meeting this challenge is the primary charge to the ecological community to confront (new) theory and develop new statistical approaches to advance our prognostic understanding.

2.5.5 Fifth Challenge

Our ability to communicate results of ecological forecasts has to be improved in order to have broad utility across sectors of society. This is everyone's responsibility: scientists, decision-makers, and the public, alike. Ecological forecasting should not attempt to provide a single "answer" to a question or be touted as a panacea. Rather, forecasts should provide a range of possible or likely outcomes or provide a trend analysis—an expected trajectory of an ecological process. In this way, a "decision space," or in engineering terms a "trade space," can bound or constrain a particular problem. Uncertainties should also be reported along with these analyses.

Reporting uncertainties goes hand-in-hand with the data in terms of informing how the data can be interpreted and how they are used, which delves into the realm of human behavior. When wishing to make science data applicable to decision-makers, interpreting uncertainties often equates to decisions based around risk. Uncertainties and risk management mean different things to scientists than to decision-makers and the public. Take, for example, the Intergovernmental Panel on Climate Change report (IPCC 2013) where the forecasting results have led scientists to refine their estimates, while convincing federal and state governments to initiate planning for food and water security, leaving average citizen disenfranchised in their own personal decision-making process by long 50–150-year forecasts, for example, why should I change my behavior if the risk may or may not hit in 100 years? New dialog and new paradigms have to be explored to make the communication of forecast results more understandable, meaningful, and actionable along all strata of society.

2.6 Future Expectations

Community engagement between the future users and NEON staff is challenging while building the observatory due to several competing pressures. Queries from future users stem from a genuine, sincere, and strong desire to understand what NEON (or other infrastructure) can provide to advance (their) science and education, what is new and exciting in the observatory design that is at the science frontier, and how they can contribute. Staff scientists are faced with applying their academic skillsets toward designing and building something that has never been done before, for example, the old paradigm of trying to building the plane while at the same time trying to fly it. At the same time, staff scientists are faced with an unfamiliar organizational structure of NEON that acts as a construction company, scientific institution, and a start-up company combined, each with its own culture, that often manifest in needs to build internal organizational function/structures for one culture. Compounding this dynamic, are the rapidly changing institutional needs, and the changing and unforeseen reporting and oversight of the sponsors themselves. The need to engage with the user community has never been greater and, at the same time, always outweighs the institutional capability to do so. This is not meant as an excuse, but rather a common, reoccurring reality seen by all research infrastructure during their construction.

Often, institutional structures are developed to foster broader engagement, such as technical working groups, advisory boards, workshops, and ad hoc position papers. While useful, they tend to be targeted efforts, or viewed by the community as having limited impact. Given the combination of culture change, the need to build institutional capacity, and the very large effort needed to report overall progress in building the infrastructure to the sponsor (i.e., NSF), it is unrealistic to think that the ability to broadly engage with future users will change in the short term (5–10 years), but rather it occurs slowly and incremental over long operational timescales, for example, decades.

Transitioning into operations (OPS) presents additional programmatic and funding challenges. Transitioning from construction to OPS is also part of a common development process (Figure 2.1). A different set of skills is needed for operations than construction. For example, the large number of PhD-level scientists needed to apply their intellectual capital for design and to assure the design has been constructed correctly, is different from the skill levels needed to execute field protocols and quality control the data. This will also mean transitioning all the ongoing relationships among the external user community, advisory groups, and staff scientists to meet the changing needs of OPS. Because workflows and protocols will be further tested and hardened during OPS, there will be a natural tendency to refine and optimize them. There will also be a natural tendency for sponsors to also require the optimization of workflows and protocols. Hence, system engineering approaches (discussed earlier) will be more than ever required during OPS to assure that the optimizations will not compromise the science being delivered.

Large federal investments in research infrastructure are often funded by "new" money, as in the case with NEON's construction funded through the MREFC program. While there may be some "new" monies being successfully secured in new annual federal budgets for an OPS budget, the majority comes from existing budgets, as in the case of the NEON OPS budget coming from the "Research and Related Activities" (R + RA) general account of the BIO Directorate at NSF. OPS budgetary decisions become a matter of balancing a portfolio of competing programs with very likely new programs emerging to directly support the science being delivered by the new research infrastructure. This is why OPS budgets are often viewed as a zero-sum game. Similarly, in Europe and elsewhere, the majority of OPS budgets for distributed research infrastructures come from budgets within member countries, each with their own political and scientific agenda, and different funding schedules. So within NSF or EU member countries, securing consistent long-term support for important environmental observations and experiments becomes a political discourse and trade-off on budgets from competing programs with very little input from the user community—and why we can expect in all circumstances to have downward pressure applied by sponsors to optimize OPS workflows and budgets. For example, the NSF MREFC construction award to NEON comes with formal assurances to support OPS, but the flow of funding has to be in the President's Budget, appropriated by Congress, transferred to NSF, and subject to internal NSF budget discussions. On the other hand, an EU infrastructure such as the Integrated Carbon Observing System (ICOS, www.icos-ri.eu) is distributed across European countries under a novel legal framework: European Research Infrastructure Consortium (ERIC). There is core ICOS OPS funding by the European Commission, but the majority of the OPS support comes from member countries, each with different National Budget cycles and priorities.

The Gartner hype cycle for technical development can be directly applied to building research infrastructures, Figure 2.5 (rf. Jarvenpaa and Makinen 2008). At the beginning, when the new concepts were being forged into the NEON design and there was broad adoption of the user community, NEON was touted as being everything to everyone (Section 2.1). Clearly, this cannot be the case and expectations were inflated. During the very difficult time of building the intellectual capital needed to design and construct NEON and faced with a tsunami of urgent tasks and reporting requirements, NEON staff were disillusioned and compromises were made. Now with NEON and similar environmental research infrastructure being close to OPS; hopefully the community has seen past inflated expectations, staff has rallied, and while the overall science capability is not what it could be, but at the same time, it is balanced with a realistic operational ability. It is also important to note that NEON was designed based on the understanding of the needs and technologies at that time with the goal of being able

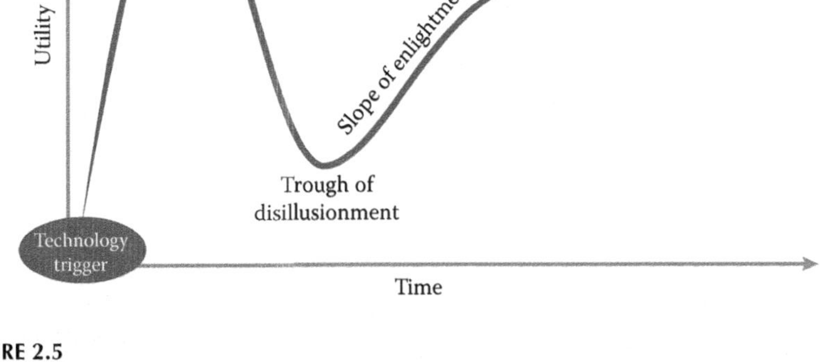

FIGURE 2.5
The abbreviated Gartner hype cycle of development.

to provide the data for ecological forecasting. It is not possible to imagine all the ways that NEON data will be used in the future, and as in the case with many large-scale federal investments in research infrastructure, the most profound discoveries are likely to be those not originally intended. The Gartner hype cycle helps communicate the development process and expectations of future science capability to scientists, decision-makers, and the public alike.

In conclusion, the success of NEON and other research infrastructures will be in how they are used, but also how well they are viewed in the constellation of all other federal (agency) programs, global initiatives, and private enterprises. Because its annual OPS budgets are subject to political (nonlinear) processes to determine funding rates, its scientific utility will also be judged in how well it can contribute to and interface with other Federal, private, and international programs, that is, USGS EROS, North American Carbon Program, USDA LTAR, DOE Atmospheric Radiation Measurement, National Aeronautics and Space Administration ROSES, NOAA Earth System Research Laboratory, and globally with Arctic Council, Group on Earth Observations, Future Earth, CoopEUS, etc. There are similar analogs with other environmental research observatories. This calls for a concerted effort in establishing and aligning efforts among these stakeholders. And even though the dovetailing with other programs was not inherently planned for, it does become part of the political discourse, funding for OPS, and metrics for success.

Acknowledgments

The authors acknowledge the National Science Foundation (NSF) for ongoing support. The NEON is a project sponsored by the NSF and managed under cooperative support agreement (EF1029808) by NEON, Inc. Any opinions, findings, and conclusions or recommendations expressed in this chapter are those of the authors and do not necessarily reflect the views of our sponsors. The authors acknowledge L. Goldman for graphics support. The authors also wish to acknowledge all those who were influential in NEON's development. Special thanks go to Drs. D. Schimel, M. Keller, A. Beasley, S. Collins, J. Ehlringer, C. Field, J. Franklin, B. Hayden, J. MacMahon, J. Melillo, and W. Michener for their vision of NEON in its nascent stages. The authors apologize for an acronym-rich chapter, unfortunately acronyms are all too commonplace in the world of project science (In Our Humble Opinion (IOHO)). The authors wish to thank the two anonymous reviewers for their constructive comments. This chapter would not have been written without decades of collegial interactions and community engagement with our peers and mentors.

References

AIBS, 2004a. Ecology and evolution of infectious diseases: Report from a NEON Science Workshop. Washington, DC: AIBS. ibrcs.aibs.org.

AIBS, 2004b. Ecological aspects of biogeochemical cycles: Report from a NEON Science Workshop. Washington, DC: AIBS. ibrcs.aibs.org.

AIBS, 2004c. Ecological impacts of climate change: Report from a NEON Science Workshop. Washington, DC: AIBS. ibrcs.aibs.org.

AIBS, 2004d. Designing NEON initiatives for invasive species: Report from a NEON Science Workshop. Washington, DC: AIBS. ibrcs.aibs.org.

AIBS, 2004e. Land use and habitat alteration: Report from a NEON Science Workshop. Washington, DC: AIBS. ibrcs.aibs.org.

AIBS, 2004f. Biodiversity, species composition, and ecosystem functioning: Report from a NEON Science Workshop. Washington, DC: AIBS. ibrcs.aibs.org.

Becknell, J.M., A. Desai, M. Dietze, G. Starr, J. Franklin, J. Hall, A. Pourmokhtarian, P. Stoy, P. Duffy, M. Binford, C.L. Staudhammer. 2015. Assessing interactions among changing climate, management, and disturbance in forests: A macrosystems approach. *BioScience* 65 (3): 263–274. doi:10.1093/biosci/biu234.

Benstead, J. P., L. A. Deegan, B. J. Peterson, A. D. Huryn, W. B. Bowden, K. Suberkropp, K. M. Buzby, A. D. Green, and J. A. Vacca, 2005. Responses of beaded Arctic stream to short-term N and P fertilization. *Freshwater Biology* 50: 277–290.

Berkes, F., J. Colding, and C. Folke, 2003. Introduction. In: *Navigating Social-Ecological Systems: Building Resilience for Complexity and Change.* Eds. F. Berkes, J. Colding, and C. Folke. Cambridge, U.K.: Cambridge University Press, pp. 1–30.

Chapin, F. S., G. P. Kofinas, and C. Folke, 2009. A framework for understanding change. In: *Principles of Ecosystem Stewardship.* New York: Springer, 401pp.

Chen, Y., and A. Hardisty, 2014. A common reference model for environmental research infrastructures. *iLEAPS Newsletter* 2014: 17–19.

Chen, D., G. Doumeingts, and F. Vernadat, 2008. Architectures for enterprise integration and interoperability: Past, present and future. *Computers in Industry* 59: 647–659.

Chen, Y., A. Hardisty, A. Preece, P. Martin, M. Atkinson, Z. Zhao, B. Magagna, H. Schentz, and Y. Legre, 2014. Analysis of common requirements for environmental science research infrastructures. In: *Proceedings of Science International Symposium on Grids and Clouds*, Taipei, Taiwan, March 17–23, 2013. v032, 11pp.

Chen, Y., P. Martin, H. Schentz, B. Magagna, Z. Zhao, A. Hardisty, A. Preece, M. Atkinson, R. Huber, and Y. Legre, 2013. A common reference model for environmental science research infrastructures. In: *Proceedings of the 27th Conference on Environmental Informatics,* Hamburg, Germany, pp. 665–673.

Collins, S. L., S. R. Carpenter, S. M. Swinton et al., 2010. An integrated conceptual framework for long-term social–ecological research. *Frontiers in Ecology and the Environment* 9(6): 351–357.

Dale, V., L. Joyce, S. McNulty et al., 2001. Climate change and forest disturbances. *BioScience* 51: 723–734. doi:10.1641/0006-3568(2001)051[0723:CCARD]2.0.CO.

Heffernan, J. B., P. A. Soranno, M. J. Angilletta Jr. et al., 2014. Macrosystems ecology: Understanding ecological patterns and processes at continental scales. *Frontiers in Ecology and the Environment* 12: 5–14. doi:10.1890/130017.

Heinz Foundation, 2006. Filling the gaps: Priority data needs and key management challenges for national reporting on ecosystem condition. www.heinzctr.org/ecosystem, accessed June, 2014.

Heinz Foundation, 2008. The state of the nation's ecosystems 2008: Measuring the lands, waters, and living resources of the United States. www.heinzctr.org/ecosystems, accessed June, 2014.

Holdren, J., T. Dickenson, G. Paulson et al., 2014. *National Plan for Earth Observations.* National Science and Technology Council, Executive Office of the President. Office of Science, Technology and Policy, Washington DC, 71pp

Holling, C., 1973. Resilience and stability of ecological systems. *Annual Review of Ecology and Systematics* 4(1): 1–23.

Hughes, R. L., K. C. Beatty, and D. L. Dinwoodie, 2014. *Becoming a Strategic Leader.* San Francisco, CA: Jossey-Bass Pub., 331pp.

Institute of Medicine (IOM), 2013. *Environmental Decisions in the Face of Uncertainty.* Washington, DC: The National Academies Press, 209pp.

Intergovernmental Panel on Climate Change (IPCC), 2013. *Climate Change 2013: The Physical Science Basis: Contribution of Working Group I to the Fifth Assessment Report on the Intergovernmental Panel on Climate Change.* Eds. T. F. Stocker, D. Qin, G.-K. Planer, M. Tignor, S.F. Allen, I Boschung, A. Nauels, Y. Xia, V. Bex, and P. M. Midgeley. Cambridge, U.K.: Cambridge University Press, 1535pp.

Iverson, L. R. and D. MacKenzie, 2013. Tree-species range shifts in a changing climate: Detecting, modeling, assisting. *Landscape Ecology,* 28: 879–889. doi:10.10007/s10980-013-9885-x.

Jain, R. K. and H. C. Triandis, 1997. *Management of Research and Development Organizations: Managing the Unmanageable.* Wiley Series in Engineering and Technology Management. New York: Wiley Press Inc., 315pp.

Jarvenpaa, H. and S. J. Makinen, 2008. Empirically detecting the hype cycle with the life cycle indicators: An exploratory analysis of three technologies. In: *Proceedings from IEEE International Conference on Industrial Engineering and Engineering Management,* Singapore, pp. 12–16. doi:10.1109/IEEM.20084737823.

Johnson, J. B. and K. S. Omland, 2004. Model selection in ecology and evolution. *Trends in Ecology and Evolution* 19: 101–108.

Katz, R. W. and A. H. Murphy, 2005. *Economic Value of Climate and Weather Forecasts.* Cambridge, U.K.: Cambridge University Press, 224pp.

Kossiakoff, A., W. N. Sweet, S. J. Seymour, and S. M. Biemer, 2011. *Systems Engineering Principles and Practice,* 2nd edn. Hoboken, NJ: Wiley & Sons Inc., 453pp.

Laurance, W. F., B. Dell, S. M. Turton et al., 2011. The 10 Australian ecosystems most vulnerable to tipping points. *Biological Conservation* 144: 1472–1480.

Loescher, H.W., Vargas, R., M. SanClements, B. Bond-Lamberty et al., 2016. Challenges in developing large scale environmental observatories: Building global ecological understanding. *Ecological Processes* (in review).

Long Term Ecological Research (LTER) Project Office, 1989. *1990's Global Change Action Plan Utilizing a Network of Ecological Research Sites.* Seattle, WA: Long Term Ecological Research Project Office, University of Washington, College of Forest Resources, AR-10, 36pp.

McDonnell, T. C., S. Belyazid, T. J. Sullivan, H. Sverdrup, W. Bowman, and E. M. Porter, 2014. Modeled subalpine plant community response to climate change and atmospheric nitrogen deposition in Rocky Mountain National Park, USA. *Environmental Pollution,* 187: 55–64. doi:10.1016/j.envpol.2013.12.021.

National Research Council (NRC), 2001. *Grand Challenges in Environmental Sciences.* Washington, DC: The National Academies Press, 106pp.

NRC, 2007. *Understanding Multiple Environmental Stresses: Report of a Workshop.* Washington, DC: The National Academies Press, 154pp.

NRC, 2011a. *Adapting to the Impacts of Climate Change; America's Climate Choices: Panel on Adapting to the Impacts of Climate Change.* Washington, DC: The National Academies Press, 272pp.

NRC, 2011b. *A Review of the U.S. Global Change Research Program's Draft Strategic Plan.* Washington, DC: The National Academies Press, 62pp.

NSF, 2000a. Report on First Workshop on the National Ecological Observatory Network, held January 10–12 at Archbold Biological Station, Lake Placid, FL. Arlington, VA: National Science Foundation.

NSF, 2000b. Report to National Science Foundation from the Second Workshop on the Development of a National Ecological Observatory Network (NEON), held on March 9–13 at San Diego Supercomputer Center, La Jolla, CA. Arlington, VA: National Science Foundation.

NSF, 2000c. Report to National Science Foundation from the Third Workshop on the Development of a National Ecological Observatory Network (NEON), held on May 3–4, 2000 at Sante Fe Institute, Santa Fe, NM. Arlington, VA: National Science Foundation.

NSF, 2002a. Report to the National Science Foundation from the Fourth Workshop on the Development of a National Ecological Observatory Network (NEON): Standard Measurements and Infrastructure Needs, held on June 4–5, 2002 at the Millenium Hotel, Boulder, CO. Arlington, VA: National Science Foundation.

NSF, 2002b. Final Report on NEON-V: CRIPTON Workshop, held on June 14–16, 2002 at Field Museum of Natural History, Chicago, IL. Arlington, VA: National Science Foundation.

NSF, 2002c. Report to the National Science Foundation from the Sixth Workshop of a National Ecological Observatory Network (NEON): Information Management, held on September 16–18 at the National Center for Ecological Analysis and Synthesis, University of California, Santa Barbara, CA. Arlington, VA: National Science Foundation.

NSF, 2002d. *National Ecological Observatory Network.* Arlington, VA: National Science Foundation.

NSF, 2002e. *Facilities Management and Oversight Guide.* Arlington, VA: National Science Foundation.

Peters, D. P. C., P. M. Groffman, K. J. Nadelhoffer, N. B. Grimm, S. L. Collins, W. K. Michener, and M. A. Huston, 2008. Living in an increasingly connected world a framework for continental scale environmental science. *Frontiers in Ecology and the Environment* 6: 229–237. doi:10.1890/070098.

Peters, D. P. C., H. W. Loescher, M. SanClements, and K. M. Havstad, 2014. Taking the pulse of a continent: Role of observatories and long-term research networks to fill critical knowledge gaps. *Ecosphere* 5(3): Article 29, 1–23. doi:10.1890/ES13–00295.1.

Presidents Council of Advisors on Science and Technology (PCSAT), 2011. Sustaining Environmental Capital: Protecting Society and the Economy. Report to the President. Office of Science Technology and Policy, Executive Branch, Washington DC. www.whitehouse.gov/ostp/pcast.

Quintero, I. and J. J. Wiens, 2012. Impacts of climate change on the future of biodiversity. *Ecology Letters* 15: 365–377.

Reckhow, K. H., 2003. Bayesian approaches in ecological analysis and modeling. In: *Models in Ecosystem Science.* Eds. C. D. Canham, J. J. Cole, W. K. Lauenroth. Princeton, NJ: Princeton University Press, pp. 168–183.

Robertson, G. P, 2008. Long-term ecological research: Re-inventing network science. *Frontiers in Ecology and the Environment* 6: 281. doi:10.1890/1540-9295.

Schimel, D. S. and M. Keller, 2015. Big questions, big science: Meeting the global ecology. *Oecologia* 177: 925–935. doi:10.1007/s00442-015-3236-3.

Schimel, D., A. Janetos, P. Backlund, J. Hatfield, M. Ryan, S. Archer, and D. Lettenmaier, 2008. *Synthesis.* In: *The Effects of Climate Change on Agriculture, Land Resources, Water Resources, and Biodiversity.* A report by the U.S. Climate Change Science Program and the Subcommittee on Global Change Research, U.S. Global Change Research Program, Washington, DC, 362pp.

Schimel, D., M. Keller, S. Berukoff, R. Kao, H. W. Loescher, H. Powell, T. Kampe, D. Moore, and W. Gram, 2011. *NEON Science Strategy; Enabling Continental-Scale Ecological Forecasting.* Boulder CO: Pub. NEON Inc., 55pp.

Shishko, R., 1995. *NASA Systems Engineering Handbook.* Pub. SP-6105. Washington, DC: National Aeronautics and Space Administration, 154pp.

Slavik, K., B. J. Peterson, L. A. Deegan, W. B. Bowden, A. E. Hershey, and J. Hobbie, 2004. Long-term responses of the Kuparuk River to phosphorus fertilization. *Ecology* 85(4): 939–954.

Smith, B. D. and M. A. Zeder, 2013. The onset of the Anthropocene. *Anthropocene* 4: 8–13.

Smith, M. D., A. K. Knapp, and S. L. Collins, 2009. A framework for assessing ecosystem dynamics in response to chronic resource alterations induced by global change. *Ecology* 30: 3279–3289.

Soranno P. A. and D. S. Schimel, 2014. Macrosystems ecology: Big data, big ecology. *Frontiers in Ecology and the Environment* 12(1): 3.

Thorpe, R. S., A. Barlow, A. Malhotra, and Y. Surget-Groba, 2015. Widespread parallel population adaptation to climate variation across a radiation: Implications for adaption to climate change. *Molecular Ecology* 24: 1019–1030. doi:10.111/mec.13093.

United States Global Change Research Program (USGCRP), 2013. Climate assessment report, Third Assessment. U.S. Global Change Research Program, Washington DC.

Wall, D., 2007. Global change tipping points: Above- and below-ground biotic interactions in a low diversity ecosystem. *Philosophical Transactions of the Royal Society B* 362: 2291–2306. doi:10.1098/rstb.2006.1950.

Zolkos, S. G., P. Jantz, T. Cornier, L. R. Iverson, D. W. McKenney, and S. J. Goetz, 2015. Projected tree species redistribution under climate change: Implications for ecosystem vulnerability across protected areas of the Eastern United States. *Ecosystems* 18: 20–220. doi:10.1007/s10021-014-9822-0.

3

Field Phenotyping: Concepts and Examples to Quantify Dynamic Plant Traits across Scales in the Field

M. Pilar Cendrero-Mateo, Onno Muller, Hendrik Albrecht, Andreas Burkart, Simone Gatzke, Benedikt Janssen, Beat Keller, Niklas Körber, Thorsten Kraska, Shizue Matsubara, Jinquan Li, Mark Müller-Linow, Roland Pieruschka, Francisco Pinto, Pablo Rischbeck, Anke Schickling, Angelina Steier, Michelle Watt, Ulrich Schurr, and Uwe Rascher

CONTENTS

Abstract

The increase in population is leading to an unprecedented demand on agriculture and natural resources. To meet human and nature needs, food production must drastically increase while, at the same time, agriculture's environmental footprint must decrease dramatically. In times of global change, a better understanding of the dynamic spatiotemporal adaptation of our crops is needed to provide the basis for crop breeding, management, and protection measures. To scientifically understand the mechanisms behind the dynamic structural and functional adaptation of plant traits, it is mandatory to phenotype plants under natural, that is, fluctuating environmental conditions in the field. Plant phenotyping aims for a quantitative description of plant traits, which is affected by genetic variation as well as by the environment.

In this chapter, we review and develop concepts for phenotyping of dynamic plant traits in the field. We give an overview on the most prominent sensors and measurement concepts that are the backbone for plant phenotyping initiatives. We focus on the quantification of most relevant traits, which are related to functional and structural root development, plant structure, water relations and transpiration, and photosynthetic energy conversion. In addition to the sensors and measurement modes, we review positioning systems that allow for a reproducible recording of experimental plots at different temporal and spatial resolution. Finally, the quantitative data on the plant's phenotype is integrated with environmental data and information on anthropogenic impact to better understand gene × environment × management interactions, which ultimately will be the basis for a sustainable and resource-efficient use of our plant resources in a future bio-based economy.

Keywords: Plant phenotyping, Root, LIDAR, Stereo cameras, Thermal imaging, Active thermography, Hyperspectral, RGB camera, Vegetation indices, Chlorophyll fluorescence, Ground positioning systems, UAV, Airplanes, Environmental monitoring

3.1 Introduction

Humankind has already entered an era where anthropogenic interference dominates large parts of our planet. The growing human population and high living standards are the driving forces for increasing demands on our Earth's ecosystem services, which in turn are leading to a greatly increased use of marine and terrestrial ecosystems. As a consequence, we are now facing four main global challenges:

1. Safeguard food, feed, and water supply to a growing global population
2. Ensure the availability and access to energy for everyone
3. Protect and sustain natural resources and the environment
4. Ensure human well-being and health to the world population

Addressing any one of these major challenges cannot be considered as a stand-alone task because of their interconnectedness. To increase food production, more land and energy may be required, thus, exacerbating already-existing pressures on nonrenewable resources. To satisfy an increasing demand for renewable energy, we may need to enhance biomass production for biofuels, thus competing for agricultural land required for food production. To protect the environment and its important services and to preserve biodiversity, we must avoid agricultural intensification potentially affecting food production.

Sustainable agriculture and primary vegetation productivity are the basis for a future bio-economy. The EU strategy and action plan named bio-economy as a key element for smart and green growth in Europe that reconciles food security with the sustainable use of renewable resources for industrial purposes, while ensuring environmental protection (McCormick and Kautto 2013).

But earth's primary productivity is constantly under the threat of several external abiotic and biotic factors, such as drought, extreme temperatures, pests, and nutrient limitation. The magnitude and impact of some of these factors may be exacerbated by global changes, which involve climate, but also major changes that are caused by a rapidly growing human population and the current unsustainable utilization of some critical, but limited and nonrenewable resources.

Thus, the sustainable and resource-efficient use of our vegetation is of outmost importance for the survival of humankind. Great advances have been made in our understanding of the potential of the genetic background of our vegetation and many agricultural crops like rice (Sequencing Project 2005) or recently wheat (Eversole et al. 2014) are sequenced. After these major steps forward, nowadays the plant phenotype is put in the center of plant research. We need to identify which genes are expressed under which environmental and management conditions (i.e., drought, extreme temperatures, pests, and nutrient limitation) resulting in a certain phenotype.

On the other hand, plant breeders, ecosystem ecologists, and agronomists need accurate and reliable sensing technologies to support their strategic decisions by detailed spatial and temporal information on plant growth (and plant growth stages) to predict future plant development and yield, as well as ecosystem functions, for example, net primary productivity and water use (Auernhammer 2001). These methods must be able to bridge spatial areas (e.g., from single plants to experimental plots, fields, and ecosystems) to deliver the necessary data to the breeder, land manager, or scientist. One key problem under field conditions is the large number of plant phenotypes as a result of almost endless permutations of genetic information and environmental conditions.

Plant phenotyping in the strict sense is not a fully new discipline. For decades agronomists and plant scientist have described the phenotype of a plant in the greenhouse and in the field. Optical approaches to detect crop status have already been developed in the 1950s and 1970s (without actually naming these approaches plant phenotyping) (Moran et al. 2003). Also, noninvasive measurements were often the method of choice to reproducibly and repeatedly quantify plant traits under greenhouse and field conditions.

Considering the recent technological developments, there are currently several phenotyping initiatives that aim to set up and improve plant phenotyping in the laboratory and also under field conditions. It is generally agreed that the discipline needs both controlled laboratory and greenhouse approaches as well as field approaches, where the plant is exposed to natural and thus varying environmental conditions (Rascher et al. 2009). Especially in the field, the environmental conditions vary by nature, and the plant phenotype has to be characterized under these nonstationary conditions, which adds an additional dimension on the requirement for sensors and the recording of time series with the appropriate spatial and temporal resolution to address salient questions. In this context, the long-term ecological research field sites (LTER), the Carbo Europe long-term study sites, various institutional field sites, and decades of ecophysiological plant research have provided the basis for modern field phenotyping activities. Knowledge from these interdisciplinary projects is the basis for sensor development, measurement protocols, and the scientific understanding of dynamic adaptation of plant traits.

The need for improved technical and scientific demands has resulted in various integrated activities worldwide, where, on the one hand, research institutions, agencies, and universities have developed phenotyping infrastructure and, on the other hand, user networks are developing the scientific tools, knowledge, and practical applications.

In this chapter, we will review the four main pillars that are needed to develop a network of plant phenotyping infrastructure. To address the challenges mentioned earlier, we will need to identify

- Which plant traits need to be monitored to understand the dynamics of the gene × environment × management interaction that determines the expression of a plant's phenotype
- A set of sensors and measurement approaches that allow fast and noninvasive quantification of relevant plant traits

- A set of positioning systems that facilitate to place the sensors in an automated and reproducible manner in the field
- A concept for environmental sensors to link the plant phenotype to the dynamic changing environment

3.2 Concept of Field Phenotyping: From Traits to Sensors and Positioning Systems

3.2.1 Relevant Traits Define Sensors for Field Prototyping

To understand how changes of environmental conditions will modulate the expression of crop genetic information, we need first to identify the relevant traits that need to be monitored. For field phenotyping, we propose four main clusters: (1) structural and functional root traits, (2) canopy and plant developmental morphology, (3) traits that determine water relations (e.g., water uptake and transpiration and water-use efficiency) of plants, and (4) functional traits that are related to photosynthetic energy conversion and carbon uptake (Figure 3.1). Depending on the scientific question of interest, those traits need to be monitored on different scales ranging from leaf, whole plant to the canopy scale. Only a good understanding of the dynamic changes at leaf, plant, and canopy level will allow us to understand field-scale

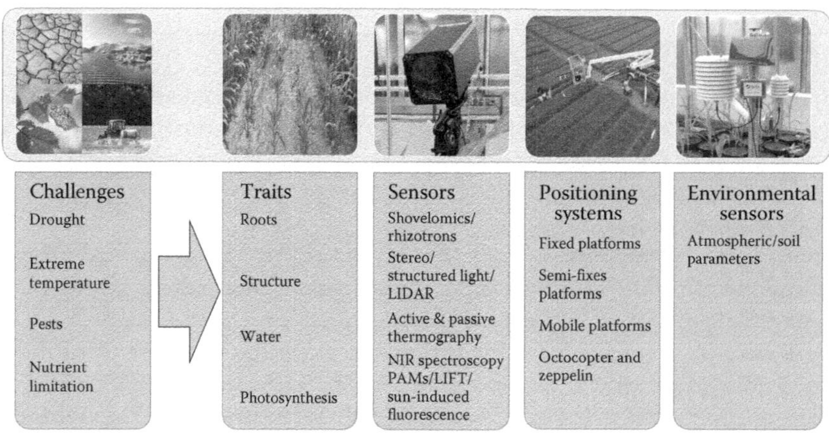

FIGURE 3.1
The pillars of field phenotyping. To evaluate relevant traits, the right sensors have to be chosen and they have to be positioned by dedicated field positioning systems. This infrastructure then can be applied in dedicated field experiments that are additionally equipped with dedicated environmental sensors.

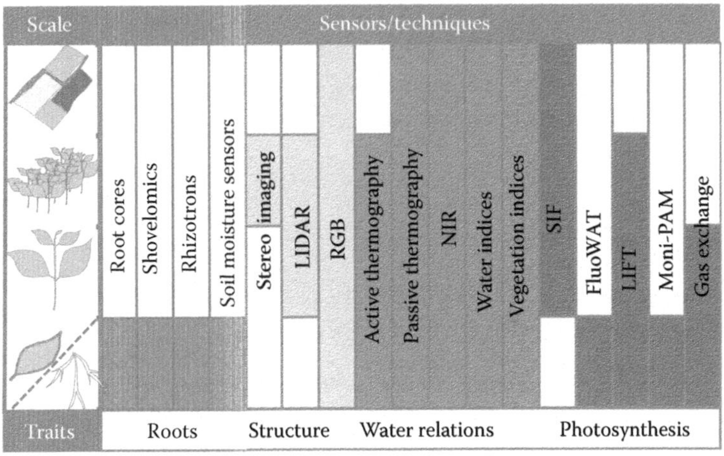

FIGURE 3.2
(See color insert.) Scheme of the sensors and techniques to characterize plant phenotypic traits for roots, structure, water relations, and photosynthesis, where the colored part of the column indicates the scale they can be employed for. These are (1) roots mostly by excavation, (2) individual leaves, for example, by clip on devices, (3) capturing whole plants or measuring organs within canopy, (4) on top of canopy mostly by mobile platforms, and (5) field from flying platforms and satellite.

measurements. In the framework of several projects, promising sensors were identified and measurement concepts to quantify traits were developed to analyze plant structure (root and shoot), photosynthesis, and water relations at different scales (Figure 3.2).

3.2.1.1 Measuring Root Traits

Many of the traits required in future crops are tightly linked to root properties and root system architecture which can strongly affect crop yield. The root system is very plastic and responds strongly to environmental influences. Sustainable plant production rely on root systems adapted to growing conditions in the field.

In the past, the majority of root phenotyping efforts were focused on establishing phenotyping platform in the green house, where the gene × environment effects are often missing (Kuijken et al. 2015). The main reason for this negligence is the technical difficulty in accessing the roots for phenotyping in the field, especially when temporal and spatial variability have to be considered too. Roots can be measured directly or indirectly in the field using destructive and nondestructive phenotyping technologies. Direct, destructive phenotyping methods are made by digging the roots out or exposing them in one point and time, such as manual coring or via shovel (Rich 1990; Trachsel et al. 2010). Both coring and "shovelomics" can be carried out rapidly to achieve the required numbers and repetition needed for field phenotyping to compare genotypes (Wasson et al. 2014; York and Lynch 2015). However, both approaches are not dynamic because they cannot be repeated over time on the same root systems.

A true dynamic approach is the use of rhizotrons, where repeated measurements of the same root systems are possible. "Mini-rhizotrons" are cylindrical clear tubes, inserted into the ground so that roots that grow on the tubes can be imaged by a camera within the tube. For instance, mini-rhizotrons have been used to measure the time course of descending root profiles of different crop species (Thorup-Kristensen and van de Boogaard 1998). To fully understand root allocation patterns, a traceable and reproducible approach to quantify the camera's technical imprecision is a prerequisite to avoid the risk of performing an improper quantification of the root system architectures (Roberti et al. 2014).

Unless the crop is very young, all technologies described earlier have the common problem of sampling a very small portion of the root systems. Coring only takes a small vertical transect through a mixture of roots below the crop canopy and generally the cores are of small diameters (4–10 cm). Shovelomics takes a single root system to a depth of 20 or 30 cm, leaving behind roots torn away during the shoveling and washing process. Mini-rhizotrons provide a small viewing window (cm range), having the added disadvantage of requiring insertion prior or during crop establishment.

Root phenotyping in the field is on a very early stage and reliable noninvasive methods to quantify root traits on the appropriate spatial and temporal level are still not available. Thus, much work is needed to develop the appropriate measurement concepts to improve root phenotyping in the field.

3.2.1.2 Measuring Structural Traits

Plant structure and function are known to be linked in natural and agronomic systems like leaf angle distribution, that in corn greatly influences light interception and yield in densely planted canopies (Ford et al. 2008). During their seasonal development, most plants display strong morphological changes, which depend on the availability of resources and on the fluctuation of abiotic and biotic factors. For instance, leaf orientation can be greatly affected by environmental factors like drought, which can be highly useful for breeders to compare drought stress tolerance between cultivars (Müller-Linow et al. 2015). Structural properties that are altered on the diurnal and seasonal scale may affect the efficiency of light interception within the canopy and thus may influence canopy light-use efficiency (Müller-Linow et al. 2015).

The most commonly used geometric measure of plant canopy is the leaf area index (LAI), which relates the projected leaf surface to the soil surface. However, the LAI does not provide information on plant architecture or the distribution of leaf orientations. For nonmanual estimation of plant architecture, several methodical approaches are available, which include structured light approaches, laser scanning techniques, and stereo imaging. In structured light techniques, the reconstruction of a plant's shape is derived by analyzing the distortion of a projected light pattern (Kjaer and Ottosen 2015). High-resolution 3D models can be achieved, which often also allow to separate objects having different colors (e.g., fruits and leaves) (Bellasio et al. 2012). Stable light conditions are needed to perform structured light measurements thus it is limited to greenhouse or growth chamber environments.

Alternatively, LIght Detection And Ranging (LIDAR) systems can measure the distance between the sensor and the objects around it very fast, enabling the construction of 3D point clouds. Appropriate algorithms can transform these point clouds into a highly precise digital reconstruction of crop structures and architecture (Sanz-Cortiella et al. 2011).

In recent years, stereo camera setups were developed (Biskup et al. 2007; Müller-Linow et al. 2015). Two cameras with a defined baseline distance are used to record simultaneously stereo images. From these stereo images, the distance of canopy elements can be computed and a full 3D representation of the canopy can be calculated. Further processing of the data can reveal relevant structural parameters such as canopy area, leaf angle distribution, the number of leaves, or the ratio between leaves and fruits, which translates impressive 3D images in quantitative data on vegetation properties (Rascher et al. 2010; Müller-Linow et al. 2015; Figure 3.3).

Dynamic changes in canopy structure were long unvalued in science and only limited implications of structural variations on ecosystem functioning were assumed. In recent years, however, structure–function relations have recently being put in the focus of breeding strategies as structural optimization may still pose some potential for yield improvements (Zhu et al. 2010). During growth, light availability becomes increasingly limited and fluctuating in the lower canopy layers. Plants may adapt to this by layer-specific distributions of leaf orientations. Structural changes are recognized to sensitively display stresses such as leaf rolling when plants are exposed to drought stress (O'Toole and Cruz 1980). Elevated atmospheric CO_2 concentrations were described to increase the total leaf area in soybean shoots (Ainsworth et al. 2002), and wind logging has been put back on the agenda of breeders to reduce the economic loss produced by wind. Thus, the quantitative understanding of the interplay of structural canopy traits with their environmental conditions is currently revisited, even though one does not want to underestimate the complexity, which is behind the mechanisms that control canopy structure.

3.2.1.3 Measuring Water Relations Traits

Breeding for drought tolerance has to consider various traits responsible for plant water status. Leaf water potential may be the most important parameter, but it is impossible to measure water potential nondestructively. Thus, various other methods have been tested that assess either leaf and canopy

FIGURE 3.3
(See color insert.) Estimation of the leaf angle distribution from a 3D reconstruction of a sugar beet canopy: (a) The stereo camera setup (see insert) has been mounted on a cherrypicker 3.5 m top of canopy; (b) shows a region of interest with ~6 plants in the original left camera RGB; (c) filtered depth (disparity) map of (b) with pixel colors indicating the object depth; (d) single leaf segmentation, which is used for further individual leaf surface modeling; (e) leaf surface models are used to derive different leaf traits like leaf area, canopy area, or leaf angle distributions; (f) the leaf angle distribution of the zenith angle of the previous reconstruction; the zenith angle ranges from 0° (flat leaf surface) to 90° (errected surface) and can derived locally or for the complete leaf.

water content or approximate transpiration rates (Farrar et al. 2011; Sampoux et al. 2011). In this context, measurements of near-infrared (NIR) and thermal imaging have been introduced as the most promising measurement approaches in the lab and in the field (Fiorani and Schurr 2013).

A qualitative approach to estimate plant water content is NIR measurements using the relative depth of the water absorption band in the NIR region (1370 and 1870 nm). This approach generally provides a good relative estimate of water content but to our knowledge there is no study available that describes the retrieval of plant water content as absolute physical values.

A second approach is exploiting thermal cameras, which are sensitive within the infrared region (9–13 μm spectral range) to evaluate plant transpiration and evapotranspiration (ET). The principle of passive thermography is that surfaces are cooled by ET, so surface temperatures are lower than ambient temperature, which is proportional to the rate of ET. However, leaf temperature (T_L) does not only depend on ET rates but also depend on the leaf boundary layer, which is a thin layer of air at the leaf surface. The thickness and composition of the leaf boundary layer determine how fast heat can be dissipated, that is, increasing leaf boundary layer decreases the transfer of heat from the leaf to the atmosphere and vice versa (Leuning et al. 1989). Furthermore, leaf boundary layer and thus T_L respond in a dynamic way to variable environmental conditions. Parameters such as solar irradiance and ambient air temperature are highly fluctuating and highly affect the leaf boundary layer.

To overcome these problems, Jackson et al. (1981) developed the crop water stress index (CWSI), which normalizes leaf temperature against the prevailing environmental conditions. This index is based on the comparison of leaf temperature to wet and dry reference surfaces. The CWSI and other indices have been shown to be sensitive to evapotranspiration and can be used to detect drought stress–induced stomatal conductance (Jackson et al. 1981; Jones 1999; Cohen et al. 2005; Grant et al. 2006; Möller et al. 2007; Alchanatis et al. 2009).

Passive thermography, and particularly the CWSI, has become a widely used tool for measuring plant evapotranspiration to analyze high numbers of plants in a short period of time. Passive thermography measures only one part of the overall plant–water relations, namely, ET. To understand plant–water relations in whole plants in response to changing environmental conditions (e.g., drought), it is essential to know how different water fluxes between roots, stems, and leaves are connected.

A promising approach is active thermography, where T_L is actively manipulated by a short heat pulse. An additional heat pulse increases T_L transiently. After a short time, T_L will decrease again approaching the former steady state. The time constants (τ) of heating or cooling can be measured. This time constant (τ) depends on the leaf heat capacity per unit area (CA_{leaf}^{-1}) and the leaf heat transfer coefficient (h_{leaf}). High leaf water content leads to a higher heat capacity and consequently higher τ. The second parameter affecting τ is h_{leaf}, which describes how fast a leaf is able to dissipate heat. h_{leaf} depends on the boundary layer and is therefore highly affected by environmental conditions. For instance, increasing wind decreases the boundary layer, which in turn increases h_{leaf} and decreases τ. Also, stomatal conductance affects h_{leaf}. High stomatal conductance accelerates leaf heat dissipation and consequently h_{leaf} increases and τ decreases. Active thermography was successfully tested in laboratory at leaf scale and under greenhouse conditions at canopy scale. At leaf scale, a linear relationship between τ and LWC and consequently CA_{leaf}^{-1} was found. This relationship changed when the leaf boundary layer and thus h_{leaf} was manipulated by wind.

We would like to highlight the need to combine passive and active measurements to fully understand the overall plant–water relations in response to drought. Intuitively, the water content limits water loss by transpiration and thus controls transpiration rates. However, neither the CWSI alone, nor τ alone are appropriate to reflect this relation. By combining τ with CWSI, one may be able to track changes in LWC, boundary layer conductance, and transpiration. This will facilitate a better understanding of the dynamic responses of plants to optimize their water relations and help to better understand the strategies to cope with drought stress.

3.2.1.4 Measuring Photosynthesis Traits

Plant growth and productivity do not only depend on the availability of nutrients and water but to a great extent on adequate sunlight and temperature. Variations of these conditions affect photosynthesis rates and consequently plant productivity. Even though photosynthesis being the primary process that determines plant growth, there is no simple link between photosynthetic CO_2 uptake rates and plant growth (Körner 2000). Nevertheless, measurements of photosynthesis are essential to either refine plant growth models or alternatively photosynthesis may be used as a sensitive indicator for environmental limitations of a plant energy metabolism.

In the past decades, different approaches have been used to estimate photosynthesis, including gas exchange measurements, spectral vegetation indices, and chlorophyll fluorescence.

Gas exchange measurements are widely used at leaf and plant level. The portable infrared gas analyzers using small leaf clip chambers allow measurements at leaf level, whereas customized whole plant chambers are built in the field to measure at canopy level (Burkart et al. 2007). These methods however are very labor intensive and may never be used for rapid screening of numerous field plots. They are nevertheless essential tools for method validation discussed in the following text, which allow noninvasive measurements of photosynthesis at field plot scale.

VIS-NIR spectroscopy has become a versatile and accessible proxy for plant photosynthetic capacity. Spectral analysis can be performed at all different scales using point and imaging spectroscopy (Chuvieco and Huete 2010). The addition of spatial information by imaging spectroscopy offers new opportunities for plant phenotyping. A high number of so-called spectral vegetation indices (VIs) have been developed to quantify pigment contents as well as structural and physiological properties at leaf, plant, and canopy level (Jackson and Huete 1991). VIs combine spectral information of two or more bands. The selected spectral bands are combined in a manner that enhances the reflectance properties of specific molecules or that could identify plant stress response (see Jansen et al. 2014 for an overview on the most commonly used VIs). Mainly due to its simplicity, VIs are widely used in research in the area of breeding, precision agriculture, and remote sensing. Nevertheless, one should not miss the potential of using the full vegetation reflectance spectra for plant phenotyping. Some of the current and most powerful approaches are partial least squares regression (Feilhauer et al. 2010), supervised and unsupervised endmember selection and unmixing, continuous support vector machines (Hostert et al. 2003), multi-block analysis (Eiden et al. 2007), or simplex volume maximization (Roemer et al. 2012). All these methods provide significantly more accurate results than the use of VIs; however, their application for phenotyping purposes requires major adaptation of computer algorithms and data processing.

However, the reflectance-based approaches only allow a quantification of photosynthesis pigments. Functional regulation of the photosynthetic apparatus may be reflected in absorption changes around 531 nm that are related to the epoxidation of the xanthophyll cycle pigments during non-photochemical energy dissipation. These changes can be measured using the photochemical reflectance index (PRI), which is widely used in phenotyping and remote sensing (Gamon et al. 1990; Garbulsky et al. 2011). It has to be noted that the PRI is greatly influenced by structural factors, such as chlorophyll/carotenoid ratio, leaf angle, and illumination geometry (Barton and North 2001), and thus this index has to be used with care.

Currently, the most widely used technique to quantify the efficiency of photosynthesis exploits the fluorescence signal of chlorophyll. Chlorophyll fluorescence is light reemission in the red (690 nm) to NIR (740 nm) following light absorption by photosynthetic pigments (chlorophylls and carotenoids) in plants. The principle underlying the use of chlorophyll fluorescence as an indicator of plant photosynthetic status is relatively straightforward. Absorbed light energy excites chlorophyll molecules and de-excitation of this energy is mainly attained through three competing processes: photosynthesis, radiative loss of photons or chlorophyll fluorescence, and non-radiative thermal energy dissipation (non-photochemical quenching, NPQ). As these three energy dissipation processes compete for excitation energy, changes in one process (e.g., photosynthesis) will affect the other two. Hence, by measuring chlorophyll fluorescence, we can derive information on NPQ and photosynthesis (Maxwell and Johnson 2000; Porcar-Castell et al. 2014). Major developments in the instrumentation for measuring chlorophyll fluorescence have been made in the last decade, and currently more than 500 scientific articles are published each year that use this method* (for recent reviews, we refer to Baker 2008, Papageorgiou and Govindjee 2004). The most widely used technique is the pulse-amplitude modulation (PAM); in this approach, chlorophyll molecules are brought to excited states by absorption of active measurement light (Schreiber 1986, 2004). The handheld Mini-PAM device allows for a quick assessment of leaf fluorescence, whereas the stationary Monitoring-PAM (e.g., Porcar-Castell et al. 2008) is developed for field measurements and can be clamped on the leaf permanently as long as weather conditions or leaf growth allow. Upscaling this method to measure the canopy from above is limited by the close vicinity needed to apply a saturating light pulse. In recent years, the light-induced fluorescence transient method (LIFT, Kolber et al. 2005; Pieruschka et al. 2010) was developed. This method allows measurements of fluorescence parameters from a distance of several meters. This so-called pump-and-probe technique has been developed and further improved for applications in the field (Kolber et al. 2005; Rascher and Pieruschka 2008; Pieruschka et al. 2014; Raesch et al. 2014).

For large-scale studies, active methods will not be applicable and thus alternative methods to passively quantify the fluorescence emission have been developed. Such passive techniques retrieve chlorophyll fluorescence emission from the solar irradiance and the vegetation-emitted radiance by using the absorption bands in surface solar irradiance (termed sun-induced fluorescence). The most important atmospheric absorption bands in the fluorescence emission region are two of the oxygen absorption bands: O_2-A at 761 nm and O_2-B at 687 nm. The Fraunhofer Line Discrimination (FLD) principle (Plascyk 1975; Plascyk and Grabriel 1975) allows to retrieve chlorophyll fluorescence emission in these absorption lines (see Meroni et al. 2009 for an in-depth review). This technique allows estimation of absolute variations in the sun-induced fluorescence intensity from leaf up to regional scales. At leaf scale, the FluoWat leaf clip (Alonso et al. 2007; Van Wittenberghe et al. 2013) has been developed to measure the whole chlorophyll fluorescence emission spectrum by clip-on approach. At canopy scale, high-resolution point spectrometers can be used to retrieve sun-induced fluorescence emission in the oxygen absorption lines (Rossini et al. 2010; Burkart et al. 2015). Today, state-of-the-art imaging spectrometers can be used to map the spatial distribution of the sun-induced fluorescence signal in the wider atmospheric oxygen absorption lines, which provides first insight into the spatiotemporal dynamics of fluorescence emission in natural canopies (Pinto et al. 2016).

An important advantage of sun-induced fluorescence is the possibility to be scaled up to larger areas. As it relies on passive detection of the emitted

* Based on a bibliographic survey in Web of Science (core collection), the following number of papers were published using the keywords "photosynthesis" *and* "fluorescence": 2011: 537 publications; 2012: 595 publications; 2013: 609 publications; 2014: 724 publications; 2015: 642+ publications.

fluorescence, this method can also be used with unmanned aerial vehicles (UAVs, e.g., multicopter), aircrafts, and even satellites. There is currently no reliable sensor for drones available yet, but one can expect that such miniaturized sensors will become available soon. Recently, an airborne sensor was released that allows the exact quantification of sun-induced fluorescence from a research aircraft (Rascher et al. 2015; Rossini et al. 2015). This sensor, called *HyPlant*, was proven to deliver novel information of large field trials. Figure 3.4 shows an exemplary *HyPlant* flight line covering a large agricultural area close to the Forschungszentrum Jülich (adapted from Rascher et al. 2015).

On the largest scale, recently the European Space Agency (ESA) selected the FLuorescence EXplorer (FLEX) mission to measure sun-induced fluorescence on a global scale in their new Earth Explorer 8.* Thus, we expect great progress in this method over the next years that will derive from the possibility to globally map vegetation health and stress.

3.2.2 Sensor Positioning Systems: A Compromise between Temporal and Spatial Resolution

To survey large fields, devices and sensors are required that deliver detailed and reproducible temporal and spatial information on the cultivated crop phenotypes on a regular base. In the following sections, we will describe and provide examples for the different positioning systems, from ground-based systems to airplanes and satellite platforms.

3.2.2.1 Ground-Based Systems

Ground-based system can be divided into four different groups: (1) handheld systems, (2) fixed platforms, (3) semi-fixed platforms, and (4) mobile platforms (Table 3.1). The main differences between these ground-based platforms are found in their compromise in temporal and spatial resolution. In other words, platforms enable either manual or semi-manual measurements with limited reproducibility or automated measurements with high reproducibility. Handheld sensors can be carried by a single person through the field, but they are excessively labor intensive and time consuming; they are not useful for high-throughput phenotyping. Fixed platforms are permanent infrastructures built over a specific field which is able to position different sensors from one experimental plot to another and offer a high flexibility regarding temporal and spatial resolution. The main drawback of this approach is that the platform movement is limited to a single field. Semi-fixed platforms, like towers, elevated platforms, or "cherry pickers," allow performing measurements in a high temporal resolution over extensive areas, which simplifies the comparison of different experimental

* http://www.esa.int/Our_Activities/Observing_the_Earth/New_satellite_to_measure_plant_health.

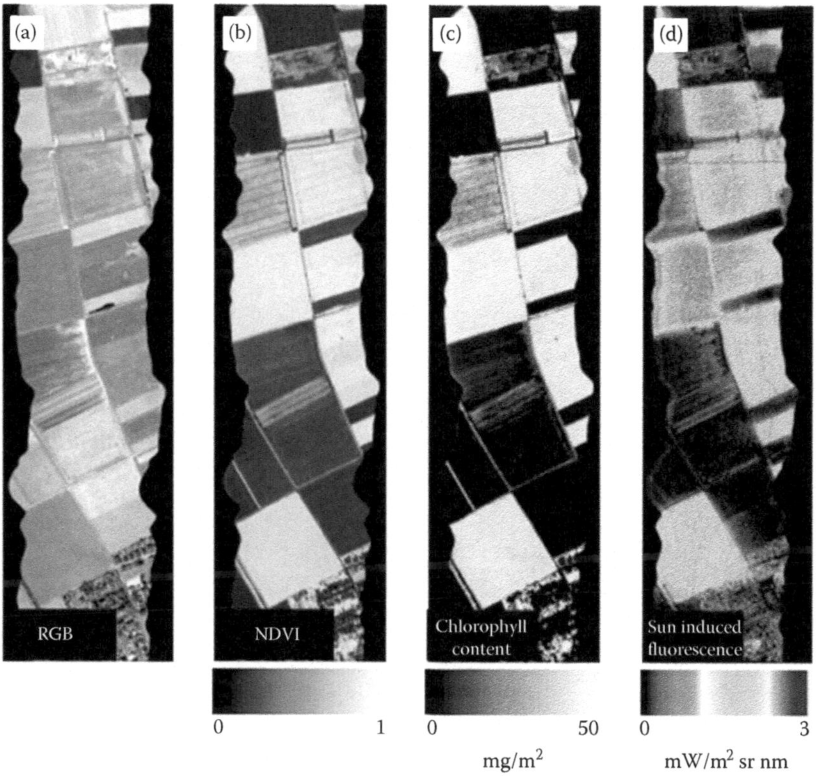

FIGURE 3.4
(See color insert.) Airborne maps of different vegetation products and sun-induced fluorescence (SIF) from an agricultural area in Western Germany. The flight line was recorded on August 23, 2012, from 600 m height, at 13:50 local time (UTC + 2 h), which was approximately 1 h after solar noon. (a) For the pseudo-RGB image, reflectance bands at 696, 708, and 677 nm are used for the red, green, and blue channel of the image. (b) The normalized difference vegetation index (NDVI) was calculated as NDVI = R758 – R670R758 + R670. (c) Leaf chlorophyll content was calculated using the Gaussian process regression method (Verrelst et al. 2012, 2013). (d) Sun-induced fluorescence (F760) was calculated using the 3FLD method with an empirical correction of effective transmittance in the relevant wavebands. Dominant green vegetation in this area at the time of observation were sugar beet and corn on one side and grassland on the other side. The two vegetation types were in contrasting periods of their seasonal development. Sugar beet and corn had a dense fully mature canopy in contrast grassland had reached senescence. A visual evaluation already indicates that fluorescence shows a different pattern compared to other vegetation variables. This indicated the added value of the sun-induced fluorescence signal that provides a new window into photosynthetic functioning that potentially can be mapped on large areas.

TABLE 3.1

Ground-Based Phenotyping Platforms

	Example	Temporal and Spatial Resolution	Disadvantages	Advantages
Handheld sensors can be carried through the field by a single person	Point spectroradiometers Thermal sensors Imagers	Low	Labor intensive and time consuming	Mostly used as a ground true reference to validate measurements performed from UAVs and airplanes
Fixed platforms permanent structures built over a specific field which are able to move different sensors from one experimental plot to another	Field Scanalyzer, Lemnatec[a] Eidgenössische Technische Hochschule, Zürich's Field Phenotyping Platform[b]	High	Limited to a single field	Possible to carry a broad number of sensors
Semi-fixed platforms towers, elevated platforms or "cherry pickers"	FieldLIFT, Forschungszentrum Jülich[c]	High	Plots further away are both smaller and have more atmosphere to traverse (relevant for both thermal and reflectance data). Difficult to move	Possible to carry a broad number of sensors

(Continued)

TABLE 3.1 (*Continued*)

Ground-Based Phenotyping Platforms

	Example	Temporal and Spatial Resolution	Disadvantages	Advantages
Mobile platforms simple hand-pushed bicycle to sophisticated platforms which can traverse the field autonomously or with a driver	Field4Cycle, Forschungszentrum Jülich[c] FieldCOP, Forschungszentrum Jülich[c] BreedVision, Osnabrucke, Germany (Busemeyer et al. 2013) Avignon system, France (Comar et al. 2012) Phenomobile Canberra, Australia (Deery et al. 2014)	High	Data handling, tons of gigabytes that need to be first processed and then combined together to generate effective measurements for a particular experimental plot	Possible to carry a broad number of sensors Constant view angle FieldCOP, work independently including night measurements

Source: Adapted from Deery, D. et al., *Agronomy*, 4(3), 349, 2014.

[a] http://www.lemnatec.com/.

[b] http://www.kp.ethz.ch/infrastructure/FIP.html.

[c] http://www.fz-juelich.de/ibg/ibg-2/EN/methods_jppc/methods_node.html.

plots. Mobile platforms ranging from simple hand-pushed bicycle system to sophisticated platforms which can traverse the field autonomously or by a driver offer the possibility to perform measurements in a high temporal and spatial resolution over extensive areas. Data handling is one of the major challenges of using mobile platforms. High temporal and spatial resolution means tons of gigabytes of data from different sensors (frame imagers, line-scan imagers, and point sensors) with their own spatial resolution. These data need to be processed first and then combined to generate effective measurements for a particular experimental plot.

3.2.2.2 UAVs and Unmanned Aircrafts

The recent developments in UAVs provide platforms for positioning lightweight sensors over agricultural fields. The UAV as a free movable platform is a "very versatile tripod" that enables the positioning of a sensor almost everywhere over agricultural fields. UAVs are capable of providing airborne data with a high spatial and temporal resolution due to their flexible and easy use. This opens up new possibilities in precision farming and management or huge breeding experiments. While common RGB cameras already provide in-depth insights, if positioned above agricultural fields (Sakamoto et al. 2012; Bendig et al. 2014), other sensor types allow complementing analyses. Recent technical developments toward small high-performance electronics allowed the production of thermal cameras that can be carried by UAVs to detect water stress (Zarco-Tejada et al., 2012). Just recently highly accurate and lightweight point spectrometers and multispectral cameras became available, which make them suitable for the use in UAVs (Burkart et al. 2014).

The retrieval of small differences in the reflectance of plants is necessary to detect relevant differences in traits. But to identify these small differences, the quality of the measurements must allow such discrimination. However, field measurements with lightweight sensors that are often used on UAVs are prone to a large variety of undesired influences increasing uncertainty. The reflection signal of plants can be altered by these uncertainties arising from variable environmental factors, such as changes in light conditions by clouds, different sun elevations, variation in temperature, or strong wind. Additionally, data retrieval depends on technical characteristics and the quality and stability of the sensor itself. Thus, to not only produce colorful UAV-based pictures, specific measurement reports have to be adapted for the combination of environmental factors and instrument characteristics to insure consistent data quality.

An alternative to the use of UAV for field phenotyping is the use of zeppelins. The high payload, extended flight time, smooth flight, and low vibrations make the airship a great platform even for sensitive scientific instruments, complementing other small UAV platforms.

3.2.2.3 Airplanes and Satellites for Field Phenotyping

Most of the sensors and measurement approaches that were described earlier are applied on the plot scale. This is the most relevant scale from a scientific perspective and for breeders and agricultural management. However, field phenotyping will be expanded to a larger scale in the near future. Demand comes not only from precision agriculture but also from research in agroecology (Diekötter et al. 2014).

In this context, aircraft- and satellite-based approaches are becoming increasingly important. On these large-scale passive measurements, new methods are becoming favorable. Current methods that can also be used with air- and spaceborne sensors include 3D surface reconstruction using LIDAR and stereo camera approaches (Asner et al. 2012), thermography (Munns et al. 2010), imaging spectroscopy (Fiorani et al. 2012) and recently sun-induced fluorescence (Rascher et al. 2015). Scaling from the aircraft to the satellite is a question of balancing spatial resolution, repeat time, and coverage. In general, airborne sensors have a higher spatial resolution, can be used upon request, but cover normally only selected flight lines of a few kilometers swath. Satellite sensors for vegetation monitoring generally start at a spatial resolution of a few meters but have the advantage of a regular and large coverage.

3.3 The Need for Environmental Monitoring

The plant phenotype is determined by the interaction of the genotype and the environmental conditions that modulates the expression of the genetic information in time and space (Fiorani and Schurr 2013). As a consequence, plant phenotyping requires an exact monitoring of the environment at the location where the plant is cultivated.

Field phenotyping depends highly on environmental monitoring at the relevant spatial and temporal scales. This may be more complicated than first anticipated. Environmental conditions could be highly variable even on small spatial and temporal scales. One prominent example is light. Light quality and quantity has an inherent diurnal cycle that is deeply imbedded in the regulation of almost all plant functions (Greenham and McClung 2015). Additionally, light constantly changes because of moving clouds, cast shadows, and wind-moved canopies. As a consequence, photosynthesis almost never operates at constant but most of the time under highly fluctuating conditions (Rascher and Nedbal 2006). These fluctuations vary on each leaf and cannot be measured appropriately by light sensors. A second example comes from soil sciences. There is a vast selection of soil moisture sensors, and recent studies demonstrate that soil moisture is very heterogeneously distributed within the soil (Hinsinger et al. 2009). This heterogeneity is important because it determines the actual water availability at fine roots. Both examples demonstrate the importance but also the complexity of environmental monitoring at the "relevant" scale.

In modern field phenotyping, a dense network of environmental sensors is installed. It is attempted to record the most relevant environmental parameters, and currently modern field phenotyping centers have numerous environmental sensors installed. These sensors are often custom made and are positioned as close to the plants as possible. Some prominent examples are the environmental sensors that were developed by the Australian field phenotyping center and Forschungszentrum Jülich (Germany). The major advantage of this concept is that the user can define the type, distribution, and density of the sensors based on the scientific question and thus ensure adequate monitoring of the environmental variability in time and space.

3.4 Conclusions and Outlook: The Future of Field Phenotyping

Currently, several national and international initiatives are on the way to developing infrastructures for field phenotyping. Networks such as the German Plant Phenotyping Network (DPPN),[*] the French Plant Phenotyping Network (FPPN),[†] the UK Plant Phenomics Network,[‡] the European Plant Phenotyping Network (EPPN),[§] or the International Plant Phenotyping Network (IPPN)[¶] do not only improve the infrastructure but also develop a new community within plant, vegetation, and agricultural sciences. For instance, the EMPHASIS (European Infrastructure for Multi-Site Plant Phenotyping and Simulation for Food Security in a Changing Climate) has been listed on the ESFRI (European Strategy Forum on Research Infrastructures) roadmap and represents an initiative to synergistically merge the plant phenotyping research infrastructure within Europe. This includes phenotyping facilities under controlled and field conditions as well as modeling and data management. Additionally, EMPHASIS will facilitate a close interaction of different stakeholders from academia and industry including breeders, technology developers, and policy makers. It can be expected that within the next years a network of instrumented field sites will become operational and that field phenotyping data will become widely available from these groups and networks.

* www.dppn.de.
† https://www.phenome-fppn.fr/.
‡ http://www.ukppn.org.uk/.
§ http://www.plant-phenotyping-network.eu/.
¶ http://www.plant-phenotyping.org/.

One of the main challenges will be to merge data from the different initiatives into one knowledge-based environment. However, up to date no standards are available to store, label, and distribute the various data. Success of previous large-scale initiatives such as the human genome project,* the FLUXNET initiative,† or the Long Term Ecological Research Network (LTER)‡ was often linked to a standardized and open data policy.

Phenotyping data are far more complex than what we have experienced before. With field phenotyping, we have to cope with a multitude of data formats and measurement protocols that finally have to be linked to the genetic, environmental, and spatial information resulting in a certain phenotype. But nevertheless linking these fundamentally different data sources is essential to increase our understanding of the gene × environment × management interaction.

Future developments in phenotyping platforms will strongly benefit from sensor arrangements, which allow the fusion of ground and remote sensing data. Particularly the fusion of 3D imaging systems with other sensors like RGB, hyperspectral, or thermal cameras may help to analyze and interpret data with respect to spatial relations between sensor and plant surface. In this context, intensively instrumented field sites, such as the field lab Campus Klein-Altendorf of Bonn University,§ where different sensors and measurement protocols can be tested in an integrated manner, will become highly valuable.

The joint measurements of plant traits will allow the analysis of functional links between phenotypic traits. This becomes especially important if we aim for a better understanding of dynamic plant stress responses. Abiotic as well as biotic stress factors generally result in a complex reaction of the plant that can often not be explained by assessing a few plant properties only. Combined sensors and the possibility to link the data from different experiments give hope that plant phenotyping will also foster our scientific understanding of plant stress physiology. This knowledge in turn is of outermost importance to breed for future plant cultivars that are adapted to the changing and more extreme environmental conditions we are inevitably facing in times of global change.

Acknowledgments

We thank institutional funds of the Helmholtz Association for research at IBG-2 Plant Sciences, DPPN (Förderkennzeichen 031A053A/B/C), EPPN

* http://www.genome.gov/10001772.
† http://fluxnet.ornl.gov/.
‡ https://www.lternet.edu/.
§ https://www.cka.uni-bonn.de/.

(Grant Agreement No. 284443), PhenoCrops project in the context of the Ziel 2-Programms NRW 2007–2013 Regionale Wettbewerbsfähigkeit und Beschäftigung by the Ministry for Innovation, Science and Research (MIWF) of the state North Rhine Westphalia (NRW), and European Union Funds for regional development (EFRE) (005-1105-0035).

References

Ainsworth, E. A., P. A. Davey, C. J. Bernacchi, O. C. Dermody, E. A. Heaton, D. J. Moore, P. B. Morgan et al. 2002. A meta-analysis of elevated [CO_2] effects on soybean (glycine max) physiology, growth and yield. *Global Change Biology* 8(8): 695–709. doi:10.1046/j.1365-2486.2002.00498.x.

Alchanatis, V., Y. Cohen, S. Cohen, M. Moller, M. Sprinstin, M. Meron, J. Tsipris, Y. Saranga, and E. Sela. 2009. Evaluation of different approaches for estimating and mapping crop water status in cotton with thermal imaging. *Precision Agriculture* 11(1): 27–41. doi:10.1007/s11119-009-9111-7.

Alonso, L., L. Gomez-Chova, J. Amoros-Lopez, L. Guanter, and J. Calpe. 2007. Sensitivity analysis of the FLD method for the measurement of chlorophyll fluorescence Using a Field Spectroradiometer. In *Proceedings of the 3rd International Workshop on Remote Sensing of Vegetation Fluorescence*, Florence, Italy.

Asner, G. P., D. E. Knapp, J. Boardman, R. O. Green, T. Kennedy-Bowdoin, M. Eastwood, R. E. Martin, C. Anderson, and C. B. Field. 2012. Carnegie airborne observatory-2: Increasing science data dimensionality via high-fidelity multi-sensor fusion. *Remote Sensing of Environment* 124(September): 454–465. doi:10.1016/j.rse.2012.06.012.

Auernhammer, H. 2001. Precision farming—The environmental challenge. *Computers and Electronics in Agriculture* 30(1–3): 31–43. doi:10.1016/S0168-1699(00)00153-8.

Baker, N. R. 2008. Chlorophyll fluorescence: A probe of photosynthesis in vivo. *Annual Review of Plant Biology* 59(1): 89–113. doi:10.1146/annurev.arplant.59.032607.092759.

Barton, C. V. M and P. R. J North. 2001. Remote sensing of canopy light use efficiency using the photochemical reflectance index: Model and sensitivity analysis. *Remote Sensing of Environment* 78(3): 264–73. doi:10.1016/S0034-4257(01)00224-3.

Bellasio, C., J. Olejníčková, R. Tesař, D. Šebela, and L. Nedbal. 2012. Computer reconstruction of plant growth and chlorophyll fluorescence emission in three spatial dimensions. *Sensors* 12(1): 1052–71. doi:10.3390/s120101052.

Bendig, J., A. Bolten, S. Bennertz, J. Broscheit, S. Eichfuss, and G. Bareth. 2014. Estimating biomass of barley using crop surface models (CSMs) derived from UAV-based RGB imaging. *Remote Sensing* 6(11): 10395–10412. doi:10.3390/rs61110395.

Biskup, B., H. Scharr, U. Schurr, and U. Rascher. 2007. A stereo imaging system for measuring structural parameters of plant canopies. *Plant, Cell & Environment* 30(10): 1299–1308. doi:10.1111/j.1365-3040.2007.01702.x.

Burkart, A., S. Cogliati, A. Schickling, and U. Rascher. 2014. A novel UAV-based ultra-light weight spectrometer for field spectroscopy. *IEEE Sensors Journal* 14(1): 62–67. doi:10.1109/JSEN.2013.2279720.

Burkart, A., A. Schickling, M. P. C. Mateo, T. J. Wrobel, M. Rossini, S. Cogliati, T. Julitta, and U. Rascher. 2015. A method for uncertainty assessment of passive sun-induced chlorophyll fluorescence retrieval using an infrared reference light. *IEEE Sensors Journal* 15(8): 4603–4611. doi:10.1109/JSEN.2015.2422894.

Burkart, S., R. Manderscheid, and H.-J. Weigel. 2007. Design and performance of a portable gas exchange chamber system for CO_2- and H_2O-flux measurements in crop canopies. *Environmental and Experimental Botany* 61(1): 25–34. doi:10.1016/j.envexpbot.2007.02.007.

Busemeyer, L., D. Mentrup, K. Möller, E. Wunder, K. Alheit, V. Hahn, H. P. Maurer et al. 2013. BreedVision—A multi-sensor platform for non-destructive field-based phenotyping in plant breeding. *Sensors* 13(3): 2830–2847. doi:10.3390/s130302830.

Chuvieco, E. and A. Huete. 2010. *Fundamentals of Satellite Remote Sensing*. CRC Press, Boca Raton, FL.

Cohen, Y., V. Alchanatis, M. Meron, Y. Saranga, and J. Tsipris. 2005. Estimation of leaf water potential by thermal imagery and spatial analysis. *Journal of Experimental Botany* 56(417): 1843–1852. doi:10.1093/jxb/eri174.

Comar, A., F. Baret, F. Viénot, L. Yan, and B. de Solan. 2012. Wheat leaf bidirectional reflectance measurements: Description and quantification of the volume, specular and hot-spot scattering features. *Remote Sensing of Environment* 121(June): 26–35. doi:10.1016/j.rse.2011.01.028.

Deery, D., J. Jimenez-Berni, H. Jones, X. Sirault, and R. Furbank. 2014. Proximal remote sensing buggies and potential applications for field-based phenotyping. *Agronomy* 4(3): 349–379. doi:10.3390/agronomy4030349.

Diekötter, T., F. Peter, B. Jauker, V. Wolters, and F. Jauker. 2014. Mass-flowering crops increase richness of cavity-nesting bees and wasps in modern agro-ecosystems. *GCB Bioenergy* 6(3): 219–226. doi:10.1111/gcbb.12080.

Eiden, M., S. Van Der Linden, J. H. Schween, C. Gerbig, B. Neininger, Y. Brunet, N. Jarosz, O. Traullé, H. Geiss, and U. Rascher. 2007. Elucidating physiology of plant mediated exchange processes using airborne hyperspectral reflectance measurements in synopsis with eddy covariance data. *International Archives of the Photogrammetry, Remote Sensing and Spatial Information Sciences* 34: 473–481.

Eversole, K., C. Feuillet, K. F. X. Mayer, and J. Rogers. 2014. Slicing the wheat genome. *Science* 345(6194): 285–287. doi:10.1126/science.1257983.

Farrar, K., D. N. Bryant, L. Turner, J. A. Gallagher, A. Thomas, M. Farrell, M. O. Humphreys, and I. S. Donnison. 2011. Breeding for bio-ethanol production in *Lolium perenne* L.: Association of allelic variation with high water-soluble carbohydrate content. *BioEnergy Research* 5(1): 149–157. doi:10.1007/s12155-011-9156-0.

Feilhauer, H., G. P. Asner, R. E. Martin, and S. Schmidtlein. 2010. Brightness-normalized partial least squares regression for hyperspectral data. *Journal of Quantitative Spectroscopy and Radiative Transfer* 111(12–13): 1947–1957. doi:10.1016/j.jqsrt.2010.03.007.

Fiorani, F., U. Rascher, S. Jahnke, and U. Schurr. 2012. Imaging plants dynamics in heterogenic environments. *Current Opinion in Biotechnology* 23(2): 227–235. doi:10.1016/j.copbio.2011.12.010.

Fiorani, F. and U. Schurr. 2013. Future scenarios for plant phenotyping. *Annual Review of Plant Biology* 64(1): 267–291. doi:10.1146/annurev-arplant-050312-120137.

Ford, E. D., A. Cocke, L. Horton, M. Fellner, and E. Van Volkenburgh. 2008. Estimation, variation and importance of leaf curvature in *Zea mays* hybrids. *Agricultural and Forest Meteorology* 148(10): 1598–1610. doi:10.1016/j.agrformet.2008.05.015.

Gamon, J. A., C. B. Field, W. Bilger, O. Björkman, A. L. Fredeen, and J. Peñuelas. 1990. Remote sensing of the xanthophyll cycle and chlorophyll fluorescence in sunflower leaves and canopies. *Oecologia* 85(1): 1–7. doi:10.1007/BF00317336.

Garbulsky, M. F., J. Peñuelas, J. Gamon, Y. Inoue, and I. Filella. 2011. The photochemical reflectance index (PRI) and the remote sensing of leaf, canopy and ecosystem radiation use efficiencies: A review and meta-analysis. *Remote Sensing of Environment* 115(2): 281–297. doi:10.1016/j.rse.2010.08.023.

Grant, O. M., M. M. Chaves, and H. G. Jones. 2006. Optimizing thermal imaging as a technique for detecting stomatal closure induced by drought stress under greenhouse conditions. *Physiologia Plantarum* 127(3): 507–518. doi:10.1111/j.1399-3054.2006.00686.x.

Greenham, K. and C. R. McClung. 2015. Integrating circadian dynamics with physiological processes in plants. *Nature Reviews Genetics* 16(10): 598–610. doi:10.1038/nrg3976.

Hinsinger, P., A. G. Bengough, D. Vetterlein, and I. M. Young. 2009. Rhizosphere: Biophysics, biogeochemistry and ecological relevance. *Plant and Soil* 321(1–2): 117–152. doi:10.1007/s11104-008-9885-9.

Hostert, P., A. Röder, and J. Hill. 2003. Coupling spectral unmixing and trend analysis for monitoring of long-term vegetation dynamics in mediterranean rangelands. *Remote Sensing of Environment* 87(2): 183–197.

Jackson, R. D. and A. R. Huete. 1991. Interpreting vegetation indices. *Preventive Veterinary Medicine* 11(3–4): 185–200. doi:10.1016/S0167-5877(05)80004-2.

Jackson, R. D., S. B. Idso, R. J. Reginato, and P. J. Pinter. 1981. Canopy temperature as a crop water stress indicator. *Water Resources Research* 17(4): 1133–1138. doi:10.1029/WR017i004p01133.

Jansen, M., F. Pinto, K. A. Nagel, D. van Dusschoten, F. Fiorani, U. Rascher, H. U. Schneider, A. Walter, and U. Schurr. 2014. Non-invasive phenotyping methodologies enable the accurate characterization of growth and performance of shoots and roots. In *Genomics of Plant Genetic Resources*. Eds. R. Tuberosa, A. Graner, and E. Frison. Springer, Dordrecht, the Netherlands, pp. 173–206. http://link.springer.com/chapter/10.1007/978-94-007-7572-5_8.

Jones, H. G. 1999. Use of thermography for quantitative studies of spatial and temporal variation of stomatal conductance over leaf surfaces. *Plant, Cell and Environment* 22(9): 1043–1055. doi:10.1046/j.1365-3040.1999.00468.x.

Kjaer, K. H. and C.-O. Ottosen. 2015. 3D laser triangulation for plant phenotyping in challenging environments. *Sensors* 15(6): 13533–13547. doi:10.3390/s150613533.

Kolber, Z., D. Klimov, G. Ananyev, U. Rascher, J. Berry, and B. Osmond. 2005. Measuring photosynthetic parameters at a distance: Laser induced fluorescence transient (LIFT) method for remote measurements of photosynthesis in terrestrial vegetation. *Photosynthesis Research* 84(1–3): 121–129. doi:10.1007/s11120-005-5092-1.

Körner, C. 2000. Biosphere responses to CO_2 enrichment. *Ecological Applications* 10(6): 1590–1619. doi:10.1890/1051-0761(2000)010[1590:BRTCE]2.0.CO;2.

Kuijken, R. C. P., F. A. van Eeuwijk, L. F. M. Marcelis, and H. J. Bouwmeester. 2015. Root phenotyping: From component trait in the lab to breeding. *Journal of Experimental Botany* June: erv239. doi:10.1093/jxb/erv239.

Leuning, R., J. Grace, J. L. Monteith, J. R. Milford, M. H. Unsworth, and D. Fowler. 1989. Leaf energy balances: Developments and applications [and discussion]. *Philosophical Transactions of the Royal Society of London Series B, Biological Sciences* 324(1223): 191–206.

Maxwell, K. and G. N. Johnson. 2000. Chlorophyll fluorescence—A practical guide. *Journal of Experimental Botany* 51(345): 659–668. doi:10.1093/jexbot/51.345.659.

McCormick, K. and N. Kautto. 2013. The bioeconomy in Europe: An overview. *Sustainability* 5(6): 2589–2608. doi:10.3390/su5062589.

Meroni, M., M. Rossini, L. Guanter, L. Alonso, U. Rascher, R. Colombo, and J. Moreno. 2009. Remote sensing of solar-induced chlorophyll fluorescence: Review of methods and applications. *Remote Sensing of Environment* 113(10): 2037–2051. doi:10.1016/j.rse.2009.05.003.

Möller, M., V. Alchanatis, Y. Cohen, M. Meron, J. Tsipris, A. Naor, V. Ostrovsky, M. Sprintsin, and S. Cohen. 2007. Use of thermal and visible imagery for estimating crop water status of irrigated grapevine. *Journal of Experimental Botany* 58(4): 827–838. doi:10.1093/jxb/erl115.

Moran, S., G. Fitzgerald, A. Rango, C. Walthall, E. Barnes, W. Bausch, T. Clarke et al. 2003. Sensor development and radiometric correction for agricultural applications. *Photogrammetric Engineering and Remote Sensing* 69(6): 705–718. doi:10.14358/PERS.69.6.705.

Müller-Linow, M., F. Pinto-Espinosa, H. Scharr, and U. Rascher. 2015. The leaf angle distribution of natural plant populations: Assessing the canopy with a novel software tool. *Plant Methods* 11(1): 1. doi:10.1186/s13007-015-0052-z.

Munns, R., R. A. James, X. R. R. Sirault, R. T. Furbank, and H. G. Jones. 2010. New phenotyping methods for screening wheat and barley for beneficial responses to water deficit. *Journal of Experimental Botany* 61(13): 3499–3507. doi:10.1093/jxb/erq199.

O'Toole, J. C. and R. T. Cruz. 1980. Response of leaf water potential, stomatal resistance, and leaf rolling to water stress. *Plant Physiology* 65(3): 428–432. doi:10.1104/pp.65.3.428.

Papageorgiou, G. C. and Govindjee. 2004. *Chlorophyll a Fluorescence: A Signature of Photosynthesis*. Springer, New York.

Pieruschka, R., H. Albrecht, O. Muller, J. A. Berry, D. Klimov, Z. S. Kolber, Z. Malenovský, and U. Rascher. 2014. Daily and seasonal dynamics of remotely sensed photosynthetic efficiency in tree canopies. *Tree Physiology*, June: tpu035. doi:10.1093/treephys/tpu035.

Pieruschka R., D. Klimov, Z. S. Kolber, and J. A. Berry. 2010. Monitoring of cold and light stress impact on photosynthesis by using the laser induced fluorescence transient (LIFT) approach. *Functional Plant Biology* 37(5): 395–402.

Pinto, F., A. Damm, A. Schickling, C. Panigada, S. Cogliati, M. Müller-Linow, A. Balvora, and U. Rascher. 2016. Sun-induced chlorophyll fluorescence from high-resolution imaging spectroscopy data to quantify spatio-temporal patterns of photosynthetic function in crop canopies. *Plant, Cell and Environment* 39: 1500–1512.

Plascyk, J. 1975. The MKII Fraunhofer line discriminator (FLD-II) for airborne and orbital remote sensing of solar stimulated luminescence. *Optical Engineering* 14: 339–346.

Plascyk, J. and F. Grabriel. 1975. The Fraunhofer line discriminator MKII—An airbone instrument for precise and standarized ecological luminescence measurements. *IEEE Transactions on Instrumentation and Measurement* 24: 306–313.

Porcar-Castell, A., E. Pfündel, J. F. J. Korhonen, and E. Juurola. 2008. A new monitoring PAM fluorometer (MONI-PAM) to study the short- and long-term acclimation of photosystem II in field conditions. *Photosynthesis Research* 96(2): 173–179. doi:10.1007/s11120-008-9292-3.

Porcar-Castell, A., E. Tyystjärvi, J. Atherton, C. van der Tol, J. Flexas, E. Pfündel, J. Moreno, C. Frankenberg, and J. A. Berni. 2014. Linking chlorophyll a fluorescence to photosynthesis for remote sensing applications: mechanisms and challenges. *Journal of Experimental Botany*, eru191.

Raesch, A. R., O. Muller, R. Pieruschka, and U. Rascher. 2014. Field observations with laser-induced fluorescence transient (LIFT) method in barley and sugar beet. *Agriculture* 4(2): 159–169. doi:10.3390/agriculture4020159.

Rascher, U., G. Agati, L. Alonso, G. Cecchi, S. Champagne, R. Colombo, A. Damm et al. 2009. CEFLES2: The remote sensing component to quantify photosynthetic efficiency from the leaf to the region by measuring sun-induced fluorescence in the oxygen absorption bands. *Biogeosciences* 6(7): 1181–1198.

Rascher, U., L. Alonso, A. Burkart, C. Cilia, S. Cogliati, R. Colombo, A. Damm et al. 2015. Sun-induced fluorescence—A new probe of photosynthesis: First maps from the imaging spectrometer HyPlant. *Global Change Biology* 21(12): 4673–4684. doi:10.1111/gcb.13017.

Rascher, U., B. Biskup, A. D. B. Leakey, J. M. McGrath, and E. A. Ainsworth. 2010. Altered physiological function, not structure, drives increased radiation-use efficiency of soybean grown at elevated CO_2. *Photosynthesis Research* 105(1): 15–25. doi:10.1007/s11120-010-9548-6.

Rascher, U. and L. Nedbal. 2006. Dynamics of photosynthesis in fluctuating light. *Current Opinion in Plant Biology, Cell Biology/Edited by L. G. Smith and U. Mayer,* 9(6): 671–678. doi:10.1016/j.pbi.2006.09.012.

Rascher, U. and R. Pieruschka. 2008. Spatio-temporal variations of photosynthesis: The potential of optical remote sensing to better understand and scale light use efficiency and stresses of plant ecosystems. *Precision Agriculture* 9(6): 355–366. doi:10.1007/s11119-008-9074-0.

Rich, P. M. 1990. characterizing plant canopies with hemispherical photographs. *Remote Sensing Reviews* 5(1): 13–29. doi:10.1080/02757259009532119.

Roberti, J. A., M. D. SanClements, H. W. Loescher, and E. Ayres. 2014. Traceable calibration, performance metrics, and uncertainty estimates of minirhizotron digital imagery for fine-root measurements. *PLoS ONE* 9(11): e112362. doi:10.1371/journal.pone.0112362.

Roemer, C., M. Wahabzada, A. Ballvora, F. Pinto, M. Rossini, C. Panigada, J. Behmann et al. 2012. Early drought stress detection in cereals: Simplex volume maximisation for hyperspectral image analysis. *Functional Plant Biology* 39(10–11): 878–890. doi:10.1071/FP12060.

Rossini, M., M. Meroni, M. Migliavacca, G. Manca, S. Cogliati, L. Busetto, V. Picchi, A. Cescatti, G. Seufert, and R. Colombo. 2010. High resolution field spectroscopy measurements for estimating gross ecosystem production in a rice field. *Agricultural and Forest Meteorology* 150(9): 1283–1296. doi:10.1016/j.agrformet.2010.05.011.

Rossini, M., L. Nedbal, L. Guanter, A. Ač, L. Alonso, A. Burkart, S. Cogliati et al. 2015. Red and far red sun-induced chlorophyll fluorescence as a measure of plant photosynthesis. *Geophysical Research Letters* 42(6): 2014GL062943. doi:10.1002/2014GL062943.

Sakamoto, T., A. A. Gitelson, A. L. Nguy-Robertson, T. J. Arkebauer, B. D. Wardlow, A. E. Suyker, S. B. Verma, and M. Shibayama. 2012. An alternative method using digital cameras for continuous monitoring of crop status. *Agricultural and Forest Meteorology* 154–155(March): 113–126. doi:10.1016/j.agrformet.2011.10.014.

Sampoux, J.-P., P. Baudouin, B. Bayle, V. Béguier, P. Bourdon, J.-F. Chosson, F. Deneufbourg et al. 2011. Breeding perennial grasses for forage usage: An experimental assessment of trait changes in diploid perennial ryegrass (*Lolium perenne* L.) cultivars released in the last four decades. *Field Crops Research* 2(123): 117–129. doi:10.1016/j.fcr.2011.05.007.

Sanz-Cortiella, R., J. Llorens-Calveras, A. Escolà, J. Arnó-Satorra, M. Ribes-Dasi, J. Masip-Vilalta, F. Camp et al. 2011. Innovative LIDAR 3D dynamic measurement system to estimate fruit-tree leaf area. *Sensors* 11(6): 5769–5791. doi:10.3390/s110605769.

Schreiber, U. 1986. Detection of rapid induction kinetics with a new type of high-frequency modulated chlorophyll fluorometer. In *Current Topics in Photosynthesis*, Eds. J. Amesz, A. J. Hoff, and H. J. Van Gorkum. Springer, Dordrecht, the Netherlands, pp. 259–270. http://link.springer.com/chapter/10.1007/978-94-009-4412-1_24.

Schreiber, U. 2004. Pulse-amplitude-modulation (PAM) fluorometry and saturation pulse method: An overview. In *Chlorophyll A Fluorescence: A Signature of Photosynthesis, Advances in Photosynthesis and Respiration*, Eds. G. C. Papageorgiou and Govindjee. Springer, Dordrecht, the Netherlands, Vol. 19, pp. 279–319.

Sequencing Project, International Rice Genome. 2005. The map-based sequence of the rice genome. *Nature* 436(7052): 793–800. doi:10.1038/nature03895.

Thorup-Kristensen, K. and R. van den Boogaard. 1998. Temporal and spatial root development of cauliflower (*Brassica oleracea* L. var. *botrytis* L.). *Plant and Soil* 201(1): 37–47. doi:10.1023/A:1004393417695.

Trachsel, S., S. M. Kaeppler, K. M. Brown, and J. P. Lynch. 2010. Shovelomics: High throughput phenotyping of maize (*Zea mays* L.) root architecture in the field. *Plant and Soil* 341(1–2): 75–87. doi:10.1007/s11104-010-0623-8.

Van Wittenberghe, S., L. Alonso, J. Verrelst, I. Hermans, J. Delegido, F. Veroustraete, R. Valcke, J. Moreno, and R. Samson. 2013. Upward and downward solar-induced chlorophyll fluorescence yield indices of four tree species as indicators of traffic pollution in Valencia. *Environmental Pollution* 173(February): 29–37. doi:10.1016/j.envpol.2012.10.003.

Wasson, A. P., G. J. Rebetzke, J. A. Kirkegaard, J. Christopher, R. A. Richards, and M. Watt. 2014. Soil coring at multiple field environments can directly quantify variation in deep root traits to select wheat genotypes for breeding. *Journal of Experimental Botany* 65(21): 6231–6249. doi:10.1093/jxb/eru250.

York, L. M. and J. P. Lynch. 2015. Intensive field phenotyping of maize (*Zea mays* L.) root crowns identifies phenes and phene integration associated with plant growth and nitrogen acquisition. *Journal of Experimental Botany* 66(18): 5493–5505. doi:10.1093/jxb/erv241.

Zarco-Tejada, P. J., V. González-Dugo, and J. A. J. Berni. 2012. Fluorescence, temperature and narrow-band indices acquired from a UAV platform for water stress detection using a micro-hyperspectral imager and a thermal camera. *Remote Sensing of Environment, Remote Sensing of Urban Environments*, 117(February): 322–337. doi:10.1016/j.rse.2011.10.007.

Zhu, X.-G., S. P. Long, and D. R. Ort. 2010. Improving photosynthetic efficiency for greater yield. *Annual Review of Plant Biology* 61(1): 235–261. doi:10.1146/annurev-arplant-042809-112206.

Section II

A New Generation of Controlled Environment, Field, and Modeling Platforms

4

Advancing Understanding of Hydrological and Biogeochemical Interactions in Evolving Landscapes through Controlled Experimentation at the Landscape Evolution Observatory

Aditi Sengupta, Luke A. Pangle, Till H. M. Volkmann, Katerina Dontsova, Peter A. Troch, Antonio A. Meira-Neto, Julia W. Neilson, Edward A. Hunt, Jon Chorover, Xubin Zeng, Joost van Haren, Greg A. Barron-Gafford, Aaron Bugaj, Nate Abramson, Michael Sibayan, and Travis E. Huxman

CONTENTS

Abstract

Understanding the multitude of processes, feedback, and interactions among water, microbes, plants, and porous geological media is crucial for obtaining better predictions about the behavior of Earth's critical zone in the face of future climatic conditions. Current studies often suffer from the limitations of the spatial scale in which they are performed. By not considering the effects brought by the heterogeneity while moving from pore to landscape scales, important feedback and integrated behavior may be missed, rendering predicted behavior different from that of the natural systems. The time span in which such experiments are executed might also not be suitable for the observation of phenomena typically occurring over years in natural settings. Studying naturally occurring phenomena in situ carries with it the uncertainty about the initial state of the system, and the fact that observations require destructive sampling, which will interfere with the processes under investigation. The investigation of hydrological and biogeochemical evolution of natural systems is thus a challenging task for Earth scientists. The Landscape Evolution Observatory (LEO), a research facility managed by the University of Arizona and located at Biosphere 2, allows for the interdisciplinary investigation of the evolution of artificial hillslopes containing an initially naive mineral assemblage that will be subjected to controlled climate experiments. The LEO's unique set of instrumentation allows for exceptional observations of energy, water, and carbon fluxes across the three 330 m^3 hillslopes. Within the time frame of 10 years of interdisciplinary research, scientists will be able to address important questions related to the interactions among hydrology, geochemistry, and ecology. The LEO project maintains a database open to scientists and practitioners from different domains to address different research questions in a collaborative way. The research done at the LEO has the potential to be a milestone in terrestrial ecosystem research infrastructures.

4.1 Introduction

The physical, chemical, and biological structures and processes controlling reaction, flow, and transport in natural landscapes interact at multiple space and timescales and are difficult to quantify. Hence, the predictions of hydrological and biogeochemical responses to natural and anthropogenic forcing at the landscape scale are highly uncertain due to the effects of heterogeneity on the scaling of reaction, flow, and transport phenomena. The current paradigm of hydrological and geochemical theory is that process descriptions derived from observations at small scales in controlled systems (e.g., the Richards equation to describe flow in porous media; first-order chemical reaction kinetics to describe reactive transport) can be applied to predict system response at much larger scales (e.g., baseflow recession at the catchment outlet, landscape chemical denudation), as long as some "equivalent" or "effective" values of the scale-dependent parameters can be identified. However, this paradigm is known to be flawed (Sivapalan, 2005; Beven, 2006) and increasingly frequent calls have been made for new theories that will better link small-scale process understanding with large-scale predictions in space and time (Troch et al., 2009).

Furthermore, how natural systems evolve in time is difficult to observe in relatively short-term laboratory experiments or in natural settings, where landscape initial conditions and time-variant forcing (e.g., changing water and energy inputs associated with changing climate regimes) are unknown. The spatial structure of flow pathways along hillslopes determines the rate, extent, and distribution of geochemical reactions and biological colonization that drives weathering, the transport and precipitation of solutes and sediments, the further evolution of soil structure, and the biotic pallet that relies on it. With feedback among all components, the resulting evolution of structures and processes, in turn, produces spatiotemporal variability of hydrological states and flow pathways. Richter and Billings (2015), in their review of Tansley's "one physical system," highlight the need to understand these interconnected processes affecting evolution of natural systems, especially in Earth's critical zone. Therefore, an integrative approach to study hydrological, geophysical, geochemical, pedological, and ecological processes stands to enhance our knowledge of coupled aboveground and belowground Earth system processes (Richter and Billings, 2015).

Hydrologists and geochemists are well equipped to make quantitative predictions of dynamic responses in relatively simple systems across space and time (e.g., uniform hillslopes with homogeneous porous media cover on top of impermeable bedrock, weathering of basalt minerals under hydrological steady-state throughflow). However, the coevolution of hydrological and (bio)geochemical processes within even simple landscapes quickly renders these predictions less accurate. For instance, the first rainfall event on a hillslope covered with homogeneous ground basalt loamy-sand soil may lead to rapid chemical weathering of the silicate minerals and possible preferential precipitation of poorly crystalline hydrated solids. Such mineral transformation can then possibly lead to local variations in particle size distribution, total porosity, and associated hydraulic properties that dictate water flow. From purely mineral and abiotic conditions, spatially heterogeneous colonization of autotrophic and heterotrophic microorganisms as well as plant establishment can further enhance such mineral transformations and hence increase the overall heterogeneity of the hydraulic structure of the subsurface. The spatial distribution of microbial communities and vascular plants across the hillslope's extent may accelerate surface and subsurface structural development, making identification of hydraulic parameters at a hillslope scale required in our hydrological models an almost impossible task. What started off as a simple homogeneous system is quickly transformed into a complex coevolving landscape that makes quantitative predictions very challenging. The complexity increases when the question of scale is considered. Researchers across the realm of hydrology (Gleeson and Paszkowski, 2013), geochemistry (Molins et al., 2012), and biology (Fierer and Lennon, 2011) agree that perception of scale and its consequent impact on coupled-earth system processes affects the outcome of such interdisciplinary studies. It thus becomes a challenge to link small-scale processes occurring at a pore-scale to the large-scale processes occurring at a landscape-scale.

Linking pore-scale and landscape-scale processes is complicated in real-world settings because of poorly constrained impacts of initial conditions, climate variability, ecosystem dynamics, and geomorphic evolution. There is a need for experimental research to improve our understanding of hydrology–biogeochemistry interactions and feedback at appropriate spatial and temporal scales (Figure 4.1), larger than laboratory soil column experiments. This need served as primary motivation for establishing LEO at Biosphere 2, which offers a unique research facility that allows real-time observations of incipient hydrological and biogeochemical response under well-constrained initial conditions and climate forcing. The LEO hillslopes are the world's largest weighing lysimeters in a controlled environment and enables elucidation of the tight coupling between the time water spends along subsurface flow paths and geochemical weathering reactions, including the feedback between water flow and pedogenesis (Huxman et al., 2009). Prior studies of hillslope-scale aqueous geochemistry have been limited to soil-mantled landscapes that have evolved over geological timescales and whose contemporary weathering processes are largely controlled by that legacy. By initiating the LEO experiment—where three convergent hillslopes mantled by a relatively uniform homogeneous and isotropic basalt porous medium are subjected to aqueous geochemical weathering—we can, for the first time, explore the impacts of hillslope-length flow paths on incipient subsurface structure development. The facility and its instrumentation also allows closure of the water, carbon, and energy budgets at hillslope scales. In this

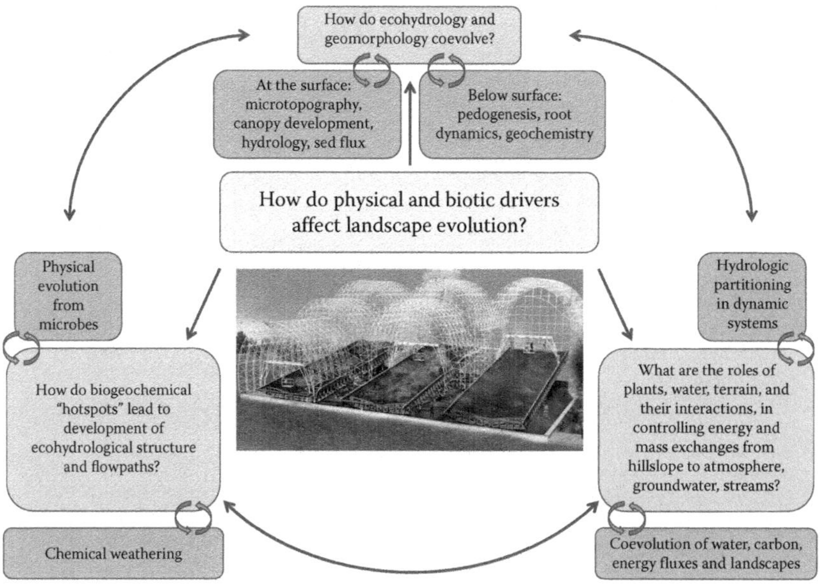

FIGURE 4.1
Conceptual framework for investigating the coevolution of life and Earth. The feedback among three principal guiding questions are central to this framework: geomorphology and ecohydrology; biogeochemistry and flow path development; and plant/water/terrain interactions with material and energy exchanges.

chapter, we discuss the infrastructural capability of LEO. We first provide a theoretical basis of LEO's instrumentation capability, followed by LEO's potential to study coupled hydrological and biogeochemical processes prevalent on a landscape scale. We conclude by highlighting long-term research goals of LEO and the consequent challenges faced, followed by a summary of LEO's unique place in terrestrial ecosystem research infrastructures.

4.2 Landscape Evolution Observatory: Description and Scope

4.2.1 The Landscape Evolution Observatory

The LEO consists of three constructed landscapes located within the climate-controlled Biosphere 2 facility of the University of Arizona, Tucson, USA (Figure 4.2a). They were designed to emulate features of zero-order basins, including a convergent topographic shape with an average slope of 10°. Maximum slope angles of approximately 17° (the angle is chosen because slope stability and hydrological response time; Hopp et al., 2009) are present near the convergence zone, located centrally with respect to the landscape

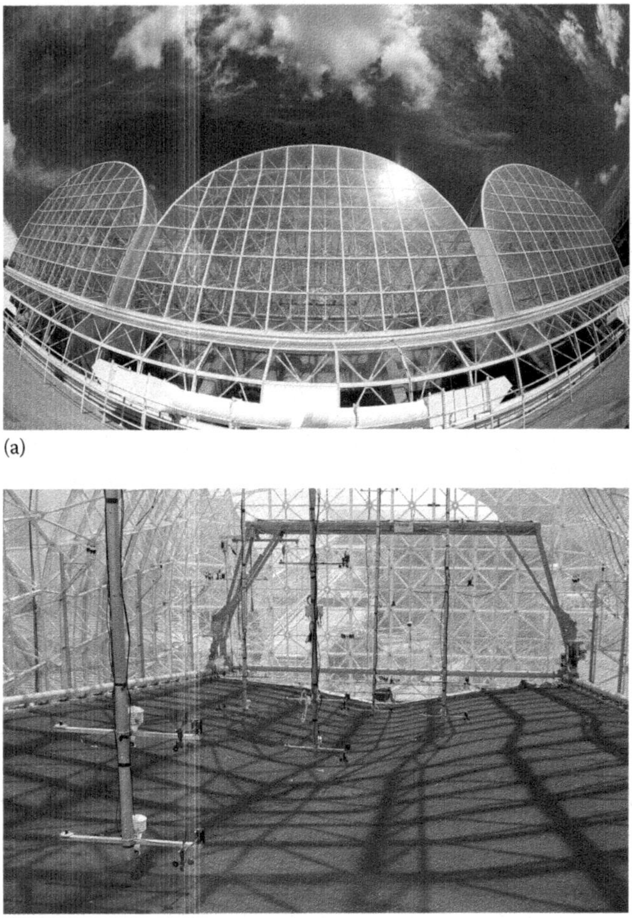

FIGURE 4.2
(a) Wide-angle photograph of the three climate-controlled bays of Landscape Evolution Observatory (LEO) at Biosphere 2, (b) Converging slope of the landscape, showing aboveground instrumentation.

width, and spanning more than half of the landscape length beginning from the downslope extent (Figure 4.2b). The convergent topography is expected to strongly control incipient coevolution of soil hydraulic, geochemical, and microbial properties, and ultimately the spatial organization of plant communities inhabiting the landscapes. The land surface is exposed to the interior atmosphere of the Biosphere 2 facility, which can be actively controlled to create specific combinations of air temperature, and wind speed—within some constraints (e.g., freezing temperatures would be cost-prohibitive, and wind speeds are limited by the flow-generating capacity of the air-handling systems). The ability to control the internal air temperature and wind speed implies that, within the stated constraint, a wide range of climatic conditions can be simulated and their effects studied. Technical details of the landscapes, sensors, samplers, instrument manufacturers, and expected precision can be found in Pangle et al. (2015).

The three LEO landscapes are experimental replicates; they have identical horizontal dimensions and nearly identical depth of parent material. Each landscape is filled with basalt tephra that was collected from a geologic deposit in northern Arizona, crushed to a loamy-sand texture, and packed to a uniform mean depth of 1 m. The packing was done by piling 30 cm of soil and compacting it to 25 cm, and repeated three times. The crushed basalt landscapes therefore represent a spatially uniform and abiotic initial condition (see also Dontsova et al., 2009; Pangle et al., 2015 for complete description of mineralogy and organic carbon content). The physical, chemical, biological, and topographical evolution of this parent material will be observed and manipulated through time. LEO is uniquely equipped to study integrated and spatially discrete measurements of (1) hydrological state and flux variables, (2) carbon cycling, (3) weathering, (4) photosynthesis and respiration, and (5) land-surface energy exchange. The facility is also capable of (1) conducting real-time isotopic measurement of water and carbon dioxide, (2) efficiently collecting and analyzing sample solutions, (3) conducting electrical resistivity tomography measurements of the landscape, and (4) detecting and monitoring spatial patterns of plant and microbial activity. In what follows, we briefly describe each of the variables noted earlier and their related theoretical basis in relation to LEO's research goals.

4.2.1.1 Integrated and Spatially Discrete Measurements of Hydrological State and Flux Variables

The water balance on the LEO landscapes is described as follows:

$$\frac{\partial S}{\partial t} = I(t) - E(t) - T(t) - Q(t) \tag{4.1}$$

where

S represents water stored within the landscape (L^3)

I represents irrigation inflow ($L^3\ T^{-1}$)

E represents the evaporative loss of water from the land surface ($L^3\ T^{-1}$)

T represents water loss from landscape to atmosphere due to transpiration ($L^3\ T^{-1}$) which is zero at present (when only bare soil exists)

Q represents discharge of water through the seepage face at the downslope extent of the landscape ($L^3\ T^{-1}$)

All terms in Equation 4.1 can be measured as integrated, landscape-scale stocks and fluxes. Temporal changes in water storage within the entire landscape are monitored via 10 load cells that are installed within the steel support structure. Each load cell is under the only load-bearing points where the main slope is connected with the supporting structure. All other connections to the surrounding structure, the rain water lines, for example, have flexible connectors to reduce load bearing. Volumetric flow rates through the irrigation system are monitored with electronic flow meters, and the specific flux (L T^{-1}) and spatial patterns associated with each of the five independent irrigation circuits per landscape were measured through a series of manual calibration tests. The term *Q* encompasses both subsurface and overland flows of liquid water from the landscape. Subsurface seepage flows are routed through a plumbing system with in-line electronic flow meters and tipping bucket gauges. Those measurements are partially redundant, though also complimentary since each instrument yields optimal precision over a different range of measured flow rates. The seepage face boundary at the downslope extent of the landscape is partitioned into six subsections. To capture spatial variability of the flow—important during high flow conditions—each subsection is measured separately. Overland flow, if present, will be measured by routing the flow over a flume-like surface, through a plumbing system, and into an open basin with known dimensions, and with a pressure transducer continuously monitoring water depth. The combined flux of water vapor associated with *E* and *T* can be estimated at the landscape scale as the residual term of Equation 4.1.

Water does not move homogeneously through the soil matrix (Sivapalan et al., 2005; Troch et al., 2009), especially when plants are present. Root uptake will cause large differences in water content across a soil area, especially in the unsaturated zones (Volkmann et al., 2016a). In order to capture the spatial variability of the total within-slope water storage, the whole landscape-scale measurements of water storage and flux are complimented by spatially resolved measurements of several hydrological variables. The volumetric water content (L^3 T^{-3}) and water gauge pressure (kPa) are measured at 496 locations within each experimental landscape (Figure 4.3). These colocated measurements are recorded at 154 horizontal (*x*–*y* plane) locations and at 3–5 different depths (0.05, 0.2, 0.35, 0.5, and 0.85 m from the land surface) at each horizontal location. The sensor locations will provide a 1 m soil moisture content grid, which we deemed detailed enough to capture most spatial variability, but coarse enough to avoid creating a slope of sensors. The depth of the perched water table, when present, is monitored by 15 pressure transducers installed within bulkhead fittings that are sealed over drilled penetrations in the underlying steel structure (i.e., at the base of the soil profile; similar penetrations exist at the 154 horizontal locations, where the water content and pressure sensors are installed). Soil surface evaporation and plant transpiration are mainly determined by a vertical gradient of atmospheric vapor pressure deficit (VPD), which is a function of air temperature and humidity. Without atmospheric measurements, the evaporation (under the current bare soil conditions) would be the only undetermined term of Equation 4.1. To capture the atmospheric variability, a network of

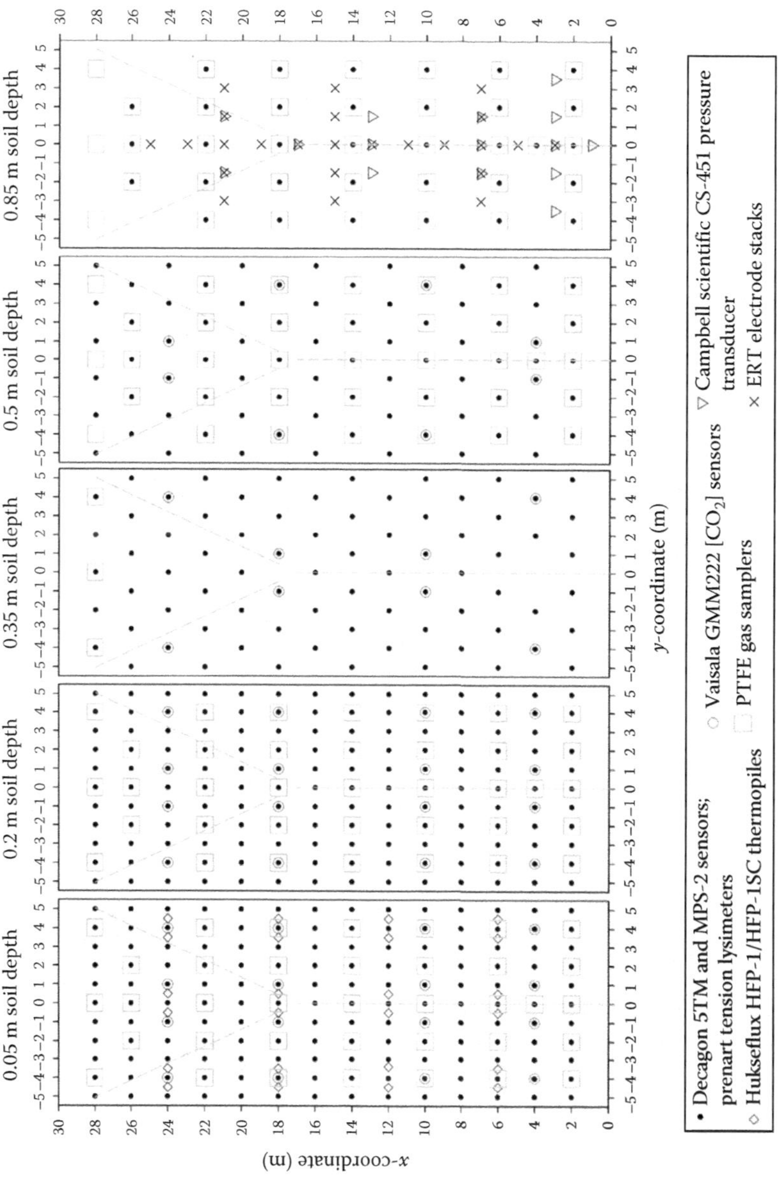

FIGURE 4.3
Diagram showing the lateral and vertical orientation of the sensor/sampler network in the LEO landscapes. For detailed description, refer to Pangle et al. (2015).

meteorological and gas sampling equipment has been installed onto vertically oriented aluminum masts that overhang each landscape. Field studies normally deploy Bowen ratio or Eddy covariance methods to determine the ecosystem carbon, energy, and water balance. However, the absence of strong above canopy mixing precludes the use of these techniques within Biosphere 2 (e.g., Arain et al. 2000). The five masts are mounted to the Biosphere 2 space frame and can be hoisted by winch, cable, and pulley system above the spray of the irrigation system to prevent dripping during experiments (Figure 4.2b). Since the atmospheric gradient will be driven by the surface exchange, we weighted the five heights on each mast stronger closer to the surface (0.25, 1, 3, 6, and 10 m above the land surface). At each height, air temperature, relative humidity, wind speed, and the water vapor concentration are measured—the former three by automated sensors, and the latter via pump-driven conveyance of air samples through a conduit system and to an infrared gas analyzer. These spatially stratified measurements, along with the closed-system nature of the Biosphere 2 facility, create additional opportunities for approximating whole-landscape evaporation based on mass balance calculations and for quantifying spatial heterogeneity of evaporative water flux (e.g., due to contrasting aspect).

4.2.1.2 Integrated and Spatially Discrete Measurements of Carbon Cycling, Weathering, Photosynthesis, and Respiration

The cycling of carbon within the LEO landscapes, and between the landscapes and atmosphere, is described as follows:

$$\frac{\partial C}{\partial t} = C_p(t) + C_a(t) + C_w(t) - C_r(t) - C_q(t) \tag{4.2}$$

where C is the mass of inorganic and organic (abiotic and biotic) carbon storage on or within the landscape (M). The subscripts attached to the carbon transfer rate terms on the right-hand side (M T^{-1}) represent carbon inputs/outputs to/from the landscape associated with precipitation p, photosynthetic assimilation a, weathering reactions w, autotrophic and heterotrophic respiration r, and water discharge from the landscape q.

The terms C_p and C_q can be quantified at the whole-landscape scale by measuring the total carbon content of inflowing irrigation water and seepage plus overland flow. Autosampling devices are in place to capture these water samples during all irrigation events on the landscapes. An on-site analytical laboratory is equipped to measure total, organic, and inorganic carbon in solution and in solid samples. A challenge for large-scale research infrastructures is to analyze huge number of samples generated at the facility. An on-site analytical laboratory proves to be advantageous here in terms of cost and time, since samples do not need to be sent out for analysis. This ensures efficient turnover time of sample analysis and reporting.

The terms C_a, C_w, C_r, and thus the derivative on the left-hand side (Equation 4.2) are approximated based on the spatial integration of distributed point-scale measurements. As mentioned in the previous section, atmospheric measurements are crucial in closing the carbon cycling budget. The aboveground network of gas sampling ports and infrared gas analyzer on the profiles, mentioned in the previous section, will be used to measure carbon dioxide (CO_2) concentration at approximately hourly interval. Atmospheric CO_2 profiles will allow for independent measurements of the whole slope carbon exchange (combined with periods of chamber closure), which is now inferred from within slope CO_2 measurements (see the following text), given the lack of autotrophic activity from vascular plants and the assumption that within-soil processes determine flux. Upon the establishment of plants, the combined soil and atmospheric profiles will be important in creating a net system flux spatially on the landscape. These measurements are complimented by estimate of stored carbon in solution and carbon exported with the solution at the seepage face.

The time resolution allows for the resolution of rapid changes following precipitation events or monitoring the expected diurnal patterns associated with radiation driving photosynthesis. The time derivative of the spatially weighted average of these measurements represents net ecosystem exchange of CO_2—the difference $C_a - C_r$, if the system is closed. At night, the same measurements represent only C_r. Those nocturnal measurements can be used to quantify functional relationships between air and/or soil temperature (measured at 25 and 496 locations, respectively) and ecosystem respiration, which, during daytime, can be used to decompose net ecosystem exchange of CO_2 into the C_a and C_r components (e.g., Phillips et al., 2011), and their source scales of spatial heterogeneity. It is well established that combinations of the approach mentioned earlier, use of CO_2 flux as a function of photosynthetically active radiation, and the periodic measurements of isotopic compositions of CO_2 flux (see the following text) provide good means for separating photosynthetic and respiratory process rates in time (Reichstein et al., 2005; Bowling et al., 2008; Lasslop et al., 2010). Additionally, 48 automated sensors measure CO_2 concentration in the soil gas phase; their spatial arrangement enables the quantification of concentration gradients within the soil, and between the soil and atmosphere, which can be used to estimate gas-phase CO_2 transfers into, and exiting from, the landscape (Barron-Gafford et al., 2011).

The carbon content of the soil solution can be monitored via 496 lysimeters that are colocated with the water content and pressure sensors (Figure 4.3) and connected to an automated vacuum system that enables relatively high-frequency sampling over the entire landscape (Section 4.3.3). These spatially and temporally intensive measurements of the soil solution carbon chemistry allow the measurement of total C storage on the landscapes and the calculation of carbon sequestration C_w. These measurements are also linked to measurements of lithogenic elements in solution that are indicators of weathering processes (Pohlmann et al.). Carbon and other solute concentrations are used further for rigorous parameterization of geochemical models that simulate the chemical evolution of the soil solid phase (Pohlmann et al.). The cumulative results of these models can be validated periodically by destructively sampling small cylindrical soil volumes that will be subsampled and analyzed to monitor inorganic and organic carbon sequestration and to determine changes in mineralogy of basalt matrix as a result of incongruent weathering using x-ray diffraction, x-ray absorption spectroscopy, and selective dissolution techniques. The same samples will undergo microbial DNA extraction, followed by high-throughput sequencing of the extracted DNA in order to determine diversity of rock-colonizing microorganisms.

4.2.1.3 Integrated and Spatially Discrete Measurements of Land-Surface Energy Exchange

The exchange of energy between the LEO landscape surfaces and their overlying atmosphere can be described as follows:

$$R_{si}(t)+R_{li}(t)+R_{so}(t)+R_{lo}(t)=H(t)+\lambda ET(t)+G(t) \tag{4.3}$$

where R terms represent radiant energy fluxes specifically associated with shortwave and longwave (s and l) radiation that is incoming or outgoing (i and o) to or from the landscape. Terms on the right-hand side represent sensible heat flux H between land and air, λET—the product of the latent heat of phase transition and the magnitude of evaporation plus transpiration, and conductive heat transport and storage into the landscape, G.

Latent heat flux is the only term in Equation 4.3 that is measured at the landscape scale, that is accomplished using the whole-landscape ET estimates, based on load-cell measurements and mass balance calculations, and the known value of latent heat of vaporization. The radiant flux terms on the left-hand side of Equation 4.3 are measured directly by a pair of four-way net radiometers that are located at 1 m height above the soil surface on the masts located over the east- and west-facing hillslope segments adjacent to the convergence zone (Figure 4.2b). Uncertainties exist in using these point measurements to represent the average at the landscape scale due to the impact of the windows and frames of the climate-controlled bay (Figure 4.2a) on solar radiation and mismatched source areas. The conductive heat flux into the ground G is measured directly by heat flux plates at 24 locations (0.08 m depth, with associated thermocouples buried at approximately 0.02 m). Those devices are arranged in uniform grid spanning most of the land surface (Figure 4.3). Finally, the sensible heat flux from land to atmosphere H can be approximated as the residual component of Equation 4.3 or possibly through application of modified gradient-flux methods that would utilize the aboveground array of meteorological instruments described earlier.

4.2.1.4 Remote Sensing of Mass and Energy Fluxes

The landscape surface is the exchange interface of carbon, energy, and water fluxes and thus warrants an even greater spatial resolution of the measurements. Remote sensing techniques can image the surface at sub centimeter resolution and thus provide detailed information on the spatial heterogeneity of evaporation (thermal imaging), water content (hyperspectral imaging), surface chemistry (potential salt precipitation during evaporation; hyperspectral imaging) on the bare soil and leaf temperature (a potential proxy for transpiration, photosynthesis, and leaf respiration; thermal imaging), and photosynthesis (hyperspectral imaging). One key challenge of remote sensing is colocation, both spatially and temporally, of the collected images, especially since no camera exists that can capture the full slope with the height constraint created by the structure. To resolve this, we mounted a 36 m long aluminum track system, including a motor-driven circulating belt with an attached climate-controlled container, below the space frame and at a constant 7 m height above each landscape. To provide greater temporal resolution and to reduce man-hours in mounting and dismounting the camera systems, one infrared imaging system and one visible-to-near-infrared hyperspectral imaging system are located within a climate-controlled box on each track system. The imaging systems and box can be rapidly and very precisely (~1 mm along path resolution and repeatability) moved along the track system, which spans the entire length of the long axis of the landscapes. Within the climate-controlled box, each imaging system is mounted on a motor-controlled rotating axle rod, which enables panning to three different angles and ultimately the integration of multiple images to provide full field-of-view coverage of the entire width of the landscape. The infrared imaging system will provide centimeter-scale image resolution, and the resulting data arrays can be used to approximate E and λE at the landscape scale via novel methods demonstrated by, for example, Shahraeeni and Or (2010). The hyperspectral imaging system is intended to enable landscape-scale estimates of C_a through novel and still experimental methods, discussed by, for example, Meroni et al. (2009).

4.2.2 Fast, Real-Time Isotope Measurement of Water and Carbon Dioxide with State-of-the-Art Laser Spectroscopic Instrumentation and Whole-Slope and Atmospheric Sampling System

A keystone to achieving the declared goal of characterizing and understanding the interactions between hydrological and biogeochemical processes is to be able to determine the pathways and residence times of water and CO_2 through the LEO landscapes and atmospheres. Both molecules are of interest by themselves with regard to understanding water resource availability and climate change but are also critical reagents for the weathering and biological colonization processes and coevolving patterns. However, measurements of states and whole-system in- and outputs of bulk water and CO_2 (see previous sections) cannot provide the required information on sources, pathways, and time spans available for chemical reactions to take place. Stable isotope analysis, in turn, is well suited for tracking fluxes and reaction processes of water and CO_2 through landscape and atmosphere (Gat, 1996; Yakir and Sternberg, 2000; Bowling et al., 2008) and can be integrated with modeling (Sprenger et al., 2015; Scudeler et al., 2016). Its application to meet the research needs at LEO is, however, challenging. While conventional isotope methodology is expensive and destructive, the space–time scales of the experiment and the highly dynamic processes under investigation demand an enormous amount and a high frequency of measurements to be performed without significantly disturbing the coevolving landscapes.

To meet the task of measuring isotope abundances associated with both water and CO_2 pools across LEO at high frequency, a state-of-the-art isotope laboratory is being implemented that makes intensive use of recent developments in laser-based analyzing (e.g., Baer et al., 2002; McManus et al., 2015) and in situ field sampling techniques (e.g., Sturm et al., 2012; Volkmann et al., 2016b). Two different isotope analyzers based on laser absorption spectroscopy (LAS) were selected. The first LAS instrument is a near-infrared gas analyzer based on off-axis integrated cavity output spectroscopy (OA-ICOS; IWA-35EP, Los Gatos Research Inc., Mountain View, CA, USA) for measurement of the hydrogen (δ^2H–H_2O) and oxygen ($\delta^{18}O$–H_2O) stable isotopic composition in injected liquid water and water vapor. The second LAS instrument is a trace gas analyzer based on quantum cascade laser absorption spectroscopy (QCLAS; TILDAS-D, Aerodyne Research Inc., Billerica, MA, USA), which measures δ^2H–H_2O and $\delta^{18}O$–H_2O and the carbon ($\delta^{13}C$–CO_2) and oxygen ($\delta^{18}O$–CO_2) isotopic composition of CO_2. Detailed descriptions of the respective technologies and instruments can be found elsewhere (Baer et al., 2002; Nelson et al., 2008; Tuzson et al., 2008; McManus et al., 2015). To facilitate stable and uninterrupted operation of the analyzers directly on-site, a custom on-site containment room was constructed below the central LEO hillslope and equipped with air-conditioning and UPS power supply. This on-site laboratory provides time-efficient analysis of the samples.

The OA-ICOS instrument will be used mainly for automated high-frequency sampling and analysis of seepage water outflow from the three LEO hillslopes. It will therefore be paired with a multiport liquid sampling system (Los Gatos Research Inc.) as described by Pangle et al. (2013). This setup was chosen because it is robust and will facilitate isotopic analysis of outflow from the three LEO hillslopes at intervals of approximately 30 min (Pangle et al., 2013). The half-hourly interval was considered sufficient to capture most of the discharge isotope dynamics associated with variable water flow path activation and transit times through the landscape given that hillslopes such as those comprising LEO act as low-pass filters to any variable input signal. Thereby, our setup should allow for reliable association

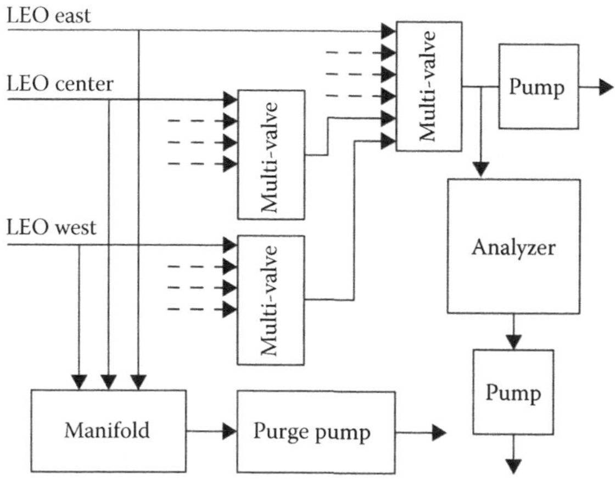

FIGURE 4.4
Schematic of the planned setup (simplified) for direct monitoring of stable isotope abundances in atmospheric water vapor and/or carbon dioxide using a laser spectroscopic analyzer at the LEO. Only 1 out of 24 available intake lines is shown for each LEO bay.

of water pathways with solute export resulting from biogeochemical processes at relevant timescales and detail.

The QCLAS instrument will be used mainly for continual monitoring of the isotopic composition of atmospheric as well as soil air CO_2 ($\delta^{13}C$–CO_2, $\delta^{18}O$–CO_2) and water vapor ($\delta^{2}H$–H_2O and $\delta^{18}O$–H_2O). In each LEO hillslope, 24 atmospheric air intake lines are available, with inlets at four to five different heights (0.25, 1, 3, 6, and 9–10 m) along each of five masts distributed over the slope surface (Figure 4.4). Subsequent sampling of the intake lines will be facilitated by stream selector valves located at the on-site isotope laboratory, with flow driven by a downstream vacuum pump (Figure 4.4). To reduce the time delay associated with gas transport from air inlet to analyzer, the intake lines upstream of the valves will be constantly purged with fresh atmospheric air using branch-off lines connected to a purge pump via manifold unions (e.g., Sturm et al., 2012).

To sample soil air for direct isotopic analysis, arrays of 151 custom sampling probes will be used that are installed along vertical profiles (0.05, 0.2, 0.35, 0.5, and 0.85 m from the land surface) within each of the model landscapes at LEO (Figure 4.3). A multivalve control system will be set up to sample automatically from the various probe locations using a closed flow-through loop approach. These measurements will not only provide insights into gas-phase fluxes and interactions such as CO_2 diffusion and consumption during weathering or evaporation; since soil temperatures are measured throughout the LEO soils, inference of the liquid water isotopic composition in the landscape's subsurface is also possible based on the soil water vapor-phase measurements (Volkmann and Weiler, 2014) due to the mainly temperature-dependent isotopic liquid–vapor equilibrium in soils

(Mathieu and Bariac, 1996). These measurements can then be used to infer flow pathways and interaction times of water throughout the hillslopes. The anticipated sampling interval is <5 min per probe location. The system will thus allow for approximately one complete measurement cycle per day, but subsets of the available soil gas probes will be selectable to attain much higher temporal resolution at selected profiles sufficient to capture even rapid processes such as preferential infiltration (Volkmann et al., 2016a). Despite the unprecedented capabilities, spatial and temporal resolution must, however, clearly be traded off with regard to the isotopic measurements.

4.2.3 Solution Collection and Analysis

The fate of organic and inorganic carbon in the soil and incongruent weathering of basalt matrix, as well as translocation of carbon and lithogenic elements on the slopes will be tracked using a combination of solution sampling and periodic coring of the soil matrix. This allows us to map carbon accumulation in soil solution and solid phases and to trace lateral flux of carbon through the soil. It also allows to identify changes in mineralogical and chemical composition of basalt as a result of dissolution of primary minerals, movement of resulting solutes through the slope, and precipitation of secondary minerals. Solution sampling will be performed through 496 samplers by Prenart samplers (Frederiksberg, Denmark) that are embedded in LEO at the same locations as moisture and water pressure sensors described in Section 4.2.1. A sampling line connects colocated Prenart samplers to 11 custom-designed sampling modules (Figure 4.5). Each sampling line has individual on/off valve. Modules consist of a Plexiglas vacuum box equipped with manual pressure regulator, vacuum gauge, and a tray capable of holding fifty-five 50 mL centrifuge tubes for sample collection, connected to a vacuum manifold. Vacuum is supplied by two single-stage rotary vane vacuum pumps common for all three slopes and located outside of the LEO area.

The vacuum system allows collection from all 496 samplers on each slope at the same time. However, when needed, the vacuum could be applied to each module or even each sampler separately, allowing sampling a section of each slope at a time as the wetting front progresses. Suction to be used in each module can be determined by water potential measured by the sensors colocated with solution samplers to ensure sufficient suction to withdraw a sample, but not high enough to withdraw solution outside of the immediate sampler vicinity and create new preferential flow pathways.

Samples will be regularly collected from all active samplers in a given hillslope and archived at –80°C. We will also collect water samples from rainfall and outflow (seepage flow and possibly some overland flow). The analytical laboratory on-site is equipped for high-throughput sample analysis for major cations and anions with capability for concurrent analysis of cations and anions, including autosampler, capillary and analytical pumps, two conductivity detectors, and eluent generators for both pumps. Electrochemical

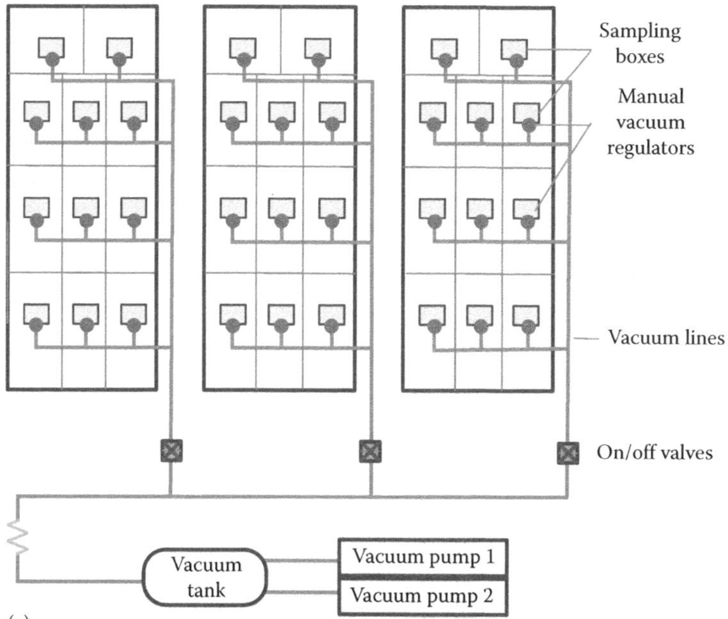

(a)

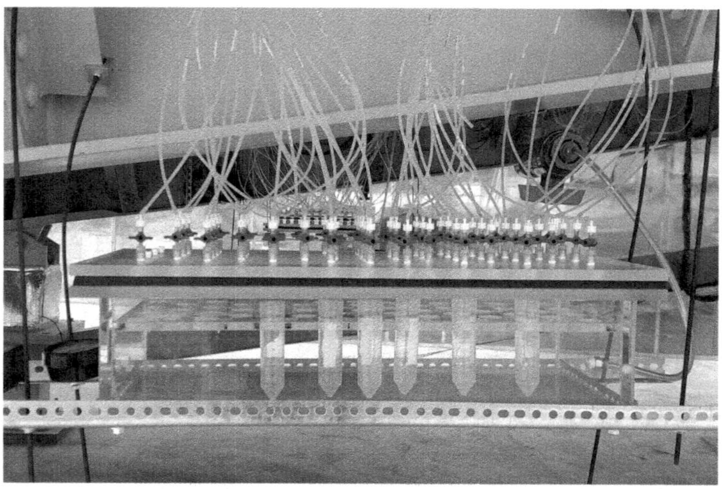

(b)

FIGURE 4.5
(a) Schematics of solution sampling module. Each hillslope will be equipped with 11 solution sampling modules. Modules include a vacuum box (b) with up to 55 sample containers connected to a vacuum line through a vacuum regulator. 496 Prenart samplers that are embedded in each LEO slope are connected through the sampling lines to the containers in the box. Vacuum is maintained through a dual pump and storage tank.

detector and a column are also available to analyze solutions for carbohydrates and amino acids; dissolved organic and inorganic carbon, and total dissolved nitrogen (Shimadzu TOC-L Series Total Organic Carbon (TOC) and nitrogen analyzer equipped with autosampler for water samples (TOC-LCSH), and an SSM-5000A unit for analysis of solid samples); and pH (VWR® sympHonymultimeter), which is necessary to determine inorganic carbon speciation in solution.

4.2.4 Electrical Resistivity Tomography

Geophysical methods have the potential to overcome limitations in spatial resolution of local point measurements, being able to measure subsurface structures and to estimate flow and transport in a minimally invasive or even noninvasive way (Vereecken et al., 2004). Among current geophysical methods used for monitoring subsurface hydrological properties, electrical resistivity tomography (ERT) has found great acceptance. Electrical resistivity within the soil can be correlated to several physical and chemical soil properties such as soil salinity, temperature, clay content, soil-water content, and solute concentration. Several studies have applied the ERT technique for obtaining high-resolution monitoring of spatial/temporal solute transport in controlled infiltration and tracer experiments (Binley et al., 1996, 2002; Slater et al., 2002) as well as in vadose zone soil water monitoring (Michot et al., 2003; Nijland et al., 2010; Beff et al., 2013). A typical ERT application consists of injecting a low-frequency electrical current through a pair of electrodes and measuring the potential at other electrode pairs. This procedure is repeated several times, with different injection locations, so that a set of redundant measured potentials is obtained, which will give rise to a distributed image of apparent electrical resistivity. The apparent electrical resistivity is further transformed into "true" resistivity through an inversion procedure (Vereecken et al., 2014).

A custom array of 24 potential-measuring electrode stacks was built inside each LEO hillslope prior to soil packing. A compromise between costs and resolution was made when installing the ERT sensors in order to arrive at the achievable resolution for electrical resistivity of nanometer scale. Each electrode stack comprises a hollow insulating acrylic cylinder, where a sequence of five stainless steel electrodes separated by different distances is placed (Figure 4.6). Each electrode stack is connected to a centralized device that fires current, reads the potential differences, and stores the results. The electrode stacks are equipped to read several potential combinations between electrodes across the hillslope, therefore, providing images of apparent electrical resistivity, which can then be interpreted after inversion algorithms are applied.

The ERT system is of importance to the LEO project, because of its key role in supporting cross-discipline research endeavors. The ERT three-dimensional scans associated with hydrological models through coupled

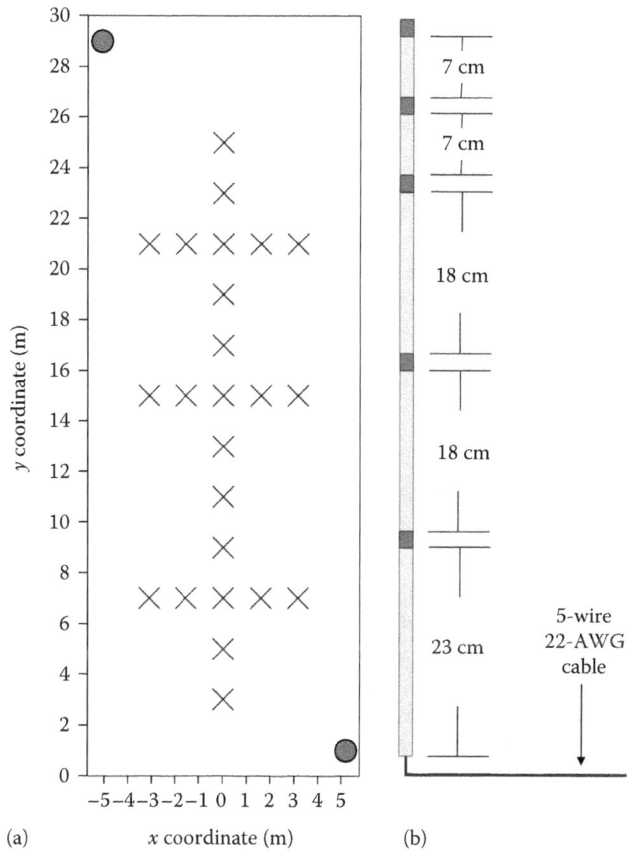

FIGURE 4.6
(a) Plan view of the LEO landscape dimensions. X symbols mark the position of potential-measuring electrode stacks. Circles mark the position of the current-injecting electrodes. (b) Diagram of potential-measuring electrode stacks: dark gray indicates stainless steel electrodes, while light gray indicates the insulating acrylic rods.

hydrogeophysical inversion (Ferré et al., 2009) will provide a continuous feed of high-resolution imaging of the flow processes at the subsurface as the LEO landscapes evolve from initially homogeneous crushed basalt rock into a complex ecosystem. Assessing evolution of flow processes within the subsurface in a spatially distributed manner is key to interpreting results of soil and water chemistry analysis along LEO's evolutionary path. Describing the location, quantity, and velocity of water flowing within the landscape at the same time that its chemical composition is being investigated, will allow investigation of feedback and interactions between coupled hydrological and biogeochemical processes. Additionally, ERT-generated data will aid the development of novel hydrogeophysical inversion methods, enabling the real-time monitoring and visualization of the movement of water, solutes, and assimilated fields of soil hydraulic properties.

4.2.5 Detection and Monitoring of Microbial Activity

In the natural world, soil and landscape dynamics are influenced by the interaction of physical, chemical, and biological processes. The organismal diversity of a site, inclusive of vascular plants, soil microorganisms, and burrowing and soil disturbing fauna, all contribute to the life–environment interactions that play a crucial role in formation of soil structure and biogeochemical cycling of nutrients. In the context of a landscape being colonized by life, microorganisms are important early pioneers that can have large impacts on nutrient cycling and through symbiosis can influence plant establishment and growth. Combined, these processes affect a number of ecosystem processes, such as net primary production, which dictates the degree to which biology may influence the evolution of a physical landscape. Consequently, during the progression of a parent material developing into soil over a landscape scale, one of the first biological features to consider is how microbial life at the microscale evolves and impacts the macro hillslope scale when subjected to varying climatic and relief conditions. Such questions of how these microbiological systems influence the physical template will be important to the continuing sequence of experimentation in LEO, where understanding plant impacts on the water balance, for example, represents another opportunity to evaluate life–environment couplings important to understanding Earth system science.

While LEO has been extensively used to study hydrological and geochemical interactions at the hillslope scale so far, the infrastructure will also allow researchers to study the diversity of microbial communities capable of exploiting the extreme oligotrophic conditions of a basalt rock hillslope and identify key environmental variables that significantly influence the diversity and activity of these communities. The basaltic soil material of LEO has low organic matter content (Pangle et al., 2015), and the absence of vegetation on the hillslopes during initial phases of the experiment eliminates the possibility of carbon input from plants. The progression of microbial life on LEO slopes is projected to be monitored from time zero through a 10-year period during which the slopes are expected to experience primary mineral weathering, release of lithogenic elements followed by secondary mineral formations. Coupled with the gradient and alternate wetting–drying of the hillslopes, a heterogeneous presence and variability of microbial occurrence across the LEO landscape is expected. It will be interesting to observe diversity patterns along the slope at multiple scales, particularly keeping in mind that species diversity in ecological systems is composed both of point estimates of species richness (what kinds), evenness (how many), and spatial patterns of heterogeneity (site-to-site turnover; Sengupta and Dick, 2015). Of particular interest are correlations between species richness of microbial communities at sample spots and the associated flow dynamics and pore-water geochemistry. Modern functional genomics will enable similarly coordinated measures of aspects of gene diversity (functional diversity) in the same context (e.g., Allison and Martiny, 2008).

Previous research has shown the presence of microorganisms in basaltic material. Rogers and Bennett (2004) reported the release of limiting nutrients from silicates by subsurface microorganisms while a study by Savostin (1972) suggested the presence of "silicate microorganisms" capable of growing on silicates in absence of organic substances. Recently, Orcutt et al. (2015) provided empirical evidence of chemoautotrophic carbon fixation by basalt-hosted microbial communities in the deep ocean environment. These findings suggest that even in oligotrophic mineral environments like the LEO slope, one can expect to find microbial life adapted to functions. The characteristic mineral composition of the LEO soil material, along with low organic matter input, can be considered as a primary ecosystem capable of harboring pioneer microorganisms with mineral-weathering abilities.

The specific challenges posed by the LEO experiment include the development of protocols to characterize the low-biomass samples that we predict will be present at time-zero in the basalt soils and the design of nondisruptive sampling strategies that will produce a suite of samples representative of the range of microenvironmental conditions present across the surface and depth of the hillslope. Using microbial DNA environmental sampling and next-generation sequencing techniques (see Chapter 8), researchers have been able to detect microbial life at regions of low biological productivity. A few of these include bacterial diversity in arid soils of the Atacama Desert, Chile (Drees et al., 2006; Neilson et al., 2012), microbes colonizing subsurface calcite cave formations in Kartchner Caverns, AZ (Ortiz et al., 2013, 2014), and microbial communities of acid-generated mine tailings (Valentín-Vargas et al., 2014). Generally, current sequencing techniques like Illumina® require minimum DNA concentration of 1 ng/μL to generate a community DNA library. However, the amount that can be retrieved from oligotrophic environments falls in the picogram range. Research groups in the Department of Soil, Water, and Environmental Science, University of Arizona, have expertise working on isolating microbial DNA in oligotrophic environments. Protocols have been developed to obtain sequence reads from 0.1 ng/μL DNA extracts (Valentin-Vargas et al., 2014), and research continues to further optimize library preparation from lower concentration DNA extracts. Currently, sampling protocols are being developed to recover samples from the LEO surface without introducing any exogenous microbial contamination. Due to the extremely low biomass levels present in the LEO landscape, slight contamination from sampling equipment or technicians can have a profound impact on the sample taxonomic profiles. The highly instrumental and environmentally controlled LEO facility provides an excellent opportunity to develop new protocols for microbial community analysis, developing trait-based approaches to representing microbial communities and integrating data generated into cutting-edge models that predict the impact of in situ microbial activity on micro- to macroscale environments (e.g., Allison, 2012; Wieder et al., 2013).

4.3 Research Foci to Advance Understanding of Interacting Hydrological and Biogeochemical Processes

4.3.1 Flow and Transport Studies at Landscape Scales

In recent years, there has been a paradigm shift in our understanding of flow and transport at watershed scales and approaches to their prediction. The complexity and heterogeneity of water movement within individual landscape units have been recognized in hillslopes (Jencso and McGlynn, 2011; Nippgen et al., 2011), riparian areas (Buttle et al., 2004; Detty and McGuire, 2010) and within streams and their hyporheic zone (Bencala et al., 2011). This has led to calls for new approaches to prediction that go beyond the traditional continuum models (i.e., Richards equation and the St Venant equations for flow and convection–dispersion/diffusion for transport—or approximations thereof) (Sivapalan, 2005; Beven, 2006; Kirchner, 2006; McDonnell et al., 2007), as these generally rely on calibrated "effective" property values to replace the spatially distributed properties of the landscape—those are essentially unknowable at catchment scales using current technology.

New approaches have sought ways to represent flow and transport directly at the scales of interest, with the expectation that the new equations may be "different in form, not just different in the parameters" (Kirchner, 2006) than the continuum-scale equations. The concept of a representative elementary watershed, or REW (Reggiani et al., 1998, 2000, 2001; Beven, 2006), provides a conceptual framework for representing flow through individual landscape elements within an REW and in a river network based on a rigorous time–space averaging of the conservation laws governing mass, energy, momentum, and entropy. These equations are not complete, however, as they require specification of "closure relations" that specify the boundary fluxes exchanged between these landscape elements in terms of their states (storage, fluxes, other boundary forcings) and parameterized by measurable properties of the landscape. These closure relations must represent the aggregate effect of the unresolved "sub-REW" heterogeneities and flow complexity without resolving them explicitly, and they have been termed the "Holy Grail" of scientific hydrology (Beven, 2006).

The understanding and prediction of the timing and partitioning of fluxes of water at landscape scales is a considerable challenge on its own, but the understanding and prediction of transport at watershed scales is an even greater challenge. Studies of using passive tracers present in precipitation (such as chloride and water molecule isotopologues) have shown that a large fraction of the water that appears in a stream in response to a storm is water that has been present in the catchment for a considerable period of time (Sklash and Farvolden, 1979; McDonnell, 1990; McDonnell et al., 2010). This water reaches the stream through a flow system that includes spatially variable infiltration and unsaturated flow (Roth, 1995), shallow transient lateral flow through perched water tables (Troch et al., 2003), preferential flow pathways that connect and disconnect uplands and riparian areas with threshold nonlinearities (Tromp-van Meerveld and McDonnell, 2006a; Jencso and McGlynn, 2011), flow through deep and shallow fractured bedrock (Montgomery and Dietrich, 2002), and other types of fine-scale dynamics (Tetzlaff et al., 2008). The challenge has been to find ways to represent the mixing of water of different ages in the discharge that agrees with observational data and accounts for the highly transient, complex nature of the transporting flow. These efforts are ultimately aimed at a complete theory of flow and transport at watershed scales suitable for use in predicting transport of solutes, contaminants, and pathogens through complex landscapes (Kirchner et al., 2000; Rinaldo et al., 2011; van der Velde et al., 2013).

Understanding the hillslope-scale structural controls on flow and transport is one of the two main research questions that LEO was built to address (Hopp et al., 2009). The LEO facility has been built to allow us to develop observations of the "closure relations" governing flow and transport at a scale approaching that of real hillslopes in a highly controlled laboratory environment. These hillslopes can be subjected to artificial rainfall events, and the subsequent flow and transport can be analyzed by over 1800 sensors and water samplers embedded in each hillslope. Analysis of deuterium and oxygen-18 stable isotope tracers can be conducted in real time using existing laser spectroscopic instrumentation. Insights obtained from the flow and transport studies can be applied to coupled biogeochemical processes occurring on the hillslope, as described in Sections 4.3.2 and 4.3.3. This forms an integral part of the second research question attached to LEO: that of landscape stabilizations and ecosystem development over time. This includes the role of biological processes and the impact of life–environment interactions. While microorganisms have soil-stabilizing properties and impact biogeochemical cycling of nutrients, vegetation controls net primary production, the order of magnitude of life in a location, along with affecting flow and transport in natural landscapes: through root-water uptake and redistribution over timescales of minutes to days; through biomass growth and canopy development over days to seasons; through the formation of preferential flow paths over weeks to years; and through controls on soil structure and geochemical weathering over years to millennia.

4.3.2 Coupled Hydrological and Geochemical Process Evolution

There exists a tight coupling between the time water spends along subsurface flow paths and geochemical weathering reactions (Steefel, 2008; Maher, 2011). Since mineral dissolution is kinetically limited, the duration of water contact time with matrix minerals along a given flow path determines the extent to which it acquires solutes from mineral dissolution and, hence, the *relative saturation* of mobile solution with respect to primary mineral dissolution and potential secondary mineral formation (Maher et al., 2009). Such reactions are especially rapid in extrusive, mafic igneous rocks such as basalts (Gislason et al., 2008). Incongruent dissolution of primary silicates to form poorly crystalline hydrated solids, as well as crystalline layer silicate clays and (oxyhydr)oxides, can affect these rates (Yokoyama and Banfield, 2002), as can the presence of complexing biogenic ligands (Neaman et al., 2005; Dontsova et al., 2014). Progressive weathering and its distribution in space leads to the development of subsurface heterogeneity (e.g., catena development), which, in turn, feeds back to affect water flow paths (Dontsova et al., 2009).

As chemical weathering drives the particle size distribution to smaller values and hydrated solids are formed, this will affect the pore size distribution, hydraulic conductivity, and hence the water transit time distribution along these flow paths (Li et al., 2008; Pacheco and Van der Weijden, 2012). Changes in particle size, surface chemistry, and porosity distributions further alter reactive interfacial area and weathering rates of the soil medium (Anbeek, 1992; Brantley and Mellott, 2000), with associated feedback on subsurface flow path evolution, bio-colonization, and nutrient and organic matter retention. At the same time, changes in flow paths and transit time distributions will affect geochemical weathering reactions through alterations of Gibbs energy of reaction (e.g., changes in the quotient of product and reactant activities relative to equilibrium) (Ganor et al., 2007). These geochemical–hydrological couplings are considered fundamental to the evolution of subsurface structure in the critical zone (Chorover et al., 2007).

Dontsova et al. (2009) used a reactive transport geochemical model coupled with a hillslope-scale hydrological model and pedotransfer functions to predict mineral transformations in a basalt soil representative of LEO (45.67% basaltic glass; 35.87% labradorite, feldspar; 11.94% forsterite, olivine; 6.53% diopside, pyroxene) and to estimate resulting changes in soil hydraulic properties. They found that significant changes in both the fraction of secondary minerals and predicted hydraulic conductivity are expected after 10 years of climate forcing representative of semiarid Southwestern United States and that observable changes are expected as early as within 3 years. Model results indicated that characteristic spatial trends in pore water chemistry, indicative of the trajectory of soil formation, arise even during initial hydrological events, well before changes in porous media solid-phase composition are likely to be detectable. It remains to be tested whether these model predictions are realistically capturing the time evolution of subsurface structure related to geochemical weathering and precipitation.

The geochemical evolution predicted by this modeling varied spatially in a way that reflected the variations in hydrological function between different parts of the hillslope. Differences in soil moisture and residence time between the unsaturated upslope areas and the persistently saturated convergent regions produced significant structure in the subsurface hydraulic heterogeneity, consistent with observations in pedogenic studies along catenas and chronosequences (Chadwick and Chorover, 2001; Jing et al., 2008).

The LEO facility will allow us to simultaneously measure soil moisture status and the associated residence times of water within the slopes, as well as the geochemical transformations that take place during and after rainfall–runoff events by chemical analysis of soil solution samples drawn from the network of soil samplers (described in Section 4.2.3). Water collections from early rainfall experiments at LEO in 2013 provided insight into the nature and distribution of secondary mineral phases formed following incipient weathering of the primary basalt material (Pohlmann et al., 2016). Samplers embedded in the 1 m profile conveyed pore water for individual collection beneath the steel frame of the hillslope. Over 900 pore water samples collected from two distinct rain events were processed and analyzed for a suite of geochemical analysis including anions, cations, inorganic carbon, and pH. Sensors also recorded the volumetric water content and temperature associated with each of these pore water samples. Initial processing and analysis for a selection of samples following the first rain event revealed solid formation in the colloidal size range (60–2000 nm) for aluminum (Al), phosphorus (P), and iron (Fe). This suggests the development of poorly crystalline mineral phases such as gibbsite (an Al hydroxide) and ferrihydrite (an Fe oxyhydroxide), in addition to phosphate (a common P-bearing anion) known to associate with such minerals. These tiny solids represent the earliest stage of soil development from our essentially unweathered basalt (Pohlmann et al.).

4.3.3 Microbiological and Biogeochemical Evolution of Landscape

Earth's land surface is the locus for coevolution of biota, soils, and landforms and is governed by processes important to soil science, geomorphology, hydrology, ecology, and atmospheric sciences (Showstack, 2012). The processes are coupled in terms that the functioning and dynamics of one affects that of the other. In the context of a colonizing landscape, microbes are at the forefront of the biotic component of landscape evolution. They play a significant role in bioweathering of minerals during soil genesis through the oxidation of primary minerals, the influence of acid-generating metabolisms that increase solubilization rates, and the production of siderophores and other organic molecules that function as metal-complexing agents to drive weathering reactions (Mapelli et al., 2012). However, interpreting microbe–mineral interactions with underlying hydrological forces raises a few fundamental questions. Focusing on geochemistry, one can ask whether microorganisms benefit by the dissolution of a mineral or if weathering is simply a coincidental by-product of basic biochemical functions. With respect to hydrology, one can question the effect of hydrological partitioning and water-transit time distribution on microbial evolution of the landscape, and conversely, the effect of microbial evolution on time-dependent change in flow paths. The answers to these questions cannot be answered by studying microbial communities in isolation. Instead, studies that focus on microbial ecology, that is, interaction of microorganisms with each other and their environment are a key variable to the landscape evolution equation.

Microorganisms colonizing bare basaltic material can potentially be categorized into three metabolic groups: (1) *chemolithoautotrophs* that obtain energy through the oxidation of diverse reduced (often mineral-bound) elemental forms (i.e., Fe, Mn, and S), thus oxidizing the inorganic element and generating acid as a by-product; (2) *phototrophic cyanobacteria and algae* that fix CO_2 and N and function as drivers of increased biodiversity in oligotrophic systems; and (3) *heterotrophic bacteria or fungi* that oxidize available organic C supplies and produce extracellular products that solubilize primary or secondary minerals and can function as complexing agents that influence pore water chemistry. In addition, microbial respiration increases soil pore CO_2 concentrations leading to enhanced pore water concentrations of carbonic acid and associated (carbonation) weathering reactions.

With the accretion of organic C from chemoautotrophic growth, heterotrophic colonization is expected to ensue. Literature suggests that the initial microbial consortium that colonizes the basalt surface takes advantage of physical conditions in the porous grain interiors (Bagshaw et al., 2011) and catalyzes some of the most important mineral transformations including those of active Fe and Mn oxidation (Edwards et al., 2003; Bailey et al., 2009). Furthermore, the strong bonding that occurs between neo-formed inorganic colloids and microbial macromolecules, lipopolysaccharides, and nucleic acids, results in the long-term persistence of otherwise labile C forms in soil and sediment (Dontsova and Bigham, 2005; Omoike and Chorover, 2004, 2006; Mikutta et al., 2011). The formation of these secondary bioinorganic solid-phase products during pedogenesis in basalt has been shown to impact soil hydrology, with potential feedback to biogeochemical processing (Lohse and Dietrich, 2005). Additionally, there is strong evidence that organic compounds produced by bacteria and fungi influence soil structure, resulting in a more porous, ordered, and aggregated microenvironment (Feeney et al., 2006) with important consequences for water and gas retention and transport. Studying microbial community diversity on LEO hillslopes will thus be able to provide supporting insights into the biotic variable of landscape evolution.

In addition to studying microbial community diversity patterns on LEO, biogeochemical and hydrological weathering of basaltic soil material of LEO is of particular relevance to global carbon (C) and nitrogen (N) cycling. Despite making up only 5% of earth's land area, basalt silicates contribute to sequestration of atmospheric carbon dioxide (CO_2) over geologic timescales, accounting for 7% of total CO_2 uptake, and nearly 30% of the uptake associated with silicate weathering in particular (Gaillardet et al., 1999; Dessert et al., 2003; Suchet et al., 2003). In fact, initial experiments conducted in LEO indicated significant intake of CO_2 from atmosphere through weathering processes (van Haren et al.). The primary mineral assemblage of basalt weathers incongruently to form poorly crystalline secondary minerals (e.g., allophane, imogolite, ferrihydrite) of high specific surface area and chemical reactivity (Chadwick et al., 2003; Chorover et al., 2004; Dontsova et al., 2014), and these weathering products have been shown to form strong interbonded complexes with biogenic organic matter, thereby, stabilizing it against heterotrophic degradation over long timescales (Torn et al., 1997). The process of primary mineral weathering followed by translocation of released chemicals and secondary mineral formation contributes to nutrient stabilization in the system. As LEO evolves over time, one can anticipate critical opportunities for understanding the component roles of plants, soil food webs, biogeochemical cycle feedback, and a range of ecological processes shaping ecosystem development.

The LEO slopes have been designed to essentially track the evolution of a nascent landscape in the presence of carefully manipulated biological diversity, to understand and illustrate the subsequent coupling of biogeochemical Earth processes with the physical system. At LEO, we expect that biogeochemical "hot spots" will emerge reproducibly within the convergent hillslope structures, and that these "hot spots" will be loci of intense weathering, microbial growth, and plant establishment. Such locations will lead to the formation of nanoparticulate, bioinorganic weathering products (e.g., organometal and organo-mineral complexes) that will lead to (1) organic C stabilization against heterotrophic degradation and (2) hillslope-scale alterations in hydrological response of the hillslope. Our initial intent is to establish a suite of vascular plants that takes advantage of model systems sufficient to understand how diverse genotypes of the same species, or alternatively different species, may establish in the context of the heterogeneity of the LEO surface. Subsequent mapping of plant performance and the surface/subsurface characteristics that influence the ecophysiology of plant growth and reproduction will lead to a better understanding of how the factors that regulate the establishment of diverse plant communities, resource extraction dynamics, and the impacts of these plants on water/resource balance and soils. All of this is guided by an ecohydrological hypothesis of "homogenization" describing the effects of plants on surface water balance in the context of a variable environment (Ivanov et al., 2010). Over time, we expect to identify clear relationships between water movement, geochemical weathering, microbial diversity, and plant activity upon the hillslopes, thereby, enabling us to answer the bigger question of how landscape and life coevolve on our planet.

4.4 Conclusions

The LEO experiment, with all its instrumentation in place, is scheduled to start in 2016, when initially the evolution of the unvegetated landscapes will be observed under identical treatment regarding climate forcing. This initial phase is estimated to last for about 2–3 years, after which vascular plants will be introduced to the slopes in order to study how complex ecosystems alter the cycling of resources through these slopes and how they affect the evolutionary trajectory of the landscapes. This second phase is anticipated to last for another 6–7 years. Initial replication and identical experimental treatment of all three slopes will allow for significant model development that will allow for nuanced experimentation in the future, such as each replicate slopes being set on different climate trajectories, to observe how identical ecosystems are perturbed by different climate forcing. The data collected during this 10-year-long experiment and the knowledge gained from it will allow us to develop more powerful coupled system models that are capable of predicting landscape evolution in the real world in the context of climate change. Such tools are urgently needed to study the fate of ecosystem goods and services in the twenty-first century.

Many challenges remain, however, with any large-scale project like LEO. First, there are always design challenges along with logistical problems associated with a large-scale research infrastructure as LEO. These issues have been addressed thoroughly from a hydrological perspective in Hopp et al. (2009) and Dontsova et al. (2009) from a hydrogeochemical perspective, Ivanov et al. (2010) from an ecohydrological perspective, and Pangle et al. (2015) from an atmospheric perspective. The additional information about LEO observation capabilities given here are needed to convey the opportunities for coupled hydro–biogeochemical research.

With a team of diverse researchers, projects have to be planned out to best serve the interests of everyone without hampering the experimental requirements of any discipline or the future value of the infrastructure. Soil sampling on LEO is also a challenge since any activity that damages the soil material compromises the integrity of the hillslopes. To overcome these challenges, LEO researchers have put multiple strategies in place. All activities are planned and executed after considerable focus group meetings that address individual research concerns. A sampling strategy for microbial analyses that require destructive sampling is being developed to best capture the spatial heterogeneity of microbial communities without compromising the hillslopes. A crucial component of large-scale research infrastructures like LEO includes handling and archiving the entire data life cycle. The ability to store, access, analyze, share, and curate such vast amounts of big data, both pre- and post-publication are vital. With real-time data acquisition and archiving servers, and the presence of on-site analytical laboratories, LEO has the ability to manage the demands of multidisciplinary data handling. As a research infrastructure committed to model itself as a tool for the scientific community, collaborative research is highly encouraged for those keen on performing experiments at LEO or use its data sets. The biggest challenge, however, is to keep this research infrastructure running. While funds to build the infrastructure is one part, significant funding is also required to conduct research.

This chapter, together with Pangle et al. (2015), describes the research infrastructure available at the LEO at Biosphere 2, The University of Arizona, which is ambitious in its scope, capacity, and intent in serving as a community asset. This research infrastructure includes a dense network of automated sensors and samplers to observe water, carbon, and nutrient spatial and temporal dynamics in large-scale convergent landscapes comprised of ground basalt, as well as aboveground instrumentation for direct observation of water, carbon, and energy fluxes at the landscape surfaces. Moreover, sophisticated analytic instrumentation (dual laser spectrometers) allow for near real-time observation of the movement of water and carbon stable isotopes in the soil, atmosphere, and vegetation populating the hillslopes. The facility is unique in the world and provides opportunities to earth scientists, representing a wide range of disciplines (from hydrology to geochemistry to microbiology and ecology) to work in an interdisciplinary setting and address fundamental questions about the coevolution of hydrological, biogeochemical, and ecological processes and their interactions.

Acknowledgments

The authors wish to acknowledge support of NSF-funded project 1344552, NSF-funded Hydrologic Synthesis Project: Water Cycle Dynamics in a Changing Environment: Advancing Hydrologic Science Through Synthesis; NSF Grant EAR-0636043, NSF Grant EAR-1340912, and Philecology Foundation of Fort Worth Texas. Additional funding support was provided by the Water, Environmental, and Energy Solutions (WEES) initiative at the University of Arizona and by the Office of the Vice President of Research at the University of Arizona.

References

Allison, S.D. 2012. A trait-based approach for modelling microbial litter decomposition. *Ecology Letters* 15(9): 1058–1070.

Allison, S.D. and J.B.H. Martiny. 2008. Colloquium paper: Resistance, resilience, and redundancy in microbial communities. *Proceedings of National Academy of Sciences of the United States of America* 105(Suppl): 11512–11519.

Anbeek, C. 1992. Surface roughness for minerals and implications for dissolution studies. *Geochimica et Cosmochimica Acta* 56: 1461–1469.

Arain, M.A., W.J. Shuttleworth, B. Farnsworth, J. Adams, and O.L. Sen. 2000. Comparing micrometeorology of rain forests in Biosphere-2 and Amazon basin. *Agricultural and Forest Meteorology* 100: 273–289.

Baer, D.S., J.B. Paul, J.B. Gupta, and A. O'Keefe. 2002. Sensitive absorption measurements in the near-infrared region using off-axis integrated-cavity-output spectroscopy. *Applied Physics B:Lasers and Optics* 75: 261–265, doi:10.1007/s00340-002-0971-z.

Bagshaw, E.A., C.S. Cockell, N. Magan, J.L. Wadham,T. Venugopalan, T. Sun, M. Mowlem, and A.J. Croxford. 2011. The microbial habitability of weathered volcanic glass inferred from continuous sensing techniques. *Astrobiology* 11: 651–664.

Bailey, B., A. Templeton, H. Staudigel, and B.M. Tebo. 2009. Utilization of substrate components during basaltic glass colonization by *Pseudomonas* and *Shewanella* isolates. *Geomicrobiology Journal* 26: 648–656.

Barron-Gafford, G.A., R.L. Scott, G.D. Jenerette, and T.E. Huxman. 2011. The relative controls of temperature, soil moisture, and plant functional group on soil CO_2 efflux at diel, seasonal, and annual scales. *Journal of Geophysical Research* 116(G1): G01023.

Beff, L., T. Günther, B. Vandoorne, V. Couvreur, and M. Javaux. 2013. Three-dimensional monitoring of soil water content in a maize field using Electrical Resistivity Tomography. *Hydrology and Earth System Sciences* 17(2): 595–609.

Bencala, K.E., M.N. Gooseff, and B.A. Kimball. 2011. Rethinking hyporheic flow and transient storage to advance understanding of stream-catchment connections. *Water Resources Research* 47(3): W00H03, doi:10.1029/2010WR010066.

Beven, K. 2006. Searching for the Holy Grail of scientific hydrology: Q t = H (S R, t) A as closure. *Hydrology and Earth System Sciences* 10(5): 609–618, doi:10.5194/hess-10-609-2006.

Binley, A., G. Cassiani, R. Middleton, and P. Winship. 2002. Vadose zone flow model parameterisation using cross-borehole radar and resistivity imaging. *Journal of Hydrology* 267(3–4): 147–159.

Binley, A., S. Henry-Poulter, and B. Shaw. 1996. Examination of solute transport in an undisturbed soil column using electrical resistance tomography. *Water Resources Research* 32(4): 763–769.

Bowling, D.R., D.E. Pataki, and J.T. Randerson. 2008. Carbon isotopes in terrestrial ecosystem pools and CO_2 fluxes. *New Phytologist* 178(1): 24–40.

Brantley, S. and N. Mellott. 2000. Surface area and porosity of primary silicate minerals. *American Mineralogist* 85: 1767–1783.

Buttle, J.M., P.J. Dillon, and G.R. Eerkes. 2004. Hydrologic coupling of slopes, riparian zones and streams: An example from the Canadian Shield. *Journal of Hydrology* 287 (1–4): 161–177, doi:10.1016/j.jhydrol.2003.09.022.

Chadwick, O.A. and J. Chorover. 2001. The chemistry of pedogenic thresholds. *Geoderma* 100: 321–353.

Chadwick, O.A., R.T. Gavenda, E.T. Kelly, K. Ziegler, C.G. Olson, W.C. Elliott, and D.M. Hendricks. 2003. The impact of climate on the biogeochemical functioning of volcanic soils. *Chemical Geology* 202: 195–223.

Chorover, J., M.K. Amistadi, and O.A. Chadwick. 2004. Surface charge evolution of mineral organic complexes during pedogenesis in Hawaiian basalt. *Geochimica et Cosmochimica Acta* 68: 4859–4876.

Chorover, J., R. Kretzschamar, F. Garcia-Pichel, and D.L. Sparks. 2007. Soil biogeochemical processes within the critical zone. *Elements* 3: 321–326.

Dessert, C., B. Dupre, J. Gaillardet, L. Francois, and C. Allegre. 2003. Basalt weathering laws and the impact of basalt weathering on the global carbon cycle. *Chemical Geology* 202(3–4): 257–273, doi:10.1016/j.chemgeo.2002.10.001.

Detty, J.M. and K.J. McGuire. 2010. Topographic controls on shallow groundwater dynamics: Implications of hydrologic connectivity between hillslopes and riparian zones in a till mantled catchment. *Hydrological Processes* 24(16): 2222–2236, doi:10.1002/hyp.7656.

Dontsova, K., C.I. Steefel, S. Desilets, A. Thompson, and J. Chorover. 2009. Solid phase evolution in the Biosphere 2 hillslope experiment as predicted by modeling of hydrologic and geochemical fluxes.Hydrology and *Earth System Sciences* 13(12): 2273–2286.

Dontsova, K., D. Zaharescu, W. Henderson, S. Verghese, N. Perdrial, E. Hunt, and J. Chorover. 2014. Impact of reduced carbon on weathering of granular basalt. *Geochimica et Cosmochimica Acta* 139: 508–526.

Dontsova, K.M. and J.M. Bigham. 2005. Anionic polysaccharide sorption by clay minerals. *Soil Science Society of America Journal* 69: 1026–1035.

Drees, K.P., J.W. Neilson, J.L. Betancourt, J. Quade, D.A. Henderson, B.M. Pryor, and R.M. Maier. 2006. Bacterial community structure in the hyperarid core of the Atacama Desert, Chile. *Applied Environmental Microbiology* 72(12): 7902–7908.

Edwards, K.J., D.R. Rogers, C.O. Wirsen, and T.M. McCollom. 2003. Isolation and characterization of novel psychrophilic, neutrophilic, Fe-oxidizing, chemolithoautotrophic α and γ-Proteobacteria from the deep sea. *Applied and Environmental Microbiology* 69: 2906–2913.

Feeney, D.S., J.W. Crawford, T. Daniell, P.D. Hallett, N. Nunan, K. Ritz, M. Rivers, and I.M. Young. 2006. Three-dimensional microorganization of the soil–root–microbe system. *Microbial Ecology* 52: 151–158.

Ferré, T. et al. 2009. Critical steps for the continuing advancement of hydrogeophysics. *EOS*, 90(23): 200, doi:10.1029/2009EO230004.

Fierer, N. and J.T. Lennon. 2011. The generation and maintenance of diversity in microbial communities. *American Journal of Botany* 98(3): 439–448.

Gaillardet, J., B. Dupre, P. Louvat, and C.J. Allegre. 1999. Global silicate weathering and CO_2 consumption rates deduced from the chemistry of large rivers. *Chemical Geology* 159 (1–4): 3–30, doi:10.1016/S0009-2541(99)00031-5.

Ganor, J., P. Lu, Z. Zheng, and C. Zhu. 2007. Bridging the gap between laboratory measurements and field estimatations of silicate weathering using simple calculations. *Environmental Geology* 53: 599–610.

Gat, J.R. 1996. Oxygen and hydrogen isotopes in the hydrologic cycle. *Annual Review of Earth and Planetary Sciences* 24: 225–262, doi: 10.1146/annurev.earth.24.1.225.

Gislason, S.R., W. Broecker, E.H. Oelkers, E. Gunnlaugsson, A. Stefánsson, J. Matter, and G. Björnsson. 2008. Mineral CO_2 sequestration into basalt: The Hellisheidi, Iceland project. *International Geological Congress*, August 6–14, Oslo, Norway.

Gleeson, T. and D. Paszkowski. 2013. Perceptions of scale in hydrology: What do you mean by regional scale? *Hydrological Science Journal* 59(1): 99–107, doi: 10.1080/02626667.2013.797581.

Hopp, L., C.J. Harman, S.L.E. Desilets, C.B. Graham, J.J. McDonnell, and P.A. Troch. 2009. Hillslope hydrology under glass: Confronting fundamental questions of soil-water-biota co-evolution at Biosphere 2. *Hydrology and Earth System Sciences* 13(11): 4411–4448, doi:10.5194/hess-13-2105-2009.

Huxman, T., P.A. Troch, J. Chorover, D.D. Breshears, S. Saleska, J. Pelletier, X. Zeng, and J. Espeleta. 2009. The hills are alive: Interdisciplinary earth science at Biosphere 2. *EOS Transactions American Geophysical Union* 90: 120.

Ivanov, V.Y., S. Fatichi, G.D. Jenerette, J.F. Espeleta, P.A. Troch, and T.E. Huxman. 2010. Hysteresis of soil moisture spatial heterogeneity and the "homogenizing" effect of vegetation. *Water Resources Research* 46(9): W09521.

Jencso, K.G. and B.L. McGlynn. 2011. Hierarchical controls on runoff generation: Topographically driven hydrologic connectivity, geology, and vegetation. *Water Resources Research* 47(11): W11527, doi:10.1029/2011WR010666.

Jing, Y., B. Zhang, A. Thimm, and A. Zepp. 2008. Anisotropy of soil hydraulic properties along arable slopes. *Pedosphere* 18(3): 353–362.

Kirchner, J.W. 2006. Getting the right answers for the right reasons: Linking measurements, analyses, and models to advance the science of hydrology. *Water Resources Research* 42(3): W03S04, doi:10.1029/2005WR004362.

Kirchner, J.W., X. Feng, and C. Neal. 2000. Fractal stream chemistry and its implications for contaminant transport in catchments. *Nature* 403(6769): 524–527, doi:10.1038/35000537.

Lasslop, G., M. Reichstein, D. Papale, A.D. Richardson, A. Arneth, A. Barr, P. Stoy, and G. Wohlfahrt. 2010. Separation of net ecosystem exchange into assimilation and respiration using a light response curve approach: Critical issues and global evaluation. *Global Change Biology* 16(1): 187–208.

Li, G., H. Li, and M.C. Boufadel. 2008. The enhancing effect of the elastic storage of the seabed aquitard on the tide-induced groundwater head fluctuation in confined submarine aquifer systems. *Journal of Hydrology* 350(1–2): 83–92.

Lohse, K.A. and W.E. Dietrich. 2005. Contrasting effects of soil development hydrological properties and flow paths. *Water Resources Research* 41: W12419.

Los Gatos Research Inc. 2015. Isotopic water analyzer (δ^2H, $\delta^{17}O$, $\delta^{18}O$)—Enhanced performance. www.lgrinc.com/documents/LGR_IWA-45EP.pdf, accessed September 6, 2015.

Maher, K. 2011. The role of fluid residence time and topographic scales in determining chemical fluxes from landscapes. *Earth and Planetary Science Letters* 312(1–2): 48–58.

Maher, K., C.I. Steefel, A.F. White, and D.A. Stonestrom. 2009. The role of reaction affnity and secondary minerals in regulating chemical weathering rates at Santa Cruz Soil Chronoseqeunce, California. *Geochimica et Cosmochimica Acta* 73(10): 2804–2831.

Mapelli, F., R. Marasco, A. Balloi, E. Rolli, F. Cappitelli, D. Daffonchio, and S. Borin. 2012. Mineral–microbe interactions: Biotechnological potential of bioweathering. *Journal of Biotechnology* 157(4): 473–481.

Mathieu, R. and T. Bariac. 1996. A numerical model for the simulation of stable isotope profiles in drying soils. *Journal of Geophysical Research* 101: 12685–12696, doi: 10.1029/96jd00223.

McDonnell, J.J. 1990. A rationale for old water discharge through macropores in a steep, humid catchment. *Water Resources Research* 26(11): 2821–2832.

McDonnell, J.J. et al. 2007. Moving beyond heterogeneity and process complexity: A new vision for watershed hydrology. *Water Resources Research* 43(7), W07301, doi:10.1029/2006WR005467.

McDonnell, J.J. et al. 2010. How old is streamwater? Open questions in catchment transit time conceptualization, modelling and analysis. *Hydrological Processes* 24(12): 1745–1754, doi:10.1002/hyp.7796.

McManus, J.B., D.D. Nelson, and M.S. Zahniser. 2015. Design and performance of a dual-laser instrument for multiple isotopologues of carbon dioxide and water. *Optics Express* 23: 6569–6586, doi: 10.1364/oe.23.006569.

Meroni, M., M. Rossini, L. Guanter, L. Alonso, U. Rascher, R. Colombo, and J. Moreno. 2009. Remote sensing of solar-induced chlorophyll fluorescence: Review of methods and applications. *Remote Sensing of Environment* 113: 2037–2051.

Michot, D., Y. Benderitter, A. Dorigny, B. Nicoullaud, D. King, and A. Tabbagh. 2003. Spatial and temporal monitoring of soil water content with an irrigated corn crop cover using surface electrical resistivity tomography. *Water Resources Research.* 39(5): 1138, doi:10.1029/2002WR001581.

Mikutta, R., G.E. Schaumann, D. Gildemeister, S. Bonneville, M.G. Kramer, J. Chorover, O.A. Chadwick, and G. Guggenberger. 2011. Biogeochemistry of mineral-organic associations across a long-term mineralogical soil gradient (0.3–4100 kyr), Hawaiian Islands. *Geochimica et Cosmochimica Acta* 73: 2034–2060.

Molins, S., D. Trebotich, C.I. Steefel, and C. Shen. 2012. An investigation of the effect of pore scale flow on average geochemical reaction rates using direct numerical simulation. *Water Resources Research* 48(3): W03527, doi: 10.1029/2011WR011404.

Montgomery, D.R. and W.E. Dietrich. 2002. Runoff generation in a steep, soil-mantled landscape. *Water Resources Research* 38(9): 7-1–7-8, doi:10.1029/2001WR000822.

Neaman, A., J. Chorover, and S.L. Brantley. 2005. Element mobility patterns record organic ligands in soils on early Earth. *Geology* 33: 117–120.

Neilson, J.W., J. Quade, M. Ortiz, W.M. Nelson, A. Legatzki, F. Tian, M. LaComb, J.L. Betancourt, R.A. Wing, C.A. Soderlund, and R.M. Maier. 2012. Life at the hyper-arid margin: Novel bacterial diversity in arid soils of the Atacama Desert, Chile. *Extremophiles* 16(3): 553–566.

Nelson, D.D., J.B. McManus, S.C. Herndon, M.S. Zahniser, B. Tuzson, and L. Emmenegger. 2008. New method for isotopic ratio measurements of atmospheric carbon dioxide using a 4.3 μm pulsed quantum cascade laser. *Applied Physics B* 90: 301–309, doi: 10.1007/s00340-007-2894-1.

Nijland, W., M. van der Meijde, E.A. Addink, and S.M. de Jong. 2010. Detection of soil moisture and vegetation water abstraction in a Mediterranean natural area using electrical resistivity tomography. *Catena* 81(3): 209–216.

Nippgen, F., B.L. McGlynn, L.A. Marshall, and R.E. Emanuel. 2011. Landscape structure and climate influences on hydrologic response. *Water Resources Research* 47(12): W12528, doi:10.1029/2011WR011161.

Omoike, A. and J. Chorover. 2004. Spectroscopic study of extracellular polymeric substances from *Bacillus subtillis*: Aqueous chemistry and adsorption effects. *Biomacromolecules* 5: 1219–1230.

Omoike, A. and J. Chorover. 2006. Adsorption to goethite of extracellular polymeric substances from *Bacillus subtilis*. *Geochimica et Cosmochimica Acta* 70: 827–838.

Orcutt, B.N., J.B. Sylvan, D.R. Rogers, J. Delaney, R.W. Lee, and P.R. Girguis. 2015. Carbon fixation by basalt-hosted microbial communities. *Frontiers in Microbiology* 6: 904. http://journal.frontiersin.org/article/10.3389/fmicb.2015.00904.

Ortiz, M. et al. 2013. Profiling bacterial diversity and taxonomic composition on speleothem surfaces in Kartchner Caverns. *Microbial Ecology* 65(2): 371–383.

Ortiz, M. et al. 2014. Making a living while starving in the dark: Metagenomic insights into the energy dynamics of a carbonate cave. *ISME Journal* 8(2): 478–491.

Pacheco, F.A.L. and C.H. Van der Weijden. 2012. Integrating topography, hydrology and rock structure in weathering rate models of spring watersheds. *Journal of Hydrology* 428–429: 32–50.

Pangle, L.A., J. Klaus, E.S.F. Berman, M. Gupta, and J.J. McDonnell. 2013. A new multi-source and high-frequency approach to measuring δ^2H and $\delta^{18}O$ in hydrological field studies. *Water Resources Research* 49: 7797–7803, doi: 10.1002/2013wr013743.

Pangle, L.A. et al. 2015. The Landscape Evolution Observatory: A large-scale controllable infrastructure to study coupled Earth-surface processes. *Geomorphology* 244: 190–203, doi:10.1016/j.geomorph.2015.01.020.

Phillips, C.L., J.W. Gregg, and J.K. Wilson. 2011. Reduced diurnal temperature range does not change warming impacts on ecosystem carbon balance of Mediterranean grassland mesocosms. *Global Change Biology* 17(11): 3263–3273, doi:10.1111/j.1365-2486.2011.02483.x.

Pohlmann, M., K. Dontsova, R. Root, P.A. Troch, J. Ruiz, and J. Chorover. 2016 Pore water chemistry reveals gradients in mineral transformation across a model basaltic hillslope. *Geochemistry, Geophysics, Geosystems* 17(6): 2054–2069, doi:10.1002/2016GC006270.

Reggiani, P., M. Sivapalan, and S.M. Hassanizadeh. 1998. A unifying framework of watershed thermodynamics: Balance equations for mass, momentum, energy and entropy and the second law of thermodynamics. *Advances in Water Resources* 22(4): 367–398.

Reggiani, P., M. Sivapalan, and S.M. Hassanizadeh. 2000. Conservation equations governing hillslope responses: Exploring the physical basis of water balance. *Water Resources Research* 36(7): 1845–1863.

Reggiani, P., M. Sivapalan, M. Hassanizadeh, and W.G. Gray. 2001. Coupled equations for mass and momentum balance in a stream network: Theoretical derivation and computational experiments. *Proceedings of the Royal Society of London A* 457: 157–189.

Reichstein, M. et al. 2005. On the separation of net ecosystem exchange into assimilation and ecosystem respiration: Review and improved algorithm. *Global Change Biology* 11(9): 1424–1439.

Richter, D. de B. and S.A. Billings. 2015. "One physical system": Tansley's ecosystem as Earth's critical zone. *The New Phytologist* 206(3): 900–912.

Rinaldo, A., K.J. Beven, E. Bertuzzo, L. Nicotina, J. Davies, A. Fiori, D. Russo, and G. Botter 2011. Catchment travel time distributions and water flow in soils. *Water Resources Research* 47(7): WR010478, doi:10.1029/2011WR010478.

Rogers, J.R. and P.C. Bennett. 2004. Mineral stimulation of subsurface microorganisms: Release of limiting nutrients from silicates. *Chemical Geology* 203(1–2): 91–108.

Roth, K. 1995.Steady-state flow in an unsaturated, two-dimensional, macroscopically homogeneous Miller-similar medium. *Water Resources Research* 31: 2127–2140.

Savostin, P. 1972. Microbial transformation of silicates. *Journal of Plant Nutrition and Soil Science* 132(1): 37–45.

Scudeler, C., L. Pangle, D. Pasetto, G. Niu, T. Volkmann, C. Paniconi, M. Putti, and P. A. Troch. 2016. Multiresponse modeling of an isotope tracer experiment at the Landscape Evolution Observatory. *Hydrology and Earth System Sciences* 20: 4061–4078, doi:10.5194/hess-20-4061-2016.

Sengupta, A. and W.A. Dick. 2015. Bacterial community diversity in soil under two tillage practices as determined by pyrosequencing. *Microbial Ecology* 70: 853–859.

Shahraeeni, E. and D. Or. 2010. Thermo-evaporative fluxes from heterogeneous porous surfaces resolved by infrared thermography. *Water Resources Research* 46(9): WR008455, doi:10.1029/2009WR008455.

Showstack, R. 2012. Emerging earth science research opportunities at U.S. National Science Foundation. *EOS, Transactions American Geophysical Union* 93(1): 3–4.

Sivapalan, M. 2005. Pattern, process and function: Elements of a unified theory of hydrology at the catchment scale, in *Encyclopedia of Hydrological Sciences*, ed. M.G. Anderson, pp. 193–219, John Wiley & Sons, New York.

Sklash, M.G. and R.N. Farvolden. 1979. The role of groundwater in storm runoff. *Journal of Hydrology* 43(1–4): 45–65, doi:10.1016/0022-1694(79)90164-1.

Slater, L., A. Binley, R. Versteeg, G. Cassiani, R. Birken, and S. Sandberg. 2002. A 3D ERT study of solute transport in a large experimental tank. *Journal of Applied Geophysics* 49: 211–229.

Sprenger M., T.H.M. Volkmann, T. Blume, and M. Weiler. 2015. Estimating flow and transport parameters in the unsaturated zone with pore water stable isotopes. *Hydrology and Earth System Sciences* 19: 2617–2635.

Steefel, C.I. 2008. Geochemical kinetics and transport, in *Kinetics of Water-Rock Interactions*, eds. S. Brantley, J.D. Kubicki, and A.F. White, pp. 545–589, Springer, New York.

Sturm, P., W. Eugster, and A. Knohl. 2012. Eddy covariance measurements of CO_2 isotopologues with a quantum cascade laser absorption spectrometer. *Agricultural and Forest Meteorology* 152: 73–82, doi:10.1016/j.agrformet.2011.09.007.

Suchet, P.A., J.L. Probst, and W. Ludwig. 2003. Worldwide distribution of continental rock lithology: Implications for the atmospheric/soil CO_2 uptake by continental weathering and alkalinity river transport to the oceans. *Global Biogeochemical Cycles* 17(2): 1038, doi:10.1029/2002GB001891.

Tetzlaff, D., J.J. McDonnell, S. Uhlenbrook, K.J. McGuire, P.W. Bogaart, F. Naef, A.J. Baird, S.M. Dunn, and C. Soulsby. 2008. Conceptualizing catchment processes: Simply too complex? *Hydrological Processes* 1730(April): 1727–1730, doi:10.1002/hyp.7069.

Torn, M.S., S.E. Trumbore, O.A. Chadwick, P.M. Vitousek, and D.M. Hendricks. 1997. Mineralcontrol of soil organic carbon storage and turnover. *Nature* 389: 170–173.

Troch, P.A., G.A. Carrillo, I. Heidbüchel, S. Rajagopal, M. Switanek, T.H.M. Volkmann, and M. Yaeger. 2009. Dealing with landscape heterogeneity in watershed hydrology: A review of recent progress toward new hydrological theory. *Geography Compass* 3(1): 375–392.

Troch, P.A., C. Paniconi, and E.E. van Loon. 2003. Hillslope-storage Boussinesq model for subsurface flow and variable source areas along complex hillslopes: 1. Formulation and characteristic response. *Water Resources Research* 39(11), WR001728, doi:10.1029/2002WR001728.

Tromp-van Meerveld, H.J. and J.J. McDonnell. 2006. Threshold relations in subsurface stormflow 1: A 147 storm analysis of the Panola hillslope trench. *Water Resources Research* 42(August 2005): W02410, doi:10.1029/2004WR003778.

Tuzson, B., J. Mohn, M.J. Zeeman, R.A. Werner, W. Eugster, M.S. Zahniser, D.D. Nelson, J.B. McManus, and L. Emmenegger. 2008. High precision and continuous field measurements of $\delta^{13}C$ and $\delta^{18}O$ in carbon dioxide with a cryogen-free QCLAS. *Applied Physics B* 92: 451–458, doi:10.1007/s00340-008-3085-4.

Valentín-Vargas, A., R.A. Root, J.W. Neilson, J. Chorover, and R.M. Maier. 2014. Environmental factors influencing the structural dynamics of soil microbial communities during assisted phytostabilization of acid-generating mine tailings: A mesocosm experiment. *Science of the Total Environment* 500–501: 314–324.

van der Velde, Y., G.H. de Rooij, J.C. Rozemeijer, F.C. van Geer, and H.P. Broers. 2013. Chloride circulation in a lowland catchment and the formulation of transport by travel time distributions. *Water Resources Research* 49(8): 4619–4632, doi:10.1002/wrcr.20309.

van Haren, J. et al. CO_2 diffusion into pore spaces limits weathering rate of an experimental basalt landscape. *Geology* (In Review).

Vereecken, H., S. Hubbard, A. Binley, and T. Ferré. 2004. Hydrogeophysics: An introduction from the guest editors. *VadoseZone Journal* 3: 1060–1062.

Vereecken, H., J.A. Huisman, Y. Pachepsky, C. Montzka, J. van der Kruk, H. Bogena, L. Weihermüller, M. Herbst, G. Martinez, and J. Vanderborght. 2014. On the spatio-temporal dynamics of soil moisture at the field scale. *Journal of Hydrology* 516: 76–96.

Volkmann, T.H.M., K. Haberer, A. Gessler, and M. Weiler. 2016a. High-resolution isotope measurements resolve rapid ecohydrological dynamics at the soil–plant interface. *New Phytologist* 210: 839–849, doi:10.1111/nph.13868.

Volkmann T.H.M., K. Kühnhammer, B. Herbstritt, A. Gessler, and M. Weiler. 2016b. A method for in situ monitoring of the isotopic composition of tree xylem water using laser spectroscopy. *Plant, Cell and Environment*. 39(9): 2055–2063, doi:10.1111/pce.12725.

Volkmann, T.H.M. and M. Weiler. 2014. Continual in situ monitoring of pore water stable isotopes in the subsurface. *Hydrology and Earth System Sciences* 18: 1819–1833, doi:10.5194/hess-18-1819-2014.

Wieder, W.R., G.B. Bonan, and S.D. Allison. 2013. Global soil carbon projections are improved by modelling microbial processes. *Nature Climate Change* 3(10): 909–912.

Yakir, D. and Sternberg, L.D.S.L. 2000. The use of stable isotopes to study ecosystem gas exchange. *Oecologia* 123: 297–311, doi:10.1007/s004420051016.

Yokoyama, T. and J. Banfield. 2002. Direct determination of the rates of rhyolite dissolution and clay formation over 52,000 years and comparison with laboratory measurements. *Geochimica et Cosmochimica Acta* 66: 2665–2681.

5

Quantifying Relationships between Biodiversity and Ecosystem Function with Experiments

Charles A. Nock, Lander Baeten, Andy Hector, Kris Verheyen, Wolfgang W. Weisser, and Michael Scherer-Lorenzen

CONTENTS

5.1 Why Study Biodiversity–Ecosystem Functioning Relationships?

In the early 1990s, it became increasingly recognized that despite mounting global biodiversity losses (Ehrlich and Ehrlich 1981; Wilson 1988), science was limited in its ability to predict the effects of these changes. Now widely referred to as the "biodiversity crisis," intensive research was triggered, seeking to answer the question: *how important are diverse biotic communities for ecosystem functioning*? (Schulze and Mooney 1993). The research underpinning this question represented a paradigm shift in ecology (Hillebrand and Matthiessen 2009). Traditionally, biodiversity research emphasized the regulation and maintenance of diversity, that is, the drivers of biodiversity, whereby biodiversity was seen as the consequence of the abiotic and biotic factors regulating a community (Chesson 2000). This emphasis has shifted toward understanding the consequences of biodiversity change on ecosystem functions, whereby biodiversity itself is a driver rather than a dependent variable in ecosystems (Schulze and Mooney 1993). Ecosystem processes sustain mankind: if biodiversity is needed for the processes, and in turn the services provided by nature to humans, then conserving biodiversity is a necessity for preserving the foundation on which our life depends (Cardinale et al. 2012; Millennium Ecosystem Assessment 2005).

In the past 20 years, the relationship between biodiversity and ecosystem functioning (BEF) has developed into a central research topic in ecology (e.g., Cardinale et al. 2012; Kinzig et al. 2002; Loreau et al. 2002; Naeem et al. 2009). Initially, individual studies generally focused on a single ecosystem function in a particular context and were of limited duration. A seminal result of the research was the demonstration of a positive effect of increasing species richness on ecosystem functioning in the short term, both in terrestrial and aquatic ecosystems (Balvanera et al. 2006; Cardinale et al. 2006, 2007; Stachowicz et al. 2007; Worm et al. 2006). As the number of studies accumulated, it became possible to bring datasets together and investigate the effects of biodiversity on multiple ecosystem functions, that is, *multifunctionality* (He et al. 2009; Hector and Bagchi 2007; Isbell et al. 2011; Maestre et al. 2012). For example, in the intensively studied grassland biodiversity experiment in Jena, Germany (Figure 5.1a), several thousand variables have been investigated in the same plots (Allan et al. 2013; Roscher et al. 2004; Scherber et al. 2010). Such analyses thus represented an achievement in broadening the scope of studies from single function studies and few explanatory variables to more comprehensive analyses. A broadening of the scope of the science in more recent studies was also made possible by a broader characterization of ecosystems by inclusion of the ecosystem consequences of a change in the diversity of taxa other than plants (insects, Cardinale et al. 2002; animals, Gagic et al. 2015). As experiments matured and evidence accumulated, it was also shown that the effects of biodiversity on diverse ecosystem functions become stronger over time (Allan et al. 2011; Cardinale et al. 2007; Reich et al. 2012).

Despite worldwide efforts to reduce rates of species loss, the biodiversity crisis is still looming (Butchart et al. 2010). This continued attrition of biodiversity is especially concerning in the context of widespread global changes. Biodiversity–ecosystem functioning experiments have now shown that when comparing the same ecosystem functions, biodiversity can have effects as large as or even greater than global change drivers such as climatic change, acidification, elevated CO_2, and nutrient pollution (Hooper et al. 2012). Because of the link of biodiversity loss to ecosystem function and in turn the provision of essential ecosystem services such as the provision of clean water and food, there is also a strong link between biodiversity conservation, ecosystem management, and poverty alleviation (Turner et al. 2012). Perhaps with greater recognition of the contribution of biodiversity to human well-being, efforts to stop the loss of species will increase. Fostering the recognition of BEF relationships is a multifaceted challenge that requires coordination from the design stage of experiments (realism and relevance) to communication of the final results of experiments to a variety of different stakeholders ranging from fellow researchers to managers and private landowners.

(a)

(b)

FIGURE 5.1
Examples from biodiversity–ecosystem functioning experiments, (a) showing a large grassland study and (b) a high-density tree plantation. In IDENT, a "synthetic assemblage experiment," a gradient of tree species diversity was created by planting plots (~3.2 m × 3.2 m) with different numbers of species and/or functional groups. ([a]: The Jena Experiment in Germany; From Roscher, C. et al., *Basic Appl. Ecol.*, 5, 107, 2004; Photo courtesy of the Jena Experiment, Jena, Germany; [b]: IDENT experiment, Freiburg, Germany; Photo by T. Kattenborn.)

5.2 How to Study BEF Relationships in Real Ecosystems?

If one is interested in answering *how does biodiversity affect ecosystem functioning,* most often the first approach coming to one's mind would involve searching for communities that differ or contrast in one aspect of biodiversity and comparing levels of ecosystem function across the sample of communities (Scherer-Lorenzen 2005). For example, in a forest ecosystem, several forest stands with contrasting community structure could be sampled with a random or grid-based selection of plots across a specific area. Then, values for measured ecosystem functions (often biomass production) would be plotted on the y-axis against the measured biodiversity, for example, number of tree species on the x-axis. This type of study, often referred to as *monitoring or sample surveys* (Figure 5.2) reflect natural conditions with respect to age structure, canopy and root architecture, food webs, or biogeochemical cycles at quasi steady state (Leuschner et al. 2009). For example, the expansive networks of permanent forest inventory plots found in many countries are increasingly valued not only for tracking timber stocks but also as a means to study BEF relationships in forest ecosystems (Ruiz-Benito et al. 2014). While the use of such data provides large advantages in terms of realism and relevance, they also present significant challenges. These challenges arise in part because different levels of biodiversity—whether defined as species richness, functional and/or phylogenetic diversity—are rarely equally represented on the landscape. Often, the majority of plots will have biodiversity values close to the mean, and fewer plots will represent low or high biodiversity levels. Alternatively, the distribution of biodiversity levels can be highly skewed, for example, with most plots at the low end of

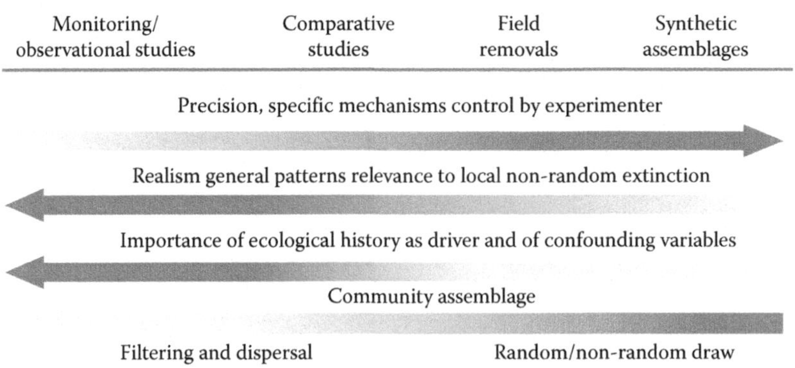

FIGURE 5.2
Comparison of scientific approaches to study the relationship between biodiversity and ecosystem functions and the different trade-offs inherent to these designs. (Modified from Díaz, S. et al., *Trends Ecol. Evol.*, 18, 140, 2003, Figure 1.)

BOX 5.1 BALANCED VERSUS UNBALANCED DESIGNS

- Balanced (orthogonal) experimental designs have the advantage that they allow the effects of two or more explanatory variables to be independently estimated.
- In contrast, with unbalanced designs, explanatory variables are partially confounded (correlated), and it is not possible to definitively separate their effects.

the diversity gradient as often found in managed ecosystems, such as plantation forests (Vilà et al. 2007).

Statistically, such unbalanced designs (Box 5.1) are less powerful to detect causal relationships among variables than those experiments with a balanced design. In addition, unless site conditions are extremely similar, such across-habitat or across-locality comparisons may hide any potential effects. This can occur when differences in diversity are partly caused by environmental differences between the sampled sites, which introduce "noise" in the diversity–function relationship. Indeed, the environmental co-variables (temperature, precipitation) themselves determine the diversity of an ecosystem and the ecosystem properties and processes. For example, an asymptotical increase of biomass production with increasing plant species richness could be due to functional differences among species leading to niche differentiation and higher relative rates of resource exploitation and, hence, higher productivity rates. However, more productive stands may simply permit the coexistence of more species. Thus, such sample surveys only represent correlations between biodiversity and ecosystem processes, which are bidirectional relationships between two dependent variables (Baruffol et al. 2013), and causal relationships can only be approximated by accounting for measured environmental co-variables with suitable statistical procedures such as path analyses. To avoid some of these problems, *comparative studies* have been designed, where a similar number of plots per diversity level (e.g., species richness = 1, 2, 4, 6) are deliberately chosen along gradients of biodiversity, ideally coupled with a maximum standardization of conditions: environmental (e.g., soils, climate), stand age, or management history (e.g., Baeten et al. 2013; Bruelheide et al. 2011; Leuschner et al. 2009) (Figure 5.2). Nevertheless, there are many potentially confounding variables influencing biodiversity, and it is difficult to ascertain if they have been adequately accounted for in analyses, which is a crucial issue for the attribution of causality. In face of such "confounding variables," experimental manipulation of plant diversity has been advocated to decouple biodiversity from environmental conditions in order to quantify causality and mechanisms of diversity effects. We will describe the basic design of such BEF experiments and their development over time in the following section.

5.3 The Basic Design of BEF Experiments and Their Development over Time

When biologists want to find out what a gene does, they develop a version of the organism without the gene of interest and grow individuals of these "knockout" organisms to see what the consequences of the loss of the gene are. Biodiversity experiments use the same basic approach to study the role of species in the functioning of ecosystems: the species to omit are chosen, experimental communities constructed from the species that are left (the species pool), and their function examined (Bell et al. 2005; Schmid et al. 2002). Such experiments fall into two main classes: "removal experiments" and "synthetic assemblage experiments" (Figure 5.2).

In the process of community assembly—the formation of a local community by a subset of a regional species pool—many potential community members are excluded or filtered by biotic or abiotic factors. For example, herbivory of palatable species by deer and the formation of a community of browsing-tolerant or unpalatable species is an example of biotic filtering. Abiotic filtering is similar but naturally involves filtering of species by abiotic factors (for plants, soil type, water availability, temperature, etc.). *Removal experiments* are conducted by manipulating, through species removals, the diversity of naturally assembled communities (species, functional groups), thus creating a gradient in species richness (or an alternative metric of diversity) ranging from natural to depauperate. These types of experiments mimic the loss of species as a result of specific pests or pathogens in the past, for instance. Removal experiments are meant to emulate natural abiotic and biotic filtering of the regional species pool, represent realistic nonrandom extinction scenarios, and usually involve a large variety of organismic and functional groups (Díaz et al. 2003; Wardle and Zackrisson 2005). However, there have been a number of limitations in studies of biodiversity effects employing removal experiments. First, appropriate experimental controls are rarely used but are important to separate the effect of local species extinction from the disturbance caused by the removal process (Guido and Pillar 2015). Second, maintaining density across the diversity gradient as the abundance of the removed species varies may not be possible. Finally, species may be clumped in space, which may lead to little or no interactions among different species in a plot even though they are defined as belonging to the same community.

Synthetic assemblage experiments establish a gradient in plant biodiversity by sowing or planting, while keeping extrinsic conditions (e.g., climate, fertility, land-use history) as constant as possible. They are conducted in microcosms and controlled environmental facilities, or in the field. For very practical reasons, these experiments mainly use fast-growing, small-sized, mainly early successional model systems such as grasslands. The first such experiments in the Ecotron facility at Imperial College, Silwood Park, U.K. (Naeem et al. 1994),

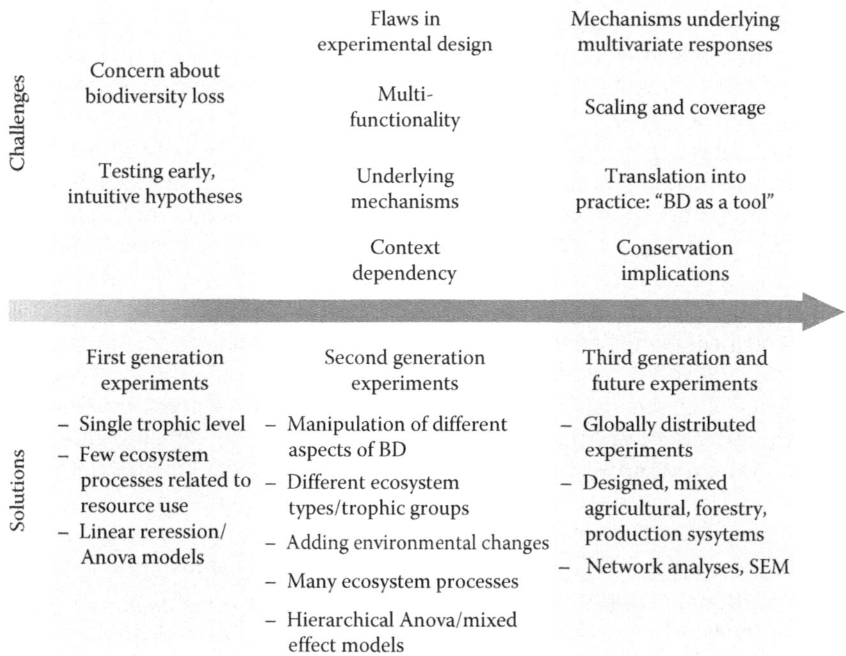

FIGURE 5.3
Schematic representation of the development of experimental biodiversity–ecosystem function research over the last few decades. Detailed are the key challenges that have emerged and the solutions that were presented in response. The development is coarsely divided into three periods representing the evolution from first-generation through to third-generation experiments.

in North American prairie systems at Cedar Creek, MN (USA, Tilman et al. 1996) in serpentine grasslands of California (Hooper and Vitousek 1997), and within the European BIODEPTH project (Hector et al. 1999) paved the road for even larger and more sophisticated experimental designs (Figure 5.3). These early biodiversity experiments tended to focus only on the effects of species richness and composition while holding other environmental factors constant. However, we now know that biodiversity loss has impacts of a similar magnitude as other global change drivers (Hooper et al. 2012). For example, in locations where species losses fall within intermediate projections (21%–40% of species loss), biomass production is expected to be reduced by 5%–10% below the most diverse mixtures, an effect that is comparable in size to the effects of global changes such as ultraviolet radiation and climate warming on plant production (Hooper et al. 2012). Given that species loss and global changes will act in concert in influencing ecosystem function, next-generation experiments have started to become multifactorial: manipulating diversity and other factors related to global change such land-use intensity (e.g., agricultural intensity, mowing and fertilization, Weigelt et al. 2009) and drought (e.g., Vogel et al. 2012), eutrophication of soil resources and elevated CO_2 levels (e.g., Reich et al. 2001), and the study of multitrophic interactions (Scherber et al. 2010; Figure 5.3).

Biodiversity experiments tend to have complex designs for several reasons. First, one key aim is to assess the effect of varying species diversity (or functional groups, genotypes, etc.); however, this was challenging in initial designs because species composition (which species are present and their traits) was sometimes confounded with the number of species. In what was probably the first formal biodiversity experiment—by Naeem et al. (1994) using the British Ecotron-controlled environment facility—a single high-, medium-, and low-diversity community were compared, and consequently, the effects of species richness (number of species) could not be separated from the effects of species composition (the particular set of species used for the high-, medium-, and low-diversity communities). In turn, later first-generation designs dealt with this issue by quantifying an average effect of changing the number of species over a sample of different possible species compositions within each diversity level (e.g., Tilman et al. 1996). For example, an experiment could start with a complete community of species and ask what is the effect (on ecosystem function) of halving the number of species. The challenge then becomes that there are many different possible ways to halve the number of species and how to do so to derive an average response and to avoid bias. In one commonly applied method, bias is avoided through random selection of the composition of the depauperate communities. However, species often do not go extinct at random in nature. Rather, depending on the cause, species may go extinct at different rates depending on their vulnerability to certain extinction drivers (Box 5.2). However, often we are not sure what drives extinction. There may be several potential extinction mechanisms that may produce different patterns of species loss. Two or more extinction drivers may even be operating at the same time and potentially interacting with each other (such as climate change and invasive species, Wolkovich and Cleland 2010). Faced with these complexities and the status of randomization in experimental design, then random species loss is a sensible starting point in the absence of better information (Schmid and Hector 2004). Nonetheless, an important goal of next-generation biodiversity experiments is to examine the consequences of particular patterns of extinction loss caused by particular extinction drivers (Bruelheide et al. 2014; Zavaleta and Hulvey 2004).

Biodiversity is more than just species richness and later generation experiments have tended to expand their focus to additional dimensions of biodiversity, such as species evenness, as well as the diversity of functional characteristics of species (a concept whose origins extend back to 300 BC and Theophrastus). In particular, many experiments focus not only on individual

BOX 5.2 RANDOM AND NONRANDOM EXTINCTION

- *Patterns of extinction*: Refers to whether extinction operates randomly or whether the probability of extinction is influenced by the traits of species.
- *Drivers of extinction*: Primarily anthropogenic-driven influences on the biosphere contributing to species extinction, such as habitat loss, harvesting, eutrophication, and climate change.
- *Random extinction*: For all members of a community, extinction is equally likely.
- *Nonrandom extinction*: Patterns of extinction in which species with similar traits are more likely to go extinct as a result of one or more extinction drivers. For example, in diverse mesic European grasslands, intensification of management has promoted species with traits favoring growth, which has led to the decline of competitively inferior species (e.g., Fuller 1986).

species but also on functional groups—species that are thought to contribute similar process rates on the overall ecosystem functions, such as biotic nutrient cycling, water usage, and productivity. These approaches are not mutually exclusive and can be combined, for example, by simultaneously manipulating the numbers of species and functional groups, albeit with practical constraints: you can have fewer groups than species but not the other way around (e.g., Roscher et al. 2004; Tilman et al. 1997). Functional diversity does not have to be approached by classifying species into groups, and, instead, various continuous measures of functional diversity (analogous to phylogenetic diversity) have been developed (Petchey and Gaston 2002; Schleuter et al. 2010). Diversity within species (i.e., genetic diversity) can also be manipulated, although this has been much less studied to date (e.g., Kotowska et al. 2010).

There has been a recent move to multisite experiments within ecology ("coordinated distributed experiments" [Fraser et al. 2013] such as Nutrient Network, Drought-Net), and this may have been partly influenced by multisite biodiversity experiments both in grasslands (Hector et al. 1999) and more recently in forests (Verheyen et al. 2015). This has implications for analysis. First-generation experiments were often analyzed with relatively simple classical least squares ANOVA and regression designs. However, later designs, particularly those with multisite experiments, require mixed-effects or multilevel models (see Box 5.3).

In summary, some of the milestones in the improvement of biodiversity experiments have been the development of designs and analyses that properly replicate both richness and composition; choice of random or directed

BOX 5.3 ANALYZING BEF EXPERIMENTS, FROM ANOVA TO MIXED-EFFECTS MODELS

The simplest of first-generation biodiversity experiments were typically analyzed using classical least squares linear regression or ANOVA models. However, as designs became more complex and also included species composition, then modern maximum likelihood mixed-effects models were a more appropriate choice (since species compositions were only a sample of the large number of possible species combinations). Mixed-effects regression models (or their Bayesian multilevel/ hierarchical model counterparts) were also ideal for quantifying the variation in the intercepts and slopes of the regressions of function on diversity among the locations within multisite studies.

species loss; multifactorial designs that address biodiversity loss and other global change drivers (and their potential interactions) and multisite experiments that can look at the variability of biodiversity effects across locations. From the earlier mentioned pros and cons, it should be evident that there is no single best way to study biodiversity–ecosystem functioning relationships. The complexity of interactions among species underlying functional diversity effects, and the multitude of abiotic and biotic factors influencing ecosystem processes require a multitude of different, albeit complementary scientific approaches. Still, biodiversity experiments allow a process-based view on the underlying mechanisms of diversity effects on ecosystem functioning (Box 5.4).

5.4 The Future of BEF Platforms: Challenges for a New Generation of Experiments

While there is broad scientific consensus that changes in biodiversity cause important ecosystem responses, functional biodiversity research faces plenty of unresolved and emerging research questions that should guide future studies and the design of new platforms.

The first-generation experiments were often designed to understand fundamental mechanisms and were therefore stripped of much ecological complexity (Figure 5.3). The major advances in this field on theoretical and empirical grounds now allow making stronger linkages with real-world settings and the applied sciences, including agricultural and forestry practices. The challenge for a new generation of functional biodiversity research

BOX 5.4 RELEVANCE OF BEF EXPERIMENTS TO THE GLOBAL PROLIFERATION OF MONO-SPECIFIC TREE PLANTATIONS

Positive diversity–stability relationships can be related to the presence of species-specific responses to changing environmental conditions and functional redundancy within groups of species ("insurance hypothesis," Yachi and Loreau 1999). Tree diversity has the potential to provide a specific type of insurance effect during replanting (or enrichment planting of selectively logged areas—we use planting as a shorthand for both practices). As with all insurance effects, it relies on ecological differences between species. In any given forestry enterprise, species are generally found to vary in their survival rates. When this is the case, there is a potential twofold cost of planting areas as monocultures (or very low diversity mixtures) that may be avoided by insurance effects of mixtures. When areas are planted as monocultures, there is the risk of recruitment and regeneration failure in the areas planted with species with low survival rates if eventual densities are insufficient to achieve the desired levels of basal area, canopy closure, and forest structure. In contrast, in the areas planted as monocultures of species with the highest survival rates, there is potential wastage of seedlings through excessive self-thinning due to intensifying intraspecific competition over time. Planting mixtures of species with different regeneration niches (conditions under which they have high and successful recruitment) may avoid this twofold cost of recruitment failure in some areas and wasteful self-thinning in others. The potential spatial insurance effect of mixtures may be particularly relevant, where forestry schemes are not able to do matching of species to sites. Most species also experience good and bad years for recruitment so that there is also a potential temporal component in which mixtures can once again bring a potential insurance effect for tree recruitment and forest regeneration.

principally involves extending both its scope and scale. In this context, we will discuss key considerations for designing future biodiversity–ecosystem functioning platforms (see also Bruelheide et al. 2014) as well as the current opportunities for doing synthesis analyses with already available data or targeted measuring campaigns in the existing infrastructures.

Novel biodiversity-oriented research platforms can extend the scope of existing infrastructures in several important ways. First, most experiments have so far created diversity gradients using random species compositions, which is a good starting point if the order of species loss in an ecosystem could follow many alternative sequences (Schmid and Hector 2004). The generality of the approach, however, is criticized for lacking relevance to real-world scenarios of biodiversity change. Recent global syntheses found dramatic changes in plant community composition, rather than a general tendency for declines in local species diversity (Dornelas et al. 2014; Vellend et al. 2013), so future work should clearly focus on the ecosystem effects of those particular species (e.g., invasive species) and species compositions that are actually becoming widespread in nature. Species have generally been the focus and few studies attempt to link intraspecific trait variation with ecosystem processes. In large part, this is because identifying individuals in many experiments is challenging (e.g., grasses). However, in many experiments, individuals can be tracked (e.g., tree diversity experiments; Potvin and Gotelli 2008; Potvin and Dutilleul 2009), and exploring the functional effects of diversity within populations of individual species also deserves further attention (e.g., Violle et al. 2012).

While we have presented numerous arguments about the design of BEF experiments, it is important to note that in many places in the world experiments have not been conducted (Verheyen et al. 2015). The global distribution of experiments is skewed, and establishing new sites in underrepresented biomes should be a priority to further increase the scope of biodiversity experiments. Experimental tree diversity plantations, for instance, are concentrated in temperate regions, and the few located in tropical systems do not represent the largest forested areas and biodiversity hotspots (Verheyen et al. 2015). One way in which the establishment of experiments in underrepresented areas could be facilitated is through the activities of research networks that can promote the exchange knowledge across communities and provide support in the establishment and operation of experiments through shared experience and sampling protocols and methodologies (e.g., TreeDivNet: www.treedivnet.ugent.be).

The concepts and mechanistic insights gained from previous functional biodiversity research are not by any means restricted to natural or seminatural ecosystems and have great potential for applications in economically important and intensively managed biological systems where one can virtually choose the desired number of species that grow together. Terrestrial ecosystems are predominantly managed. Management alternatives should thus be included in a greater number of biodiversity experiments in order to increase their relevance to the "real world" (Figure 5.3). Previous experiments have done so through the use of split-plot designs: for example, in Jena, a split-plot design was utilized to test the effects of agricultural management on biodiversity effects (fertilizer addition and mowing) (e.g., Weigelt et al. 2009). In BIOTREE, which has a very-long-term horizon, the influence of thinning on tree diversity effects on biomass production will be tested using a split-plot design (Scherer-Lorenzen et al. 2007). To maintain the diversity gradient that was created, experiments often require management, so careful consideration of management alternatives, developed together with stakeholders, can represent a valuable way forward. As argued by Bullock et al. (2001) based on positive biodiversity–productivity relationships documented in British hay fields, the outreach of new experiments would dramatically increase if they were more relevant for economically important systems. In addition to production grasslands, agricultural production systems, agroforestry, and short-rotation coppices deserve more attention (Verheyen et al. 2015). Based on such studies, one could use biodiversity as a tool to achieve multifunctional and sustainable production systems. The functional biodiversity field also has clear conceptual insights and mechanisms to offer in the context of disease and pest control. For instance, plant diversity generally reduces the impact of pests and diseases through associational resistance, that is, via reduced host colonization and increased top-down regulation by natural enemies (e.g., Barbosa et al. 2006; Castagneyrol et al. 2013).

The scale of most existing BEF platforms is usually relatively small. Plots are typically only few square meters to one hectare for grasslands and forests, respectively, and are clustered in small and environmentally quite homogeneous areas. Using smaller plots generally allows for more realizations of the diversity and composition treatments in a single experiment, increasing the power to detect effects at the plot scale (Bruelheide et al. 2014), but limits opportunities to determine the context dependency of BEF relationships within these homogeneous areas, because the environmental variation covered is necessarily reduced. For instance, the relative importance of species identity effects and complementarity effects is likely to vary along environmental gradients. Indeed, the strength of diversity–functioning relationships is expected to increase with environmental stress, with stronger positive effects in stressed habitats (e.g., Jucker and Coomes 2012). Within a management context, this is important because high diversity may appear to be redundant in one context, but then increase in importance in response to a perturbation (Isbell et al. 2011). Multisite experiments (BIODEPTH, Hector et al. 1999; or FORBIO, Verheyen et al. 2013), experiments that incorporated environmental heterogeneity in their design (BEF China, Bruelheide et al. 2014) and projects that use complementary research platforms covering different scales from experiments to regional inventories (FunDivEUROPE, Baeten et al. 2013), are producing highly relevant data for understanding the relative importance of environmental drivers in directly affecting ecosystem function as well as in modifying the biodiversity effects. Plot-based approaches, even at multiple sites or along gradients, are less appropriate though to make inferences about ecosystem processes that operate at larger spatial scales such as water purification and nutrient cycles. Larger, watershed-scale platforms would be required to study those processes, which also provide opportunities to study landscape-level disturbances, including fragmentation, mammalian herbivory, and fires. Finally, biodiversity–functioning relationships appear to grow stronger with time, for instance, because complementarity effects accumulate when species are replaced by functionally dissimilar ones (Reich et al. 2012). Extending the monitoring of experiments over temporal scales that exceed typical project lives is challenging but critically needed to fully understand diversity-dependent ecosystem responses.

Biodiversity effects on individual ecosystem processes including productivity and nutrient-use efficiency have been extensively studied (although, as argued earlier, there are numerous places on the globe where this important groundwork is still lacking). Now, the time is ripe to work on syntheses that quantify how diversity influences multiple ecological functions simultaneously (Figure 5.3; Byrnes et al. 2014). Such analyses can lead to guidelines about which species or species combinations will lead to the optimization of multiple ecosystem functions or rather create important trade-offs between functions (Gamfeldt et al. 2008). Context dependencies can be integrated in multifunctionality studies, by quantifying whether the variation in BEF relationships across environments is consistent across multiple functions. For example, increasingly strong biodiversity effects with increasing abiotic stress for both productivity (e.g., forests, Jucker et al. 2014) and drought tolerance (e.g., forests, Grossiord et al. 2014) would result in a positive covariation between the BEF relationships across environments. Comprehensive data for multifunctionality studies should come from groups of researchers from very different disciplines measuring responses on a common set of sites. Integrative research with close collaboration between researchers with various backgrounds working on different spatiotemporal scales and levels of taxonomic resolution is challenging but is now facilitated by recently developed data platforms that allow harmonizing, annotating, and sharing data in BEF research (Nadrowski et al. 2013).

References

Allan, E., Weisser, W.W., Fischer, M. et al., 2013. A comparison of the strength of biodiversity effects across multiple functions. *Oecologia* 173, 223–237.

Allan, E., Weisser, W.W., Weigelt, A., Roscher, C., Fischer, M., Hillebrand, H., 2011. More diverse plant communities have higher functioning over time due to turnover in complementary dominant species. *Proceedings of the National Academy of Sciences of the United States of America* 108, 17034–17039.

Baeten, L., Verheyen, K., Wirth, C. et al., 2013. A novel comparative research platform designed to determine the functional significance of tree species diversity in European forests. *Perspectives in Plant Ecology Evolution and Systematics* 15, 281–291.

Balvanera, P., Pfisterer, A.B., Buchmann, N. et al., 2006. Quantifying the evidence for biodiversity effects on ecosystem functioning and services. *Ecology Letters* 9, 1146–1156.

Barbosa, P., Hines, J., Kaplan, I., Martinson, H., Szczepaniec, A., Szendrei, Z., 2006. Associational resistance and associational susceptibility: Having right or wrong neighbors. *Annual Review of Ecology, Evolution, and Systematics* 40, 1–20.

Baruffol, M., Schmid, B., Bruelheide, H. et al., 2013. Biodiversity promotes tree growth during succession in subtropical forest. *PLoS ONE* 8, e81246.

Bell, T., Newman, J.A., Silverman, B.W., Turner, S.L., Lilley, A.K., 2005. The contribution of species richness and composition to bacterial services. *Nature* 436, 1157–1160.

Bruelheide, H., Böhnke, M., Both, S. et al., 2011. Community assembly during secondary forest succession in a Chinese subtropical forest. *Ecological Monographs* 81, 25–41.

Bruelheide, H., Nadrowski, K., Assmann, T. et al., 2014. Designing forest biodiversity experiments: General considerations illustrated by a new large experiment in subtropical China. *Methods in Ecology and Evolution* 5, 74–89.

Bullock, J.M., Pywell, R.F., Burke, M.J.W., Walker, K.J. 2001. Restoration of biodiversity enhances agricultural production. *Ecology Letters* 4, 185–189.

Butchart, S.H., Walpole, M., Collen, B. et al., 2010. Global biodiversity: Indicators of recent declines. *Science* 328, 1164–1168.

Byrnes, J.E.K., Gamfeldt, L., Isbell, F. et al., 2014. Investigating the relationship between biodiversity and ecosystem multifunctionality: Challenges and solutions. *Methods in Ecology and Evolution* 5, 111–124.

Cardinale, B.J., Duffy, J.E., Gonzalez, A. et al., 2012. Biodiversity loss and its impact on humanity. *Nature* 486, 59–67.

Cardinale, B.J., Palmer, M.A., Collins, S.L., 2002. Species diversity enhances ecosystem functioning through interspecific facilitation. *Nature* 415, 426–429.

Cardinale, B.J., Srivastava, D.S., Duffy, E.J. et al., 2006. Effects of biodiversity on the functioning of trophic groups and ecosystems. *Nature* 443, 989–992.

Cardinale, B.J., Wright, J.P., Cadotte, M.W. et al., 2007. Impacts of plant diversity on biomass production increase through time because of species complementarity. *Proceedings of the National Academy of Sciences of the United States of America* 104, 18123–18128.

Castagneyrol, B., Giffard, B., Péré, C., Jactel, H., 2013. Plant apparency, an overlooked driver of associational resistance to insect herbivory. *Journal of Ecology* 101, 418–429.

Chesson, P., 2000. Mechanisms of maintenance of species diversity. *Annual Review of Ecology and Systematics* 31, 343–366.

Díaz, S., Symstad, A.J., Stuart Chapin, I., F., Wardle, D.A., Huenneke, L.F., 2003. Functional diversity revealed by removal experiments. *Trends in Ecology and Evolution* 18, 140–146.

Dornelas, M., Gotelli, N.J., McGill, B. et al., 2014. Assemblage time series reveal biodiversity change but not systematic loss. *Science* 344, 296–299.

Ehrlich, P., Ehrlich, A., 1981. *Extinction: The Causes and Consequences of the Disappearance of Species*. Random House, New York.

Fraser, L.H., Henry, H.A.L., Carlyle, C.N. et al., 2013. Coordinated distributed experiments: An emerging tool for testing global hypotheses in ecology and environmental science. *Frontiers in Ecology and the Environment* 11, 147–155.

Fuller, R.M., 1986. The changing conservation interest of lowland grasslands in England and Wales: A review of grassland surveys 1939–1984. *Biological Conservation* 40, 281–300.

Gagic, V., Bartomeus, I., Jonsson, T. et al., 2015. Functional identity and diversity of animals predict ecosystem functioning better than species-based indices. *Proceedings of the Royal Society of London B: Biological Sciences* 282, 20142620.

Gamfeldt, L., Hillebrand, H., Jonsson, P.R., 2008. Multiple functions increase the importance of biodiversity for overall ecosystem functioning. *Ecology* 89, 1223–1231.

Grossiord, C., Granier, A., Ratcliffe, S. et al., 2014. Tree diversity does not always improve resistance of forest ecosystems to drought. *Proceedings of the National Academy of Sciences of the United States of America* 111, 14812–14815.

Guido, A., Pillar, V.D., 2015. Are removal experiments effective tools for assessing plant community resistance and recovery from invasion? *Journal of Vegetation Science* 26, 608–613.

He, J.Z., Ge, Y., Xu, Z.H., Chen, C.R., 2009. Linking soil bacterial diversity to ecosystem multifunctionality using backward-elimination boosted trees analysis. *Journal of Soils and Sediments* 9, 547–554.

Hector, A., Bagchi, R., 2007. Biodiversity and ecosystem multifunctionality. *Nature* 446, 188–190.

Hector, A., Schmid, B., Beierkuhnlein, C. et al., 1999. Plant diversity and productivity experiments in European grasslands. *Science* 286, 1123–1127.

Hillebrand, H., Matthiessen, B., 2009. Biodiversity in a complex world: Consolidation and progress in functional biodiversity research. *Ecology Letters* 12, 1405–1419, doi:10.1111/j.1461-0248.2009.01388.x.

Hooper, D.U., Adair, E.C., Cardinale, B.J. et al., 2012. A global synthesis reveals biodiversity loss as a major driver of ecosystem change. *Nature* 486, 105–108.

Hooper, D.U., Vitousek, P.M., 1997. The effects of plant composition and diversity on ecosystem processes. *Science* 277, 1302–1305.

Isbell, F., Calcagno, V., Hector, A. et al., 2011. High plant diversity is needed to maintain ecosystem services. *Nature* 477, 199–202.

Jucker, T., Bouriaud, O., Avacaritei, D., Coomes, D.A., 2014. Stabilizing effects of diversity on aboveground wood production in forest ecosystems: Linking patterns and processes. *Ecology Letters* 17, 1560–1569.

Jucker, T., Coomes, D.A., 2012. Comment on plant species richness and ecosystem multifunctionality in global dryland. *Science* 337, 155.

Kinzig, A.P., Pacala, S.W., Tilman, D., 2002. *The Functional Consequences of Biodiversity: Empirical Progress and Theoretical Extensions.* Princeton University Press, Princeton, NJ.

Kotowska, A.M., Cahill Jr, J.F., Keddie, B.A., 2010. Plant genetic diversity yields increased plant productivity and herbivore performance. *Journal of Ecology* 98, 237–245.

Leuschner, C., Jungkunst, H.F., Fleck, S., 2009. Functional role of forest diversity: Pros and cons of synthetic stands and across-site comparisons in established forests. *Basic and Applied Ecology* 10, 1–9.

Loreau, M., Naeem, S., Inchausti, P., 2002. *Biodiversity and Ecosystem Functioning.* Oxford University Press, New York.

Maestre, F.T., Quero, J.L., Gotelli, N.J. et al., 2012. Plant species richness and ecosystem multifunctionality in global drylands. *Science* 335, 214–218.

Millennium Ecosystem Assessment, 2005. *Ecosystems and Human Well-Being: Biodiversity Synthesis.* World Resources Institute, Washington, DC.

Nadrowski, K., Ratcliffe, S., Bönisch, G. et al., 2013. Harmonizing, annotating and sharing data in biodiversity-ecosystem functioning research. *Methods in Ecology and Evolution* 4, 201–205.

Naeem, S., Bunker, D.E., Hector, A., Loreau, M., Perrings, C., 2009. *Biodiversity, Ecosystem Functioning, and Human Wellbeing: An Ecological and Economic Perspective*. Oxford University Press, Oxford, U.K.

Naeem, S., Thompson, L.J., Lawler, S.P., Lawton, J.H., Woodfin, R.M., 1994. Declining biodiversity can alter the performance of ecosystems. *Nature* 368, 734–737.

Petchey, O.L., Gaston, K.J., 2002. Functional diversity (FD), species richness and community composition. *Ecology Letters* 5, 402–411.

Potvin C., Gotelli N.J., 2008. Biodiversity enhances individual performance but does not affect survivorship in tropical trees. *Ecology Letters* 11, 217–223.

Potvin C., Dutilleul P., 2009. Neighborhood effects and size-asymmetric competition in a tree plantation varying in diversity. *Ecology* 90, 321–327.

Reich, P.B., Knops, J., Tilman, D. et al., 2001. Plant diversity enhances ecosystem responses to elevated CO_2 and nitrogen deposition. *Nature* 410, 809–812.

Reich, P.B., Tilman, D., Isbell, F. et al., 2012. Impacts of biodiversity loss escalate through time as redundancy fades. *Science* 336, 589–592.

Roscher, C., Schumacher, J., Baade, J. et al., 2004. The role of biodiversity for element cycling and trophic interactions: An experimental approach in a grassland community. *Basic and Applied Ecology* 5, 107–121.

Ruiz-Benito P., Gómez-Aparicio L., Paquette A., Messier C., Kattge J., Zavala M.A., 2014. Diversity increases carbon storage and tree productivity in Spanish forests. *Global Ecology and Biogeography* 23, 311–322.

Scherber, C., Eisenhauer, N., Weisser, W.W. et al., 2010. Bottom-up effects of plant diversity on multitrophic interactions in a biodiversity experiment. *Nature* 468, 553–556.

Scherer-Lorenzen, M., 2005. Biodiversity and ecosystem functioning: Basic principles. In *Encyclopedia of Life Support Systems (EOLSS)*, Vol. 1: Biodiversity: Structure and Function, eds. W. Barthlott, K.E. Linsenmair, S. Porembski. EOLSS Publisher, Oxford, U.K.

Scherer-Lorenzen, M., Schulze, E.-D., Don, A., Schumacher, J., Weller, E. 2007. Exploring the functional significance of forest diversity: A new long-term experiment with temperate tree species (BIOTREE). *Perspectives in Plant Ecology, Evolution and Systematics* 9, 53–70.

Schleuter, D., Daufresne, M., Massol, F., Argillier, C., 2010. A user's guide to functional diversity indices. *Ecological Monographs* 80, 469–484.

Schmid, B., Hector, A., 2004. The value of biodiversity experiments. *Basic and Applied Ecology* 5, 535–542.

Schmid, B., Hector, A., Huston, M.A. et al., 2002. The design and analysis of biodiversity experiments, In *Biodiversity and Ecosystem Functioning: Synthesis and Perspectives*, eds. M. Loreau, S. Naeem, P. Inchausti, pp. 61–75. Oxford University Press, Oxford, U.K.

Schulze, E.-D., Mooney, H.A., 1993. *Biodiversity and Ecosystem Function*. Springer, Berlin, Germany.

Stachowicz, J.J., Bruno, J.F., Duffy, J.E., 2007. Understanding the effects of marine biodiversity on communities and ecosystems. *Annual Review of Ecology, Evolution and Systematics* 38, 739–766.

Tilman, D., Knops, J., Wedin, D., Reich, P., Ritchie, M., Siemann, E., 1997. The influence of functional diversity and composition on ecosystem processes. *Science* 277, 1300–1302.

Tilman, D., Wedin, D., Knops, J., 1996. Productivity and sustainability influenced by biodiversity in grassland ecosystems. *Nature* 379, 718–720.

Turner, W.R., Brandon, K., Brooks, T.M. et al., 2012. Global biodiversity conservation and the alleviation of poverty. *BioScience* 62, 85–92.

Vellend, M., Baeten, L., Myers-Smith, I.H. et al., 2013. Global meta-analysis reveals no net change in local-scale plant biodiversity over time. *Proceedings of the National Academy of Sciences of the United States of America* 110, 19456–19459.

Verheyen, K., Ceunen, K., Ampoorter, E. et al. 2013. Assessment of the functional role of tree diversity: The multi-site FORBIO experiment. *Plant Ecology and Evolution* 146, 26–35.

Verheyen, K., Vanhellemont, M., Auge, H. et al., 2015. Contributions of a global network of tree diversity experiments to sustainable forest plantations. *Ambio* 45, 1–13.

Vilà, M., Vayreda, J., Comas, L., Ibáñez, J.J., Mata, T., Obón, B., 2007. Species richness and wood production: A positive association in Mediterranean forests. *Ecology Letters* 10, 241–250.

Violle C., Enquist B.J., McGill B.J., Jiang L., Albert C.H., Hulshof C., Jung V., Messier J., 2012. The return of the variance: intraspecific variability in community ecology. *Trends in Ecology & Evolution* 27, 244–252.

Vogel, A., Scherer-Lorenzen, M., Weigelt, A., 2012. Grassland resistance and resilience after drought depends on management intensity and species richness. *PLoS ONE* 7, e36992.

Wardle, D.A., Zackrisson, O., 2005. Effects of species and functional group loss on island ecosystem properties. *Nature* 435, 806–810.

Weigelt, A., Weisser, W.W., Buchmann, N., Scherer-Lorenzen, M., 2009. Biodiversity for multifunctional grasslands: Equal productivity in high-diversity low-input and low-diversity high-input systems. *Biogeosciences* 6, 1695–1706.

Wilson, E.O., 1988. *Biodiversity*. National Academy Press, Washington, DC.

Wolkovich, E.M., Cleland, E.E., 2010. The phenology of plant invasions: A community ecology perspective. *Frontiers in Ecology and the Environment* 9, 287–294.

Worm, B., Barbier, E.B., Beaumont, N. et al., 2006. Impacts of biodiversity loss on ocean ecosystem services. *Science* 314, 787–790.

Yachi, S., Loreau, M., 1999. Biodiversity and ecosystem productivity in a fluctuating environment: The Insurance Hypothesis. *Proceedings of the National Academy of Sciences of the United States of America* 96, 57–64.

Zavaleta, E.S., Hulvey, K.B., 2004. Realistic species losses disproportionately reduce grassland resistance to biological invaders. *Science* 306, 1175–1177.

6

Frontiers of Ecosystem Modeling and Large-Scale Experiments

Lifen Jiang, Jiang Jiang, Junyi Liang, Kevin R. Wilcox, Scott L. Collins, Alan K. Knapp, William T. Pockman, Melinda D. Smith, and Yiqi Luo

CONTENTS

6.1 Introduction

Observational studies, ecological experiments, and modeling are the three foundational approaches in ecological research. With increasing data availability from ecological observations and experiments, data–model integration becomes a critical tool to reduce model uncertainty and gain more reliable predictions of future ecosystem dynamics. In this chapter, we review the need, history, and current status of data–model integration to improve model simulations. Data from long-term manipulative experiments (e.g., free-air CO_2 enrichment experiments), observation networks of eddy covariance measurements (e.g., FLUXNET), and global databases (e.g., IGBP-DIS soil C database) have been successfully incorporated into ecosystem or global biogeochemical models to improve model projections. We also take two ongoing projects, Spruce and Peatland Responses Under Climatic and Environmental Change Experiment (SPRUCE) and Extreme Drought in Grasslands Experiment (EDGE), as examples to illustrate how experiment–model integration approaches are designed to achieve different research goals. In SPRUCE, the operational forecasting system is developed to assimilate data streams in real time to predict ecosystem responses to two global change factors, warming and elevated CO_2, whereas in EDGE, the application of data assimilation is designed to disentangle the role of environmental context versus ecosystem attributes in response to drought and eventually to scale findings obtained at distributed sites to regional scales. Finally, we review the challenges in data–model integration and propose a few strategies to move forward.

6.2 Experiment–Model Integration: Needs, History, and Current Status

In recent decades, tremendous efforts have been made to improve the predictability of ecosystem states and processes using three independent approaches: carefully designed long-term observational studies, ecological experimentation, and process-based modeling. The state of this knowledge is becoming increasingly important in light of global climate change. Ecological experiments, including in situ observations, have been conducted at many individual sites globally, and through these, we have obtained large amounts of observational and experimental data. Manipulative experiments, by creating novel conditions of projected climate, have improved our understanding of how ecosystems respond to climate change. While site-based experiments are still an important way to explore ecological patterns and underlying mechanisms, coordinated distributed experiments have become an emerging tool to test hypotheses at the global scale (Fraser et al. 2012). More and more networks and cross-site studies at regional and global scales are generating huge amounts of data on daily or annual time scales such as FLUXNET, LTER Network, Nutrient Network, Drought-Net, NGEE-Aritic, NGEE-Tropics, and the National Ecological Observatory Network (NEON). Ecological research is now in a data-rich era (Luo et al. 2011b), and a major challenge is to know how to use those data in an efficient and appropriate manner to understand the ecological pattern and process and, based on that understanding, to forecast how ecosystems will change under global environmental change.

Empirical data from observational and experimental studies are useful in exploring general patterns of ecosystem phenomena and studying underlying mechanisms. However, ecological observations and experiments have inherent limitations to predict future changes, such as (1) difficulties in observing or measuring certain critical variables (e.g., turnover of different nutrient or carbon pools in soils), (2) large costs of conducting observations and manipulations at high temporal and spatial resolutions, (3) expense associated with conducting experiments over a time period long enough to detect changes in slow processes (Luo et al. 2011a), and (4) logistical constraints when conducting experiments across broad spatial scales; many difficulties remain for using empirical data to assess the impacts caused by changes in environmental conditions at multiple spatial and temporal scales. In addition, observational data and manipulative experiments with certain treatments alone lack power to forecast future dynamics of ecosystems in response to global change, which is challenging because we now live in a world that is undergoing rapid environmental change and highly altered disturbance regimes (Clark et al. 2001, Luo et al. 2011a,b, Niu et al. 2014).

Ecological models, particularly process-based models, can supplement limitations inherent in observations and experiments (Luo et al. 2011b). Process-based models have advanced quickly in recent years, which can partly be attributed to the dramatic increase of computational capability. Global models of the terrestrial carbon (C) cycle, developed about two decades ago, have been widely used to predict changes in terrestrial C storage under increasing atmospheric CO_2 concentration and climate change (IPCC 2013). Models have become increasingly complex due to the need for incorporating additional ecosystem processes, such as multiple vertical soil layers in the Community Land Model, CLM4.5 (Koven et al. 2013) and dynamic vegetation model (e.g., Sitch et al. 2003). However, with more processes added to models, models are more complex and generate larger uncertainties in their projections of terrestrial carbon storage and uptake, thus poorly fitting observations as shown in a number of Model Intercomparison Projects (MIP, Friedlingstein et al. 2006, Jones et al. 2013, Todd-Brown et al. 2013, 2014, Carvalhais et al. 2014, Jiang et al. 2015, Tian et al. 2015). On one hand, models need to include critical processes that predict future dynamics correctly; on the other hand, models should lower their uncertainty in predictions. The large uncertainties associated with projections among different models as well as between models and observations strongly limit their usefulness for informing policy makers. This makes experiment–model integration improve model behaviors and forecasting an urgent task (Clark et al. 2001, Luo et al. 2011a,b). The research community is still struggling to improve predictability of models (Luo et al. 2016). Informed by empirical data, models can obtain more reliable projections (Niu et al. 2014).

Empirical data can help model development and improvement through a few pathways: providing mechanistic understanding behind ecosystem processes (e.g., photosynthesis and litter decomposition), generalizing patterns (e.g., vegetation distributions along temperature and precipitation gradients), identifying ranges and patterns of key model parameters, and constraining models. In practice, experiment–model integration can improve model performance by evaluating the goodness-of-fit of models using data (i.e., benchmarking, Luo et al. 2012, Walker et al. 2014), model parameterization with data, assimilating multiple streams of data to constrain models (Figure 6.1, Luo et al. 2011b, Niu et al. 2014), and representing unresolved processes with new algorithms supported by data (Williams et al. 2009).

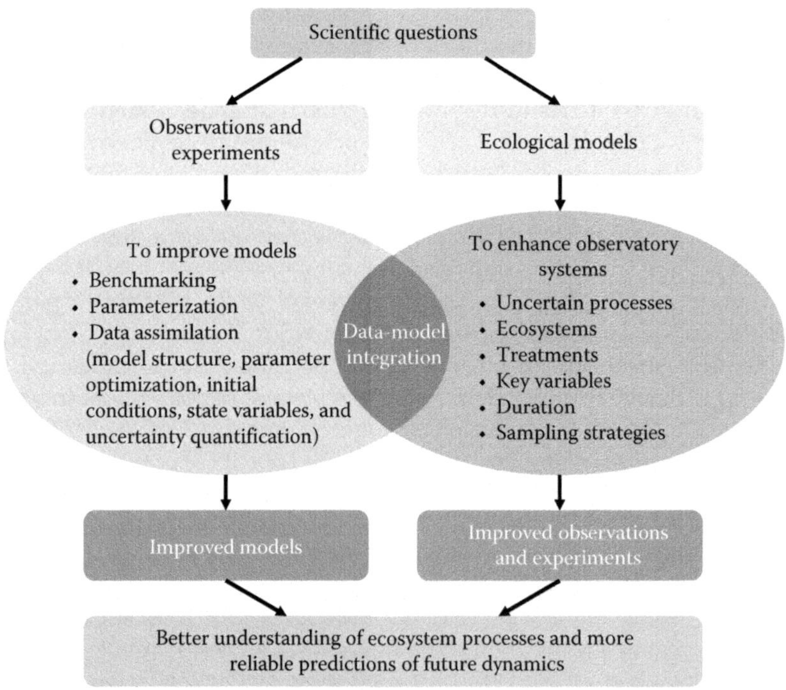

FIGURE 6.1
Conceptual flow of ecological research via experiment–model integration.

Modeling studies, on the other hand, can guide observations and experiments (e.g., Wang et al. 2009, Keenan et al. 2012). Modeling results may inform observational and experimental studies about the most uncertain processes that experiments should be conducted to explore, where or what ecosystems need more research, what treatments need to be considered, what measurements should be added (i.e., key variables needed to improve modeling), how long a manipulative experiment should last, and appropriate sampling strategies such as sampling frequency and spatial resolution for sampling (Figure 6.1).

Observational and experimental data have often been used for evaluating process-based models since the early stages of their development (e.g., Parton et al. 1987, 1993, Pacala et al. 1993, Luo et al. 2012). It is still common practice for modelers to tune their models to match specific data or reflect general patterns. However, multiple factors could result in bias between model simulations and data, and it becomes frustrating to tune models if a model has lots of interacting parameters. Experiment–model integration via rigorous statistical methods such as the Bayesian approaches provides a new way to enable simultaneous analysis of diverse sources of data to train a process-based model (Ogle 2009). Data assimilation is such a cyber-enabled process, which incorporates data into models to determine initial values and state variables, estimate model parameters (Richardson et al. 2010), and evaluate alternative model structures (Figure 6.1; Keenan et al. 2012, Liang et al. 2015b). As a result, data assimilation can substantially improve model simulations and gain more reliable predictions for future ecosystem dynamics (Raupach et al. 2005, Williams et al. 2005, 2009, Luo et al. 2011b, Peng et al. 2011, Niu et al. 2014). Indeed, data assimilation is recognized as a top priority for near-term modeling research to reduce systematic biases in modeling soil carbon dynamics by Earth system models, making data assimilation an essential tool in ecological research in a data-rich era (Luo et al. 2016). Also, data assimilation allows quantification of uncertainties stemming from data, model structures, parameters, boundary conditions, and statistical method selection (Luo et al. 2003, 2011b, Weng and Luo 2011, Dietze et al. 2013, LeBauer et al. 2013).

While data assimilation has been applied in biogeochemical cycles for a relatively short period, it has long been used in weather forecasts since the 1950s when the advances in computational power allowed data assimilation to be done within a reasonable time frame (LeBauer et al. 2013, Niu et al. 2014). The application of data assimilation in process-based models has been largely motivated by the need for ecological forecasting to inform decision-making processes toward better management of natural resources in a world undergoing rapid global change (Clark et al. 2001, 2003, Luo et al. 2011a,b). Since 2000, the application of data assimilation in process-based models has incorporated a variety of sources of data. For example, data from manipulative experiments (Luo et al. 2011a, Shi et al. 2015b,c), particularly those Free-Air CO_2 Enrichment Experiments (FACE, Luo and Reynolds 1999,

Parton et al. 2007, Weng and Luo 2011, Walker et al. 2014, Medlyn et al. 2015, Norby et al. 2016, Xu et al. 2016), have been frequently assimilated to constrain model parameters and state variables to improve predictions; data contributed by the observational networks such as FLUXNET and AmeriFlux have been used to train ecosystem or land surface models (e.g., White et al. 2006, Williams et al. 2009, Wu et al. 2009, Zhang et al. 2010, Kuppel et al. 2012); compiled continental and global data sets of biomass and soil C content were found to be useful in improving models (Zhou and Luo 2008, Zhou et al. 2009, 2013, 2015, Hararuk and Luo 2014, Hararuk et al. 2014, 2015); soil incubation data were collected and used to evaluate model structures and responses of soil C decomposition to warming and the factors regulating soil C decomposition (Li et al. 2013, Schädel et al. 2013, Liang et al. 2015a, Xu et al. 2016).

Many data assimilation techniques have been employed to constrain models. Usually, model parameters are specified and set a priori according to knowledge of processes. Optimal solutions and uncertainty analyses are determined by searching and selecting within realistic ranges of the chosen parameter values. The applied optimization techniques include the adjoint method (White and Luo 2002), the Levenberg–Marquardt method in combination with quasi-Monte Carlo algorithm (Luo et al. 2003), genetic algorithms (Zhou and Luo 2008), Markov chain Monte Carlo (MCMC) with the Metropolis–Hastings algorithm (Xu et al. 2006, Zhou et al. 2010, Hararuk et al. 2015, Shi et al. 2016), conditional inversion (Wu et al. 2009), and the ensemble Kalman filter (Gao et al. 2011). These optimization techniques identify sets of parameters, which result in model output that best fits the data. One of the most important applications of data assimilation is addressing the initial condition problems. Traditionally, Earth system models usually use spin-up cycles with preindustrial forcing to determine the initial condition (i.e., steady state), which may overestimate terrestrial C pools (Wutzler and Reichstein 2007). By relaxing the steady-state assumption, data assimilation can significantly increase model efficiency and reduce normalized average error (Carvalhais et al. 2008, Williams et al. 2009).

However, applications of data assimilation, even at a regional or global scale, have mostly been conducted with simplified models that had a limited number of parameters to be constrained by data, typically from several to less than 20 parameters (e.g., Braswell et al. 2005, Williams et al. 2009, Zhou et al. 2009, 2012). Data assimilation with complex models containing thousands of parameters are not feasible until a breakthrough in semi-analytic spin-up of global land model (Xia et al. 2012) and traceability analysis of model structures (Xia et al. 2013). High-fidelity emulators that reproduce original complex models are often used to make data assimilation computationally feasible. By assimilating a global soil carbon data set into an emulator of the Community Land Model coupled with Carnegie-Ames-Stanford Approach biogeochemistry submodel (CLM-CASA) to optimize the parameters, Hararuk et al. (2014) were able to substantially improve soil carbon simulations, increasing the correlation between model simulations and the observations from an R^2 of 0.27 before calibration of parameters to an R^2 of 0.41. Using the emulator of CLM-CASA and a combination of global soil carbon and soil microbial biomass data sets, model simulations after parameter optimization via data assimilation detected stronger soil carbon responses to climate change (Hararuk et al. 2015).

6.2.1 Experiment–Model Integration with Data from Global Change Experiments and Observational Networks

As mentioned earlier, traditionally, process-based models are usually validated by comparing model simulations with data from ecological observations or experiments. This validation, however, can be biased by limited availability of observations or experiments. Recently, large-scale benchmarking, which includes as many available observations and experiments as possible, has been proposed and developed to evaluate the performance of models by scoring models based on how well the models reproduce data (Luo et al. 2012). While this approach provides comprehensive evaluations of model performance, benchmarking itself rarely leads to model improvement because reasons for model performance are not typically identified (Medlyn et al. 2015).

Over the past decades, an increasing number of global change experiments have been initiated due to the need for information concerning ecosystem responses to a rapidly changing climate. Data from these global change experiments are valuable in helping models identify ecosystem responses to long-term climate change. Enormous efforts have been made to elucidate ecosystem dynamics in response to elevated CO_2 and increased temperature. For example, a recent FACE Model-Data Synthesis (FACE-MDS) project was sponsored to assess terrestrial models by using observational data collected in two temperate forest FACE experiments in the United States (Walker et al. 2014, Medlyn et al. 2015). In this project, 11 terrestrial models were evaluated by comparing with data from the Oak Ridge National Laboratory FACE (ORNL FACE) and Duke FACE sites. In addition to comparing model simulations with direct measurements, they also focused on assumptions resulting in disagreement among models (Medlyn et al. 2015). For example, the 11 terrestrial models generally reproduced the responses of NPP to CO_2 enrichment in the early years of the experiment, but they were not able to predict NPP responses in later years. By further decomposing NPP to nitrogen (N) use efficiency (NUE) and N uptake, they found that even if correctly estimating the response of NPP to CO_2 enrichment in the early years, models achieved that by overestimating NUE response but underestimating the response of N uptake (Zaehle et al. 2014, Medlyn et al. 2015).

In addition to NPP, other processes and ecosystem properties, such as stomatal conductance, water use efficiency, NPP allocation, N cycling, stoichiometry, and leaf mass per area (LMA), were also studied in the FACE-MDS project (De Kauwe et al. 2013, 2014, Walker et al. 2014, Zaehle et al. 2014, Medlyn et al. 2015). This project has identified specific aspects for improving model performance and experimental design. The identified aspects for model improvement include stomatal conductance, NPP allocation, LMA, plant stoichiometry, and priming (Medlyn et al. 2015). It is suggested that models should employ a proportional relationship between the ratio of photosynthetic C assimilation to stomatal conductance and atmospheric CO_2 concentration (De Kauwe et al. 2013, Medlyn et al. 2015). Dependence of NPP allocation to plant tissues (e.g., leaves, wood, and fine roots) on resource availability is necessary to match observations (De Kauwe et al. 2014, Medlyn et al. 2015). Moreover, priming of soil N release seems important to reproduce N transfer from soil organic matter to vegetation (Zaehle et al. 2014, Medlyn et al. 2015). In addition to these mechanisms on which clear suggestions have been proposed, there are some others on which new theories are needed. For example, LMA is usually set constant in models, leading to a stronger response of leaf area index (LAI)—a key property that determines photosynthetic C assimilation in both models and reality—to CO_2 enrichment in models than identified in observations (Medlyn et al. 2015). Additionally, new theory of flexibility of plant stoichiometry is needed to better simulate the balance of N supply and demand (Zaehle et al. 2014, Medlyn et al. 2015).

Aspects of improvement of experimental design have also been identified based on findings in the FACE-MDS project. First, more empirical evidence is needed to determine the relative importance of electron transport and Rubisco limitations to photosynthetic C assimilation on the leaf and canopy scales (Medlyn et al. 2015). Additionally, observations of sensitivity of transpiration to stomatal conductance, precipitation interception by canopy, the interactive effect of drought, turnover of plant tissues, and ecosystem's N losses (e.g., leaching) are also needed to provide a scientific basis for model assumptions. However, one potential limitation of the FACE-MDS project is that only two FACE experiments, both located in temperate regions, are involved. Extensive meta-analyses (e.g., Liang et al. 2015b) and experiments in other regions, such as boreal and tropical systems, are necessary to help evaluate and improve the performance of Earth system models on the global scale (Norby et al. 2016).

In addition to site-level manipulative experiments, observations at larger scales are valuable resources that can be used to improve model performance. One of the successful data set networks that has been assimilated to improve terrestrial model performance is the FLUXNET observations (Williams et al. 2009). FLUXNET is an international observation network based on the eddy covariance (EC) technique, which can provide continuous measurements of land surface–atmosphere exchanges of C, water, and energy (Verma 1990). There are many regional EC networks across the globe, such as AmeriFlux (http://ameriflux.ornl.gov/), ChinaFlux (http://www.chinaflux.org/), JapanFlux (http://www.japanflux.org/), and Swiss Fluxnet (http://www.swissfluxnet.ch/). Each of the regional networks consists of a number of EC observation sites. These regional networks then form the global network FLUXNET (http://fluxnet.ornl.gov/), which provides rich collection of information concerning the exchanges of C, water, and energy between land surface and the atmosphere across spatial and temporal scales. FLUXNET observations have been widely used in data assimilation techniques to improve model performance. For example, Zhu and Zhuang (2015) assimilated five sites of AmeriFlux carbon flux data to Terrestrial Ecosystem Model (TEM) to reduce uncertainty of 10 posterior parameters and improve simulations of regional carbon dynamics.

Those global change experiments and observational networks are usually designed for examining patterns of ecosystem responses to environmental changes and underlying mechanisms. The resulting data sets are then found useful for model improvements. Recently, a number of projects have been designed and initiated to conduct both experiment and data–model integration for forecasting or upscaling site-level knowledge to a regional scale. In the following text, we take two ongoing projects, SPRUCE and EDGE, as examples to illustrate how data assimilation is designed to achieve different research goals.

6.2.2 Experiment–Model Integration: A Case Study with the SPRUCE Project

SPRUCE is a climate change manipulative experiment supported by Terrestrial Ecosystem Science Scientific Focus Area of ORNL's Climate Change Program. The experiment aimed toward integration of experiments with ecosystem modeling, data assimilation, and model structure evaluation to yield reliable model projections. Although the SPRUCE experiment only initiated the Deep Peatland Heating treatments in June of 2014 and the whole-ecosystem warming and elevated CO_2 treatments will not start until June 2016, the integrated model-experiment approach based on pretreatment data sets and modeling activities have promoted an interactive and mutually beneficial engagement between modelers and experimentalists to advance predictions from experiments and models.

6.2.2.1 Infrastructure Challenges in the SPRUCE Experiment

The SPRUCE experiment is being operated as the first whole-ecosystem, forest-scale experiment to increase temperature and CO_2 concentrations from deep soil to the tops of tree canopies. This decade-long experiment is conducted in a weakly ombrotrophic peatland with a perched water table that has little regional groundwater influence and is located in northern Minnesota in the USDA Forest Service Marcell Experimental Forest (MEF). The 8.1 ha experimental site (S1-Bog) is dominated by *Picea mariana* (black spruce), regenerated from strip cuts in 1969 and 1974. Located at the southern margin of the boreal peatland, this ecosystem is anticipated to be approaching its tipping point with high vulnerability in response to climate change. Shifts of plant communities at the southern margin of boreal ecosystems under climate change is not fully understood, so large-scale long-term experiments like SPRUCE are needed to improve mechanistic representation of unresolved processes in understudied ecosystems in Earth system models.

To achieve the overall goal of assessing ecosystem-level biological responses of vulnerable, high carbon terrestrial ecosystems to climate change, the SPRUCE experiment faces many infrastructure challenges including, but not limiting to, plot facilities, sensors and instruments, data acquisition and control system, automated data monitoring system, data management, model–data integration, and synthesis of model outputs. In this section, we do not attempt to describe all the research infrastructure challenges facing large-scale experiments like SPRUCE, but we highlight those key elements necessary for model–data integration.

Long-term monitoring of ecosystem dynamics is an important data source for data–model integration. But more powerful manipulative ecosystem experiments are needed to distinguish future climate change impacts from those inherent responses to natural variability. Climate change–caused warming scenarios predicted by the Intergovernmental Panel on Climate Change (IPCC 2013) are much higher than observed variation in mean annual temperatures (±2°C) under current climate. The SPRUCE experiment provides a platform to understand physiological and biogeochemical processes under future climate through a combination of multiple levels of warming up to +9°C at ambient or elevated CO_2 levels. Air warming is achieved with heating infrastructures enclosed in 10 plots of 12 m diameter by 8 m high open-top enclosures (Figure 6.2). The open-top enclosures can keep warming air around the enclosed plots by limiting air turnover, while still allowing natural precipitation to fall on experimental plots. The design also enables maintenance of high concentration of CO_2 (800–900 ppm) in the elevated CO_2 treatment. One unprecedented data set that the SPRUCE experiment can provide for modeling activities is the Deep Peat Heating (DPH, –2 m) with belowground temperatures being consistent with future aboveground warming scenarios (Hanson et al. 2011). Belowground deep warming evaluates responses of deep peat C stocks, microbial communities, and biogeochemical cycling processes, which can be used for evaluating model projections.

The SPRUCE program also provides data from real-time automated monitoring systems, which advance data–model integration approach in an interactive fashion. All data collected from sensors or instruments are recorded by dataloggers and then transferred to a data storage server every 30 min. Data are then published via a web server and used for remote monitoring and control (Krassovski et al. 2015). The standardized data streams are then immediately available to feed and inform models.

(a) (b) (c)

FIGURE 6.2
(See color insert.) View of SPRUCE experiment infrastructure with (a) exterior view of experimental chamber, (b) interior view of experimental chamber, and (c) aerial view of the S1 Bog site. (Pictures a and c are Oak Ridge National Laboratory Images from: http://mnspruce.ornl.gov. Image b is a PHENOCAM network SPRUCE image from http://phenocam.sr.unh.edu/webcam/gallery/.)

6.2.2.2 Informing Models Using the SPRUCE Experiment

Current Earth system models predict high variability of terrestrial carbon uptake with uncertainty in the direction and magnitude of global carbon cycling under future climate change (Friedlingstein et al. 2006, Arora et al. 2013). Northern peatlands contain 200–400 Pg of carbon, about 30% of that contained in the global soil carbon pool. These peat carbon pools are sensitive to climate change due to their tight dependence on hydrology and temperature (Frolking et al. 2013). In spite of the importance of peat carbon, the dynamics are still less represented in most Earth system models. Current models often change parameter values related to wetland and peatland regions assuming the same formulation as other regions (Luo et al. 2016), and therefore underestimate peat carbon storage (Limpens et al. 2008).

Manipulative experiments are needed to improve the unresolved processes in terrestrial carbon cycle models to refine projections of the net carbon balance of the northern peatlands in the face of a warming global environment. The SPRUCE experiment interacts with multiple modeling activities to improve the representation of terrestrial C processes required to reduce uncertainty in forecasting from ESMs. For example, the Community Land Model currently has poor representations of vegetated peatlands, where the carbon fluxes are strongly influenced by water table height. Future warming could drop the water table due to increased evapotranspiration, thus influencing the stability of carbon stocks. To improve the representations of peatland water tables, Shi et al. (2015a) modified CLM to include fully prognostic water table algorithms to improve projections of peatland responses to future climate change. The model was parameterized and evaluated against half-hourly measurements of daily water table levels for 3 years from the SPRUCE experiment. Those observations include pretreatment data sets and long-term peatland hydrology studies on the MEF, where the SPRUCE site located. To further quantify how warming may influence water table and carbon fluxes, however, we need data sets from manipulative experiments to validate models, and further refine related processes.

6.2.2.3 Toward an Interactive Experiment–Model Approach

With more realistic representations of ecosystem processes, models become more complex and more parameters are required to be constrained. Developing an operational forecasting system is an interactive way to integrate experiments and models to accomplish this. The forecasting system would assimilate various data streams into models so as to improve model predictions. Forecasting outcomes can provide feedback to experiments from which data sets are needed to further improve model predictions. In turn, new data sets can then be fed back into models further constraining and improving model predictions.

In the SPRUCE project, we are developing a data assimilation and operational ecological forecasting system called ECOlogical Platform for Assimilation of Data (ECOPAD). Pretreatment data sets from different field campaigns are compiled, including large-collar in situ CO_2 flux measurements across 4 years, aboveground NPP and carbon pool sizes from sampled vegetation, phenological data derived from PhenoCam imagery, and peat carbon from core samples. The data sets are then assimilated into the Terrestrial ECOsystem (TECO) model using a Markov chain Monte Carlo technique to constrain parameters. The TECO model is used because it simulates processes of canopy photosynthesis, plant growth, carbon transfer among compartments, and soil water dynamics. Unlike Earth system models, the TECO model is simple enough to overcome computational cost. With data assimilation, we can quantify how much uncertainty of forecasting could be reduced as more data become available. The relative contributions of external forcing versus model parameters to the uncertainty of forecasting can also be estimated. The projections of carbon cycles will be compared to future data streams to refine model structure, update model parameters, generate new scientific questions, and test competing hypotheses. The new data sets are then assimilated to enable new projections. This flow of work should be done regularly and automated in an operational system such as ECOPAD.

Although researchers have collected enormous amounts of data to understand various ecosystem processes over past decades, a major challenge is to combine understanding of multiple processes together to form a complete picture of how ecosystems will respond as a whole. Usually, empirical data from observations and manipulative experiments are scattered among individual teams and a significant proportion of them is not published in a timely manner, or is not available to modelers even after published. In the SPRUCE experiment, while individual teams collect data to answer questions related to specific ecosystem processes, they also work as a large group to confront models with data. ORNL is developing and deploying data and information management, and integration capability required for the collection, storage, processing, discovery, access, and delivery of data, including experimental data and model outputs. These capabilities and systems are designed to facilitate uncertainty associated with characterization and quantification. The systems will also be developed for assimilation of available measurements, synthetic analysis of results, model forcing and boundary condition data sets, and model results. Such an information system facilitates data–model integration and provides accessibility to model output, benchmarking analysis, visualization, and synthesis activities.

6.2.3 Experiment–Model Integration: A Case Study with EDGE Project

Site-level studies (e.g., a temperature manipulative experiment in a forest or measurements made in a grassland across years with different precipitation) can provide excellent information concerning ecological responses to climate change on small spatial scales. However, substantial limitations exist when trying to scale up to form regional and global predictions. Indeed, we know much more about how climate drivers are likely to be altered at various spatial scales (Murphy 2000, Schoof et al. 2010) than how the resulting impacts on ecosystem processes will play out to affect ecosystem services.

There are a number of major challenges associated with extending field-based findings to larger spatial scales. Two major ones are (1) understanding the mechanisms driving ecosystem responses to alterations in environmental variables and (2) obtaining knowledge about how these mechanisms vary across ecosystems. These challenges lead to an important overarching question: what are the relative strengths of climatic context (e.g., wet versus dry systems) versus ecosystem attributes (e.g., types of plant community) in driving how sensitive an ecosystem will be to changes in climate? For example, an arid system may have high sensitivity to drought because conditions are already dry and water is highly limiting. Yet on the other hand, xeric plant species may buffer primary production due to their having plant traits enabling growth even under stressful conditions. Indeed, empirical evidence has been shown for both increased sensitivity in drier systems (Huxman et al. 2004) and ecosystem attributes moderating responses to altered precipitation (Wilcox et al. 2015), yet methodological differences among studies obscure the relative strengths of these drivers in controlling ecosystem responses to climate change.

6.2.3.1 Coordinated Experiments across Space in the EDGE Project

EDGE is a multisite study being conducted in six grasslands spanning climate gradients but also varying in their ecosystem attributes (Figure 6.3). The approach of this experiment solves three major problems typically confronting single-site studies, meta-analyses, and data synthesis approaches

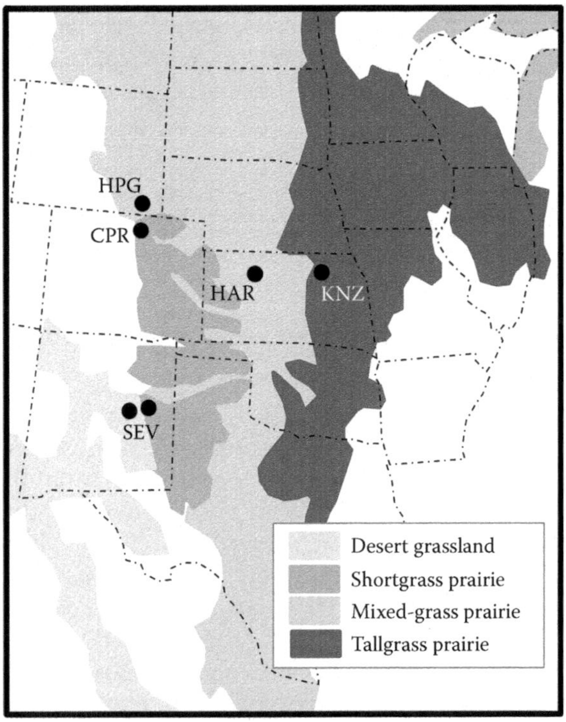

FIGURE 6.3
(See color insert.) Map of the Midwest United States showing locations of the six experimental sites of the EDGE project. SEV, Sevilleta National Wildlife Refuge, NM; CPR, Central Plains Experimental range; HPG, High Plains Grassland Research Center; HAR, Hays Agricultural Research Center; KNZ, Konza Prairie Biological Center.

to predict large-scale responses to climate change. First, consistent methodology among sites removes the uncertainty associated with meta-analyses that often compare experiments having different techniques for measuring ecosystem responses or implementing treatments; this allows for better detection of patterns of sensitivity across sites. For example, if a researcher has to compare results from two experiments, both simulating drought by removing 50% of ambient rainfall, but one study did this by removing rainfall from May to September and the other by removing rainfall from April to August, the researcher would not be able to know how much of the difference in responses was due to differential sensitivities of the systems and how much was due to the differences of drought timing. Second, the EDGE study is conducting extensive monitoring and sampling of ecosystem attributes at each site before and during the experiment. Meta-analyses or data syntheses usually get partial information from each single study. This provides mechanistic understanding into why sensitivity differs across ecosystems and tracks changes in ecosystem attributes as well as overall responses. Third, the EDGE experiment is designed to have three pairs of sites (Figure 6.3), with each pair existing along a precipitation gradient ranging from ~240 to 860 mm of mean annual rainfall and a temperature gradient ranging from 7.6°C to 13.3°C of mean annual temperature. Yet each pair represents two different plant community types existing within a particular climate envelope. For example, the two arid sites at SEV; both receive ~250 mm of precipitation on average and are quite close in space. However, one is a Chihuahuan Desert grassland dominated by *Bouteloua eriopoda* (black grama) while the other is more like the shortgrass steppe dominated by *Bouteloua gracilis* (blue grama). Comparison of differential responses between these two sites versus responses across the broad climate gradient will provide insight into how plant communities drive sensitivity of ecosystems to drought within the context of climate.

6.2.3.2 Integrating Experimental Findings and Process-Based Models

In EDGE, modelers and experimental ecologists are working in close collaboration to incorporate field data into an integrated experiment–modeling framework (Figure 6.4), where data are used to estimate parameters of the process-based model, TECO. In turn, this process identifies information gaps, for which additional sampling efforts can be conducted within the experiment. For example, some variables being measured in EDGE concerning C cycling processes are aboveground and belowground standing crop biomass, aboveground litter, soil respiration, and total soil C content. By mapping and comparing TECO output to these measured variables using a Markov chain Monte Carlo technique (Metropolis–Hastings algorithm; Hastings 1970), multiple model parameters are able to be estimated for each site. Additionally, model parameters not well constrained using the current level of data inform where additional sampling efforts should be applied.

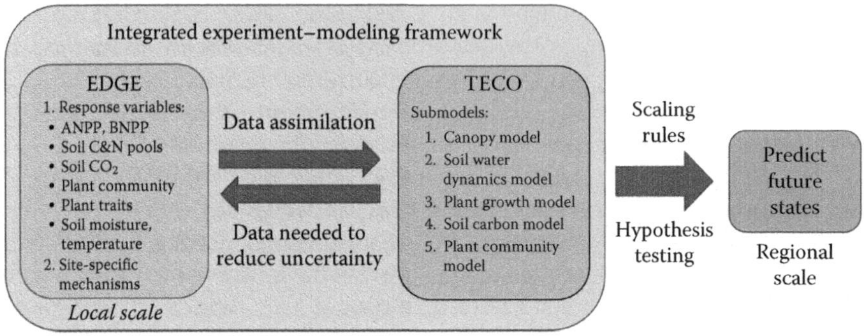

FIGURE 6.4
Conceptual representation of the integrated experiment–modeling framework employed by EDGE to assess mechanisms behind differential ecosystem sensitivity to drought across grasslands spanning climatic gradients and having various ecosystem attributes. Data obtained from the multisite experiment are integrated into the process-based model, TECO, to identify scaling rules for model parameters. These, in turn, are incorporated into regional-scale projections that allow for broader testing of hypotheses related to future ecosystem status under altered climate regimes.

Once parameters are able to be properly estimated, they can then be linked with the climatic conditions across sites to provide insight into how ecosystem processes as represented by model parameters and climate should covary. Second, examination of parameter estimates in different sites, but within the same climate envelope, gives insight into how parameters should be assigned based on ecosystem attributes separate from climatic context. For example, despite both having mean annual precipitation close to 350 mm and mean annual temperature around 9°C, paired sites in northern Colorado and southern Wyoming have very different functional composition with one site being dominated by C_4 warm-season grasses and the other by C_3 cool-season grasses. These different functional groups have very different growth strategies and water use efficiencies, which will impact how these grasslands respond to alterations in rainfall. Variation of estimated parameters in these grasslands provides insight into how mechanisms driving ecosystem responses to drought may differ along with plant functional types and how model projections should be scaled across ecosystems varying in plant functional types. Using the information about how model parameters vary across space due to environmental context and ecosystem attributes allows for model projections to be scaled up from single-sites to the regional level. Once scaling rules are established and incorporated into process-based models, hypotheses are then able to be tested at larger spatial scales.

Both the design of the EDGE study and integration of experiment and process-based models are extremely important for this study's success. We suggest that, to provide much needed predictive power under altered climate regimes at regional scales, future studies examining ecosystem sensitivity to environmental drivers should consider implementing consistent methodology across study sites at least when experimental sites exist within the same biome. Assimilating data across those regional sites would improve model performance upscaling from distributed experiments.

6.3 Challenges and Strategies to Promote Integrated Experiment–Model Approaches

Integrated experiment–model approaches will continue to serve a critical role for better understanding the natural world and more reliable predictions of future ecosystem states. Despite that experiment–model integration has the potential to significantly improve both model performance and experimental design, several challenges need to be addressed when conducting experiment–model integration. To move the experiment–model integration approach forward, some key actions need to be considered and stimulated.

6.3.1 Data Set Development

Incomplete use of empirical data in model parameterization is a major cause of model–model differences (Luo et al. 2016). More observational and experimental data are needed to constrain model parameters (Bonan 2008, Luo et al. 2015, 2016), potentially helping address equifinality, an issue that models yield similar outputs with different combinations of parameters (Williams et al. 2009). Therefore, it is necessary to evaluate whether model parameters can be well constrained by the available data. Different types of observations and experiments may be helpful to reduce equifinality. Including various types of observational and experimental data, instead of a single database, can lead to more accurate and efficient data assimilation.

An enormous amount of site-level ecological data has been generated over many decades, and ongoing projects and research networks will increasingly produce even more high-quality, high-resolution data. These data need to be synthesized, compiled, and often transformed before they can be used in model–data integration. To facilitate this process, it is vital that data are saved in a usable format, appropriately archived and well documented, and easily discoverable for use in synthesis and modeling.

Another challenge related to data set development is to fully address measurement uncertainty when documenting data set contents and protocols. It is critically important to identify data bias due to techniques and/or spatial-temporal distributions. Larger-scale data sets are essential to improve land surface models (Williams et al. 2009, Luo et al. 2016). A big challenge in compiling regional and global data sets is to harmonize data processing.

The schemes to process data should be documented clearly in the literature (Williams et al. 2009). Scaling site measurements to represent the whole grid is also problematic and the uncertainty related to scaling should be well specified. More well-designed experiments like EDGE could help to solve the scaling issue.

Moreover, new observations and experiments need to be designed to reveal mechanisms behind observed patterns and to discover some unresolved or missing processes. From the perspective of model improvements, future observational and experimental research should be focused on those model components whose predictability is poorly understood (Luo et al. 2015). These components may include ecosystem state transitions, changes in disturbance regimes, disturbance events and recovery trajectories, and ecosystem response functions.

6.3.2 Development of High-Fidelity Emulators and Traceability Analysis

Technical challenges of experiment–model integration impede the widespread use of models (LeBauer et al. 2013). Models are becoming more and more complex, such as land surface models in Earth system models. It is infeasible for the complex models to assimilate multiple data streams with traditional data assimilation techniques due to computational cost. High-fidelity emulators of complex models provide new ways to solve the problem. For a complex C model, an emulator is a simplified version of the model to describe the carbon flows and pools exactly as in simulations with the original model (Luo et al. 2016). Fox example, built upon the traceability framework proposed by Xia et al. (2013), Hararuk et al. (2014) developed an emulator of CLM-CASA, which enabled data assimilation with a global soil C data set. They first extracted the C cycle component of CLM-CASA into a set of C input–output equations that described C transfer among pools and then the set of equations was encoded into MATLAB® to perform data assimilation. The steady-state soil C produced by the MATLAB version matched closely to that simulated by the original CLM-CASA despite that the former used the semi-analytic spin-up approach developed by Xia et al. (2012).

As model complexity increases, traceability becomes more and more difficult. It is still a big challenge to explain differences among models in model intercomparisons (Friedlingstein et al. 2006, Jones et al. 2013, Luo et al. 2015, 2016). A clear traceability framework allows modelers to trace uncertainty back to each key component of a complex model (Xia et al. 2013, Ahlström et al. 2015) and thus lead to more practical strategies for model improvement.

6.3.3 Infrastructure Development

Both data processing and archiving require significant computing resources. The demand for computing resources will increase as more data become available for public access through repositories such as DataONE (https://www.dataone.org/). The computational costs of data assimilation are already high and will become more so as data availability increases. The requirements for computational power to develop new data assimilation techniques might even be higher. In addition, we need to develop online tools to assimilate real-time data streams including model outputs and to visualize data assimilation for forecasting as illustrated by SPRUCE, so more researchers can easily use the online tools to employ data assimilation even though they do not have much technical training. The online tools for forecasting may also help policy makers and land managers to make decisions. Both computational costs and development of online tools require improvements of existing cyber infrastructure or building of new cyber infrastructure.

On the other hand, new experiments that simultaneously manipulate multiple global change factors are critically needed to determine how interactions among factors shape ecosystem responses to global change, to explore synergisms among variables and resolve complex processes, or to study highly sensitive ecosystems such as tundra and tropical forests (Seddon et al. 2016). Multifactor experiments are complicated to design and expensive to implement building and maintaining the required cyber infrastructure. Therefore, development of cyber infrastructure will be a key determinant for the success of experiment–model integration.

6.3.4 Communications between Experimentalists and Modelers

Due to the strengths and limitations of both models and experiments, the advance of experiment–model integration depends on the effectiveness of communication between experimentalists and modelers. Opportunities should be created and funded to facilitate interactions among these research communities to design useful and appropriate sampling regimes (also see Luo et al. 2011b) to provide useful parameters for models and for modelers to improve their models. This will not only maximize the efficiency of monetary investment into research but will also allow for rapid advancement and improvement of projections concerning future states of ecosystem services on a dynamic planet.

6.4 Conclusions

We are living in a world that is undergoing rapid environmental changes. Ecological research is now in a data-rich era, with massive amounts of accumulated data distributed globally and large volumes of incoming data produced by many ongoing research projects and networks. Fortunately, advances in computational capacity continue to develop at an unprecedented rate. However, our ability to predict future states still remains limited, as indicated by large uncertainties in current model projections. Enhancing the accuracy of predictions of future ecosystem states under global change remains a challenge that must be resolved to better manage resources for sustainable development.

Ecological observations, experiments, and models remain the three foundational approaches for ecologists to tackling scientific problems. Field research, no matter how complex, only provides partial information about the system under study, whereas models, no matter how comprehensive, are always imperfect. To maximize gain from imperfect data and models, integrated experiment–model approaches are becoming a high priority. In the next a few decades, data–model integration to enhance observatory systems and to improve model projections will be critical to advance ecological understanding of dynamic systems. Closer collaborations between experimentalists and modelers will likely enhance data collection while new data assimilation techniques will improve model projections, resulting in better decisions for sustainable management of natural resources.

References

Ahlström, A., J. Xia, A. Arneth, Y. Luo, and B. Smith. 2015. Importance of vegetation dynamics for future terrestrial carbon cycling. *Environmental Research Letters* **10**:054019.

Arora, V. K., G. J. Boer, P. Friedlingstein, M. Eby, C. D. Jones, J. R. Christian, G. Bonan et al. 2013. Carbon–concentration and carbon–climate feedbacks in CMIP5 earth system models. *Journal of Climate* **26**:5289–5314.

Bonan, G. B. 2008. Forests and climate change: Forcings, feedbacks, and the climate benefits of forests. *Science* **320**:1444–1449.

Braswell, B. H., W. J. Sacks, E. Linder, and D. S. Schimel. 2005. Estimating diurnal to annual ecosystem parameters by synthesis of a carbon flux model with eddy covariance net ecosystem exchange observations. *Global Change Biology* **11**:335–355.

Carvalhais, N., M. Forkel, M. Khomik, J. Bellarby, M. Jung, M. Migliavacca, M. Mu, S. Saatchi, M. Santoro, and M. Thurner. 2014. Global covariation of carbon turnover times with climate in terrestrial ecosystems. *Nature* **514**:213–217.

Carvalhais, N., M. Reichstein, J. Seixas, G. J. Collatz, J. S. Pereira, P. Berbigier, A. Carrara et al. 2008. Implications of the carbon cycle steady state assumption for biogeochemical modeling performance and inverse parameter retrieval. *Global Biogeochemical Cycles* **22**:GB2007.

Clark, J. S., S. R. Carpenter, M. Barber, S. Collins, A. Dobson, J. A. Foley, D. M. Lodge et al. 2001. Ecological forecasts: An emerging imperative. *Science* **293**:657–660.

Clark, J. S., M. Lewis, J. S. McLachlan, and J. HilleRisLambers. 2003. Estimating population spread: What can we forecast and how well? *Ecology* **84**:1979–1988.

De Kauwe, M. G., B. E. Medlyn, S. Zaehle, A. P. Walker, M. C. Dietze, T. Hickler, A. K. Jain et al. 2013. Forest water use and water use efficiency at elevated CO_2: A model-data intercomparison at two contrasting temperate forest FACE sites. *Global Change Biology* **19**:1759–1779.

De Kauwe, M. G., B. E. Medlyn, S. Zaehle, A. P. Walker, M. C. Dietze, Y.-P. Wang, Y. Luo et al. 2014. Where does the carbon go? A model–data intercomparison of vegetation carbon allocation and turnover processes at two temperate forest free-air CO2 enrichment sites. *New Phytologist* **203**:883–899.

Dietze, M. C., D. S. Lebauer, and R. O. B. Kooper. 2013. On improving the communication between models and data. *Plant, Cell and Environment* **36**:1575–1585.

Fraser, L. H., H. A. Henry, C. N. Carlyle, S. R. White, C. Beierkuhnlein, J. F. Cahill Jr., B. B. Casper, E. Cleland, S. L. Collins, and J. S. Dukes. 2012. Coordinated distributed experiments: An emerging tool for testing global hypotheses in ecology and environmental science. *Frontiers in Ecology and the Environment* **11**:147–155.

Friedlingstein, P., P. Cox, R. Betts, L. Bopp, W. von Bloh, V. Brovkin, P. Cadule et al. 2006. Climate–carbon cycle feedback analysis: Results from the C4MIP model intercomparison. *Journal of Climate* **19**:3337–3353.

Frolking, S., N. Roulet, and D. Lawrence. 2013. Issues related to incorporating northern peatlands into global climate models. In A. J. Baird, L. R. Belyea, X. Comas, A. S. Reeve, and L. D. Slater, eds., *Carbon Cycling in Northern Peatlands*, pp. 19–35. American Geophysical Union, Washington, DC.

Gao, C., H. Wang, E. Weng, S. Lakshmivarahan, Y. Zhang, and Y. Luo. 2011. Assimilation of multiple data sets with the ensemble Kalman filter to improve forecasts of forest carbon dynamics. *Ecological Applications* **21**:1461–1473.

Hanson, P. J., K. W. Childs, S. D. Wullschleger, J. S. Riggs, W. K. Thomas, D. E. Todd, and J. M. Warren. 2011. A method for experimental heating of intact soil profiles for application to climate change experiments. *Global Change Biology* **17**:1083–1096.

Hararuk, O. and Y. Luo. 2014. Improvement of global litter turnover rate predictions using a Bayesian MCMC approach. *Ecosphere* **5**:1–13.

Hararuk, O., M. J. Smith, and Y. Luo. 2015. Microbial models with data-driven parameters predict stronger soil carbon responses to climate change. *Global Change Biology* **21**:2439–2453.

Hararuk, O., J. Xia, and Y. Luo. 2014. Evaluation and improvement of a global land model against soil carbon data using a Bayesian Markov chain Monte Carlo method. *Journal of Geophysical Research: Biogeosciences* **119**:403–417.

Hastings, W. K. 1970. Monte Carlo sampling methods using Markov chains and their applications. *Biometrika* **57**:97–109.

Huxman, T. E., M. D. Smith, P. A. Fay, A. K. Knapp, M. R. Shaw, M. E. Loik, S. D. Smith et al. 2004. Convergence across biomes to a common rain-use efficiency. *Nature* **429**:651–654.

IPCC. 2013. *Climate Change 2013: The Physical Science Basis. Contribution of Working Group I to the Fifth Assessment Report of the Intergovernmental Panel on Climate Change.* Cambridge University Press, Cambridge, U.K.

Jiang, L., Y. Yan, O. Hararuk, N. Mikle, J. Xia, Z. Shi, J. Tjiputra, T. Wu, and Y. Luo. 2015. Scale-dependent performance of CMIP5 earth system models in simulating terrestrial vegetation carbon. *Journal of Climate* **28**:5217–5232.

Jones, C., E. Robertson, V. Arora, P. Friedlingstein, E. Shevliakova, L. Bopp, V. Brovkin, T. Hajima, E. Kato, and M. Kawamiya. 2013. Twenty-first-century compatible CO_2 emissions and airborne fraction simulated by CMIP5 earth system models under four representative concentration pathways. *Journal of Climate* **26**:4398–4413.

Keenan, T. F., E. A. Davidson, J. W. Munger, and A. D. Richardson. 2012. Rate my data: Quantifying the value of ecological data for the development of models of the terrestrial carbon cycle. *Ecological Applications* **23**:273–286.

Koven, C., W. Riley, Z. Subin, J. Tang, M. Torn, W. Collins, G. Bonan, D. Lawrence, and S. Swenson. 2013. The effect of vertically resolved soil biogeochemistry and alternate soil C and N models on C dynamics of CLM4. *Biogeosciences* **10**:7109–7131.

Krassovski, M. B., J. S. Riggs, L. A. Hook, W. R. Nettles, P. J. Hanson, and T. A. Boden. 2015. A comprehensive data acquisition and management system for an ecosystem-scale peatland warming and elevated CO_2 experiment. *Geoscientific Instrumentation, Methods and Data Systems* **4**:203–213.

Kuppel, S., P. Peylin, F. Chevallier, C. Bacour, F. Maignan, and A. Richardson. 2012. Constraining a global ecosystem model with multi-site eddy-covariance data. *Biogeosciences* **9**:3757–3776.

LeBauer, D. S., D. Wang, K. T. Richter, C. C. Davidson, and M. C. Dietze. 2013. Facilitating feedbacks between field measurements and ecosystem models. *Ecological Monographs* **83**:133–154.

Li, D., C. Schädel, M. L. Haddix, E. A. Paul, R. Conant, J. Li, J. Zhou, and Y. Luo. 2013. Differential responses of soil organic carbon fractions to warming: Results from an analysis with data assimilation. *Soil Biology and Biochemistry* **67**:24–30.

Liang, J., D. Li, Z. Shi, J. M. Tiedje, J. Zhou, E. A. Schuur, K. T. Konstantinidis, and Y. Luo. 2015a. Methods for estimating temperature sensitivity of soil organic matter based on incubation data: A comparative evaluation. *Soil Biology and Biochemistry* **80**:127–135.

Liang, J., X. Qi, L. Souza, and Y. Luo. 2015b. Processes regulating progressive nitrogen limitation under elevated carbon dioxide: A meta-analysis. *Biogeosciences Discussions* **12**:16953–16977.

Limpens, J., F. Berendse, C. Blodau, J. G. Canadell, C. Freeman, J. Holden, N. Roulet, H. Rydin, and G. Schaepman-Strub. 2008. Peatlands and the carbon cycle: From local processes to global implications—A synthesis. *Biogeosciences* **5**:1475–1491.

Luo, Y., A. Ahlström, S. D. Allison, N. H. Batjes, V. Brovkin, N. Carvalhais, A. Chappell et al. 2016. Toward more realistic projections of soil carbon dynamics by Earth system models. *Global Biogeochemical Cycles***30**:40–56.

Luo, Y., T. F. Keenan, and M. Smith. 2015. Predictability of the terrestrial carbon cycle. *Global Change Biology* **21**:1737–1751.

Luo, Y., J. Melillo, S. Niu, C. Beier, J. S. Clark, A. T. Classen, E. Davidson et al. 2011a. Coordinated approaches to quantify long-term ecosystem dynamics in response to global change. *Global Change Biology* **17**:843–854.

Luo, Y., K. Ogle, C. Tucker, S. Fei, C. Gao, S. LaDeau, J. S. Clark, and D. S. Schimel. 2011b. Ecological forecasting and data assimilation in a data-rich era. *Ecological Applications* **21**:1429–1442.

Luo, Y., J. T. Randerson, G. Abramowitz, C. Bacour, E. Blyth, N. Carvalhais, P. Ciais, D. Dalmonech, J. B. Fisher, and R. Fisher. 2012. A framework for benchmarking land models. *Biogeosciences* **9**:3857–3874.

Luo, Y. and J. F. Reynolds. 1999. Validity of extrapolating field CO_2 experiments to predict carbon sequestration in natural ecosystems. *Ecology* **80**: 1568–1583.

Luo, Y., L. W. White, J. G. Canadell, E. H. DeLucia, D. S. Ellsworth, A. Finzi, J. Lichter, and W. H. Schlesinger. 2003. Sustainability of terrestrial carbon sequestration: A case study in Duke Forest with inversion approach. *Global Biogeochemical Cycles* **17**:1021.

Medlyn, B. E., S. Zaehle, M. G. De Kauwe, A. P. Walker, M. C. Dietze, P. J. Hanson, T. Hickler et al. 2015. Using ecosystem experiments to improve vegetation models. *Nature Climate Change* **5**:528–534.

Murphy, J. 2000. Predictions of climate change over Europe using statistical and dynamical downscaling techniques. *International Journal of Climatology* **20**:489–501.

Niu, S., Y. Luo, M. C. Dietze, T. F. Keenan, Z. Shi, and J. Li. 2014. The role of data assimilation in predictive ecology. *Ecosphere* **5**:1–16.

Norby, R. J., M. G. De Kauwe, T. F. Domingues, R. A. Duursma, D. S. Ellsworth, D. S. Goll, D. M. Lapola, K. A. Luus, A. R. MacKenzie, and B. E. Medlyn. 2016. Model–data synthesis for the next generation of forest free-air CO_2 enrichment (FACE) experiments. *New Phytologist* **209**:17–28.

Ogle, K. 2009. Hierarchical Bayesian statistics: Merging experimental and modeling approaches in ecology. *Ecological Applications* **19**:577–581.

Pacala, S. W., C. D. Canham, and J. Silander Jr. 1993. Forest models defined by field measurements: I. The design of a northeastern forest simulator. *Canadian Journal of Forest Research* **23**:1980–1988.

Parton, W. J., J. A. Morgan, G. Wang, and S. Del Grosso. 2007. Projected ecosystem impact of the Prairie Heating and CO_2 enrichment experiment. *New Phytologist* **174**:823–834.

Parton, W. J., D. S. Schimel, C. Cole, and D. Ojima. 1987. Analysis of factors controlling soil organic matter levels in Great Plains grasslands. *Soil Science Society of America Journal* **51**:1173–1179.

Parton, W., J. Scurlock, D. Ojima, T. Gilmanov, R. Scholes, D. S. Schimel, T. Kirchner, J. C. Menaut, T. Seastedt, and E. Garcia Moya. 1993. Observations and modeling of biomass and soil organic matter dynamics for the grassland biome worldwide. *Global Biogeochemical Cycles* **7**:785–809.

Peng, C., J. Guiot, H. Wu, H. Jiang, and Y. Luo. 2011. Integrating models with data in ecology and palaeoecology: Advances towards a model–data fusion approach. *Ecology Letters* **14**:522–536.

Raupach, M., P. Rayner, D. Barrett, R. DeFries, M. Heimann, D. Ojima, S. Quegan, and C. Schmullius. 2005. Model–data synthesis in terrestrial carbon observation: Methods, data requirements and data uncertainty specifications. *Global Change Biology* **11**:378–397.

Richardson, A. D., M. Williams, D. Y. Hollinger, D. J. Moore, D. B. Dail, E. A. Davidson, N. A. Scott, R. S. Evans, H. Hughes, and J. T. Lee. 2010. Estimating parameters of a forest ecosystem C model with measurements of stocks and fluxes as joint constraints. *Oecologia* **164**:25–40.

Schädel, C., Y. Luo, R. D. Evans, S. Fei, and S. M. Schaeffer. 2013. Separating soil CO_2 efflux into C-pool-specific decay rates via inverse analysis of soil incubation data. *Oecologia* **171**:721–732.

Schoof, J. T., S. C. Pryor, and J. Surprenant. 2010. Development of daily precipitation projections for the United States based on probabilistic downscaling. *Journal of Geophysical Research: Atmospheres* **115**:D13106.

Seddon, A. W. R., M. Macias-Fauria, P. R. Long, D. Benz, and K. J. Willis. 2016. Sensitivity of global terrestrial ecosystems to climate variability. *Nature* **531**:229–232.

Shi, X., P. E. Thornton, D. M. Ricciuto, P. J. Hanson, J. Mao, S. D. Sebestyen, N. A. Griffiths, and G. Bisht. 2015a. Representing northern peatland microtopography and hydrology within the Community Land Model. *Biogeosciences* **12**:6463–6477.

Shi, Z., X. Xu, O. Hararuk, L. Jiang, J. Xia, J. Liang, D. Li, and Y. Luo. 2015b. Experimental warming altered rates of carbon processes, allocation, and carbon storage in a tallgrass prairie. *Ecosphere* **6**:1–16.

Shi, Z., Y. Yang, X. Zhou, E. Weng, A. C. Finzi, and Y. Luo. 2016. Inverse analysis of coupled carbon–nitrogen cycles against multiple datasets at ambient and elevated CO_2. *Journal of Plant Ecology*. **9**:285–295.

Sitch, S., B. Smith, I. C. Prentice, A. Arneth, A. Bondeau, W. Cramer, J. Kaplan, S. Levis, W. Lucht, and M. T. Sykes. 2003. Evaluation of ecosystem dynamics, plant geography and terrestrial carbon cycling in the LPJ dynamic global vegetation model. *Global Change Biology* **9**:161–185.

Tian, H., C. Lu, J. Yang, K. Banger, D. N. Huntzinger, C. R. Schwalm, A. M. Michalak, R. Cook, P. Ciais, and D. Hayes. 2015. Global patterns and controls of soil organic carbon dynamics as simulated by multiple terrestrial biosphere models: Current status and future directions. *Global Biogeochemical Cycles* **29**:775–792.

Todd-Brown, K. E. O., J. T. Randerson, F. Hopkins, V. Arora, T. Hajima, C. Jones, E. Shevliakova, J. Tjiputra, E. Volodin, T. Wu, Q. Zhang, and S. D. Allison. 2014. Changes in soil organic carbon storage predicted by Earth system models during the 21st century. *Biogeosciences* **11**:2341–2356

Todd-Brown, K., J. Randerson, W. Post, F. Hoffman, C. Tarnocai, E. Schuur, and S. Allison. 2013. Causes of variation in soil carbon simulations from CMIP5 Earth system models and comparison with observations. *Biogeosciences* **10**:1717–1736.

Verma, S. B. 1990. Micrometeorological methods for measuring surface fluxes of mass and energy. *Remote sensing reviews* **5**:99–115.

Walker, A. P., P. J. Hanson, M. G. De Kauwe, B. E. Medlyn, S. Zaehle, S. Asao, M. Dietze, T. Hickler, C. Huntingford, and C. M. Iversen. 2014. Comprehensive ecosystem model-data synthesis using multiple data sets at two temperate forest free-air CO_2 enrichment experiments: Model performance at ambient CO_2 concentration. *Journal of Geophysical Research: Biogeosciences* **119**:937–964.

Wang, Y.-P., C. M. Trudinger, and I. G. Enting. 2009. A review of applications of model–data fusion to studies of terrestrial carbon fluxes at different scales. *Agricultural and Forest Meteorology* **149**:1829–1842.

Weng, E. and Y. Luo. 2011. Relative information contributions of model vs. data to short- and long-term forecasts of forest carbon dynamics. *Ecological Applications* **21**:1490–1505.

White, L. and Y. Luo. 2002. Estimation of carbon transfer coefficients using Duke Forest free-air CO_2 enrichment data. *Applied Mathematics and Computation* **130**:101–120.

White, L., F. White, Y. Luo, and T. Xu. 2006. Estimation of parameters in carbon sequestration models from net ecosystem exchange data. *Applied Mathematics and Computation* **181**:864–879.

Wilcox, K. R., J. C. Fischer, J. M. Muscha, M. K. Petersen, and A. K. Knapp. 2015. Contrasting above- and belowground sensitivity of three Great Plains grasslands to altered rainfall regimes. *Global Change Biology* **21**:335–344.

Williams, M., A. Richardson, M. Reichstein, P. Stoy, P. Peylin, H. Verbeeck, N. Carvalhais, M. Jung, D. Hollinger, and J. Kattge. 2009. Improving land surface models with FLUXNET data. *Biogeosciences* **6**:1341–1359.

Williams, M., P. A. Schwarz, B. E. Law, J. Irvine, and M. R. Kurpius. 2005. An improved analysis of forest carbon dynamics using data assimilation. *Global Change Biology* **11**:89–105.

Wu, X., Y. Luo, E. Weng, L. White, Y. Ma, and X. Zhou. 2009. Conditional inversion to estimate parameters from eddy-flux observations. *Journal of Plant Ecology* **2**:55–68.

Wutzler, T. and M. Reichstein. 2007. Soils apart from equilibrium? consequences for soil carbon balance modelling. *Biogeosciences* **4**:125–136.

Xia, J., Y. Luo, Y.-P. Wang, and O. Hararuk. 2013. Traceable components of terrestrial carbon storage capacity in biogeochemical models. *Global Change Biology* **19**:2104–2116.

Xia, J., Y. Luo, Y.-P. Wang, E. Weng, and O. Hararuk. 2012. A semi-analytical solution to accelerate spin-up of a coupled carbon and nitrogen land model to steady state. *Geoscientific Model Development* **5**:1259–1271.

Xu, X., Z. Shi, D. Li, A. Rey, H. Ruan, J. M. Craine, J. Liang, J. Zhou, and Y. Luo. 2016. Soil properties control decomposition of soil organic carbon: Results from data-assimilation analysis. *Geoderma* **262**:235–242.

Xu, T., L. White, D. Hui, and Y. Luo. 2006. Probabilistic inversion of a terrestrial ecosystem model: Analysis of uncertainty in parameter estimation and model prediction. *Global Biogeochemical Cycles* **20**:1–15. GB2007.

Zaehle, S., B. E. Medlyn, M. G. De Kauwe, A. P. Walker, M. C. Dietze, T. Hickler, Y. Luo et al. 2014. Evaluation of 11 terrestrial carbon–nitrogen cycle models against observations from two temperate free-air CO_2 enrichment studies. *New Phytologist* **202**:803–822.

Zhang, L., Y. Luo, G. Yu, and L. Zhang. 2010. Estimated carbon residence times in three forest ecosystems of eastern China: Applications of probabilistic inversion. *Journal of Geophysical Research: Biogeosciences* **115**:G01010.

Zhou, T. and Y. Luo. 2008. Spatial patterns of ecosystem carbon residence time and NPP-driven carbon uptake in the conterminous United States. *Global Biogeochemical Cycles* **22**:1–15. GB3032

Zhou, X., Y. Luo, C. Gao, P. S. J. Verburg, J. A. Arnone, A. Darrouzet-Nardi, and D. S. Schimel. 2010. Concurrent and lagged impacts of an anomalously warm year on autotrophic and heterotrophic components of soil respiration: A deconvolution analysis. *New Phytologist* **187**:184–198.

Zhou, T., P. Shi, D. Hui, and Y. Luo. 2009. Global pattern of temperature sensitivity of soil heterotrophic respiration (Q10) and its implications for carbon-climate feedback. *Journal of Geophysical Research: Biogeosciences* **114**:1–9. G02016.

Zhou, T., P. Shi, G. Jia, Y. Dai, X. Zhao, W. Shangguan, L. Du, H. Wu, and Y. Luo. 2015. Age-dependent forest carbon sink: Estimation via inverse modeling. *Journal of Geophysical Research: Biogeosciences* **120**:2473–2492.

Zhou, T., P. Shi, G. Jia, and Y. Luo. 2013. Nonsteady state carbon sequestration in forest ecosystems of China estimated by data assimilation. *Journal of Geophysical Research: Biogeosciences* **118**:1369–1384.

Zhou, X., T. Zhou, and Y. Luo. 2012. Uncertainties in carbon residence time and NPP-driven carbon uptake in terrestrial ecosystems of the conterminous USA: A Bayesian approach. *Tellus B* **64**:17223.

Zhu, Q. and Q. Zhuang. 2015. Ecosystem biogeochemistry model parameterization: Do more flux data result in a better model in predicting carbon flux? *Ecosphere* **6**:art283.

Section III

New Tools to Meet New Challenges

Emerging Technologies for Exploring Unknown Ecosystem Processes

7

Large-Scale Sequence-Based Information: Novel Understanding of Ecology and Novel Avenues to Test Ecological Hypotheses

Achim Quaiser, Alexis Dufresne, Sophie Michon-Coudouel, Marine Biget, and Philippe Vandenkoornhuyse

CONTENTS

Abstract

Ecology entered a new era with growing evidences of the huge impact of microorganisms on all existing ecosystems. Molecular approaches, such as the SSU rRNA gene analysis and environmental genomic approaches, identified an unexpected microbial taxonomic and functional diversity in all kinds of habitats. This diversity and the low cultivability suggest functional interactions within microbial communities as well as between microbial communities and plants or animals in dependence of the environmental conditions. These findings shape the groundwork of new revolutionary ecological concepts away from isolated characteristics of organisms to broad *holistic**/integrated views. Aiming to capture and to understand the functioning of interacting microbial communities, they must be considered as a whole, in constant evolution and in adaptation to environmental factors. Since there is no all-inclusive approach combining all targeted aspects of analysis, results obtained from particular molecular ecology approaches are still rather tesserae. Nevertheless, combining different approaches fills the mosaic leading to more comprehensive reproductions of in situ community functions.

In this chapter, we highlight the findings and limitations of environmental genomics that paved the way for new *holistic** concepts of *community assemblages** as well as perspectives. Combined with insights into molecular techniques, the reader should be well fitted to apply and develop adequate research strategies that imply environmental genomic approaches combined with high-throughput sequencing.

Keywords: Metagenomics, Metatranscriptomics, Metabarcoding, Taxonomic and functional diversity, Culture-independent ecology

7.1 Use of Sequence Data in Ecology: An Overview

The major aims of ecology are the study and understanding of the processes that foster biodiversity and that biodiversity contributes toward the understanding of ecosystem functioning. Biodiversity is defined as the species assemblage within a community, the genetic variability within the species, and the diversity of ecosystems formed by these species assemblages (Wilson 1989). Thus, the species level does not intrinsically summarize the biodiversity but corresponds to one key level. However, defining the species level is not straightforward. It depends on the organism observed and the definition of species that is adopted. A good *species concept** allows to group *evolutionary-related organisms**. This systematic view of life based on evolutionary relationships among organisms allows to formally describe diversity that can lead to modifications of taxonomy (e.g., the delineation a third domain of life, the Archaea, besides Bacteria and Eukarya). This systematic view is possible from the analysis of nucleic acid sequences, DNA and RNA, because these vertically inherited sequences (i.e., parents to offspring transmission) have recorded traces of evolution (i.e., mutation, insertion, deletion). Besides

* Definition of terms in *italic* with the sign "*": are found in the glossary.

the taxonomic criteria, which could be sometimes subjective, this *diachronic information**, contained in DNA and RNA, is a fundamental material to detect and classify the organisms and microorganisms, known or unknown. The use of DNA-/RNA-based methods has revolutionized our understanding of the extent of the diversity especially considering microorganisms.

Microorganisms are unicellular, and due to their small size, they are difficult to observe and describe. Microorganisms belong to the three domains of life: Eukarya, Bacteria, and Archaea. For a long time, microbiology was more regarded as a discipline of medical interest. Often, microbiologists analyzed one single cultured species from different characteristics such as phenotype, physiology, metabolism, and genetics. This culture-based approach is disconnected from the natural world and selects only a tiny fraction of the microorganisms. The development of molecular ecology approaches has changed the perception of the diversity of organisms and revealed that in the microbial world the cultivated and known microorganisms represent only a drop in an ocean of diversity. For example, 1 kg of soil contains about 10^{13} microorganisms in average (Whitman et al. 1998), thus, about a number 100 times higher than the number of stars in the Milky Way (Curtis and Sloan 2005), while waters (oceans + continental water) contain in total about $1.2 \cdot 10^{29}$ prokaryote cells (Whitman et al. 1998) and the human gut contains 100 times more unique genes than the human genome (Qin et al. 2010). This huge diversity of microorganisms plays a key role in ecosystem functioning and services (e.g., Vandenkoornhuyse et al. 2010).

Considering the historic background, one milestone was the development of culture-independent technologies, notably the SSU rRNA gene approaches allowing the detection and classification of uncultured microorganisms (Pace et al. 1986; Figure 7.1). The SSU rRNA gene is amplified from extracted DNA from environmental samples, cloned in an adequate *vector**, amplified in *Escherichia coli* cells, sequenced and classified by phylogenetic analysis (Figure 7.2). With this approach, the quantities of available SSU rRNA sequences, deposited in dedicated databases, were growing. The quality of the phylogenetic signal contained within the SSU rRNA gene has allowed its use as *molecular clock**, starting with the pioneering work of Woese (1987). Cloning, sequencing, and analysis of bulk genomic DNA without targeted amplification was applied shortly thereafter (Schmidt et al. 1991; Figure 7.1). Subsequently, the analysis of the first cloned and sequenced large genomic fragments (>35 kb) from marine plankton microbial assemblages further allowed to link taxonomic classification and functional potential together (Stein et al. 1996). Cultivation-independent surveys revealed that in natural ecosystems at most about 0.1% of the microbial species could be cultivated. Today, the SSU rRNA sequences generated by cultivation-independent molecular approaches combined with new sequencing technologies lead to the exponential growth of dedicated databases reaching more than five billions of SSU rRNA sequences in 2015 (e.g., www.arb-silva.de).

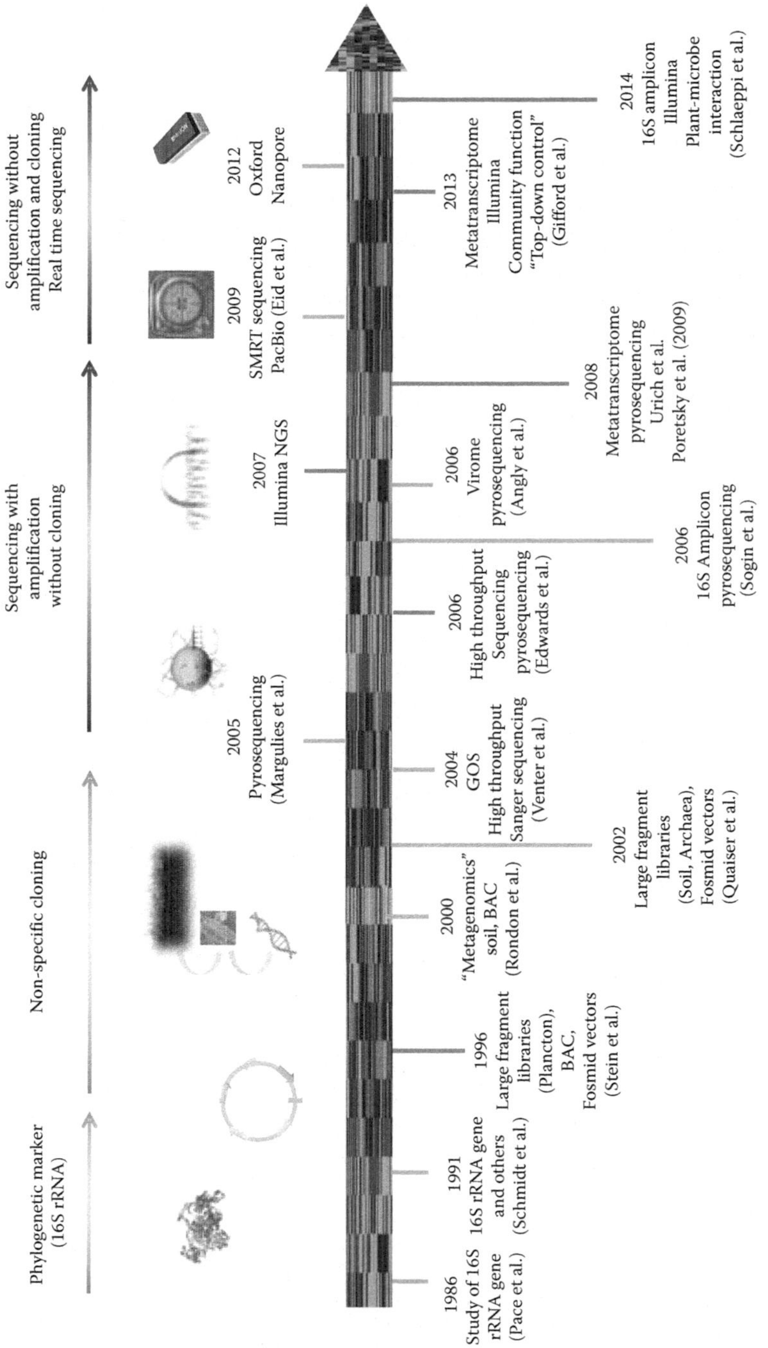

FIGURE 7.1
Historical overview of major advances in molecular ecology.

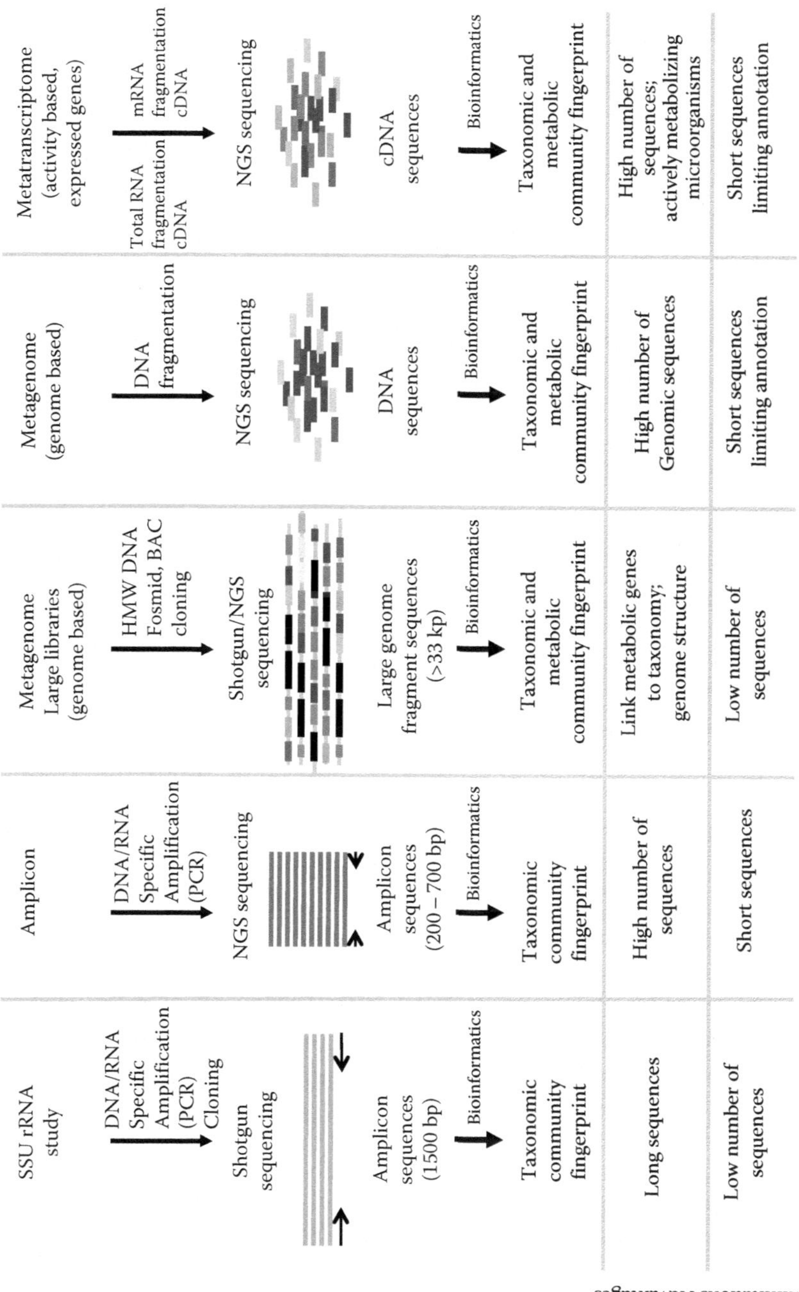

FIGURE 7.2
Overview of the different nucleic acid approaches.

Molecular ecology has entered the genomic era (e.g., Vandenkoornhuyse et al. 2010). New understanding of evolution from genomic comparisons is an important field of research supported by the increasing number of available genomes (i.e., in October 2015, 42,323 Bacteria, 997 Archaea, and 6,572 Eukarya genomes sequenced, https://gold.jgi.doe.gov). Environmental genomics currently drives important discoveries in biology and a cortege of new ideas, concepts, and theories (Vandenkoornhuyse et al. 2010). Developments of genomic and metagenomic approaches to microorganisms and microbial communities have demonstrated a high number of *horizontal gene transfers** even among evolutionarily distant microbial species, allowing the acquisition of new physiological functions (Nelson et al. 1999; Quaiser et al. 2003; Gladyshev et al. 2008; Keeling and Palmer 2008; Deschamps et al. 2014). In addition, viruses as mobile genetic elements involved in HGT were identified as the most abundant biological entity and are equipped with fast evolving genetic diversity (Fuhrman 1999; Stern and Sorek 2011). These attributes combined with the low cultivability of prokaryotes inevitably lead to new views of microbial *community assemblages** and evolutionary trajectories (Woese 2004; Goldenfeld and Woese 2007).

Current new frontiers in ecology are related to the recent knowledge and understanding of microorganisms in the environment. We underline urgent research questions considering the understanding of the complexity of the microbial communities and associated ecological functions. It is essential to understand the rules of microbial *community assembly** and the associated theories to describe these rules since the selection and genetic drift impact individual microorganisms, population and community, and expressed function.

The focus of this chapter is thus to provide an overview of ideas and frontiers together with past, current, and upcoming strategies related to environmental genomics.

7.2 Mass Sequencing: Current and Upcoming Technologies

The first sequencing technology was developed by Sanger (Sanger and Coulson 1975) and in parallel by Maxam and Gilbert using a more complex technology (Maxam and Gilbert 1977). Starting with fluorescence-based labeling of the four nucleotides and manual analysis of radiographies, the first automated sequencer was introduced in 1987 using Sanger technology. This technology was scaled up for the sequencing of the first human genome (Collins et al. 2003) costing ~US$2.7 billion. This was followed by second-generation sequencing using the sequencing-by-synthesis technology, thereby avoiding the cloning steps (Schuster 2008). Pyrosequencing improved the throughput over Sanger sequencing approaches (Margulies et al. 2005) with >99% accuracy and much lower cost per base pair. Complete microbial genomes could be sequenced and assembled in several hours with sequence *coverage** allowing the assembly of the complete genomes. Due to the continuous development and improvement of pyrosequencing technology, the sequence length reached up to 700 base pairs in average and up to 1.5 million sequences per run (i.e., 454 GS flx Plus instrument, Roche, Switzerland). The current dominant technology of mass sequencing can produce in a single run 25 million reads of up to 300 bp per read (MiSeq instrument, Illumina Inc., United States). With HiSeq X series instrument (Illumina Inc., United States), even more spectacular numbers are obtained, reaching up to six billion sequences per run and 2 × 150 base pairs per read. Besides these methods, sequencing where nucleotides are read in real time without the bias-prone amplification step during sequencing is currently being developed (i.e., third-generation sequencing [TGS]) aiming to perform single-molecule sequencing (see Section 7.6.1). New technologies are currently developed. For example, the newest TGS technology targets the abilities to increase read length simplifying subsequent sequence data analyses. Since significantly lower number of reads and lower *coverage** is necessary, this goes along with reduced analytical computation and storage costs (see section in the following text for more detail). Among these methods, three technologies can be underlined: (1) single-molecule synthesis by DNA polymerase with real-time recording (Sequel instrument by Pacific Bioscience, United States); (2) nanopore technology, in which the nucleotides are identified by the modification of electric potential during the transfer of a single DNA molecule through a nanopore (minion by Oxford Nanopore Technologies, United Kingdom); and (3) direct imaging of labeled DNA molecules in real time (ongoing developments by ZS Genetics Inc., United States). These TGS developments have all the same focus, long sequences, and high accuracy at low cost.

The current dominant sequencing technologies allow only producing short sequences. These short stretches of data form a giant puzzle to reassemble. The reconstruction of complete genomes from these sequences necessitates strong computational and laboratory efforts. Major difficulties to assemble a full genome from short reads are related to occurring sequence repeats and natural genetic variations (*haplotypic variations**). The generation of long sequences, as targeted by TGS technologies, allows reducing these difficulties and thereby reducing the number of sequences needed. In other words, if the sequencing length would not be limited, only one read would be necessary to complete one *haploid** genome or chromosome.

7.3 Diversity, Molecular Barcoding, and New Prospects

Prior to addressing communities as a whole, the first key issue is to properly describe the organismal inventory (diversity) (Figure 7.3). For detailed

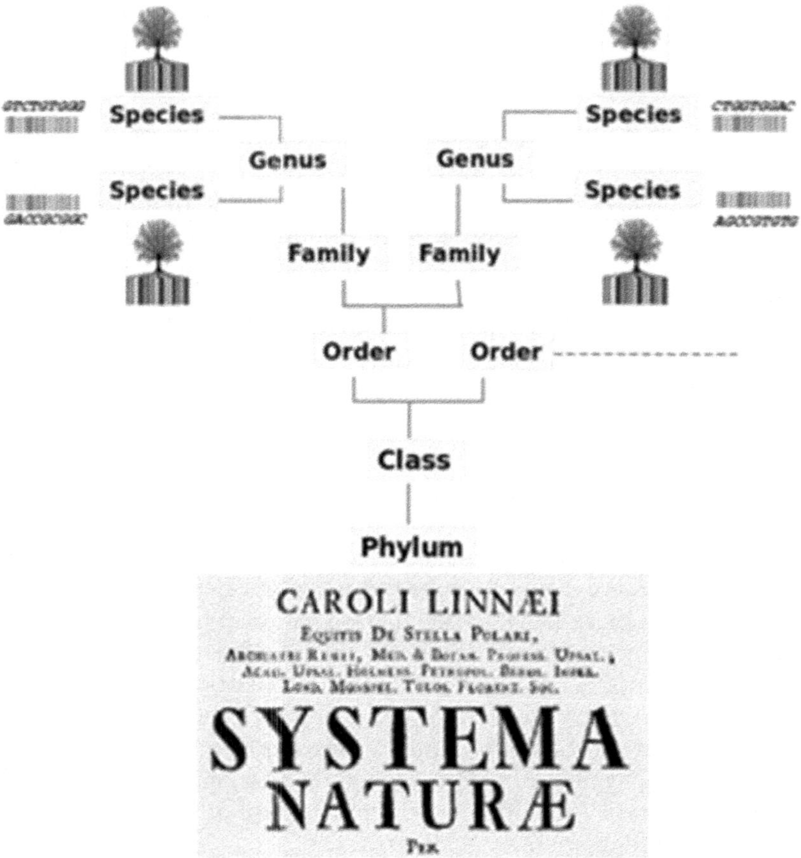

FIGURE 7.3
The hierarchy of biological classification by Carl Linneaus, and related data, metadata, and environmental factors to consider in a holistic perception of "systems ecology." The bar coding of life is based upon this taxonomic diversity.

distinction of the organisms and to describe evolutionary relationships, molecular identification of organisms has been developed by so-called molecular barcoding approaches. This important focus of current research is developed in the following text.

7.3.1 Molecular Barcoding of Plants and Animals

The DNA barcoding aims at identifying any animal or plant (i.e., multicellular organisms) using a short genetic sequence (i.e., the molecular barcode). Thus, molecular barcoding implicitly accept the existing Linnaean taxonomy (Figure 7.3). Molecular barcoding was developed with the broad rationale of (1) identification of all species, (2) rapid identification of species, (3) linkage of the species to environmental conditions, (4) discovery of new species, and (5) increasing the use and utility of biological collections. For molecular comparison and distinction between organisms, adequate gene or sequence targets must be determined. To be considered as adequate, these DNA fragments should contain conserved regions and enough genetic variations to distinguish species. A consensual DNA barcoding region for animals, the mitochondrial Cytochrome Oxidase 1 (CO1) gene, seems to fit well with the aims of molecular barcoding. Conversely, the selection of a good DNA barcode was more challenging for plants. In this quest for a universal plant DNA barcode, two short coding regions of *plastid DNA**, *rbcLa* and *matK*, are used most often, while other combinations of molecular barcode, such as *trnH-psbA*, are successful in some cases (Ghahramanzadeh et al. 2013). A key difficulty is that there are relatively few nucleotides that are detectable that can determine robust differences among similar or related species (Chase and Fay 2009). It is known that the gene fragments of CO1, *rbc*L and *mat*K, contain an important amount of homoplasy (i.e., inherited homology due to the high rate of mutation within these genomic regions); thus, the quality of the phylogenetic information is moderate. Despite this fact, the successful development of the barcoding of life must be underlined (iBol and related BOLD database URL http://www.ibol.org/). The use of molecular barcoding has brought new depth to our understanding of traditional plant and animal systematic taxonomy and suggests insights that we otherwise would not have obtained. Considering the well-known example of two existing elephant species (i.e., African and Indian), DNA barcoding has allowed to reveal the likely coexistence of three African species (Blaxter 2003). In another example, invasive alien plant species threaten the structure and function of aquatic ecosystems. Barcoding tools have been able to distinguish these invasive plants, which is of particular importance when managing the trade of aquatic plants (Ghahramanzadeh et al. 2013). The DNA barcoding tools are also used to manage the conservation of animal species, such as the large and illegal influx of snapper turtle (*Macroclemys temminckii*) in the food chain (Roman and Bowen 2000). As demonstrated by these examples, robust barcoding tools are continually being developed to further diagnose species in the environment from DNA traces. From an applied aspect, this knowledge and associated tools are developing to better manage illegal use, trade, and will likely contribute to the conservation and preservation of species.

7.3.2 Microorganisms and Molecular "Barcoding"

For microorganisms, detection and taxonomy-based community descriptions within an environmental sample mostly rely on the analyses of the SSU rRNA gene fragment from an *amplicon** mass sequencing approach applied on all kind of ecosystems (e.g., Sogin et al. 2006; Lundberg et al. 2012; Ben Maamar et al. 2015; de Vargas et al. 2015; Nunes et al. 2015) (Figure 7.2). Other options are the recruitment of SSU rRNA sequences from metatranscriptomes bypassing the bias-prone amplification step (e.g., Urich et al. 2008; Quaiser et al. 2014). From these SSU rRNA sequence analyses, a number of unknown phyla in Bacteria, Archaea, and Eukarya were revealed, and a large proportion of them are still without cultivated representatives (e.g., López-García and Moreira 2008). These SSU rRNA approaches can be assimilated to barcoding or metabarcoding methods, since the sequence analysis tends to the detection and identification of the microorganisms. It must be underlined that one key difference in comparison to plant or animal barcoding approaches is that the analyses of the SSU RNA gene sequences of microorganisms are not necessarily driven by the known taxonomy allowing the detection of undetermined lineages. The use of mass sequencing strategies to study SSU rRNA gene fragments from environmental samples has allowed major advances in the knowledge and understanding of the gut microbiota (e.g., Turnbaugh et al. 2009), the soil microbiota (e.g., Fierer et al. 2012), the plant microbiota (e.g., Bulgarelli et al. 2012; Lundberg et al. 2012), and the pico-eukaryotes in oceans (e.g., Rodríguez-Martínez et al. 2013; de Vargas et al. 2015).

Since the SSU rRNA gene sequence is highly conserved, albeit much less than eukaryotic barcoding targets, one of the limits of SSU rRNA *amplicon** approaches is the resolution level especially when using short sequence fragments. On the other hand, in contrast to PCR-independent approaches such as metagenomic and metatranscriptomic on bulk nucleic acids, *amplicon** approaches necessitate previous knowledge about the targets inevitably discriminating taxonomic lineages.

7.4 Meta-Omics to Assess the Different Dimensions of Diversity

Here, we define meta-omics as the culture-independent analyses beyond the SSU rRNA gene–based analyses of microbial assemblages. These meta-omics approaches are of different types. For a given environmental sample, the metagenomics use the extracted DNAs, while the metatranscriptomics focus on the extracted RNAs and multi-omics approaches aim to combine metagenomics, metatranscriptomics, and metaproteomics. In this section, we describe these different strategies and highlight the advantages and disadvantages.

7.4.1 Metagenomics

Environmental genomics (metagenomics) developed from the adoption of genomic techniques applied to complex microbial communities from natural ecosystems. Typically, microbial sequencing studies targeted complete genomes, while in environmental genomics, due to the large diversity of microbial communities, this was not feasible. Large genomic fragments, above 32 kb, from marine (Stein et al. 1996; Béjà et al. 2000, 2001) and from soil (Rondon et al. 2000; Quaiser et al. 2002, 2003; Liles et al. 2003; Treusch et al. 2004) microbial communities were cloned, sequenced, and analyzed (Figure 7.2). Rondon et al. (2000) fixed the term "metagenome" for environmental genomic studies. These studies allowed getting first insights into the genomes from uncultured microorganisms. The large genomic fragments revealed the gene content and genome structure as well as the potential function mostly linked to the taxonomic origin contributing significantly to the understanding of microorganisms in the environment. For example, Beja et al. (2000) identified a novel type of bacterial *rhodopsin**, named *proteorhodopsin**, which has been shown to be a very abundant marine plankton (Béjà et al. 2001), and highlights an important previously unknown phototrophic ocean process. This functional identification has helped subsequent isolation and genome sequencing of phototrophic microorganisms from different bacterial phyla (Fuchs et al. 2007; Gómez-Consarnau et al. 2007). Analogical examples of fundamental functions discovered from a metagenomic approach also exist for soils. For instance, the identification of genes encoding for the key enzyme for *nitrification**, the ammonium monooxygenase on an archaeal genome (Treusch et al. 2005; Leininger et al. 2006) suggested key roles of *mesophilic** archaea in nitrogen cycle in soils. Combined with the detection of ammonium monooxygenases in marine plankton samples (Venter et al. 2004) and large genome fragment analyses of *Cenarchaeum symbiosum* (Hallam et al. 2006), mesophilic Archaea (Thaumarchaeota) are to date recognized as keystone players of the *nitrification** within the nitrogen cycle previously thought to be performed exclusively by a restricted number of Bacteria. These two prime examples revealed the strength of environmental genomics. Isolation or enrichment cultures can demonstrate direct information about phenotype, metabolism, physiology, and genomics, while culture-independent approaches can show the characteristics within the environment, such as distribution, diversity, dynamics, and interactions. These "classical" metagenomic approaches permitted also to get first insights into the genomes from species of other major phyla, such as Acidobacteria (Liles et al. 2003; Quaiser et al. 2003, 2008; Kielak et al. 2010), Poribacteria (Fieseler et al. 2006), and marine Thaumarchaeota (Martín-Cuadrado et al. 2008). Interestingly, evolutionary information provided findings about *horizontal gene transfers** among different groups, for example, between Alphaproteobacteria and Acidobacteria (Quaiser et al. 2003), and most impressively massive *horizontal gene transfers**, up to 29.7%, from Bacteria to marine Thaumarchaeota and Euryarchaeota (Deschamps et al. 2014).

The environmental DNA sequencing or metagenomic shotgun sequencing is a powerful method to analyze the microbial diversity. Shotgun *fosmid** end sequencing was applied to reveal stratified microbial assemblages in the ocean (DeLong et al. 2006) or to explore the diversity and metabolic potential of soil communities (Treusch et al. 2004). *Fosmid** sequencing approaches (Sanger sequencing) provided, at this time, an affordable method to produce relatively high number (5,000–10,000 sequences in these examples) of environmental sequences representing a DNA fingerprint of the microbial community.

To the contrary of large genome fragment analysis, one of the major challenges applying high-throughput sequencing, which goes along with short sequences, is to link the diversity or taxonomic origin with function. The first study applying mass sequencing (high-throughput Sanger sequencing) was the Global Ocean Sequencing Project (Venter et al. 2004; Rusch et al. 2007; Yooseph et al. 2007) (Figure 7.1). This sequencing effort (about 700 Mbp), at the time, doubled the number of existing nucleotide and protein databases (NCBI), allowing the study of the genetic potential of surface ocean water. Several follow-up studies revealed interesting results about the ribotype diversity (Rusch et al. 2007), protein families (Yooseph et al. 2007), prokaryotic genomes (Biers et al. 2009), and others. Analogously, the comparison of whale fall (3 × 25 Mbp) and soil microbial communities (100 Mbp) (Tringe et al. 2005), applying cloning and short fragment Sanger sequencing, revealed habitat-specific fingerprints and, due to the complexity of the microbial communities, evidence for the impossibility to assemble individual genomes. Only about 1% of the reads had overlaps with other reads.

In contrast, the application of the same approach combined with *fluorescence in situ hybridization** was applied on a low-complexity microbial community from an acid mine drainage biofilm (Tyson et al. 2004). Interestingly, with about the same sequencing effort (76 Mbp), several near-complete draft genomes could be reconstituted, and based on the gene content, a metabolic map and pathways were proposed. A recent review details draft or complete genome reconstructions from metagenomic analyses (Gasc et al. 2015).

Higher number of sequences at lower cost can be obtained applying next-generation mass sequencing on bulk DNA. In contrast to gene-targeted *amplicon** approaches described earlier (i.e., Section 7.3), no prior knowledge about the target is necessary and potential *primer** biases are avoided. Otherwise, the bioinformatics analysis is more complex targeting in general numerous genes or gene families, while homologs of the majority of the sequences are missing. Several studies yielded important information about the metabolic potential of microbial communities representing unique fingerprints useful to highlight particular differences (e.g., Ghai et al. 2010; Mackelprang et al. 2011; Quaiser et al. 2011; Shi et al. 2011). For example, recent metagenomic analyses addressing the human microbiota have revealed the role of the microbiome in health and disease (Korem et al. 2015). Assuming that patterns of sequencing read *coverage** in metagenomic samples reflects the microbial growth rate within the microbiota, Korem et al. (2015) used the ratio of copy number to show differences between virulent and avirulent strain dynamics and correlations to bowel disease. Gut microbiome datasets generated by next-generation sequencing are growing, allowing a comprehensive view of the functional diversity in humans, animals, and invertebrates (Yoon et al. 2015). These metagenomic analyses helped to get insights into the *pleiotropic** roles of the gut microbiome in the nutritional processing, the microbial metabolisms and interactions, the modulation of the host physiology, and the impact of the microbiota on the host defense system (e.g., Doré and Blottière 2015; Yoon et al. 2015). In addition, the likely mutualistic functions of the microbiota emerged and opened new perspectives for the understanding of metazoa along with plants as meta-organisms (Vandenkoornhuyse et al. 2015). Despite this growing understanding and knowledge, and considering the current high research effort to analyze the gut microbiota, an obvious limit of the metagenomic approach is that a large proportion of *open reading frames** are cataloged as uncharacterized and/or novel gene with unknown function (e.g., Qin et al. 2010; Yoon et al. 2015). Thus, the use of the metagenomic approaches has a lot of advantages revealing important discoveries that lead to modification of paradigms (i.e., discovery of archaeal *nitrification**). Nevertheless, it is clear that the more the metagenome is complex, the more the metagenomic analyses are descriptive. In this case, large parts of the metagenome, due to the huge microbial diversity, are not interpretable because a high number of genes and proteins are still unknown and uncharacterized, and in consequence no homologs are deposited in international databases. Altogether, all the genomes sequenced and all the sequences yet produced represent an estimation of about only $1 \times 10^{-19}\%$ of the total DNA sequenced, the ongoing "earth microbiome project" being included in this estimation (Microbiology by Numbers 2011).

7.4.2 Metatranscriptomics

Metatranscriptomics targets the expressed genes in a microbial community at the time of sampling. Compared to metagenomics, this approach targets only active modules from intact cells of the microbial community. Since the half-life of mRNA transcripts in general is short (i.e., seconds to minutes, in general), there is no doubt that these sequences are derived from actively metabolizing cells. Otherwise, gene expression adapts rapidly to changing environmental conditions, requiring rapid and elaborated sampling strategies, in order to fix cells as fast as possible assuring the cells' in situ conditions (Feike et al. 2011). In environmental metatranscriptomic studies, the ribosomal RNAs represent about 80%–97% of the total transcripts, while messenger RNAs count only for up to 20% depending on the ecosystem analyzed (He et al. 2010; Quaiser et al. 2014; Tveit et al. 2014). Therefore, new approaches for mRNA enrichment, through mRNA amplification and by rRNA depletion, were developed (Carvalhais et al. 2012). Other studies favor the extraction and analysis of total RNA, thereby preserving the possibility to exploit the full potential of the metatranscriptome sequences that include mRNA and rRNA from all three kingdoms (e.g., Urich et al. 2008, 2014; Radax et al. 2012; Quaiser et al. 2014). The first metatranscriptome study analyzed freshwater bacterioplankton communities by mRNA enrichment, cloning of reverse-transcribed RNA, Sanger sequencing, and linking the transcripts to environmental processes (Poretsky et al. 2005). Even though the *sequencing depth** was limited, the first signature transcripts of an environmental community were obtained. In another study, pyrosequencing of metatranscriptome and metagenome from ocean surface water community revealed gene- and taxon-specific expression patterns and surprisingly a high number of unique sequences in the metatranscriptome absent in the metagenome were identified (Frias-Lopez et al. 2008).

The full potential of the metatranscriptomic approach was demonstrated on different ecosystems such as marine sponges (Radax et al. 2012), marine sediments (Urich et al. 2014), peat soils (Tveit et al. 2012, 2014), and iron-rich microbial mats (Quaiser et al. 2014). The stratification of an iron-rich microbial mat could be shown based on taxonomic relevant rRNA as well as mRNA analysis, highlighting the strong expression level of iron oxidizers affiliated to *Leptothrix* species particularly near the surface and surprisingly type I *methanotrophs** in deeper layers (Quaiser et al. 2014). The presence of *methanotrophs** coincided with about 10–100 times higher methane concentrations within the microbial mat compared to surrounding waters. The *methanotrophs'* key enzyme coincided with the pattern of rRNA sequences of *methanotrophs,** confirming the stratification of the taxonomic analysis and linking diversity to function. Comparison of surface marine plankton metatranscriptomes over the seasons using Illumina sequencing suggested strong top-down ecological control for some faster-growing microbial groups (Gifford et al. 2013).

For the analysis of community diversity and functioning, several observations from metatranscriptome studies can be highlighted, especially as follows: (1) mRNA pools provide precious information about ongoing microbial community processes but only for the time of sampling, (2) replication and comparative metatranscriptome analysis are essential for emerging reliable results, (3) metatranscriptome analysis represents a highly sensitive tool to display the instantaneous metabolic state of the microbial community, (4) metatranscriptomes based on total RNA extraction reduces methodological biases but can limit the functional information included in mRNA, and (5) total RNA metatranscriptomics has a strong potential to link diversity to function including representatives from the three domains. However, the same limit as for metagenomic analyses also exists for metatranscriptomic approaches related to the high number of detected genes without homologs and with unknown function in databases.

The major advantage of metagenomic and metatranscriptomic analyses is the possibility to address the different levels of the ecological hierarchy at the same time (i.e., individual, population, community, ecosystem; Figure 7.4) as

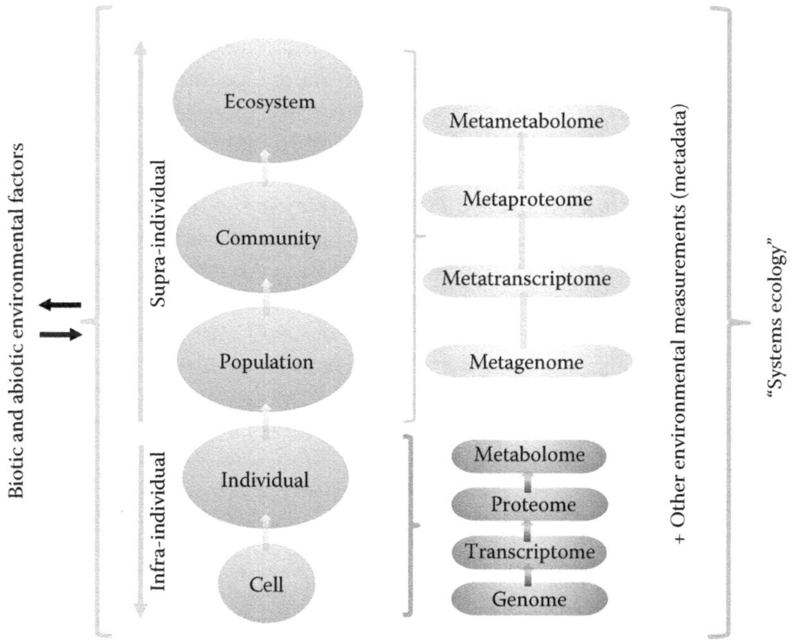

FIGURE 7.4
The ecological hierarchy and related data. Metadata and environmental factors to consider in a holistic perception of "systems ecology."

exemplified recently by analyzing the community and the population level of an iron-rich microbial mat (Quaiser et al. 2014). However, most studies published so far focus on the individual and/or the community level while the population level is omitted. This is certainly related to the *species concept** used for microorganisms but also to the *sequencing depth** and rate of sequencing errors.

7.4.3 Multi-Omics

It is currently possible to integrate heterogeneous data sources including (meta)genomic, (meta)transcriptomic, (meta)proteomic, and metabolomic to provide a *holistic* top-down overview of an organism or an environmental sample. This data cascade could allow deciphering the thin functioning of a tissue, an organism, or an environmental sample. This kind of strategy is still challenging but is important to model or validate in silico biological systems (Yoon et al. 2012) or "ecological system" (see Section 7.6).

The "multi-omics profiling" refers specifically to an expanding multi-omics resource that supports transcriptomics and proteomics information from publicly available studies on model organisms. This developing tool aims to improve proteomic data analysis from RNA sequence annotation predictions for translated RNAs. We can anticipate that this "multi-omics profiling" will also be developed for non-model organisms to be useable for metaproteomic analyses especially with the development of the next-generation proteomic instruments.

Up to date, multi-omics approaches have intended to identify expressed genes (transcriptomic approach) to understand the genome functioning of a cell/tissue/organism under different life constraints in a community. This multi-omics approach has also been viewed from a combination of metagenomic–metatranscriptomic approaches in an aim to understand the metagenome functioning both (1) at the gene level and (2) at the individual level by differentiating present versus active cells within the analyzed environmental sample.

Compared to mRNA that have short half-time lives and a high variability in copy numbers, the use of (meta)proteomics is promising as proteins are more stable and more abundant than transcripts (Moran et al. 2013). Nevertheless, the identification of peptides revealed by mass spectrometry is difficult, since the variability of protein sequences is very high and the direct comparison of peptides to general protein databases inefficient, since the large majority of observed peptides do not match to the databases. A combined approach, metagenomics, metatranscriptomics, and metaproteomics, was applied on deep-sea hydrothermal vent sediments (Urich et al. 2014), which proved effective to detect the presence and estimated the abundance of *methanotrophs* affiliated to *Methylobacter* sp. at the level of the three molecules, rRNA, mRNA, and protein, thereby clearly linking the diversity to function (i.e., *Methylobacter* species perform aerobic *methanotrophy** in hydrothermal vents). A similar multi-omics approach was applied to the rhizosphere and phyllosphere of rice plants linking *methanogenesis** and *methanotrophy** to particular microbial genera (Knief et al. 2012). A more recent work questioning the carbon storage/emission from permafrost at different states of thaw elegantly combined metagenomics, metatranscriptomics, and metaproteomics approaches to address the phylogenetic composition of the microbial communities and their potential functions and activity (Hultman et al. 2015). The study showed clear linkages between metagenomics/metatranscriptomics data and functioning. In more details, the work provided a new understanding into adaptation strategies for life in frozen conditions, and actors and conditions for *methanogenesis** (Hultman et al. 2015) in relation to the question of permafrost carbon–climate feedback and consequences of global warming (Koven et al. 2011).

7.4.4 Single-Cell Genomics

Single-cell genomic (SCG) approaches proved to be a powerful tool to access genomes from uncultured microorganisms. A high number of complete or draft genomes even from low-abundance microorganisms ("microbial dark matter") were obtained from all kind of habitats (for an overview, see Gasc et al. 2015). For example, the draft genome of the candidate phylum TM7 was reconstructed from the human mouth microbial community (Marcy et al. 2007) and from soil (Podar et al. 2007) revealing first insights into the metabolic potential and structure of representatives of this candidate phylum. The combination of SCG with other -omics tools applied on the same samples can have strong impacts on the understanding of the microbial community diversity and functioning. Combined with metagenomics and metatranscriptomics, draft genomes can be confirmed or completed as shown recently, where 35 draft genomes from 15 phyla were generated from a methanogenic bioreactor community (Nobu et al. 2015). Recently, more than 200 uncultivated archaeal and bacterial single-cell genomes were generated from nine diverse marine habitats. In this study, 29 major taxonomical lineages of the microbial dark matter could be enriched with genomic sequences (Rinke et al. 2013). In both cases, concrete community functions and interactions were deduced applying metagenome or metatranscriptome recruitment. For further details about the potential of SCG, see Gasc et al. (2015).

7.5 Linking Diversity to Functions

To determine the mechanistic links and combined theory between species diversity and community/ecosystem function is still an ecological grand challenge. This is particularly true for the most numerous fraction of life on the planet, the microorganisms, and is an active area of research. Metagenomic analyses can allow linking the predicted function to a given organism if sequence assembly into large scaffolds is successful (cf. Section 7.4). However, the within-sample microbial diversity is often too high to generate good-quality assemblages of sequences that allow deducing functions when applying bulk sequencing of microbial community DNA or RNA (*cDNA**). In other words, the signals hinting to community functions are too low or hidden by the overwhelming majority of other sequences. Alternative methods to focus on a specific fraction of the diversity involved in a specific process exist. Stable isotope probing (SIP) represents a powerful technique for targeting species involved in specific ecological functions. Typically, from a ^{13}C-enriched substrate, the microorganism able to metabolize the labeled molecule will integrate the ^{13}C leading to the enrichment in ^{13}C of its lipids, proteins, RNAs, and DNAs. The ^{13}C-enriched DNA or RNA of microorganisms active in the studied process can be separated from the others after isopycnic ultracentrifugation on the basis of a density modification. SIP-RNA strategy has the advantage to be more direct and more sensitive than SIP-DNA, the latter being dependent on a cell division and the ^{13}C enrichment signal being diluted due to the semiconservative DNA replication (Dumont et al. 2011). To assess hypotheses about evolution of plant–microorganisms interactions, SIP-RNA was applied on plant roots after incubation of the plant using the stable isotope ^{13}C–CO_2 at atmospheric concentration (Vandenkoornhuyse et al. 2007; Kiers et al. 2011). Interestingly, the results showed the ability of plants to sanction symbiotic cheaters (Kiers et al. 2011). SIP-RNA was also used to identify fractions of microbial diversity involved in the degradation of organic molecules (Monard et al. 2008) including xenobiotic pollutant in soil, the atrazine (Monard et al. 2011). Authors showed that atrazine degradation likely relied on keystone microbial species. Combined with high-throughput sequencing, SIP-DNA was applied on lake sediment microbial communities targeting single-carbon oxidizing compounds (Kalyuzhnaya et al. 2008). Samples of the same community were incubated with different labeled one-carbon substrates separately, and the different metagenomes containing only the labeled DNA were sequenced and compared. The results showed specific sequence enrichments in response to different carbon-one substrates, thereby identifying the ecological roles of individual phylotypes. In addition, due to the enrichment, via compositional binning, a nearly complete genome of a novel methylotroph, *Methylotenera mobilis*, could be reconstituted.

Alternative labeling can be performed with labeled bromodeoxyuridine, an analog of thymidine that incorporates in the DNA in actively replicating microorganisms. Combined with pyrosequencing, this was applied on coastal ocean plankton–targeting species that utilize dissolved organic carbon (Mou et al. 2008). This study revealed that the coastal microbial communities are populated by taxa capable of metabolizing a wide variety of organic carbon compounds.

Simpler approaches to link diversity to function can be applied if genes encoding key enzymes are characteristic for particular taxa. A good example is the gene amoA encoding for one of three subunits of the ammonium monooxygenase, the key enzyme of *nitrification*.* The diversity of nitrifying microorganisms is very reduced, and they are found within the bacteria affiliated to beta- and gamma-proteobacteria and recently identified within the Archaea from the phylum Thaumarchaeota. Both genes, the bacterial and archaeal amoA are sufficiently different that they can be discriminated by *primers** used in PCR. Therefore, simple PCR or qPCR methods can be applied to detect or quantify the two types of nitrifiers separately (Leininger et al. 2006; Agogué et al. 2008; and others). Combined with 16S rRNA applying qPCR, the distribution and repartition of bacterial and archaeal nitrifiers can be detected, representing a very efficient tool linking the diversity to function targeted on an essential step of the nitrogen cycle.

7.6 Future

7.6.1 Sequencing Technologies: Read Length Matters

The future is in part dependent on new instruments and technologies for sequencing. The throughput is no longer a problem for sequencing. The main expected innovation is the increase of sequencing read length, simplifying genome assembly, and in consequence data analyses and at the same time requiring lower sequence *coverage**. This innovation should facilitate the linkage between diversity and function in ecosystems processes to analyze the rules of *community assembly** and the evolution of host-associated microbiota. Different companies are working on such innovation. Nanopore technology is based on measurement of electric potential differences induced when a native DNA molecule pass through the nanopore (Oxford Nanopore Technologies, United Kingdom). The main advantage of this strategy is that there is no instrument required, the sequencing being done on a small device from which the data are collected directly. To date, the still relatively high error rate (i.e., >10%) can be combined with high-quality short reads (i.e., Illumina sequencing). The other most advanced technology is based on light emission optically detected within a small chamber each with a zero-mode waveguide (Pacific Bioscience, United States). The light emission generated while copying the target DNA is measured within the chamber. In addition to the low error rate, this technology allows the direct analysis of particular epigenetic modifications (i.e., bacterial DNA modifications and methyltransferase recognition motifs; eukaryotic hyper- and hypo-methylated CpG islands) (see Section 7.6.3).

7.6.2 Metaviromes

Viruses are the most abundant biological entities exceeding the number of microorganisms by a factor of 5–25 (Fuhrman 1999). Metaviromes, that is, application of metagenomics on viral communities, were generated already in 2002 applying Sanger sequencing (Breitbart et al. 2002) and later by pyrosequencing (Edwards and Rohwer 2005; Thurber et al. 2009). Since viruses do not possess a common conserved molecular marker as the SSU rRNA genes, the viral diversity is essentially assessed by metagenomic approaches searching primarily to reconstruct complete viral genomes leading to the discovery of novel viral lineages (i.e., Roux et al. 2012; Quaiser et al. 2015). Metaviromes from all kind of habitats are available, for example, from stromatolites (Desnues et al. 2008), Antarctic lakes (López-Bueno et al. 2009), freshwater lakes (Roux et al. 2012), and peatlands (Ballaud et al. 2016). One of the major outcomes of these studies is the enormous diversity of viral sequences making their genomes the largest resource of genetic information on earth. This is in accordance with the high rates of evolutionary changes in viral genomes that are much less conserved than microbial genes (Duffy et al. 2008). Combined with the widespread horizontal gene transfers in the viral world, their abundance, and their diversity, viruses strongly impact the microbial world whose extent is still unclear.

7.6.3 Sequencing the Epigenome

In the Neo-Darwinian synthesis of evolution, phenotypes are seen as a consequence of genotypes. Nevertheless, we know that other drivers can control phenotypes. The epigenetic mechanisms (i.e., DNA methylation, histone modifications, histone variants, and small RNA mainly) can affect gene expression and phenotypes. If selection acts on phenotypes that are not necessarily genetically controlled, the new phenotypes, arising from adaptive plasticity, are not random variants and the trait frequency can be seen as a consequence of genetic accommodation or the transgenerational inheritance of epigenetic mark. Considering that both Eukarya and Prokaryotes have epigenetic marks in their DNAs, the understanding of genome functioning and evolution needs to consider more carefully these epigenetic mechanisms within a multi-omics approach.

7.6.4 Modeling: Toward "Systems Ecology"?

Research works published so far are generally focused and limited to a single level of ecological organization (Figure 7.4). They have allowed identifying a number of interactions but are limited to this level of ecological organization. In the same framework of system biology, one important prospect is to develop a "systems ecology" integrating the different levels of ecological hierarchy. The reductionist approach, still perfectly valid and justified, has however the disadvantage of not being able to provide convincing concepts or methods to understand the emergent properties of an ecological system.

It is now necessary to develop a *holistic** approach taking into account all the levels of the ecology and integrating the physical and chemical processes at different spatial and temporal scales. The pluralism of causes and effects on the organization and evolution of ecological systems should be integrated simultaneously by different approaches (i.e., multi-omics, spatio-temporal dynamics, interaction with environmental conditions) (Figure 7.4). Recently, a new theory based on the simple idea that flow of carbon through ecosystems, as a consequence of metabolisms of individual organisms, has been developed (Schramski et al. 2015). Other modeling strategies of multi-organisms are currently developed on the basis of metabolic networks with the idea of parameterization of pathways using stoichiometric coefficients (i.e., flux balance analysis) and by developing multi-objective approaches to integrate a (micro)organism network. This kind of modeling requires a very precise knowledge, and stoichiometric matrixes of reaction should be tuned by multi-omics data.

From this complex system, we can anticipate new understandings and new hypotheses from the description of emerging properties. We can also speculate the possibility that this "system ecology" will be a new tool to make up scaling inferences.

Glossary

Amplicon: The result of the in vitro synthesis of a targeted DNA using PCR.

cDNA: Double-stranded DNA with a sequence that is complementary to the RNA template from which it is synthesized in a reaction catalyzed by reverse transcriptase.

Community assemblage: A group of populations of different species co-occurring within a given habitat.

Coverage/sequence coverage: The average numbers of sequenced nucleotides that contribute to a portion of an assembly.

Diachronic information in DNA: Sequence variations due to mutation representing evolutionary information.

Evolutionary-related organisms: Organisms sharing a phylogenetic ancestor.

Fluorescence in situ hybridization: Method that uses labeled oligonucleotide probes binding to a specific DNA target to visualize particular cells or microorganisms.

Fosmid: Particular vector designed for the insertion of large-size DNA fragments.

Genome/sequence assembly: The aligning and merging of sequence fragments aiming to reconstruct the original sequence (i.e., full-length chromosome or genome).

Haploid: A single set of chromosome or a cell having a single copy genome.

Haplotype/haplotypic variation: A set of genetic variations, combination of allele on the same chromosome.

Holistic: Integrative perception; parts are interconnected to represent the whole.

Horizontal gene transfers: Transfers of DNA fragments from the environment into the genome of a given organism. This transfer differs from vertical gene transmission, that is, parent to offspring.

Large insert libraries: Collections of cloned DNA inserts of long size (>20,000 bp).

Methanogenesis: Metabolic pathway of methane production by microorganisms (methanogens).

Methanotrophy/methanotroph: The capacity to metabolize methane as a source of carbon and energy by microorganisms, that is, the methanotrophs.

Molecular clock: It is possible to detect and count mutations among homologous DNA regions (i.e., comparison). DNA sequences accumulate a certain number of mutations per unit of time. The more the organism is ancient in term of evolution, the more mutations accumulate.

Mesophile/mesophilic: Ability to grow at moderate temperature, that is, 20°C–40°C.

Metagenomics/metatranscriptomics: Sequencing and analysis of bulk DNA or cDNA (RNA) from the mixed genomes of microbial communities.

Monophyletic group: Group of organisms sharing the same common ancestor.

Nitrification: Key biological process of the nitrogen cycle by which a microorganism transforms ammonium to nitrite by oxidation.

Open reading frames (ORF): An ORF is a continuous stretch of codons that is not interrupted by stop codon.

Phylogeny: Sequence-based analyses of the evolutionary relationship between a set of organisms.

Plastid DNA: DNA contained in the chloroplast.

Pleiotropic/pleiotropy: Multiple effects from a single origin, for example, multiple phenotypes driven by a single gene.

Primer: Refers to a specific oligonucleotide designed by the user for in vitro synthesis of a targeted DNA for use in polymerase chain reaction.

Pyrosequencing: DNA sequencing method based on "sequencing by synthesis."

Sanger technology: First DNA sequencing method that incorporates chain-terminating dideoxynucleotides by copying the target DNA to sequence.

Scaffolds/sequence scaffolds: Portion of the genome sequence reconstructed from contigs. A contig is a shorter genomic sequence assembled from smaller sequences/reads.

Read length: The number of nucleotide sequenced from a sequenced DNA molecule.

Rhodopsin/proteorhodopsin: Molecule at the basis of photoheterotrophy capturing light energy.

Sequencing depth: See coverage.

Species concepts: Concept applied to group individual organisms within a species. In animals, the biological species concept defines a species as organisms that can sexually reproduce. However, most of the (micro)organisms have clonal reproduction and the biological species concept cannot be used. The phylogenetic species concepts defining a species as a *monophyletic group** of individuals is often adopted.

Vector: DNA molecule that carries foreign genetic material in cells.

References

Agogué, H, M Brink, J Dinasquet, and G J Herndl. 2008. Major gradients in putatively nitrifying and non-nitrifying archaea in the deep North Atlantic. *Nature* 456 (7223): 788–791.

Angly, F E, B Felts, M Breitbart, et al. 2006. The marine viromes of four oceanic regions. *PLoS Biology* 4 (11): e368.

Ballaud, F C, A Dufresne, A-J Francez et al. 2016. Dynamics of viral abundance and diversity in a sphagnum-dominated peatland: Temporal fluctuations prevail over habitat. *Frontiers in Microbiology* 6: 1494.

Béjà, O, L Aravind, E V Koonin et al. 2000. Bacterial rhodopsin: Evidence for a new type of phototrophy in the sea. *Science* 289 (5486): 1902–1906.

Béjà, O, E N Spudich, J L Spudich, M Leclerc, and E F DeLong. 2001. Proteorhodopsin phototrophy in the ocean. *Nature* 411 (6839): 786–789.

Ben Maamar, S, L Aquilina, A Quaiser et al. 2015. Groundwater isolation governs chemistry and microbial community structure along hydrologic flowpaths. *Frontiers in Microbiology* 6: 1457.

Biers, E J, S Sun, and E C Howard. 2009. Prokaryotic genomes and diversity in surface ocean waters: Interrogating the global ocean sampling metagenome. *Applied and Environmental Microbiology* 75 (7): 2221–2229.

Blaxter, M. 2003. Molecular systematics: Counting angels with DNA. *Nature* 421 (6919): 122–124.

Breitbart, M, P Salamon, B Andresen et al. 2002. Genomic analysis of uncultured marine viral communities. *Proceedings of the National Academy of Sciences of the United States of America* 99 (22): 14250–14255.

Bulgarelli, D, M Rott, K Schlaeppi et al. 2012. Revealing structure and assembly cues for *Arabidopsis* root-inhabiting bacterial microbiota. *Nature* 488 (7409): 91–95.

Carvalhais, L C, P G Dennis, G W Tyson, and P M Schenk. 2012. Application of metatranscriptomics to soil environments. *Journal of Microbiological Methods* 91 (2): 246–251.

Chase, M W and M F Fay. 2009. Barcoding of plants and fungi. *Science* 325 (5941): 682–683.

Collins, F S, E D Green, A E Guttmacher, M S Guyer, and US National Human Genome Research Institute. 2003. A vision for the future of genomics research. *Nature* 422 (6934): 835–847.

Curtis, T P and W T Sloan. 2005. Exploring microbial diversity—A vast below. *Science* 309 (5739): 1331–1333.

DeLong, E F, C M Preston, T Mincer et al. 2006. Community genomics among stratified microbial assemblages in the ocean's Interior. *Science* 311 (5760): 496–503.

Deschamps, P, Y Zivanovic, D Moreira, F Rodríguez-Valera, and P López-García. 2014. Pangenome evidence for extensive interdomain horizontal transfer affecting lineage core and shell genes in uncultured planktonic thaumarchaeota and euryarchaeota. *Genome Biology and Evolution* 6 (7): 1549–1563.

Desnues, C, B Rodriguez-Brito, S Rayhawk et al. 2008. Biodiversity and biogeography of phages in modern stromatolites and thrombolites. *Nature* 452 (7185): 340–343.

de Vargas, C, S Audic, N Henry et al. 2015. Eukaryotic plankton diversity in the sunlit ocean. *Science* 348 (6237): 1261605.

Doré, J and H Blottière. 2015. The influence of diet on the gut microbiota and its consequences for health. *Current Opinion in Biotechnology* 32: 195–199.

Duffy, S, L A Shackelton, and E C Holmes. 2008. Rates of evolutionary change in viruses: Patterns and determinants. *Nature Reviews Genetics* 9 (4): 267–276.

Dumont, M G, B Pommerenke, P Casper, and R Conrad. 2011. DNA-, rRNA- and mRNA-based stable isotope probing of aerobic methanotrophs in lake sediment. *Environmental Microbiology* 13 (5): 1153–1167.

Edwards, R A and F Rohwer. 2005. Viral metagenomics. *Nature Reviews Microbiology* 3 (6): 504–510.

Edwards, R A, B Rodriguez-Brito, L Wegley et al. 2006. Using pyrosequencing to shed light on deep mine microbial ecology. *BMC Genomics* 7 (1): 57.

Eid, J, A Fehr, J Gray et al. 2009. Real-time DNA sequencing from single polymerase molecules. *Science (New York, NY)* 323 (5910): 133–138.

Feike, J, K Jürgens, J T Hollibaugh et al. 2011. Measuring unbiased metatranscriptomics in suboxic waters of the central Baltic Sea using a new in situ fixation system. *The ISME Journal* 6 (2): 461–470.

Fierer, N, J W Leff, B J Adams et al. 2012. Cross-biome metagenomic analyses of soil microbial communities and their functional attributes. *Proceedings of the National Academy of Sciences of the United States of America* 109 (52): 21390–21395.

Fieseler, L, A Quaiser, C Schleper, and U Hentschel. 2006. Analysis of the first genome fragment from the marine sponge-associated, novel candidate phylum poribacteria by environmental genomics. *Environmental Microbiology* 8 (4): 612–624.

Frias-Lopez, J, Y Shi, G W Tyson et al. 2008. Microbial community gene expression in ocean surface waters. *Proceedings of the National Academy of Sciences of the United States of America* 105 (10): 3805–3810.

Fuchs, B M, S Spring, and H Teeling. 2007. Characterization of a marine gammaproteobacterium capable of aerobic anoxygenic photosynthesis. *Proceedings of the National Academy of Sciences of the United States of America* 104 (8): 2891–2896.

Fuhrman, J A. 1999. Marine viruses and their biogeochemical and ecological effects. *Nature* 399 (6736): 541–548.

Gasc, C, C Ribière, N Parisot et al. 2015. Capturing prokaryotic dark matter genomes. *Research in Microbiology* 166: 814–830.

Ghahramanzadeh, R, G Esselink, L P Kodde et al. 2013. Efficient distinction of invasive aquatic plant species from non-invasive related species using DNA barcoding. *Molecular Ecology Resources* 13 (1): 21–31.

Ghai, R, A-B Martin-Cuadrado, A G Molto et al. 2010. Metagenome of the Mediterranean deep chlorophyll maximum studied by direct and fosmid library 454 pyrosequencing. *The ISME Journal* 4 (9): 1154–1166.

Gifford, S M, S Sharma, M Booth, and M A Moran. 2013. Expression patterns reveal niche diversification in a marine microbial assemblage. *The ISME Journal* 7 (2): 281–298.

Gladyshev, E A, M Meselson, and I R Arkhipova. 2008. Massive horizontal gene transfer in bdelloid rotifers. *Science* 320 (5880): 1210–1213.

Goldenfeld, N and C R Woese. 2007. Biology's next revolution. *Nature* 445 (7126): 369.

Gómez-Consarnau, L, J M González, M Coll-Lladó et al. 2007. Light stimulates growth of proteorhodopsin-containing marine flavobacteria. *Nature* 445 (7124): 210–213.

Hallam, S J, T J Mincer, C Schleper et al. 2006. Pathways of carbon assimilation and ammonia oxidation suggested by environmental genomic analyses of marine crenarchaeota. *PLoS Biology* 4 (4): e95.

He, S, O Wurtzel, K Singh et al. 2010. Validation of two ribosomal RNA removal methods for microbial metatranscriptomics. *Nature Methods* 7 (10): 807–812.

Hultman, J, M P Waldrop, R Mackelprang et al. 2015. Multi-omics of permafrost, active layer and thermokarst bog soil microbiomes. *Nature* 521 (7551): 208–212.

Kalyuzhnaya, M, A Lapidus, N Ivanova et al. 2008. High-resolution metagenomics targets specific functional types in complex microbial communities. *Nature Biotechnology* 26: 1029–1034.

Keeling, P J and J Palmer. 2008. Horizontal gene transfer in eukaryotic evolution. *Nature Reviews Genetics* 9 (8): 605–618.

Kielak, A M, J A van Veen, and G A Kowalchuk. 2010. Comparative analysis of acidobacterial genomic fragments from terrestrial and aquatic metagenomic libraries, with emphasis on acidobacteria subdivision 6. *Applied and Environmental Microbiology* 76 (20): 6769–6777.

Kiers, E T, M Duhamel, Y Beesetty et al. 2011. Reciprocal rewards stabilize cooperation in the mycorrhizal symbiosis. *Science* 333 (6044): 880–882.

Knief, C, N Delmotte, S Chaffron et al. 2012. Metaproteogenomice. *The ISME Journal* 6 (7): 1378–1390.

Korem, T, D Zeevi, J Suez et al. 2015. Growth dynamics of gut microbiota in health and disease inferred from single metagenomic samples. *Science* 349 (6252): 1101–1106.

Koven, C D, B Ringeval, P Friedlingstein et al. 2011. Permafrost carbon-climate feedbacks accelerate global warming. *Proceedings of the National Academy of Sciences of the United States of America* 108 (36): 14769–14774.

Leininger, S, T Urich, M Schloter et al. 2006. Archaea predominate among ammonia-oxidizing prokaryotes in soils. *Nature* 442 (7104): 806–809.

Liles, M R, B F Manske, S B Bintrim, J Handelsman, and R M Goodman. 2003. A census of rRNA genes and linked genomic sequences within a soil metagenomic library. *Applied and Environmental Microbiology* 69 (5): 2684–2691.

López-Bueno, A, J Tamames, D Velázquez et al. 2009. High diversity of the viral community from an Antarctic lake. *Science* 858 (November): 1–25.

López-García, P and D Moreira. 2008. Tracking microbial biodiversity through molecular and genomic ecology. *Research in Microbiology* 159 (1): 67–73.

Lundberg, D S, S L Lebeis, S H Paredes et al. 2012. Defining the core *Arabidopsis thaliana* root microbiome. *Nature* 488 (7409): 86–90.

Mackelprang, R, M P Waldrop, K M DeAngelis et al. 2011. Metagenomic analysis of a permafrost microbial community reveals a rapid response to thaw. *Nature* 480 (7377): 368–371.

Marcy, Y, C Ouverney, E M Bik et al. 2007. Dissecting biological "dark matter" with single-cell genetic analysis of rare and uncultivated TM7 microbes from the human mouth. *Proceedings of the National Academy of Sciences of the United States of America* 104 (29): 11889–11894.

Margulies, M, M Egholm, W E Altman et al. 2005. Genome sequencing in microfabricated high-density picolitre reactors. *Nature* 437 (7057): 376–380.

Martín-Cuadrado, A-B, F Rodríguez-Valera, D Moreira et al. 2008. Hindsight in the relative abundance, metabolic potential and genome dynamics of uncultivated marine archaea from comparative metagenomic analyses of bathypelagic plankton of different oceanic regions. *The ISME Journal* 2 (8): 865–886.

Maxam, A M and W Gilbert. 1977. A new method for sequencing DNA. *Proceedings of the National Academy of Sciences of the United States of America* 74 (2): 560–564.

Microbiology by Numbers. 2011. *Nat Rev Micro* 9 (9). Editorial: 628.

Monard, C, F Binet, and P Vandenkoornhuyse. 2008. Short-term response of soil bacteria to carbon enrichment in different soil microsites. *Applied and Environmental Microbiology* 74 (17): 5589–5592.

Monard, C, P Vandenkoornhuyse, B Le Bot, and F Binet. 2011. Relationship between bacterial diversity and function under biotic control: The soil pesticide degraders as a case study. *The ISME Journal* 5 (6): 1048–1056.

Moran, M A, B Satinsky, S Gifford et al. 2013. Sizing up metatranscriptomics. *The ISME Journal* 7 (2): 237–243.

Mou, X, S Sun, R A Edwards, R E Hodson, and M A Moran. 2008. Bacterial carbon processing by generalist species in the coastal ocean. *Nature* 451 (7179): 708–711.

Nelson, K E, R A Clayton, S R Gill et al. 1999. Evidence for lateral gene transfer between archaea and bacteria from genome sequence of thermotoga maritima. *Nature* 399 (6734): 323–329.

Nobu, M K, T Narihiro, C Rinke et al. 2015. Microbial dark matter ecogenomics reveals complex synergistic networks in a methanogenic bioreactor. *The ISME Journal* 9 (8): 1710–1722.

Nunes, F L D, L Aquilina, J de Ridder et al. 2015. Time-scales of hydrological forcing on the geochemistry and bacterial community structure of temperate peat soils. *Scientific Reports* 5: 14612.

Pace, N R, D A Stahl, D J Lane, and G J Olsen. 1986. The analysis of natural microbial populations by ribosomal RNA sequences. *Advances in Microbial Ecology* 9: 1–55.

Podar, M, C B Abulencia, M Walcher et al. 2007. Targeted access to the genomes of low-abundance organisms in complex microbial communities. *Proceedings of the National Academy of Sciences of the United States of America* 73 (10): 3205–3214.

Poretsky, R S, N Bano, A Buchan et al. 2005. Analysis of microbial gene transcripts in environmental samples. *Proceedings of the National Academy of Sciences of the United States of America* 71 (7): 4121–4126.

Qin, J, R Li, J Raes et al. 2010. A human gut microbial gene catalogue established by metagenomic sequencing. *Nature* 464 (7285): 59–65.

Quaiser, A, X Bodi, A Dufresne et al. 2014. Unraveling the stratification of an iron-oxidizing microbial mat by metatranscriptomics. *PLoS One* 9 (7): 1–9.

Quaiser, A, A Dufresne, F Ballaud et al. 2015. Diversity and comparative genomics of microviridae in sphagnum-dominated peatlands. *Frontiers in Microbiology* 6: 375.

Quaiser, A, P López-García, Y Zivanovic et al. 2008. Comparative analysis of genome fragments of acidobacteria from deep Mediterranean plankton. *Environmental Microbiology* 10 (10): 2704–2717.

Quaiser, A, T Ochsenreiter, H-P Klenk et al. 2002. First insight into the genome of an uncultivated crenarchaeote from soil. *Environmental Microbiology* 4 (10): 603–611.

Quaiser, A, T Ochsenreiter, C Lanz et al. 2003. Acidobacteria form a coherent but highly diverse group within the bacterial domain: Evidence from environmental genomics. *Molecular Biology* 50 (2): 563–575.

Quaiser, A, Y Zivanovic, D Moreira, and P López-García. 2011. Comparative metagenomics of bathypelagic plankton and bottom sediment from the Sea of Marmara. *The ISME Journal* 5 (2): 285–304.

Radax, R, T Rattei, A Lanzen et al. 2012. Metatranscriptomics of the marine sponge *Geodia barretti*: Tackling phylogeny and function of its microbial community. *Environmental Microbiology* 14 (5): 1308–1324.

Rinke, C, P Schwientek, A Sczyrba et al. 2013. Insights into the phylogeny and coding potential of microbial dark matter. *Nature* 499 (7459): 431–437.

Rodríguez-Martínez, R, G Rocap, G Salazar, and R Massana. 2013. Biogeography of the uncultured marine picoeukaryote MAST-4: Temperature-driven distribution patterns. *The ISME Journal* 7 (8): 1531–1543.

Roman, J and B W Bowen. 2000. The mock turtle syndrome: Genetic identification of turtle meat purchased in the south-eastern United States of America. *Animal Conservation* 3 (01): 61–65.

Rondon, M R, P R August, A D Bettermann et al. 2000. Cloning the soil metagenome: A strategy for accessing the genetic and functional diversity of uncultured microorganisms. *Applied and Environmental Microbiology* 66 (6): 2541–2547.

Roux, S, F Enault, A Robin et al. 2012. Assessing the diversity and specificity of two freshwater viral communities through metagenomics. *PLoS ONE* 7 (3): e33641.

Rusch, D B, A L Halpern, G Sutton et al. 2007. The sorcerer II global ocean sampling expedition: Northwest Atlantic through eastern tropical Pacific. *PLoS Biology* 5 (3): e77.

Sanger, F and A R Coulson. 1975. A rapid method for determining sequences in DNA by primed synthesis with DNA polymerase. *Journal of Molecular Biology* 94 (3): 441–448.

Schlaeppi, K, N Dombrowski, R G Oter, E V L van Themaat, and P Schulze-Lefert. 2014. Quantitative divergence of the bacterial root microbiota in *Arabidopsis thaliana* relatives. *Proceedings of the National Academy of Sciences of the United States of America* 111 (2): 585–592.

Schmidt, T M, E F DeLong, and N R Pace. 1991. Analysis of a marine picoplankton community by 16S rRNA gene cloning and sequencing. *Journal of Bacteriology* 173 (14): 4371–4378.

Schramski, J R, A I Dell, J M Grady, R M Sibly, and J H Brown. 2015. Metabolic theory predicts whole-ecosystem properties. *Proceedings of the National Academy of Sciences of the United Staes of America* 112 (8): 2617–2622.

Schuster, S C. 2008. Next-generation sequencing transforms today's biology. *Nature Methods* 5 (1): 16–18.

Shi, Y, G W Tyson, J M Eppley, and E F DeLong. 2011. Integrated metatranscriptomic and metagenomic analyses of stratified microbial assemblages in the open ocean. *The ISME Journal* 5 (6): 999–1013.

Sogin, M L, H G Morrison, J A Huber et al. 2006. Microbial diversity in the deep sea and the underexplored "rare biosphere". *Proceedings of the National Academy of Sciences of the United States of America* 103 (32): 12115–12120.

Stein, J L, T L Marsh, K Y Wu, H Shizuya, and E F DeLong. 1996. Characterization of uncultivated prokaryotes: Isolation and analysis of a 40-kilobase-pair genome fragment from a planktonic marine archaeon. *Journal of Bacteriology* 178 (3): 591–599.

Stern, A and R Sorek. 2011. The phage-host arms race: Shaping the evolution of microbes. *BioEssays* 33 (1): 43–51.

Thurber, R V, M Haynes, M Breitbart, L Wegley, and F Rohwer. 2009. Laboratory procedures to generate viral metagenomes. *Nature Protocols* 4 (4): 470–483.

Treusch, A H, A Kletzin, G Raddatz et al. 2004. Characterization of large-insert DNA libraries from soil for environmental genomic studies of archaea. *Environmental Microbiology* 6 (9): 970–980.

Treusch, A H, S Leininger, A Kletzin et al. 2005. Novel genes for nitrite reductase and amo-related proteins indicate a role of uncultivated mesophilic crenarchaeota in nitrogen cycling. *Environmental Microbiology* 7 (12): 1985–1995.

Tringe, S G, C von Mering, A Kobayashi et al. 2005. Comparative metagenomics of microbial communities. *Science* 308 (5721): 554–557.

Turnbaugh, P J, M Hamady, T Yatsunenko et al. 2009. A core gut microbiome in obese and lean twins. *Nature* 457 (7228): 480–484.

Tveit, A, R Schwacke, M M Svenning, and T Urich. 2012. Organic carbon transformations in high-Arctic peat soils: Key functions and microorganisms. *The ISME Journal* 7 (2): 299–311.

Tveit, A T, T Urich, and M M Svenning. 2014. Metatranscriptomic analysis of Arctic peat soil microbiota. *Applied and Environmental Microbiology* 80 (18): 5761–5772.

Tyson, G W, J Chapman, P Hugenholtz et al. 2004. Community structure and metabolism through reconstruction of microbial genomes from the environment. *Nature* 428 (6978): 37–43.

Urich, T, A Lanzén, J Qi et al. 2008. Simultaneous assessment of soil microbial community structure and function through analysis of the meta-transcriptome. *PLoS One* 3 (6): e2527.

Urich, T, A Lanzén, R Stokke et al. 2014. Microbial community structure and functioning in marine sediments associated with diffuse hydrothermal venting assessed by integrated meta-omics. *Environmental Microbiology* 16 (9): 2699–2710.

Vandenkoornhuyse, P, A Dufresne, A Quaiser et al. 2010. Integration of molecular functions at the ecosystemic level: Breakthroughs and future goals of environmental genomics and post-genomics. *Ecology Letters* 13 (6): 776–791.

Vandenkoornhuyse, P, S Mahé, P Ineson et al. 2007. Active root-inhabiting microbes identified by rapid incorporation of plant-derived carbon into RNA. *Proceedings of the National Academy of Sciences of the United States of America* 104 (43): 16970–16975.

Vandenkoornhuyse, P, A Quaiser, M Duhamel, A Le Van, and A Dufresne. 2015. The importance of the microbiome of the plant holobiont. *The New Phytologist* 206 (4): 1196–1206.

Venter, J C, K Remington, J F Heidelberg et al. 2004. Environmental genome shotgun sequencing of the Sargasso Sea. *Science* 304 (5667): 66–74.

Whitman, W B, D C Coleman, and W J Wiebe. 1998. Prokaryotes: The unseen majority. *Proceedings of the National Academy of Sciences of the United States of America* 95 (12): 6578–6583.

Wilson, E O. 1989. Threats to biodiversity. *Scientific American* 261: 108–116.

Woese, C R. 1987. Bacterial evolution. *Microbiological Reviews* 51 (2): 221–271.

Woese, C R. 2004. A new biology for a new century. *Microbiology and Molecular Biology Reviews* 68 (2): 173–186.

Yoon, S H, M-J Han, H Jeong et al. 2012. Comparative multi-omics systems analysis of *Escherichia coli* strains B and K-12. *Genome Biology* 13 (5): R37.

Yoon, S S, E-K Kim, and W-J Lee. 2015. Functional genomic and metagenomic approaches to understanding gut microbiota-animal mutualism. *Current Opinion in Microbiology* 24: 38–46.

Yooseph, S, G Sutton, D B Rusch et al. 2007. The Sorcerer II global ocean sampling expedition: Expanding the universe of protein families. *PLoS Biology* 5 (3): 432–466.

8

Characterization of Biogeochemical Processes at the Microscale: Concepts and Applications of NanoSIMS

Carsten W. Mueller, Laurent Remusat, and Cornelia Rumpel

CONTENTS

Abstract

Nanoscale secondary ion mass spectrometry (NanoSIMS) is a state-of-the-art analytical technique allowing for the visualization of the distribution of up to seven isotopes in environmental samples at high lateral resolution. The use of this technique combined with stable isotope labeling and other microscale methods will advance our understanding of biochemical cycles. Particularly, NanoSIMS has a high potential to generate new knowledge concerning functioning of microbial communities and biogeochemical interfaces because it allows for studying biogeochemical processes at the relevant scales, where they occur. In this chapter, we present this technique and its limitations and applications in recent biogeochemical research. Moreover, we show through some examples of significant advances achieved by the use of NanoSIMS how this may lead to changing paradigms in environmental sciences.

8.1 Introduction

Large-scale biogeochemical processes (e.g., C cycling) are driven by micro- to nanoscale interactions among microorganisms, mineral constituents, and natural organic matter (Schmidt et al., 2011). At the same time, small-scale processes often show faster reaction rates in isolated zones ("hot spots"), in comparison to adjacent areas. Such hot spots occur at terrestrial–aquatic interfaces (McClain et al., 2003) or, in terms of microbial activity, at specific biophysical microenvironments, for example, the rhizosphere (Kuzyakov and Blagodatskaya, 2015; Young and Crawford, 2004). In soils and sediments, there are many biotic hot spots located within the mineral matrix, including those generated by bioturbation and microbial or plant activity (Hinsinger et al., 2009; Young et al., 1998). Recently, Falconer et al. (2015) simulated the effect of structural heterogeneity on microbial respiration. Using mathematical simulations, the authors showed that larger particulate organic matter was metabolically easier to decompose than smaller, unevenly distributed particulate organic matter, due to a so-called insulation of the fungal colony within the soil structure (Falconer et al., 2015). This and other works imply the great need for more refined data, obtained at the relevant scale, where microbial transformations take place. But most biogeochemical research is conducted using large bulk (>1 g) samples, which are usually strongly altered prior to analysis (e.g., grinding, solvent extraction). Furthermore, the investigation of intact soil/sediment structures and microbial ecosystems at the scale of so-called microhabitats, for example, hot spots of increased organic matter turnover and higher microbial activity, is usually prevented by the destruction of intact soil/sediment structures by most bulk sampling techniques. A lack in progress concerning the deeper understanding of organic carbon dynamics is mainly due to the restricted answers obtained by macroscopic measurements (Baveye, 2015). Hence, to gain a deeper understanding of the elemental fluxes (C, N, P, and S) among plants, microorganisms, and minerals in complex environments, it is necessary to study these elemental constituents at the spatial scale relevant for biogeochemical processes. Characterizing the spatial heterogeneity at soil microenvironments with respect to microbial ecology and upscaling of their importance for global cycles was named as one of the major challenges in ecological science (Baveye, 2015). It was further suggested that classical concepts in biogeochemical research would substantially evolve in the future by adapting recent approaches used in modern microbiology (Baveye, 2015). Consequently, the preparation of samples should clearly aim to maintain the structural integrity that keeps microorganisms and microstructures intact and avoids their dislocation in order to be able to study biogeochemical processes in intact microenvironments. Thus, to gain a real comprehensive understanding of how biogeochemical processes are determined by organic matter composition, microbial activity, and soil/sediment physical structure, it is crucial to integrate biogeochemistry, microbiology, and physics (Baveye and Laba, 2015; Dijkstra et al., 2013). Great technical developments over the last years now provide access to a number of analytical instruments (e.g., x-ray tomography, x-ray spectromicroscopy) to study microenvironments at previously unresolved scales (Baveye, 2015). Especially, the nanoscale secondary ion mass spectrometry (NanoSIMS), capable of the visualization of stable isotopes up to a lateral resolution of 50 nm, has revolutionized our view on microscale biogeochemical processes. The use of NanoSIMS to explore soil processes in particular is the most recent application for this rather new SIMS technique, which was previously mainly used in cosmochemistry, geology, microbiology, and life science (Hoppe et al., 2013; Moore et al., 2011; Musat et al., 2009). By using NanoSIMS in concert with other imaging techniques (e.g., scanning transmission x-ray microscopy [STXM], scanning electron microscopy [SEM]), classic analytical analyses (isotopic and elemental analysis) and biochemical methods (NMR spectroscopy, GC-MS), it is now possible to exhibit a more complete picture of environmental processes at the microscale, which already yielded results allowing to challenge existing paradigms (Keiluweit et al., 2012; Liu et al., 2013; Moore et al., 2011; Mueller et al., 2013; Remusat et al., 2012; Vogel et al., 2014).

8.2 SIMS Principle, Technical Requirements, and Limitations

NanoSIMS is a destructive spectromicroscopic technique in which a primary ion beam is sputtering the surface of a solid sample yielding detectable secondary ions. The acceleration of primary ions (about 16 keV) results in a depth resolution of approximately 10 nm; hence, NanoSIMS can be considered as a technique to assess surface chemistry. It allows for the study of a much smaller sample volume in comparison to other microscopic techniques, like SEM, which integrate often over several μm in depth due to larger-scale interaction between the incident particle and the sample atoms and molecules. Together with the ability to visualize the distribution of isotopes at a high lateral resolution, NanoSIMS is an unprecedented tool for microscale studies in biogeochemistry (Hoppe et al., 2013). The improved primary ion optics and secondary ion transmission at high mass resolving power, in contrast to large geometry SIMS (e.g., Cameca 1280), enable the Cameca NanoSIMS to image isotopes with high sensitivity (lower than ppm levels) at the submicron scale. Using cesium as primary ion (Cs^+), NanoSIMS analysis is able to reach a theoretical spatial resolution of ~50 nm using very low primary currents (less than 1 pA). However, in order to still yield a statistically sufficient amount of secondary ions from natural samples, the primary ion beam is set to a few pA, resulting in ~100–150 nm lateral resolution. This is, for example, sufficient to differentiate between specific sulfur-rich domains on extracellular polysaccharides–covered goethite needles with a size of a few hundred nanometers (Liu et al., 2013). This study indicates nicely the elemental imaging capabilities of organomineral sample material at the upper spatial resolution boundary. Due to the coaxial setup of the NanoSIMS, allowing for the sharp focusing of the primary ion beam and thus the high lateral resolution, primary and secondary ions have to be of different polarity. NanoSIMS is theoretically capable of detecting masses from 1 to 400 amu. Up to seven secondary ions can be simultaneously detected in the multicollection chamber, with some limitations imposed by its design: there is a maximum factor of 21 between the lightest and heaviest ions that can be collected simultaneously. This enables investigating a wide range of biogeochemical interesting isotopic systems by the detection of species released from the main constituents of organic matter (Cs^+ as primary ion: $^{12}C^-$, $^{13}C^-$, $^{12}C^{14}N^-$, $^{12}C^{15}N^-$, $^{31}P^-$, $^{32}S^-$) or minerals (Cs^+ as primary ion: $^{28}Si^-$, $^{27}Al^{16}O^-$, $^{56}Fe^{16}O^-$). Nitrogen is detected as cyanide (CN^-, molecular ion) due to the very low ionization of the nitrogen atom under the Cs^+ (McMahon et al., 2006). Aluminum and iron (and elements like Ca, Na, or metals being more akin for positive ionizing) can be detected as molecular ion together with oxygen when using the Cs^+ primary ion beam. It must be noted that for studying the distribution of metals, for instance, aluminum or iron, an O^- source, producing cations like $^{27}Al^+$ and $^{56}Fe^+$, is more suitable than a Cs^+ source. However, the conventional O^- source has low brightness hampering the lateral resolution at approximately 300 nm. Very recent developments on a new design of the oxygen source have led to significant improvements of the lateral resolution (down to 50 nm), unleashing great opportunities for the study of the distribution of Mg, Ca, or K. Nevertheless, the O^- source is not suitable to investigate organic matter as C and N ionize poorly with O^- primary ions.

Although analyses of natural abundance ^{13}C and ^{15}N in organic matter are widely used in biogeochemistry, very few NanoSIMS studies investigate the distribution of isotopic ratios from untreated natural samples because the NanoSIMS precision for isotope ratios is limited to ~1‰ in imaging analytical conditions (Slodzian, 2004). To obtain the high precision (<0.5‰) required to disentangle biogeochemical processes, large amounts of sample material have to be retrieved from the sample surface (nanograms), which requires large primary spots of about ~10 μm; this is achieved by large SIMS instruments like the IMS 1280. But such spot size is beyond the relevant scale for most biogeochemical processes (e.g., microorganisms, primary minerals). Consequently, the use of isotopic tracers to study microscale processes can be seen as the most promising way to obtain relevant results with NanoSIMS in biogeochemistry as described in this chapter. Adding constituents (for instance, organic compounds) enriched in specific stable isotopes results in contrast enhancement in isotope images because the range of isotopic composition in the locations of the label is significantly different from those of unaffected areas.

Aside from the technical constraints of the instrument itself, collection of meaningful data to elucidate environmental processes depends on the selection and preparation of suitable samples. Identifying and isolating a nanomaterial being represented at a larger scale and addressing its spatial heterogeneity may be crucial for the interpretation of results. Moreover, instrumental constraints require that the samples be vacuum stable, as flat as possible and conductive (Mueller et al., 2013). The fundamental requirement is the stability under high vacuum (up to 10^{-11} mbar), which excludes samples that either strongly degas (e.g., thick resin-embedded sections) or have higher amounts of residual water (e.g., fresh biological tissues). Due to the high sensitivity of the NanoSIMS, another key challenge is how to deal with the occurrence of contaminations, which is an important issue especially for the classical sample embedding using epoxy or polyester resins. Such resins contain C and N, which can dilute isotopic signals from the sample itself and hamper the spatial differentiation between resin and C and CN in the organic matter of the sample. Systematic errors can be avoided by using adequate embedding materials, such as the resin Araldite 502 (Figure 8.1; Herrmann et al., 2007) or sulfur (Lehmann et al., 2005). Other studies used special computer techniques to differentiate resin-derived elements and those from the actual samples (Rennert et al., 2014). A possible alternative is avoiding embedding agents during sample preparation, for example, by applying cryofixation and cryosectioning (Mueller et al., 2013) or an erbium oxide marker layer to avoid any mixing with the organic matter of the soil sample (Hoeschen et al., 2015).

Systematic errors through resin embedding and sectioning can also be circumvented by using sedimentation and deposition of fine grain samples (e.g., clay-sized fractions) on Si wafers (Vogel et al., 2014) or gold foil (Hatton et al., 2015). As Si wafers are very flat and conductive, they are optimal as sample substrate for minerals but also thin sections from biological tissues. Nevertheless, the deposition of particles on a flat carrier material does not prevent the occurrence of topography issues. The use of the technique reported by Hoeschen et al. (2015) to produce flat sample sections can yield promising results at substantially reduced topography. Especially in microbiology, gold-coated polycarbonate filters are used as sample substrate for microbial cells (Miot et al., 2015; Musat et al., 2009). These allow for the combination of molecular techniques with subsequent visualization using microscopic techniques in combination with NanoSIMS (Musat et al., 2009).

Other systematic errors may include the representativeness of the sample to be analyzed as well as the reproducibility of the obtained data. Especially with respect to large-scale phenomena, it will be crucial to put more efforts on upscaling approaches. This is a major issue, when NanoSIMS analyses are carried out on complex material like soil. In this case, particles of a specific fraction are selected, allowing to address specific questions, for instance, with regard to soil organic matter stabilization. For this kind of analyses, sample choice and preparation are extremely important in order to avoid errors.

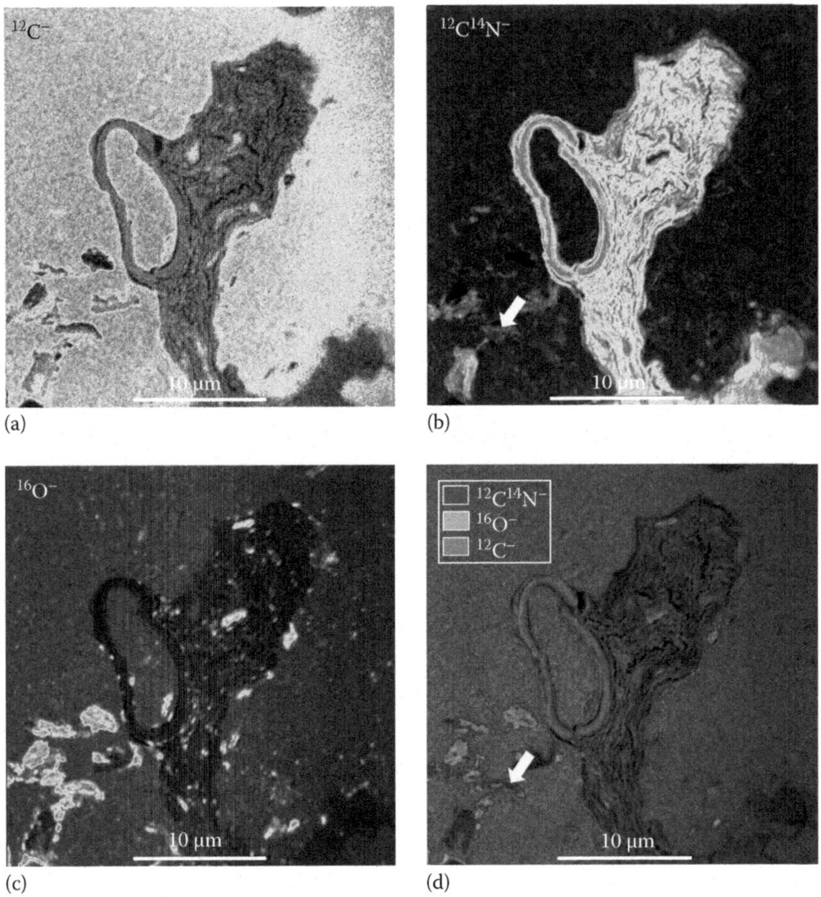

FIGURE 8.1
(See color insert.) NanoSIMS images of the distribution of $^{12}C^-$ (a), $^{12}C^{14}N^-$ (b), $^{16}O^-$ (c), and a composite image of the three secondary ions (d) derived from a resin-embedded section of a particulate organic matter–rich permafrost soil structure (Typic Aquiturbel, Barrow, Alaska). The distinctive different distributions between secondary ion counts of $^{12}C^-$ and $^{12}C^{14}N^-$ (d) indicate the good spatial differentiation between the used epoxy resin (a, Araldite 502) and the soil organic matter (b). The $^{16}O^-$ shows the distribution of mineral particles; the arrow indicates mineral-associated organic matter. (From Mueller, C.W., TU Munich, Germany, unpublished data.)

This concerns not only the specific sample preparation as outlined earlier but also the choice of the actual sample and the region of interest to be analyzed. Users of the technique should be clear about the mechanisms to be analyzed, which will determine sample choice and preparation. For complex systems like soils or sediments, the sample preparation protocol has to be adjusted almost for every new set of samples, trying to minimize artifacts and contaminations while aiming for preferably flat conductive samples. The analysis of microbial cells in soils is usually hampered due to their dispersal in a large background of particles. Recently, an efficient sample preparation technique for the extraction of single microbial cells from soils for NanoSIMS analysis was introduced (Eichorst et al., 2015).

In order to upscale NanoSIMS results, it will be tremendously important to combine these nano- to microscale explorations with techniques suitable for use at larger scales (Baveye, 2015). Techniques like laser ablation ICP-MS but also imaging techniques like VIS-NIR imaging might represent constructive tools for upscaling. Further development in digital image analyses, as recently demonstrated by Rennert et al. (2014), and geostatistical approaches (Mueller et al., 2012, 2013) could constitute key techniques to set NanoSIMS results in a larger perspective. These techniques allow for quantitative exploitation of the visual distribution of the isotopes detected by NanoSIMS. The colocation of labeled and unlabeled components at the microscale may thus be evaluated statistically and compared to their distribution at larger scales. This is a first step to upscale the NanoSIMS results. Another approach may include the integration and testing of concepts derived from microscale observations by modeling to verify if they improve the prediction of large-scale phenomena.

8.3 Application in Biogeochemistry

The NanoSIMS offers the capabilities to image stable isotopic ratios at the scale, where biogeochemical processes occur in a broad range of samples ranging from soils to sediments and living organisms. However, most of the natural variance in stable isotopic quantities occurs within a few per mil. As stated earlier, the current instrument limitations prevent for measuring sub-per mil precision isotopic ratios in micron-scale objects using the imaging mode of the NanoSIMS instrument. There is nevertheless a huge field of applications when considering isotopic labeling experiments, for example, element uptake, sorption, and microbial use. Isotopic labeling is a powerful tool to shed light on biogeochemical processes, as it permits to track a specific element of interest (C, N, etc.) without modification of its concentration or induction of radioactivity. Overall, when performing an isotopic labeling experiment, a labeled substrate is introduced and behaves in the same way as an unlabeled substrate. Therefore, all biogeochemical processes will happen in the same way but they are observable via the isotopic label. In pulse-chase experiments, for example, the label is brought into the system at the beginning of the experiment, during short-term exposure of the system to isotopically labeled compounds (isotopic pulse). Thereafter, the transfer of the labeled elements through different compartments of the system can be followed: transformations of the initial compounds and their chemical pathways can be elucidated. By sampling at different times after the isotopic pulse, it becomes possible to assess timescales for the investigated processes, allowing for the analyses of reaction rates of biogeochemical processes. NanoSIMS imaging is the ideal tool to reveal the location of this isotopic label at the microscale, allowing for visualization of the labeled compounds in compartments (microaggregates, plant or microbial cells) of the studied biogeochemical system. Studies with labeled material are useful to elucidate microbial functioning under varying environmental conditions. Isotopic labeling of the microbes itself may allow for studying their ecology, networks, and their interaction with organic and inorganic surfaces.

Many studies in the field of environmental science and biogeoscience have reported insights gained from NanoSIMS imaging of samples labeled with stable isotopes. Such experiments may be performed in order to visualize the uptake of ^{13}C- or ^{15}N-labeled nutrients (Musat et al., 2009), like acetate, or a chemical element relevant for the system, like Mg or Ca in organisms precipitating carbonate minerals, bringing constraints on transfers and cycling in biogeochemical systems. In soil science, stable isotopic labeling was employed to reveal N uptake in the rhizosphere (Clode et al., 2009). Double-labeling experiments have revealed that N and S cycling rates differ in their fate during organic matter digestion by earthworms (Gicquel et al., 2013) or that C and N cycles are rapidly decoupled by microbial activity in top soils (Hatton et al., 2015). Experiments with labeled organic matter can provide compelling advances on the processes that govern organomineral associations (Keiluweit et al., 2012; Vogel et al., 2014). In marine environment, Lechene et al. (2007) have followed the transfer of N inside animal cells subsequent to N_2 fixation by endosymbiotic bacteria. Organic matter transfers can therefore be evaluated thanks to stable isotopic labeling and imaging of the label down to processes occurring within single cells.

Recent developments on the combination of NanoSIMS with stable isotopic labeling have been fruitful in the studies of interactions between life and minerals. Living systems can be subjected to biomineralization: cells are then associated with the precipitation of various kinds of phosphate, carbonates, or oxides. In some cases, this is achieved to get a benefit, as for coral biomineralization. Brahmi et al. (2012) developed a methodology using ^{86}Sr pulse-chase labeling experiments to determine the extension rate of specific components of the coral skeleton. The combination of ^{86}Sr, ^{44}Ca, or ^{26}Mg isotopic labeling with NanoSIMS imaging sheds new light on the processes governing the growth of calcareous structures formed by organisms (Domart-Coulon et al., 2014; Gorzelak et al., 2014; Nehrke et al., 2013). Corals are very sensitive to concentration variations in their environment, but they do not notice isotope variations. In these studies, the addition of the stable isotope label did not significantly alter the concentrations in the growing media. Coral organisms were then grown in conditions very close to the natural environment thus avoiding disturbing their physiological development. Stable isotopic labeling is also a valuable tool to assess symbiotic relations in reef corals. Pernice et al. (2012) studied the ammonium assimilation in a similar association for the *Acropora aspera* reef coral, using ^{15}N-rich ammonium. Kopp et al. (2015) investigated inorganic C and N uptake, as bicarbonate and nitrate, respectively, in the symbiotic reef coral *Pocillopora damicornis*. In both cases, they were able to show that the dinoflagellate symbionts are responsible for the inorganic C and N uptake. In addition, they could prove that symbionts are transferring C and N as organic compounds to the reef coral. In other cases, biomineralization can be induced as a reaction to the environment composition due to cell activity or/and super saturation of phosphates or carbonates. This process, however, has often a deleterious consequence on cells, by preventing cell division or transmembrane transport. As such, Miot et al. (2015) investigated how *Acidovorax* sp. strain BoFeN1, a bacteria known to reduce nitrate and oxidize Fe(II), can overcome death due to biomineralization. The assimilation of ^{13}C-labeled acetate was used as a marker for viability (Figure 8.2). It appears from this experiment that biomineralization leads to cell death and that a few individuals remain mineral-free, due to some phenotypic heterogeneity, and hence alive. These individuals will divide once the conditions are favorable, ensuring the community viability.

As NanoSIMS cannot characterize molecular properties, a significant step forward may be achieved when NanoSIMS imaging is combined with other imaging techniques, or any other means to identify the type of organic compounds or microorganisms that are sampled by the ion beam. However, techniques used should work at a comparably meaningful scale(s). Transmission electron microscopy (TEM) is addressing similar scales and can be used to provide additional structural identification of cells or tissues (Lechene et al., 2007; Vidal et al., 2016). Hybridization techniques, like fluorescence in situ hybridization (FISH), HISH, or Card-FISH, can identify microbial species at a single-cell level by the tracking of specific DNA sequences (Dekas et al., 2009; Musat et al., 2008). The molecular signature of organic material can be investigated at the micron scale by STXM or Raman microspectroscopy in order to distinguish plant- or bacteria-derived organic matter (Keiluweit et al., 2012; Liu et al., 2013; Remusat et al., 2012). By combining the NanoSIMS with such techniques, an isotopic composition can be attributed to a type of cell or organic matter, shedding lights on the origin of the component trapping the stable isotope label (e.g., Figure 8.3). Using TEM, microorganisms are identified without ambiguity, and it is possible to distinguish bacteria from fungi or plant residues. Such identifications may be challenging with NanoSIMS images, alone. With joint comparison of the sample using both NanoSIMS and TEM approaches, it becomes possible to specifically ascribe an isotopic ratio to a type of microorganism. In a study by Vidal et al. (2016), a labeled litter was amended to a soil in the presence of earthworms. Twenty-four weeks after the beginning of the experiment, casts produced by earthworms were sampled and mounted for both TEM and NanoSIMS imaging. Results

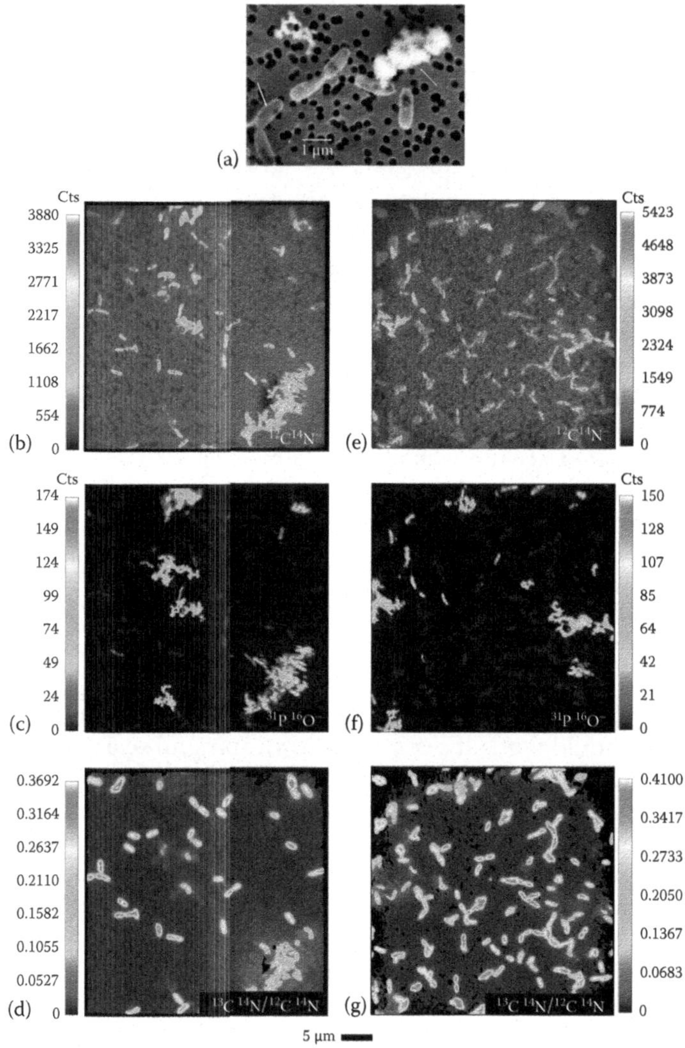

FIGURE 8.2
(See color insert.) *Acidovorax* sp. strain BoFeN1 is inducing Fe mineralization (here Fe-phosphate, image a, orange arrow) as a consequence of Fe oxidation in the presence of large amount of dissolved phosphate in the environment. A cell culture was exposed to the mineralization media for 4 days before being transferred to a media with ^{13}C-labeled acetate and no dissolved phosphate. Cells were then harvested after 1 day (NanoSIMS images b–d) or 4 days (e–g) to visualize the evolution of the bacterial colony. CN images (b, e) show the cells location, PO images (c, f) the precipitated minerals. The ^{13}C-enrichment (d, g) reveals the assimilation of acetate by living cells; this is a marker of active bacteria (e.g., living cells). Cells showing significant mineralization are free of ^{13}C-enrichment, showing that mineralization induces cell death. (Modified from Miot, J. et al., *Front. Microbiol.*, 6, 2015.)

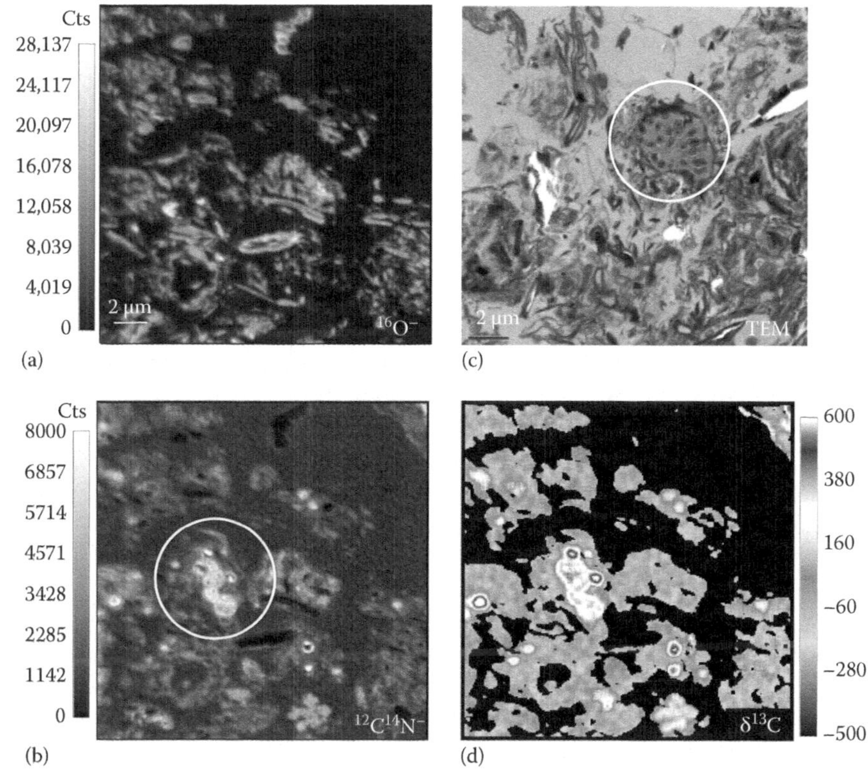

FIGURE 8.3
(See color insert.) NanoSIMS and TEM (top right) images of an earthworm cast recovered after a 24-week labeling experiment. ^{16}O image (a) is showing minerals, the $^{12}C^{14}N$ image (b) the organic matter. Thanks to the TEM image (c), we can identify the typical morphology for bacterial colonies and we observe that bacteria have trapped C from the labeled litter (shown by the $\delta^{13}C$ image (d)). (Modified from Vidal, A. et al., *Soil Biol. Biochem.*, 93, 8, 2016.)

showed that bacteria living within the cast had fixed C from the litter, demonstrating that earthworms incorporated fresh organic matter from the litter and facilitate the delivery of the substrate to decomposer microorganisms.

8.4 Main Achievements in Soil Science Thanks to NanoSIMS Analyses

8.4.1 Characterization of Microbial Communities

Biogeochemical cycles are intimately linked with microbial ecology and microbial ecophysiology. In microbial ecology, great progress has been made in the characterization of microbial community dynamics with the development of 16S and 18S rRNA sequencing and environmental genomics (Tringe et al., 2005). These techniques have revolutionized microbial ecology by identifying a huge genetic diversity of microorganisms in environmental samples, mainly bacteria, associated with functional potential. However, these approaches were successful in determining the presence of microorganisms but not in discerning their function or whether they were active or not. Thus, making functional relationships with other environmental parameters (soil type, water availability, season, substrate, temperature, etc.) was hardly possible (Wagner, 2009). This relationship may now be studied at the single-cell level and therefore at the scale at which microbial processes occur, thanks to the high lateral resolution of NanoSIMS (Li et al., 2008). NanoSIMS analyses, especially when combined with stable isotope labeling, provide deep insight into the functioning of single cells (Musat et al., 2012; Wagner, 2009). With this analytical breakthrough, characterization of microbial processes driving global biogeochemical cycles has become possible and thus changed some of the existing paradigms (Behrens et al., 2008, 2012; Musat et al., 2012; Wagner, 2009). For example, it has been shown that phylogenetics are in many cases not related to function and that therefore the well-established assumption of linkages between phylogenetic and functional prediction need to be revised even for closely related taxonomic groups (Mayali et al., 2012). This illustrates that NanoSIMS may be able to advance our understanding about the relationship between taxonomy/systematics on the one hand and function/activity on the other. Other studies successfully investigated important ecophysiological processes at the single cell level, such as uptake of elements (Behrens et al., 2008; Popa et al., 2007; Zimmermann et al., 2015). They indicate a large heterogeneity of the temporal and spatial nature of element uptake within and among individual microbial cells due to specific metabolic rates of single cells of the same species (Behrens et al., 2008; Popa et al., 2007; Zimmermann et al., 2015). Moreover, NanoSIMS analyses also revealed that species present in minor abundances may contribute greatly to biogeochemical processes (Musat et al., 2008).

Progress was also made at the process level through the visualization of carbon and nitrogen fluxes within the atmosphere–soil–plant continuum at the single-cell level (Clode et al., 2009). NanoSIMS now allows distinguishing elemental uptake at biological interfaces such as root and mycorrhiza hyphae (Nuccio et al., 2013). Thus, new experiments were set up to analyze in detail plant–microbial interactions, including mycorrhizal fungi. The results allowed for a better understanding of the quantitative importance of organic and inorganic N forms for plant nutrition. It was confirmed that NH_4^+ is the major contributor to plant N nutrition, whereas organic N forms are mainly used by C-limited soil microorganisms (Jones et al., 2013). NanoSIMS results further indicated that *arbuscular mycorrhizae* could control litter decay by influencing the litter–decaying microbial communities and through N export (Nuccio et al., 2013). *Arbuscular mycorrhizae* is also an effective distributor of recently assimilated plant-derived C to soil microorganisms (Kaiser et al., 2015). By this process, plants may be able to actively control the microbial communities involved in the decomposition and hence nutrient release from soil organic matter.

8.4.2 Biogeochemical Functioning of Soil Primary Particles and Microaggregates

Soil comprises a complex hierarchical organization characterized by a heterogeneous spatial arrangement of mineral particles, organic matter, and biota. This structural organization as well as its biogeochemical functioning control water, gas, and energy flow in soils. As already mentioned earlier, the understanding of the biogeochemical functioning of soils has long been hampered by the analyses of bulk soil samples, ignoring the intrinsic complexity of soil, that is, the 3D arrangement of minerals, organic matter, soil (micro-) organisms, water, and air. Many microbial processes determining important soil properties and ecosystem services such as carbon storage and nutrient release occur at distinctive hot spots at micro- to nanoscale biogeochemical interfaces (McClain et al., 2003; Totsche et al., 2010). Scientific advances were made with NanoSIMS using isotopic labeling to visualize the incorporation of organic C, N, and S into the soil matrix. In particular, the heterogenic distribution and composition of organic matter (OM) has been evidenced at soil mineral surfaces. With this technique, major advances concerning the stabilization processes of OM have been made. The paradigm of C storage potential of soils depending on the availability of mineral surface area free of OM has been questioned due to recent results with NanoSIMS (Vogel et al., 2014). By using short-term incubation and isotopic labeling, it has been shown that new OM is sorbed on already preexisting clusters, indicating that only a limited proportion of the total mineral surface area may be available for C sequestration (Hatton et al., 2015; Vogel et al., 2014). These results could strongly impact the way that C storage potential of soils is determined.

NanoSIMS has led to great progress in the understanding of the origin of OM stabilized at mineral surfaces. Soil incubation studies using labeled organic material, supported the preferential binding of microbial-derived OM at mineral surfaces by combining NanoSIMS isotopic analyses with compositional information derived from STXM measurements (Keiluweit et al., 2010; Remusat et al., 2012). The increased binding of microbial-derived OM known from bulk analyses (Kleber et al., 2007) was also supported by the approach of Hatton et al. (2012), who used the elemental composition (^{12}C, $^{12}C^{14}N$) of OM at mineral surfaces measured by NanoSIMS in comparison to bulk C/N ratios from classical approaches and showed that microbial-processed material is stabilized preferentially (Hatton et al., 2015). In situ studies, using long-term field experiments in combination with NanoSIMS, showed that microbial recycling of OM is spatially heterogeneous (Remusat et al., 2012) and depending on soil depth (Rumpel et al., 2015). It is assumed that this may be more intense in microbial hot pots and in subsoil horizons as compared to topsoil (Rumpel et al., 2015). Moreover, different OM compounds may be affected by this process in different depths of the soil profile (Rumpel et al., 2015). Decoupling of C and N during the stabilization process, as indicated for mineralization experiments from bulk analyses (Bimueller et al., 2014), has been noted at the microscale by several studies (Hatton et al., 2015; Rumpel et al., 2015). These findings show that phenomena existing in bulk soil also occur at the microscale. The application of these more detailed analyses to generate process information led to understanding of the coupling of biotic and abiotic controls on biogeochemical cycling. For example, very recently the paradigm of long-term stability of mineral-associated OM was challenged using a combination of NanoSIMS and STXM (Keiluweit et al., 2015). It seems that mineral-associated OM may be liberated through the addition of labile substrates such as those present in root exudates. Other studies addressed biogeochemical processes at biological interfaces created by earthworm activity. While earthworms were found to be of great importance for physical soil properties, such as aeration and water infiltration, only little is known about their impact on small-scale biogeochemical cycling of elements between plant, microorganisms, and soil. As stated earlier, NanoSIMS analyses showed the tight interaction between the OM uptake of the earthworm and the resulting enrichment of associated microorganisms in the mineral-dominated earthworm casts (Figure 8.3) (Gicquel et al., 2013; Vidal et al., 2016). These findings show that great progress in the understanding of the coupling of biotic and abiotic controls on biogeochemical cycling has already been made, and is to be expected by application of the NanoSIMS technique to environmental samples.

8.5 Summary and Outlook

Due to the growing number of facilities and easier access to spectroscopic and spectrometric imaging techniques, we see an advent of the study of biogeochemical processes at the microscale. Some of these studies led to a change of paradigms concerning our understanding of microbial ecology and elemental cycling in marine as well as terrestrial environments. Even if these proof-of-concept studies show that process information concerning microbial functioning as well as elemental cycles at biogeochemical interfaces may be obtained at the relevant scales, many issues concerning the use of NanoSIMS in environmental studies remain unresolved. There are still some technical and conceptual developments required to enhance the scientific outcome of microscale explorations. As stated earlier, the quantification of soil/sediment chemical, physical, and biological complexity is an urgent need to underpin large-scale elemental cycles. Nanoscale information on the interactions between the different living and nonliving parts of ecosystems are crucial to understand the processes behind natural phenomena. We need the combination of several classical and modern spectromicroscopic tools in order to address all these different aspects. NanoSIMS is one of these tools, which can provide additional information on compound dynamics thanks to the measure of stable isotopes. Upscaling of the information obtained at microscale is a critical issue, even if there is some evidence that processes occurring at the small scale may also be seen at larger scales. The understanding of environmental systems, as a whole, will only benefit if microscale measurements are incorporated in models considering a multitude of factors of the "3D porous medium." There is a need for new scientific concepts, experimental designs, and analytical protocols to enable the future exploration of intact soil and sediment samples at the process scale. The available techniques at different scales ranging, for example, from x-ray computed micro-tomography (μ-CT) and x-ray photoelectron spectroscopy to electron microscopy (SEM and TEM) and synchrotron-based analyses (STXM) will, together with spatially high-resolved isotopic information (NanoSIMS), set the base for a new generation of scientific concepts of biogeochemical processes and enable the development of more accurate multifunctional process models. In order to fully understand the complexity of biogeochemical microenvironments and their structures, traditional macroscopic and new microscopic techniques have to be used in scientific sound concert. In order to advance our understanding of the functioning of natural ecosystems, NanoSIMS and other microscale techniques need to be part of ecosystem research infrastructures, allowing for meaningful experiments to generate process information. This new information may then be tested in ecosystem models addressing large-scale processes with measured data obtained from monitoring facilities within ecosystem research infrastructure networks.

Glossary

Sputtering: Physical process resulting from the interaction of incident primary ions with the surface of the sample. Primary ions impact the surface and induce a cascade of collisions between the atoms of the sample. This process leads to the formation of secondary ions.

TEM (transmission electron microscopy): Transmission microscopy technique using a focused electron beam for nanometer-scale structural characterization and elemental analyses.

STXM (scanning transmission x-ray microscopy): Synchrotron-based x-ray imaging technique used to characterize, for instance, organic matter at the nanometer scale (spatial resolution 50 nm). It provides spatially resolved XANES (x-ray absorption near-edge structure) spectra. A XANES spectrum reports the absorption of x-rays of specific energies for electronic transitions and provides information about the chemical environment of a desired chemical element (e.g., functional groups in the case of C-XANES).

FISH (fluorescence in situ hybridization): Cytogenetic technique with fluorescence probes to mark specific DNA sequences, which are detected by fluorescence microscopy. CARD-FISH is its combination with catalyzed reporter deposition for signal amplification. In some experiments, the fluorescent tag can be replaced by a halogen-based tag that is easily detected by NanoSIMS (FISH-SIMS).

Raman microspectroscopy: Vibrational technique used to characterize the structure of organic matter and minerals. A monochromatic laser beam interacts with the molecular vibrations in the sample and the energy of the laser photons is shifted. This shift provides the vibrational modes in the sample, hence hints on the structural properties. Raman spectroscopy is complimentary to infrared spectroscopy.

LA-ICP-MS (laser ablation inductively coupled plasma mass spectrometry): In situ technique using a laser beam to scavenge the sample surface to analyze its elemental or isotopic composition with a high sensitivity. Typical lasers drill holes of 10–20 μm in diameter. Cannot analyze H, N, and O.

μ-CT (x-ray computed micro tomography): Radiographic imaging technique that uses x-ray to analyze the internal structure of materials. It can be used to produce 3D images at a spatial resolution >1 μm.

XPS (x-ray photoelectron spectroscopy): Surface analysis technique providing information on elemental composition and chemical state of surfaces (<5 nm depth). The lateral spatial resolution is < 7.5 μm.

References

Baveye PC (2015) Grand challenges in the research on soil processes. *Frontiers in Environmental Science*, 3(10). doi: 10.3389/fenvs.2015.00010.

Baveye PC, Laba M (2015) Moving away from the geostatistical lamppost: Why, where, and how does the spatial heterogeneity of soils matter? *Ecological Modelling*, **298**, 24–38.

Behrens S, Kappler A, Obst M (2012) Linking environmental processes to the in situ functioning of microorganisms by high-resolution secondary ion mass spectrometry (NanoSIMS) and scanning transmission x-ray microscopy (STXM). *Environmental Microbiology*, **14**, 2851–2869.

Behrens S, Loesekann T, Pett-Ridge J et al. (2008) Linking microbial phylogeny to metabolic activity at the single-cell level by using enhanced element labeling-catalyzed reporter deposition fluorescence in situ hybridization (EL-FISH) and NanoSIMS. *Applied and Environmental Microbiology*, **74**, 3143–3150.

Bimueller C, Mueller CW, Von Luetzow M et al. (2014) Decoupled carbon and nitrogen mineralization in soil particle size fractions of a forest topsoil. *Soil Biology and Biochemistry*, **78**, 263–273.

Brahmi C, Domart-Coulon I, Rougee L et al. (2012) Pulsed Sr-86-labeling and NanoSIMS imaging to study coral biomineralization at ultra-structural length scales. *Coral Reefs*, **31**, 741–752.

Clode PL, Kilburn MR, Jones DL, Stockdale EA, Cliff JB, Herrmann AM, Murphy DV (2009) In situ mapping of nutrient uptake in the rhizosphere using nanoscale secondary ion mass spectrometry. *Plant Physiology*, **151**, 1751–1757.

Dekas AE, Poretsky RS, Orphan VJ (2009) Deep-sea archaea fix and share nitrogen in methane-consuming microbial consortia. *Science*, **326**, 422–426.

Dijkstra FA, Carrillo Y, Pendall E, Morgan JA (2013) Rhizosphere priming: A nutrient perspective. *Frontiers in Microbiology*, **4**, 216.

Domart-Coulon I, Stolarski J, Brahmi C, Gutner-Hoch E, Janiszewska K, Shemesh A, Meibom A (2014) Simultaneous extension of both basic microstructural components in scleractinian coral skeleton during night and daytime, visualized by in situ Sr-86 pulse labeling. *Journal of Structural Biology*, **185**, 79–88.

Eichorst SA, Strasser F, Woyke T, Schintlmeister A, Wagner M, Woebken D (2015) Advancements in the application of NanoSIMS and Raman microspectroscopy to investigate the activity of microbial cells in soils. *FEMS Microbiology Ecology*, 91(10). doi: 10.1093/femsec/fiv106.

Falconer RE, Battaia G, Schmidt S, Baveye P, Chenu C, Otten W (2015) Microscale heterogeneity explains experimental variability and non-linearity in soil organic matter mineralisation. *PLoS ONE*, **10**, e0123774.

Gicquel A, Francez A-J, Delhaye T, Gruau G, Hallaire V, Binet F (2013) Understanding the fate and linkage of N and S in earthworm-engineered peat soil by coupling stable isotopes and nano-scale secondary ion mass spectrometry. *Biogeochemistry*, **112**, 165–177.

Gorzelak P, Stolarski J, Dery A, Dubois P, Escrig S, Meibom A (2014) Ultrascale and microscale growth dynamics of the cidaroid spine of *Phyllacanthus imperialis* revealed by Mg-26 labeling and nanoSIMS isotopic imaging. *Journal of Morphology*, **275**, 788–796.

Hatton P-J, Remusat L, Zeller B, Brewer EA, Derrien D (2015) NanoSIMS investigation of glycine-derived C and N retention with soil organo-mineral associations. *Biogeochemistry*, **125**, 303–313.

Hatton P-J, Remusat L, Zeller B, Derrien D (2012) A multi-scale approach to determine accurate elemental and isotopic ratios by nano-scale secondary ion mass spectrometry imaging. *Rapid Communications in Mass Spectrometry*, **26**, 1363–1371.

Herrmann AM, Clode PL, Fletcher IR, Nunan N, Stockdale EA, O'donnel AG, Murphy DV (2007) A novel method for the study of the biophysical interface in soils using nano-scale secondary ion mass spectrometry. *Rapid Communications in Mass Spectrometry*, **21**, 29–34.

Hinsinger P, Bengough AG, Vetterlein D, Young IM (2009) Rhizosphere: Biophysics, biogeochemistry and ecological relevance. *Plant and Soil*, **321**, 117–152.

Hoeschen C, Hoeschen T, Mueller CW, Lugmeier J, Elgeti S, Rennert T, Koegel-Knabner I (2015) Novel sample preparation technique to improve spectromicroscopic analyses of micrometer-sized particles. *Environmental Science & Technology*, **49**, 9874–9880.

Hoppe P, Cohen S, Meibom A (2013) NanoSIMS: Technical aspects and applications in cosmochemistry and biological geochemistry. *Geostandards and Geoanalytical Research*, **37**, 111–154.

Jones DL, Clode PL, Kilburn MR, Stockdale EA, Murphy DV (2013) Competition between plant and bacterial cells at the microscale regulates the dynamics of nitrogen acquisition in wheat (*Triticum aestivum*). *New Phytologist*, **200**, 796–807.

Kaiser C, Kilburn MR, Clode PL et al. (2015) Exploring the transfer of recent plant photosynthates to soil microbes: Mycorrhizal pathway vs direct root exudation. *New Phytologist*, **205**, 1537–1551.

Keiluweit M, Bougoure JJ, Nico PS, Pett-Ridge J, Weber PK, Kleber M (2015) Mineral protection of soil carbon counteracted by root exudates. *Nature Climate Change*, **5**, 588–595.

Keiluweit M, Bougoure JJ, Zeglin LH et al. (2012) Nano-scale investigation of the association of microbial nitrogen residues with iron (hydr)oxides in a forest soil O-horizon. *Geochimica et Cosmochimica Acta*, **95**, 213–226.

Keiluweit M, Nico PS, Zeglin LH, Pett-Ridge J, Weber P, Myrold DD, Kleber M (2010) C and N dynamics in soil microstructures: A joined STXM/NEXAFS and NanoSIMS approach. *Geochimica et Cosmochimica Acta*, **74**, A502.

Kleber M, Sollins P, Sutton R (2007) A conceptual model of organo-mineral interactions in soils: Self-assembly of organic molecular fragments into zonal structures on mineral surfaces. *Biogeochemistry*, **85**, 9–24.

Kopp C, Domart-Coulon I, Escrig S, Humbel BM, Hignette M, Meibom A (2015) Subcellular investigation of photosynthesis-driven carbon assimilation in the symbiotic reef coral *Pocillopora damicornis*. *mBio*, 6(1):e02299-14. doi:10.1128/mBio.02299-14.

Kuzyakov Y, Blagodatskaya E (2015) Microbial hotspots and hot moments in soil: Concept & review. *Soil Biology and Biochemistry*, **83**, 184–199.

Lechene CP, Luyten Y, Mcmahon G, Distel DL (2007) Quantitative imaging of nitrogen fixation by individual bacteria within animal cells. *Science*, **317**, 1563–1566.

Lehmann J, Liang B, Solomon D, Lerotic M, Luizao F, Schafer T, Wirick S (2005) Near-edge x-ray absorption fine structure (NEXAFS) spectroscopy for mapping nano-scale distribution of organic carbon forms in soils: Application to black carbon particles. *Global Biogeochemical Cycles*, **19**, GB1013.

Li T, Wu T-D, Mazeas L, Toffin L, Guerquin-Kern J-L, Leblon G, Bouchez T (2008) Simultaneous analysis of microbial identity and function using NanoSIMS. *Environmental Microbiology*, **10**, 580–588.

Liu X, Eusterhues K, Thieme J et al. (2013) STXM and nanoSIMS investigations on EPS fractions before and after adsorption to goethite. *Environmental Science & Technology*, **47**, 3158–3166.

Mayali X, Weber PK, Brodie EL, Mabery S, Hoeprich PD, Pett-Ridge J (2012) High-throughput isotopic analysis of RNA microarrays to quantify microbial resource use. *ISME Journal*, **6**, 1210–1221.

Mcclain ME, Boyer EW, Dent CL et al. (2003) Biogeochemical hot spots and hot moments at the interface of terrestrial and aquatic ecosystems. *Ecosystems*, **6**, 301–312.

Mcmahon G, Saint-Cyr HF, Lechene C, Unkefer CJ (2006) CN-secondary ions form by recombination as demonstrated using multi-isotope mass spectrometry of 13C- and 15N-labeled polyglycine. *Journal of American Society for Mass Spectrometry*, **17**, 1181–1187.

Miot J, Remusat L, Duprat E, Gonzalez A, Pont S, Poinsot M (2015) Fe biomineralization mirrors individual metabolic activity in a nitrate-dependent Fe(II)-oxidizer. *Frontiers in Microbiology*, 6, 879. doi: 10.3389/fmicb.2015.00879.

Moore KL, Lombi E, Zhao F-J, Grovenor CRM (2011) Elemental imaging at the nanoscale: NanoSIMS and complementary techniques for element localisation in plants. *Analytical and Bioanalytical Chemistry*, **402**, 3263–3273.

Mueller CW, Kölbl A, Hoeschen C, Hillion F, Heister K, Herrmann AM, Kögel-Knabner I (2012) Submicron scale imaging of soil organic matter dynamics using NanoSIMS—From single particles to intact aggregates. *Organic Geochemistry*, **42**, 1476–1488.

Mueller CW, Weber PK, Kilburn MR, Hoeschen C, Kleber M, Pett-Ridge J (2013) Advances in the analysis of biogeochemical interfaces: NanoSIMS to investigate soil microenvironments. *Advances in Agronomy*, **121**, 1–46.

Musat N, Foster R, Vagner T, Adam B, Kuypers MMM (2012) Detecting metabolic activities in single cells, with emphasis on nanoSIMS. *FEMS Microbiology Reviews*, **36**, 486–511.

Musat N, Halm H, Winterholler B et al. (2008) A single-cell view on the ecophysiology of anaerobic phototrophic bacteria. *Proceedings of the National Academy of Sciences of the United States of America*, **105**, 17861–17866.

Musat N, Halm H, Winterholler B et al. (2009) Linking microbial activity to cell identity in the environment using nanoSIMS. *Geochimica et Cosmochimica Acta*, **73**, A923.

Nehrke G, Keul N, Langer G, De Nooijer LJ, Bijma J, Meibom A (2013) A new model for biomineralization and trace-element signatures of Foraminifera tests. *Biogeosciences*, **10**, 6759–6767.

Nuccio EE, Hodge A, Pett-Ridge J, Herman DJ, Weber PK, Firestone MK (2013) An arbuscular mycorrhizal fungus significantly modifies the soil bacterial community and nitrogen cycling during litter decomposition. *Environmental Microbiology*, **15**, 1870–1881.

Pernice M, Meibom A, Van Den Heuvel A, Kopp C, Domart-Coulon I, Hoegh-Guldberg O, Dove S (2012) A single-cell view of ammonium assimilation in coral-dinoflagellate symbiosis. *ISME Journal*, **6**, 1314–1324.

Popa R, Weber PK, Pett-Ridge J et al. (2007) Carbon and nitrogen fixation and metabolite exchange in and between individual cells of *Anabaena oscillarioides*. *ISME Journal*, **1**, 354–360.

Remusat L, Hatton P-J, Nico PS, Zeller B, Kleber M, Derrien D (2012) NanoSIMS study of organic matter associated with soil aggregates: Advantages, limitations, and combination with STXM. *Environmental Science & Technology*, **46**, 3943–3949.

Rennert T, Haendel M, Hoeschen C, Lugmeier J, Steffens M, Totsche KU (2014) A NanoSIMS study on the distribution of soil organic matter, iron and manganese in a nodule from a Stagnosol. *European Journal of Soil Science*, **65**, 684–692.

Rumpel C, Baumann K, Remusat L et al. (2015) Nanoscale evidence of contrasted processes for root-derived organic matter stabilization by mineral interactions depending on soil depth. *Soil Biology and Biochemistry*, **85**, 82–88.

Schmidt MWI, Torn MS, Abiven S et al. (2011) Persistence of soil organic matter as an ecosystem property. *Nature*, **478**, 49–56.

Slodzian G (2004) Challenges in localized high precision isotope analysis by SIMS. *Applied Surface Science*, 231–**232**, 3–12.

Totsche KU, Rennert T, Gerzabek MH, Koegel-Knabner I, Smalla K, Spiteller M, Vogel H-J (2010) Biogeochemical interfaces in soil: The interdisciplinary challenge for soil science. *Journal of Plant Nutrition and Soil Science*, **173**, 88–99.

Tringe SG, Von Mering C, Kobayashi A et al. (2005) Comparative metagenomics of microbial communities. *Science*, **308**, 554–557.

Vidal A, Remusat L, Watteau F, Derenne S, Quenea K (2016) Incorporation of 13C labelled shoot residues in *Lumbricus terrestris* casts: A combination of transmission electron microscopy and nanoscale secondary ion mass spectrometry. *Soil Biology and Biochemistry*, **93**, 8–16.

Vogel C, Mueller CW, Hoeschen C et al. (2014) Submicron structures provide preferential spots for carbon and nitrogen sequestration in soils. *Nature Communications*, **5**, 2947.

Wagner M (2009) Single-cell ecophysiology of microbes as revealed by Raman microspectroscopy or secondary ion mass spectrometry imaging. *Annual Review of Microbiology*, **63**, 411–429.

Young IM, Crawford JW (2004) Interactions and self-organization in the soil-microbe complex. *Science*, **304**, 1634–1637.

Young IM, Blanchart E, Chenu C et al. (1998) The interaction of soil biota and soil structure under global change. *Global Change Biology*, **4**, 703–712.

Zimmermann M, Escrig S, Huebschmann T et al. (2015) Phenotypic heterogeneity in metabolic traits among single cells of a rare bacterial species in its natural environment quantified with a combination of flow cell sorting and NanoSIMS. *Frontiers in Microbiology*, **6**, 243.

9

Climate Warming Experiments: Selecting the Appropriate Technique

Hans J. De Boeck and Ivan Nijs

CONTENTS

Abstract

Experiments allow the testing of elaborate hypotheses to gain mechanistic understanding of how ecosystems respond to climate change. Establishing warming experiments is an integral part of this quest for knowledge. Many methods exist to increase temperatures, some of which use energy from the environment (passive warming) while others rely on electric power (active warming). We discuss strengths and weaknesses of each of the methods, also focusing on recent improvements. Passive warming methods are generally less costly and require less technological know-how compared to active methods, but this comes at the expense of control over the achieved temperatures. Active warming methods provide more control, but their reliance on electric power limits their use in remote locations. Weaknesses differ among the methods: asymmetric warming, unwanted co-occurring changes in other environmental variables, disturbances, etc. No warming method is free of flaws, yet every method also has distinct advantages. The fact that strengths and weaknesses differ among the methods also implies that researchers can pick the most optimal method for their specific experiment. This would avoid mismatches between the warming method and the study at hand and should thus lower the chances of conclusions being skewed by experimental artifacts.

Keywords: Active, artifacts, climate change, heating, passive, state-of-the-art, technology, temperature

9.1 Introduction

Experimental approaches are invaluable tools to promote better understanding of ecological processes in the context of a changing climate. They enable researchers to test specific hypotheses and elucidate unknown processes and the effects of their drivers (see Chapters 1 and 2), with the ultimate goal of gaining fundamental mechanistic knowledge, so that case-specific results can be extrapolated to other systems and situations (Beierkuhnlein and Nesshöver, 2006). One of the most pressing issues today is the impact of ongoing human-induced changes to the climate system. Several consequences of climate change have already been observed in ecosystems around the world, such as upward and poleward migration of species (Parmesan and Yohe, 2003), drying of tropical rainforests (Zhou et al., 2014), and earlier green-up of deciduous tree species (Menzel et al., 2006). These observations demonstrate that even small changes in the global climate can significantly alter the structure and functioning of ecosystems, with important potential feedbacks to the climate, primarily via changes in the carbon balance (e.g., Heimann and Reichstein, 2008). As deviations from the preindustrial climate are fast becoming more pronounced, an increasing number of studies that attempt to mimic certain aspects of (potential) future environmental conditions are being conducted, in order to gain understanding of likely impacts of further changes to the local and global climate.

As increases in surface temperatures are a key aspect of climate change, studies considering warming as an explicit experimental treatment are highly relevant in global change research. Warming studies may focus on various aspects of climate change. In some experiments, year-round temperature increases are considered, which ideally need to be imposed for a long time to avoid noise generated by natural interannual variability from obfuscating responses to the treatment (cf. Hollister and Webber, 2000; Norby and Zak, 2011). Other research endeavors are more concerned with the effect of temperature increases in certain seasons, such as warm winters

(Schuerings et al., 2014), or earlier onset of spring and the subsequent impacts on phenology and/or legacy effects in the rest of the year (e.g., Fu et al., 2014). Finally, some studies focus on large but short-lived deviations from normal temperatures (heat waves), which can have significant impacts on ecosystem structure and functioning both in the short and in the long term (Dreesen et al., 2015). Not all warming methods are suitable for every of such research goals, as will become apparent throughout this chapter.

Broadly, two categories of warming methods can be distinguished: those using passive mechanisms that harness energy from the environment and those where temperature is changed and controlled actively, relying mostly on electric power. In this chapter, we present an overview of widely used methods of either category and discuss the advantages and disadvantages of each technique (which often result from trade-offs between cost, control, practicality, and ecosystem disturbance). We also highlight recent developments that resolve some of the weaknesses or artifacts involved. The ultimate aim is to help researchers in selecting the most appropriate technique for each experiment.

9.2 Passive Warming Methods

9.2.1 Open-Top Chambers

Many of the first warming experiments made use of passive warming in closed chambers (e.g., Chapin and Shaver, 1985). Unfortunately, this approach came with a number of important concerns, chief of which were high-temperature peaks and significant, unwanted alteration of environmental variables such as humidity, light, and wind speed (Kennedy, 1995). Over the course of the years, open top chambers (OTCs) were increasingly preferred as these avoided the most excessive artifacts associated with their closed predecessors (Marion, 1996). The passive warming in OTCs is achieved by slowing heat dissipation, as the sides of the chamber reduce the influence of outside airflows (which carry away warmed air and cool down the canopy and soil). They have been used in different climates (e.g., De Frenne et al., 2010; Cavieres and Sierra-Almeida, 2012; Bokhorst et al., 2013), with the warming effect sometimes being an unintended side effect in early atmospheric CO_2 enrichment studies (Drake et al., 1989). The construction of OTCs requires no special technical skills, and the materials used (mostly transparent PVC plates and simple connections to keep these together) are commonly available and inexpensive. In accessible areas, more complex OTCs with active support via ventilated air in combination with heating and cooling systems (cf. Pelini et al., 2011) can be employed to increase control over the achieved temperatures, making them no longer strictly passive (see Section 9.3.2). OTCs that are only warmed by passive means have been used extensively especially in cool and remote regions such as the arctic (Wahren et al., 2005; Strebel et al., 2010) where their essentially power- and maintenance-free nature provides a major logistic advantage. Their appeal is therefore understandable, although already early on researchers acknowledged a number of weaknesses.

The achieved passive increases in air and soil temperature are generally modest in the regions and ecosystems where they are most employed (in the range of 0.5°C–1.5°C), and the amount of warming is also variable in time (Marion et al., 1997; De Frenne et al., 2010). Because the warming signal tends to be small in comparison to interannual variability (Hollister and Webber, 2000), experiments using passive OTCs are required to run for many years before being able to confidently attribute changes in ecosystem functioning or composition to the (small) increases in temperature (Hollister et al., 2005; Hudson and Henry, 2010). Nevertheless, a recent study demonstrated that at the level of the aboveground plant tissues, the daytime warming was more substantial than reported increases in air temperature because of significantly lower wind speeds (De Boeck et al., 2012a). Apart from increasing canopy temperatures, calmer conditions can also alter exchange of important gasses like CO_2 (Kimball et al., 1997) and can render wind pollination more difficult (cf. Molau and Shaver, 1997). Moreover, the decreased wind speeds inside the chambers also increase temporal variability of temperature, that is, they tend to increase maximum and decrease minimum temperatures (especially under clear-sky conditions), which differs from the general reduction in diurnal temperature range projected under climate change (Lindvall and Svensson, 2015). The variability of the achieved level of warming is also more pronounced between days and seasons, because the amount of warming is dependent on incoming energy, therefore leading to lower increases under cloudy conditions and in winter time.

Next to temporal variability of the achieved warming, spatial variation is also an issue inside OTCs, with substantial changes in surface temperature observed in the space of less than a meter (Figure 9.1). Another reported artifact is that snow tends to get trapped within the OTCs (Marion et al., 1997), which can change soil temperatures because snow cover serves as an insulation layer (cf. Wahren et al., 2005) and which can alter the growing season length if the date of snowmelt is affected. Finally, many of the commonly used passive OTCs are of a fairly small size (1–2 m across), meaning that island effects (cf. De Boeck et al., 2015), which tend to dilute experimentally imposed treatments, are likely. For example, soils will be cooler at the plot edge due to contact with the surrounding (unwarmed) soil, and roots are likely to grow outside of the chamber's zone of influence.

The artifacts and drawbacks associated with passive OTCs can be avoided to an extent by taking known artifacts into account. Researchers could, for example, only consider plants from the central patch, as changes in the microclimate (radiation, air circulation, moisture, etc.) are less acute there. Also, acquiring data of variables such as snowmelt dates, tissue temperatures, and

FIGURE 9.1
(See color insert.) Infrared image of a passive OTC covering a lawn, taken on a sunny day. (Copyright H.J. De Boeck.) Note the significant differences in surface temperature within the chamber.

irradiance would allow more precise linking of cause and effect, therefore increasing the mechanistic understanding of observed biotic responses to experimental warming, as illustrated by Bokhorst et al. (2013). Other technical improvements have also been proposed. For example, one of the most promising refinements is the use of thermally inert mass in the form of a liquid contained inside plastic pipes at the edge of the OTC (Godfree et al., 2011). This creates a more stable warming, topping off both hot and cold peaks by about 1°C, thus avoiding some of the reported increases in temporal variability in normal OTCs. Moreover, some of the island effects may also be decreased by introducing such a warm rim around the plots. Unfortunately, this improvement may in turn create new artifacts, such as increased shading. Although clearly not free of drawbacks, the major advantage of being nearly maintenance- and cost-free will ensure that passive OTCs will continue to be used, especially in remote and inhospitable environments.

9.2.2 Thermal Screens

A second method that utilizes energy from the environment to experimentally warm plots is the use of thermal screens. This method was developed in the 1990s, based on reducing the amount of net radiation lost from the land surface at night to reduce nighttime cooling. To that end, infrared reflective curtains (thermal screens) are suspended above the vegetation, which reflect a portion of the outgoing infrared radiation back to the vegetation and soil. In one of the first of such experiments, Zeiher et al. (1994) covered a cotton crop with thermal screens during 42 consecutive nights, which led to a 1°C–6°C increase in canopy temperatures compared to controls. Based on such early work, this method was used in the pan-European CLIMOOR project, where the reflective screens were applied using automated shelters, creating a modest 0.4°C–1.2°C increase in minimum air and soil temperatures (Beier, 2004).

A major advantage of using thermal screens is that there is no light attenuation (which causes a reduction of photosynthesis and can decrease surface temperatures; cf. De Boeck et al., 2012a). One aspect of the method that has been criticized is the asymmetric nature of the warming. Amthor et al. (2010) note that nighttime-only warming may be "of limited value to understanding effects of future warming on terrestrial ecosystems" as they argue that further reductions of the diurnal temperature range (i.e., more warming at night than during day) are uncertain and geographically varied. However, the most recent model projections still indicate that the diurnal temperature range is likely to decrease further (Lindvall and Svensson, 2015). Furthermore, applying the screens only at night actually also results in limited daytime warming because of lag effects via the warmer soils (Bruhn et al., 2013).

A more pertinent critique is that, like for passive OTCs, the achieved warming depends greatly on the accumulated radiation (Bruhn et al., 2013). This leads to similar issues of increased variability through time, as the weather conditions are so important in determining the realized temperature increase. An additional factor to be considered in the case of thermal screens (a horizontal surface, contrary to the more or less vertical plates in OTCs) is the more pronounced influence of wind speeds, which can again increase variation within a day and between days. On the other hand, the more natural wind environment reduces a number of artifacts linked to OTCs such as abnormal buildup of water vapor or other gasses. Indeed, Bruhn et al. (2013) reported that artifacts related to changes in air humidity or dew formation are generally minimal or insignificant. Another clear advantage compared to passive OTCs is that warming can be applied to much larger plots (20 m^2 in the case of the CLIMOOR project; Beier, 2004), so that artifacts associated with small plot sizes such as edge effects are likely to be reduced. On the downside, construction and maintenance is more complicated and costly than OTCs, and although the warming in itself is passive, automated screens require an external power source.

In order to remedy some of the weaknesses associated with this technique, several improvements were tested by Bruhn et al. (2013). Higher-temperature increases (adding a few extra tenths of degrees) were achieved by using screens with higher infrared radiation reflection and installing shields to reduce lateral wind speeds (although this could lead to other artifacts; see Section 9.2.1). The addition of thermal mass, much like in improved OTCs, was not effective, likely because of the large plot size. In spite of the limited temperature increases possible when using thermal screens (although this depends on the climate; cf. Zeiher et al., 1994), issues with asymmetric and variable changes in temperature, and the technical and logistical challenges involved, warming by thermal screens may be appealing in certain cases as it allows large plot sizes to be covered with no or minimal disturbance. Although the method has been used to cover short-statured vegetation, Luxmoore et al. (1998) developed a theoretical setup with thermal screens deployed with a cable and pulley systems mounted on a tower and scaffolding structure to warm forest sites. Such an experiment would require electric power to open and close the screens, however, and has to the best of our knowledge not been put into practice so far.

9.2.3 Translocation

A method that requires no outside power output to achieve warming but that allows more control is the translocation of monoliths (i.e., patches of vegetation with intact soil in containers or lysimeters) to a different climate zone. This can encompass relocating across hundreds of kilometers (especially to lower latitudes) or across mere kilometers (to lower elevations) when the landscape has suitable topography (hills or mountains). As an example of the former, Backhaus et al. (2014) moved *Fagus sylvatica* seedlings across more than 250 km, achieving an average air warming of 3.2°C. Regarding experiments in mountainous regions, there are studies that look into impacts of mild changes in temperature sustained for multiple years, like in Switzerland where monoliths were excavated from 2100 m a.s.l. and translocated to elevations ranging from 1680 to 2360 m a.s.l. (Seraina Bassin, pers. comm.). On the other hand, large vertical gradients have also been used in short-term studies on heat waves, for example, in the study of De Boeck et al. (2016), where an 1800 m vertical gradient led to average warming up to 11°C compared to the reference site (Figure 9.2). This illustrates that monolith translocation allows a wide range of warming. Moreover, the temperature increase is fairly predictable, especially if sites are not located too far apart and if the experiment lasts longer.

A major advantage is that a number of environmental parameters change in a manner that is realistic for a future climate. If site selection is carefully made, changes in air humidity, cloudiness, and wind speed can be consistent with climate projections. For example, relative humidity is expected to change little on average with climate warming, but if heat waves are the focus of the study, selecting a site with lower air humidity would improve realism as air humidity tends to decrease during heat waves (cf. De Boeck et al., 2010). Further "free" improvements of experimental realism using translocation compared to other warming methods are the increased percentage of precipitation as rainfall instead of snow and the change in partial pressure of CO_2 in studies using an elevational gradient. Indeed, warming by translocation encompasses a relative enrichment with CO_2, as a 100 m

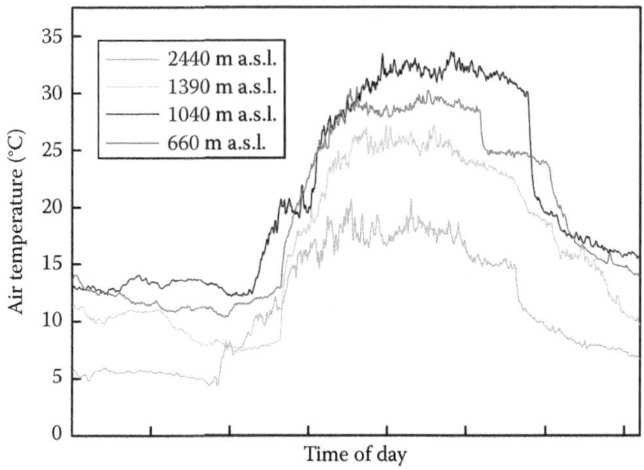

FIGURE 9.2
(**See color insert.**) Course of air temperatures during one day (July 31, 2013) measured at 40 cm height (just above the vegetation) at four sites used in a translocation experiment. Remarkably, the lowest site was not the warmest, due to specific local microclimates. (From De Boeck, H.J. et al., *New Phytol.*, 209, 531, 2016.)

difference (= ±0.7°C warming) in altitude approximately corresponds to a 1% change in the pressure of CO_2 (Körner, 2007). This amount of CO_2 fertilization is about an order of magnitude lower than what could be expected in a future environment, however.

Translocation also comes with a number of disadvantages. First, some conditions may change in an unwanted manner. Litter inputs (quantity and quality) originating from the environment can differ from the site of origin (e.g., when moving monoliths from above to below the tree line), as can radiation, pollution, herbivory, and pollination. Some of these artifacts, such as changes in pollution, could be mitigated by careful site selection, while others, such as those related to plant–animal interactions (e.g., with potential impacts on community composition), are largely unavoidable. Second, several issues are specifically associated with the use of monoliths. During the extraction of the soil blocks, some plant roots are inevitably cut, which can affect subsequent growth and induce additional edge effects (as roots from plants near the sides are trimmed to a larger extent than plants toward the center). As a minimal precautionary measure, data collection should be focused on plants in the center of the plot. Furthermore, the lysimeters containing the monoliths tend to warm up more than the outside soil, even when buried (Klaus Butterbach-Bahl, pers. comm.). To reduce this soil warming artifact, the lysimeter construction can be adapted, for example, by providing passive cooling (via evaporation of a water reservoir) at the bottom and by isolating the rim at the top. Third, the spatial separation of the warming treatments complicates measurements. To survey essential environmental variables such as temperature, humidity, and radiation, sensors and data loggers need to be installed at each site, which can imply a significant additional cost compared to when treatments are imposed at the same site. Furthermore, manual measurements cannot be carried out simultaneously, unless multiple collaborators and multiple precisely cross-calibrated measurement devices are available. As many variables of interest, such as gas exchange, differ within and between days, measurement schemes and data analyses have to be adapted to reflect the fact that individual data recordings cannot be compared one-on-one between two warming treatments. This encompasses spreading out measurements within one day to avoid intraday variability skewing results. For example, if stomatal conductance would only be measured before noon at site A and only in the afternoon at site B, this could affect the interpretation of the responses, as stomatal conductance is known to vary throughout the day (and to a larger extent in case of warming and/or drought; cf. Nijs et al., 1997). If data are acquired on many days during the experiment, data analysis can be focused on the general trends in time, which incorporate day-to-day variability so that the problem that not every measurement can be made at every site on each day is significantly reduced (e.g., De Boeck et al., 2016).

The spatial separation of treatments, which lies at the basis of the monolith translocation method, causes a number of difficulties, as discussed earlier. These may make precise attribution of cause and effect more difficult and can complicate logistics and data analysis. Yet in spite of this, using translocation as a warming method has the distinct advantage that it is very flexible (the amount of warming can be chosen relatively freely), open air (which avoids a number of artifacts), and relatively precise. A basic requirement is that sites are chosen carefully to avoid issues with locally differing conditions such as microclimate, litterfall, and pollution and that the measurement and data analysis protocols are designed with the specific characteristics of translocation studies in mind.

9.2.4 Geothermal Hot Spots

Harnessing the possibilities provided by natural anomalies is not new. Several studies have made use of locally increased levels of CO_2 (e.g., Tognetti et al., 2000). A similar approach also exists for temperature by using hot spots found in geologically active regions such as New Zealand (Burns, 1997) and Iceland (O'Gorman et al., 2014). Locally, gradients can be found where the soil is heated in different amounts without concomitant factors such as changes in pH and/or nutrients and pollutants interfering with straightforward comparisons along the warming gradient. Employing such a gradient and including extreme ranges brings the added benefit of being able to detect nonlinear trends and tipping points better (Kayler et al., 2015).

Because this type of passive warming is fully *in situ*, artifacts associated with excavation or shielding described earlier are avoided. The plots can often be chosen close to each other, with a gradient from +0°C (control) to

FIGURE 9.3
(See color insert.) FORHOT site in Iceland, where an earthquake in 2008 led to parts of the forest and grassland being exposed to marked increases in soil temperature. A gradient with plots warmed with 1°C–20°C at 10 cm depth was established here. (Copyright H.J. De Boeck.)

+20°C in soil temperature (at 10 cm depth) present within approximately 100 m at the Icelandic "FORHOT" grassland site (Figure 9.3). This allows measurements to be performed in rapid succession and under very similar conditions. One of the main other advantages of the use of hot spots is the fact that the warming is free and maintenance costs are mainly limited to loggers. Such loggers are evidently essential to keep track of temperatures during but also prior to the experiments. With good a priori quantification of isotherms, the desired warming levels can be chosen with relative confidence. Finally, interesting comparisons can be made between sites where the warming has been present for a long time and sites where soil warming is new. Such is the case in the FORHOT experiment in Iceland, where seismic shifts caused a grassland and forest to become affected by warming since 2008, while elsewhere in the area there is another grassland where the geothermal warming has been present for at least several decades (and likely for several centuries). By comparing responses to similar levels of warming, transient responses and long-term acclimation can be identified.

Of course, the warming is ultimately not controlled, so new geological activity may change the temperature gradient (or provide new opportunities). A more pronounced concern is that the warming originates from the soil, which implies that the aboveground part experiences temperatures that are close to ambient, especially in higher vegetation. This asymmetrical nature of the warming renders interpretation of its effects less straightforward, as important feedbacks exist between above- and belowground systems

(Van der Putten et al., 2009). Therefore, hypotheses have to be very well delineated to avoid extrapolation of trends that would change if the warming was symmetrical. As the heat source originates from deep within the soil, there is a vertical gradient of higher to lower temperatures in the direction of the soil surface, regardless of the season. In winter, such a situation is natural (soil temperature is fairly constant near the annual average in deeper soil layers), but in summer a reversed thermal profile would be expected under ambient conditions because of stronger irradiation and high air temperatures. Finally, even when having existed for hundreds of years, local hot spots still remain islands in a wider landscape that does not experience the same warming. As a result, mobile organisms may be attracted more to those warm patches than they would if the entire environment was warmed (De Boeck et al., 2015). This may affect, for example, pollination or grazing and hence alter processes and ecosystem characteristics down the line, such as community composition.

Such island effects are a phenomenon that occurs to some degree in all warming experiments and that is largely unavoidable, however. In general, making use of natural hot spots seems to provide some unique advantages. One is the relative ease with which gradients can be found where temperature changes while other environmental variables remain unaffected. Another important asset is the possibility of comparing locations where the temperature change is new with sites where the warming has been present for a considerable time. Nevertheless, hypotheses need to explicitly acknowledge the asymmetrical nature of the warming so that extrapolation of responses beyond the inherent limitations associated with soil-only warming is avoided.

9.3 Active Warming Methods

9.3.1 Soil Heating Cables

The use of soil heating cables is the least costly and the least technically demanding of the three active warming methods discussed here. Major studies applying this technique were started up in the 1990s (Peterjohn et al., 1993; Bergh and Linder, 1999), but heating cables have been used in ecological research even well before that (e.g., McBee et al., 1968). The method makes use of thin cables laid out in several rows (usually in one horizontal plane) with spacing and burial depth varying between experiments (Patil et al., 2013). The amount of warming is then controlled by modulation to preserve a constant difference between soil temperature sensors deployed in warmed and control plots. Usually, this control is precise and malfunctions are rare (Strömgren and Linder, 2002; Patil et al., 2013, but see Schindlbacher et al., 2009). Another major advantage is that the warming can be imposed also in forests, as the technical challenges of warming tall vegetation are not applicable here. The relative cost-effectiveness is a major advantage when performing research on long-living organisms such as trees, as responses found in short experiments may be especially misleading in such systems (cf. Norby and Zak, 2011). A prime example of a long-term warming study is found in Flakaliden (Sweden), where soil warming has been imposed for 18 consecutive growing seasons (Leppälammi-Kujansuu et al., 2013).

The temperature at Flakaliden was increased to simulate earlier soil thawing in spring, thus advancing the onset of the growing season by several weeks. Soil warming releases this boreal system from belowground cold limitation (due to the frozen soils) while cold limitation for the aboveground structures is largely absent in this season. Hence, the asymmetrical nature of the warming is less of a concern in this case. The relevance of the results when using soil heating cables depends very much on carefully developing assumptions that are well grounded in theory and avoiding a focus on processes that are highly dependent on aboveground warming, much like for soil warming through passive means (Section 9.2.4). In contrast with warming via passive methods, the installation of heating cables causes soil disturbance through the digging of slits (unless the cables are deployed at the soil surface), which cut through roots. Obviously, the same procedure needs to be followed in control plots so that the disturbance is similar. The system tends to recover from this fairly rapidly (Sune Linder, pers. comm.), implying that longer-term results should not be affected significantly by the initial perturbation. The technique creates both horizontal and vertical temperature gradients in the soil (as temperature decreases away from the cables), which is another drawback. These gradients differ from those found in naturally warmer soils especially if the cables are buried neither at the soil surface (e.g., Reth et al., 2009), which would be reminiscent of summer warming, nor in deep soil layers (Hanson et al., 2011), which would mimic winter warming. Finally, the increased soil temperatures inevitably increase evaporation, which can lead to drought. Depending on the study's goals, this drought effect may be part of the experimental design, or it can be avoided by providing irrigation (like was done in the Flakaliden experiment).

In general, soil warming provided via heating cables comes with a number of clear limitations and artifacts, of which asymmetric nature of the warming is the most important. Its major appeal is that it currently is the only technique that has been used to actively warm mature forests (not merely single trees). The relative cost-effectiveness allows for long-term warming experiments, which is an additional benefit when testing responses to temperature increases in long-lived ecosystems.

9.3.2 Climate-Controlled Chambers

Early CO_2 enrichment experiments made use of OTCs and were confronted with concurrent warming as an unwanted additional treatment (e.g., Drake et al., 1989). In response to this warming side effect, researchers added systems to dissipate warmer air by ventilation with outside air (e.g., Norris et al., 1996). Although this reduces the temperature increase measured within the OTCs, such an approach cannot keep temperatures close to those measured outside the chambers under all circumstances (Leadley and Drake, 1993), especially in late spring and summer when radiation loads on the system are very high. This is why active cooling systems were developed that are able to maintain the air temperature inside the chambers closer to ambient temperatures (e.g., Van Oijen et al., 1999) and that are often complemented with heaters so that warming could be included as an additional factor in a full-factorial design. Similar climate controls that cool or warm the air on demand have also been used in complete enclosures (e.g., Naudts et al., 2013).

Creating a fully climate-controlled environment places substantial demands on technology (sensors, control algorithms, etc.) and requires a high amount of external energy (cf. Pelini et al., 2011). A major benefit, apart from the tight control, is that the air is the medium that is warmed, which matches the characteristics of natural temperature increases. Nevertheless, some dampening may occur in the soil due to the influence of outside, unwarmed soil (another island effect). In complete enclosures, other climate variables such as atmospheric CO_2 concentrations and precipitation can be varied in a straightforward manner. Some setups also allow varying air humidity (Barton et al., 2010), which is highly relevant to avoid conditions that are drier than they would be in a future climate. Indeed, warming depletes the air of humidity in a relative sense while climate projections indicate that relative humidity would be generally unchanged in a warmer climate (IPCC, 2013). Installing air humidifiers can remedy this potential artifact and also increases the range of climate scenarios that can be imposed experimentally. The most sophisticated systems, sunlit ecotrons (e.g., Milcu et al., 2014), enable continuous sampling of gasses so that whole-system CO_2 and H_2O fluxes can be followed almost in real time. The size of climate-controlled greenhouses usually allows for experiments on short-statured systems (grassland, crops, heather, etc.) and young trees (Bauweraerts et al., 2013), but even single adult trees have been used (Medhurst et al., 2006).

Enclosing a plant or ecosystem offers some advantages, such as increased control and the possibility to measure gas fluxes continuously, but also leads to a number of artifacts. Because the radiation environment inside is different from outside (not only because of radiation reduction but also because the cover materials are usually warmer than the sky and therefore radiate more heat), there is the possibility that leaf temperatures would deviate. Results obtained using an energy balance model show that this different radiation environment does not lead to large leaf temperature changes (under ambient air temperatures) compared to outside (De Boeck et al., 2012a). The same model demonstrated that avoiding low wind speeds is crucial as these affect leaf temperatures substantially (by several degrees). Adjustable ventilation should therefore be a priority in the technical design of greenhouse experiments. Light attenuation is impossible to avoid, with reductions of 15% of photosynthetic active radiation commonly reported. This could be remedied by installing additional lighting, especially in climates where light reduction is expected to have important impacts. Another concern is that many materials block (part of the) UV wavelengths, which can additionally affect plant growth and functioning (Kittas et al., 2006). Proper material choice (Milcu et al., 2014) can avoid this. Finally, the (partially) enclosed nature of climate-controlled chambers also changes or blocks out interactions with many higher trophic groups (such as pollinators and herbivores). Apart from the targeted introduction of certain species (e.g., Stevnbak et al., 2012), no real solution exists (cf. De Boeck et al., 2015).

Climate-controlled chambers require a fairly big budget and technical know-how, and some of the downsides associated with enclosing ecosystems may mean that they are not the proper choice for certain studies (e.g., if a large spatial scale is required). Nevertheless, they provide unique research opportunities through their flexibility regarding potential treatments, their level of control, and the possibility to instantly measure whole-system gas exchange in the most sophisticated systems.

9.3.3 Infrared Heating Systems

The use of infrared heaters in ecological research was introduced in the 1990s to be able to warm ecosystems under free-air conditions, first by Harte et al. (1995) in a field experiment where commercially available IR lamps were suspended above montane meadow plots. The lamps were set to emit a constant flux, effectively increasing the amount of incoming radiation, resulting in temperature increases in vegetation and soil. Later, Nijs et al. (1996) adapted this technique by adding a modulator so that a constant difference between the surface temperatures of warmed and control plots could be maintained. Such modulation greatly decreases variability of the amount of warming caused by fluctuations in wind speed and other environmental variables that influence canopy temperatures.

However, maintaining a fixed temperature difference between plots also implies that normal physical and physiological feedbacks are negated. In reality, warmed plots tend to dry out more, and drought-stressed vegetation should warm up more than well-watered vegetation (Larcher, 2003), leading to potential differences in heat stress. Control systems that aim at a constant temperature difference between heated and unheated plots will eliminate this natural drought-associated heat by lowering the emitted infrared radiation, however. A control method was recently developed that takes this natural feedback into account while still allowing a great amount of control (De Boeck and Nijs, 2011). By using energy balance equations supported by instantaneous measurements of all the relevant environmental variables, realistic canopy temperatures are achieved while allowing input of target air temperatures. The latter is another advantage compared to conventional infrared heating, which relies on canopy temperature inputs, while these are not generally explicit in climate projections. A variation on this method, but with the same general advantages, was proposed by Kimball (2015).

One of the major benefits of using IR heaters is that it does not require enclosing the plants, therefore avoiding most artifacts associated with enclosures (reduced light and wind, absence of pollinators, etc.). This also means IR warming can be combined with the well-established free-air CO_2 enrichment technique (LeCain et al., 2015). In principle, the surface area of plots is unlimited as new hexagons (the normal shape of infrared heater arrays) can be added (Kimball et al., 2012). The fact that warming is direct because infrared lamps heat the surface without having to overcome a boundary layer resistance makes the technique very responsive (Nijs et al., 1996; Kimball et al., 2008). Obviously, this implies that the air is only warmed indirectly and generally to a lesser extent than the surface (a characteristic of infrared heating that is sometimes misinterpreted). Finally, large deviations from the ambient (canopy) temperatures can be achieved (+8°C and more), allowing heat wave research (e.g., Dreesen et al., 2014).

A drawback of infrared irradiation is that it works well for low-statured vegetation (e.g., tundra, grassland, heather) but that multilayered vegetation poses serious challenges. Although theoretically not impossible, warming a forest would require infrared heaters suspended throughout the canopy, resulting in substantial technological complexity (e.g., differential energy outputs in different canopy layers) and high running costs. Similarly, the fact that the technique warms surfaces also results in artifacts in less complex canopies. First, especially in dense canopies, soil warming can be lower than could be expected under air warming (Rich et al., 2015), although this problem is not universal (Nijs et al., 1996). Second, the gradient for water vapor from surface to air is increased, resulting in higher evapotranspiration. This artifact was recently quantified (De Boeck et al., 2012), with results suggesting a 12%–15% increase in transpiration for a 1°C canopy warming compared to 1°C air warming at constant relative humidity. Potential resolutions are adding supplemental irrigation water (Kimball, 2005) or making use of free-air humidification systems (Kupper et al., 2011). Importantly, the artifact is less of a concern when simulating heat waves as the higher atmospheric water demand underneath the heaters reflects naturally occurring increases of potential evapotranspiration during heat waves resulting from atmospheric feedback, that is, drier and warmer air (De Boeck et al., 2010).

Infrared heating seems most suitable for warming small statured vegetation and especially for imposing short but intense temperature increases as this turns an artifact of the heating method (exacerbating the atmospheric demand for water) into a benefit as it mimics naturally occurring feedbacks. New control methods (De Boeck and Nijs, 2011; Kimball, 2015) also incorporate other natural feedbacks, which together with the open air nature of the method further increases its realism. Like with actively climatized greenhouses, technological know-how is a necessity, however.

9.4 Conclusions

The methods we discussed here are the most commonly used techniques to warm ecosystems under sunlit conditions. Passive warming methods are generally cheaper to install and require less technology, but the main drawback is that the amount of warming is generally uncontrolled and dependent on ambient conditions, which means that not only the mean is changed but also the variability. Active warming methods generally rely on electric power to provide the temperature increase, confining their applicability in remote areas especially if power demands are high (i.e., when the use of generators would not meet demands or would prove too costly). In places that are connected to mains power, such as on campus research sites, active warming methods may be preferred because of their high degree of temperature control and the ability to create stable and large temperature increases year-round. Active methods also facilitate the creation of warming gradients, an approach that allows for better detection of general trends, thresholds, and tipping points (Kreyling et al., 2014). The recent SPRUCE experiment with climate-controlled OTCs (Krassovski et al., 2015), for example, features five levels of warming (from +0°C to +9°C). Note that such a gradient approach is also possible through monolith translocation and in geothermal hot spots, although these passive methods come with the limitations described earlier.

Some studies have combined warming methods in order to profit from the advantages each technique offers. For example, Sun et al. (2013) used OTCs supplemented with heating cables suspended in the air to provide a more continuous warming. Rich et al. (2015) combined an infrared heater array with soil heating cables to provide a more uniform warming above- and belowground. While such combined approaches can indeed reduce some artifacts (e.g., temperature variability), they also can create new issues (e.g., the need for electricity in the first example and soil disturbance in the second). Combinations with alternative, niche methods not discussed here, such as the addition of black carbon to decrease albedo (De Simon et al., 2013), installing heating cables within the tree canopy (Nakamura et al., 2010), or the soil warming side effect of belowground high-pressure natural gas ducts (De Frenne, 2015), could also be considered. In some cases, combinations of different techniques may be used to unravel the impact of asymmetric warming on ecosystem functioning (e.g., soil heating and warmed OTCs; Bronson et al., 2008), thereby reducing uncertainties associated with the method(s) in question.

Our overview of the main current warming methods indicates clearly that no method is without flaws and disadvantages. Luckily, weaknesses vary among the different methods (Table 9.1), which allows researchers to carefully consider what would work best for their research goals and their ecosystem. A clear understanding of peculiarities, advantages, and disadvantages of a

TABLE 9.1

Overview of General Characteristics of Different Warming Methods as Described in This Chapter

Method	Technological Demands	Cost	Intrusiveness	Plot Size	Potential ΔT	Control	Main Artifact
Passive types							
Open top chambers	Low	Low	Low	Small	Small	Low	Reduced wind
Thermal screens	Medium	Medium	Low	Large	Small	Low	Asymmetric warming
Translocation	Low	Medium	High	Small	High	Medium	Soil disturbance
Geothermal hot spots	Low	Low	Low	Unlimited	High	Medium	Asymmetric warming
Active types							
Soil heating cables	Medium	Medium	High	Large	Medium	High	Asymmetric warming
Climate-controlled chambers	High	High	Medium	Medium	High	High	Reduced light
Infrared heating	Medium/high	Medium	Low	Medium/large	High	High	Increased ET

technique could avoid unwanted consequences (e.g., erroneous extrapolation, fatal flaws) of mismatches between the warming method and the envisaged study.

Acknowledgment

The authors wish to thank S. Bassin, B. Sigurdsson, and S. Linder for much appreciated feedback.

References

Amthor, J.S. et al. 2010. A comment on "Appropriate experimental ecosystem warming methods by ecosystem, objective, and practicality" by Aronson and McNulty. *Agricultural and Forest Meteorology* 150:497–498.

Backhaus, S. et al. 2014. A transplantation experiment along climatic gradients suggests limitations of experimental warming manipulations. *Climate Research* 60:63–71.

Barton, C.V.M. et al. 2010. Whole-tree chambers for elevated atmospheric CO_2 experimentation and tree scale flux measurements in south-eastern Australia: The Hawkesbury Forest Experiment. *Agricultural and Forest Meteorology* 150:941–951.

Bauweraerts, I. et al. 2013. The effect of heat waves, elevated CO_2 and low soil water availability on northern red oak (*Quercus rubra* L.) seedlings. *Global Change Biology* 19:517–528.

Beier, C. 2004. Climate change and ecosystem function—Full-scale manipulations of CO_2 and temperature. *New Phytologist* 162:243–245.

Beierkuhnlein, C. and C. Nesshöver. 2006. Biodiversity experiments—Artificial constructions or heuristic tools? *Progress in Botany* 67:486–535.

Bergh, J. and S. Linder. 1999. Effects of soil warming during spring on photosynthetic recovery in boreal Norway spruce stands. *Global Change Biology* 5:245–253.

Bokhorst, S. et al. 2013. Variable temperature effects of open top chambers at polar and alpine sites explained by irradiance and snow depth. *Global Change Biology* 19:64–74.

Bronson, D.R. et al. 2008. Response of soil surface CO_2 flux in a boreal forest to ecosystem warming. *Global Change Biology* 14:856–867.

Bruhn, D. et al. 2013. Improving the performance of infrared reflective night curtains for warming field plots. *Agricultural and Forest Meteorology* 173:53–62.

Burns, B. 1997. Vegetation change along a geothermal stress gradient at the Te Kopia steamfield. *Journal of the Royal Society of New Zealand* 27:279–293.

Cavieres, L.A. and A. Sierra-Almeida. 2012. Facilitative interactions do not wane with warming at high elevations in the Andes. *Oecologia* 170:575–584.

Chapin, F.S. III and G.R. Shaver. 1985. Individualistic growth response of tundra plant species to manipulation of light, temperature, and nutrients. *Ecology* 66:564–576.

De Boeck, H.J., T. De Groote, and I. Nijs. 2012a. Leaf temperatures in glasshouses and open-top chambers. *New Phytologist* 194:1155–1164.

De Boeck, H.J. and I. Nijs. 2011. An alternative approach for infrared heater control in warming and extreme event experiments. *Journal of Ecology* 99:724–728.

De Boeck, H.J. et al. 2010. Climatic characteristics of heat waves and their simulation in plant experiments. *Global Change Biology* 16:1992–2000.

De Boeck, H.J. et al. 2012b. Quantification of excess water loss in plant canopies warmed with infrared heating. *Global Change Biology* 18:2860–2868.

De Boeck, H.J. et al. 2015. Global change experiments: Challenges and opportunities. *Bioscience* 65:922–931.

De Boeck, H.J. et al. 2016. Simulated heat waves affected alpine grassland only in combination with drought. *New Phytologist* 209:531–541.

De Frenne, P. et al. 2010. The use of open-top chambers in forests for evaluating warming effects on herbaceous understorey plants. *Ecological Research* 25:163–171.

De Frenne, P. 2015. Innovative empirical approaches for inferring climate-warming impacts on plants in remote areas. *New Phytologist* 205:1015–1021.

De Simon, G. et al. 2013. Short-term cropland responses to temperature extreme events during late winter. *Biogeosciences* 10:6493–6515.

Drake, B.G. et al. 1989. An open top chamber for field studies of elevated atmospheric CO_2 concentration on saltmarsh vegetation. *Functional Ecology* 3:363–371.

Dreesen, F.E. et al. 2014. Do successive climate extremes weaken the resistance of plant communities? An experimental study using plant assemblages. *Biogeosciences* 11:109–121.

Dreesen, F.E. et al. 2015. Recovery dynamics and invasibility of herbaceous plant communities after exposure to experimental climate extremes. *Basic and Applied Ecology* 16:583–591.

Fu, Y.S.H. et al. 2014. Variation in leaf flushing date influences autumnal senescence and next year's flushing date in two temperate tree species. *Proceedings of the National Academy of Sciences of the United States of America* 111:7355–7360.

Godfree, R. et al. 2011. An improved hexagon open-top chamber system for stable diurnal and nocturnal warming and atmospheric carbon dioxide enrichment. *Global Change Biology* 17:439–451.

Hanson, P.J. et al. 2011. A method for experimental heating of intact soil profiles for application to climate change experiments. *Global Change Biology* 17:1083–1096.

Harte, J. et al. 1995. Global warming and soil microclimate—Results from a meadow-warming experiment. *Ecological Applications* 5:132–150.

Heimann, M. and M. Reichstein. 2008. Terrestrial ecosystem carbon dynamics and climate feedbacks. *Nature* 451:289–292.

Hollister, R.D. and P.J. Webber. 2000. Biotic validation of small open-top chambers in a tundra ecosystem. *Global Change Biology* 6:835–842.

Hollister, R.D., P.J. Webber, and C. Bay. 2005. Plant response to temperature in Northern Alaska: Implications for predicting vegetation change. *Ecology* 86:1562–1570.

Hudson, J.M.G. and G.H.R. Henry. 2010. High Arctic plant community resists 15 years of experimental warming. *Journal of Ecology* 98:1035–1041.

IPCC. 2013. Climate change 2013: The physical science basis. In: Stocker, T.F. et al. (eds.), *Contribution of Working Group I to the Fifth Assessment Report of the Intergovernmental Panel on Climate Change*. Cambridge, U.K.: Cambridge University Press.

Kayler, Z.E. et al. 2015. Experiments to confront the environmental extremes of climate change. *Frontiers in Ecology and the Environment* 13:219–225.

Kennedy, A.D. 1995. Simulated climate change: Are passive greenhouses a valid microcosm for testing the biological effects of environmental perturbations? *Global Change Biology* 1:29–42.

Kimball, B.A. 2005. Theory and performance of an infrared heater for ecosystem warming. *Global Change Biology* 11:2041–2056.

Kimball, B.A. 2015. Using canopy resistance for infrared heater control when warming open-field plots. *Agronomy Journal* 107:1105–1112.

Kimball, B.A., M.M. Conley, and K.F. Lewin. 2012. Performance and energy costs associated with scaling infrared heater arrays for warming field plots from 1 to 100 m. *Theoretical and Applied Climatology* 108:247–265.

Kimball, B.A. et al. 1997. Comparisons of responses of vegetation to elevated carbon dioxide in free-air and open-top chamber facilities. In: Allen, L.H. et al. (eds.), *Advances in Carbon Dioxide Effects Research*, pp. 113–130. ASA/CSSA/SSSA, Madison, WI.

Kimball, B.A. et al. 2008. Infrared heater arrays for warming ecosystem field plots. *Global Change Biology* 14:309–320.

Kittas, C. et al. 2006. Effect of two new UV-absorbing greenhouse-covering films on growth and yield of an eggplant soilless crop. *Scientia Horticulturae* 110:30–37.

Körner, C. 2007. The use of "altitude" in ecological research. *Trends in Ecology and Evolution* 22:569–574.

Krassovski, M.B. et al. 2015. A comprehensive data acquisition and management system for an ecosystem-scale peatland warming and elevated CO_2 experiment. *Geoscientific Instrumentation, Methods and Data Systems* 4:203–213.

Kreyling, J., A. Jentsch, and C. Beier. 2014. Beyond realism in climate change experiments: Gradient approaches identify thresholds and tipping points. *Ecology Letters* 17:125-e1.

Kupper, P. et al. 2011. An experimental facility for free air humidity manipulation (FAHM) can alter water flux through deciduous tree canopy. *Environmental and Experimental Botany* 72:432–438.

Larcher, W. 2003. *Physiological Plant Ecology*, 4th edn. Berlin, Germany: Springer-Verlag.

Leadley, P.W. and B.G. Drake. 1993. Open top chambers for exposing plant canopies to elevated CO_2 concentration and for measuring net gas exchange. *Vegetatio* 104:3–15.

LeCain, D. et al. 2015. Microclimatic performance of a free-air warming and CO_2 enrichment experiment in windy Wyoming, USA. *PLoS ONE* 10:e0116834.

Leppälammi-Kujansuu, J. et al. 2013. Effects of long-term temperature and nutrient manipulation on Norway spruce fine roots and mycelia production. *Plant and Soil* 366:287–303.

Lindvall, J. and G. Svensson. 2015. The diurnal temperature range in the CMIP5 models. *Climate Dynamics* 44:405–421.

Luxmoore, R.J. et al. 1998. Passive nighttime warming facility for forest ecosystem research. *Tree Physiology* 18:615–623.

Marion, G.M. 1996. Temperature enhancement experiments. In: Molau, U. and Mølgaard, P. (eds.), *ITEX Manual*, pp. 17–22. Copenhagen, Denmark: Danish Polar Center.

Marion, G.M. et al. 1997. Open-top designs for manipulating field temperature in high-latitude ecosystems. *Global Change Biology* 3:20–32.

McBee, G.G., W.E. McCune, and K.R. Beerwinkle. 1968. Effect of soil heating on winter growth and appearance of Bermudagrass and St. Augustinegrass. *Agronomy Journal* 60:228–232.

Medhurst, J. et al. 2006. A whole-tree chamber system for examining tree-level physiological responses of field-grown trees to environmental variation and climate change. *Plant, Cell and Environment* 29:1853–1869.

Menzel, A. et al. 2006. European phenological response to climate change matches the warming pattern. *Global Change Biology* 12:1969–1976.

Milcu, A. et al. 2014. Functional diversity of leaf nitrogen concentrations drives grassland carbon fluxes. *Ecology Letters* 17:435–444.

Molau, U. and G.R. Shaver. 1997. Controls on seed production and seed germinability in Eriophorum vaginatum. *Global Change Biology* 3:80–88.

Nakamura, M. et al. 2010. Experimental branch warming alters tall tree leaf phenology and acorn production. *Agricultural and Forest Meteorology* 150:1026–1029.

Naudts, K. et al. 2013. Combined effects of warming and elevated CO_2 on the impact of drought in grassland species. *Plant and Soil* 369:497–507.

Nijs, I. et al. 1996. Free air temperature increase (FATI): A new tool to study global warming effects on plants in the field. *Plant, Cell and Environment* 19:495–502.

Nijs, I. et al. 1997. Stomatal regulation in a changing climate: A field study using Free Air Temperature Increase (FATI) and Free Air CO_2 Enrichment (FACE). *Plant, Cell and Environment* 20:1041–1050.

Norby, R.J. and D.R. Zak. 2011. Ecological lessons from free-air CO_2 enrichment (FACE) experiments. *Annual Review of Ecology, Evolution, and Systematics* 42:181–203.

Norris, T. et al. 1996. Performance of a controlled-ventilation open-top chamber for climate change research. *Agricultural and Forest Meteorology* 78:239–257.

O'Gorman, E.J. et al. 2014. Climate change and geothermal ecosystems: Natural laboratories, sentinel systems, and future refugia. *Global Change Biology* 20:3291–3299.

Parmesan, C. and G. Yohe. 2003. A globally coherent fingerprint of climate change impacts across natural systems. *Nature* 421:37–42.

Patil, R.H. et al. 2013. Soil temperature manipulation to study global warming effects in arable land: Performance of buried heating-cable method. *Environment and Ecology Research* 1:196–204.

Pelini, S.L. et al. 2011. Heating up the forest: Open-top chamber warming manipulation of arthropod communities at Harvard and Duke Forests. *Methods in Ecology and Evolution* 2:534–540.

Peterjohn, W.T. et al. 1993. Soil warming and trace gas fluxes: Experimental design and preliminary flux results. *Oecologia* 93:18–24.

Reth, S. et al. 2009. Sustained stimulation of soil respiration after 10 years of experimental warming. *Environmental Research Letters* 4:024005.

Rich, R.L. et al. 2015. Design and performance of combined infrared canopy and belowground warming in the B4WarmED (Boreal Forest Warming at an Ecotone in Danger) experiment. *Global Change Biology* 21:2334–2348.

Schindlbacher, A., S. Zechmeister-Boltenstern, and R. Jandl. 2009. Carbon losses due to soil warming: Do autotrophic and heterotrophic soil respiration respond equally? *Global Change Biology* 15:901–913.

Schuerings, J. et al. 2014. Winter warming pulses differently affect plant performance in temperate heathland and grassland communities. *Ecological Research* 29:561–570.

Stevnbak, K. et al. 2012. Interactions between above- and belowground organisms modified in climate change experiments. *Nature Climate Change* 2:805–808.

Strebel, D. et al. 2010. Cold-season soil respiration in response to grazing and warming in High-Arctic Svalbard. *Polar Research* 29:46–57.

Strömgren, M. and S. Linder. 2002. Effects of nutrition and soil warming on stemwood production in a boreal Norway spruce stand. *Global Change Biology* 8:1194–1204.

Sun, S.Q. et al. 2013. An improved open-top chamber warming system for global change research. *Silva Fennica* 47:960.

Tognetti, R., P. Cherubini, and J.L. Innes. 2000. Comparative stem growth rates of Mediterranean trees under background and naturally enhanced ambient CO_2 concentrations. *New Phytologist* 146:59–74.

Van der Putten, W.H. et al. 2009. Empirical and theoretical challenges in aboveground–belowground ecology. *Oecologia* 161:1–14.

Van Oijen, M. et al. 1999. Do open-top chambers overestimate the effects of rising CO_2 on plants? An analysis using spring wheat. *Global Change Biology* 5:411–421.

Wahren, D.H.A., M.D. Walker, and M.S. Bret-Harte. 2005. Vegetation responses in Alaskan Arctic tundra after 8 years of a summer warming and winter snow manipulation experiment. *Global Change Biology* 11:537–552.

Zeiher, C.A. et al. 1994. The effect of night temperature on cotton reproductive development. In: Silvertooth, J. (ed.), *Cotton: A College of Agriculture Report*, pp. 89–96. Tucson, AZ: The University of Arizona.

Zhou, L. et al. 2014. Widespread decline of Congo rainforest greenness in the past decade. *Nature* 509:86–90.

10

Remote Sensing in the Reflective Spectrum: A Powerful and Applied Technology for Terrestrial Ecosystem Science

Arnon Karnelli

CONTENTS

Abstract

Combining remote sensing techniques and methods with ecosystem science is challenging since the former has several notable benefits that offer a great number of advanced applications to terrestrial ecology in both spatial and temporal domains. Understanding the spectroscopy foundations and principles can open new horizons to ecologists by introducing a large variety of data, along with their algorithms and processing techniques, which are not be able to be achieved by traditional data collection methods. In this regard, this chapter, which was compiled by a remote sensing specialist, reviews remote sensing capabilities for terrestrial ecology and provides the readers with detailed information on remote sensing means, algorithms, and techniques. The review contains four sections on (1) past, current, and future spaceborne broadband and hyperspectral instruments and platforms; (2) spectral indices, including broad and narrow spectral indices for vegetation, soil, and other environmental substrates; (3) spectral analysis methods, including multivariable analysis, multitemporal data processing, phenology studies, and multisource data fusion; and (4) a tabular summary of the key ecological variables with respect to their analysis methods, main spectral bands, and the preferred type of remote sensing platform. The summary notes several fundamental challenges in the path toward better integrating the two disciplines. The review is focused on the reflective part of the spectrum, that is, the visible, near-, and shortwave infrared regions; however, other regions, such as the thermal and microwave, are excluded. Moreover, the review mainly describes the spectral dimension of remote sensing, leaving the related image analysis techniques for another document.

Keywords: Spaceborne platforms and instrumentation, Spectral indices and algorithms, Narrowbands, Hyperspectral, Spectral analysis, Multivariable analysis, Multitemporal analysis, Phenological studies, Ecological variables

10.1 Introduction

Remote sensing is the science and technology of deriving information about a target (material, object, or phenomenon) from measurements made at a distance and analyzing the interrelationships between the target and the reflected or emitted electromagnetic radiation. The electromagnetic spectrum comprises the full range of all radiation types from very short waves (e.g., gamma rays) to very long ones (e.g., radio and television), which characterize light. Although, in the wide sense, remote sensing can be used by a large number of disciplines with numerous applications (e.g., biomedical engineering, astrophysics), this chapter focuses only on earth observation remote sensing. In this regard and despite the fact that the electromagnetic spectrum is continuous, the most applicative spectral regions for terrestrial ecosystem observations are the near-ultraviolet (NUV, 0.3–0.4 μm); visible (VIS, 0.4–0.7 μm), subdivided into blue (B, 0.4–0.5 μm), green (G, 0.5–0.6 μm), and red (R, 0.6–0.7 μm); red-edge (RE, 0.7–0.8 μm); near-infrared (NIR, 0.8–1.1 μm); shortwave infrared (SWIR, 1.1–4.5 μm); thermal infrared (TIR, 4.5–14 μm); and microwave (0.3–100 cm). These wavelength intervals, called spectral bands, were defined according to the physical property and the specific application of each region. Therefore, different technologies and sensors are needed to acquire data from a particular region and in conjunction with a specific application.

The sun is the main source of radiation to *passive remote sensing sensors*; thus, *reflected radiation* refers to light that is returned from the detected target in the *reflective* spectral region (0.3–4.5 μm). However, since all materials at temperatures above absolute zero, 0 K, send out radiation, *emitted radiation* is connected to the thermal properties of the substance. In contrast to passive sensors, *active sensors* both transmit pulses of energy to the target and record backscattered pulses. Thus, active systems, typically radars and lasers, control the length, timing, intensity, and polarization of the transmitted signal.

According the previously mentioned definition of remote sensing, it can be implemented from any distance as long as there is no physical contact between the measuring device and the target. Consequently, remote sensing measurements can be carried out in the laboratory, in the field, from elevated platforms, and from airborne systems at different heights in the atmosphere, as well as from spaceborne systems in space. Historically, earth observations were filmed by cameras with spectral widths limited only to the *photographic region* (0.3–0.9 μm). In the last decades, analog films were replaced by digital instruments that have notable advantages. First, they enlarged the applicative spectral region using a variety of detectors. Furthermore, digital data can be stored, manipulated, transmitted, and archived by computers.

Remote sensing data are characterized by four types of resolutions that constitute the smallest observable (measurable) differences, namely spatial, radiometric, spectral, and temporal resolutions. *Spatial resolution* refers to the finest angular (termed *instantaneous field of view, IIFOV*) or linear separation (determined by the pixel size) between two objects that can be resolved by the sensor. *Radiometric resolution* is the number of digital levels that a sensor can use to express the variability of gray levels (brightness values) within the data. Radiometric resolution determines the information content of the image such that the more the levels, the more details can be expressed. The radiometric resolution is determined by the number of bits within which the digital information is encoded and expressed in binary notation. For example, the acquisition level of an 8-bit system has (2^8 =) 256 gray levels confined within a dynamic range between 0 and 256. *Spectral resolution* characterizes the number and width of spectral bands used by the remote sensing system. More numerous and narrower spectral bands enable diagnostic absorption features to be observed, thus facilitating the identification of biophysicochemical processes in vegetation, soils, rock, and minerals (see Section 10.2.8 for more details). Finally, *temporal resolution* is the revisit time of a specific sensor to acquire data from a specific location of earth. Revisit time can range from several minutes to several days, depending on the orbital characteristics of the satellite, the latitude of the target, the swath width of the sensor, and the pointing (tilting) capability of the system.

Remote sensing has several notable benefits that offer a great number of advanced applications in both spatial and temporal domains. First, remote sensing can be applied to spectral regions that are beyond the human sense of sight, thus enabling the discovery of extrasensory materials, objects, and phenomena. Measurements are unobtrusive and do not destroy the substrate, allowing periodic and repetitive sampling processes. From the spatial point of view, traditional data collection in the field is limited to the plot/local scales. On the other hand, spaceborne remote sensing covers regional to global views, even over inaccessible sites and under extreme conditions, thus enabling the causes and effects of climate and environmental changes to be assessed. Spaceborne observations have a constant revisit time that allows repetitive geo-referenced looks at the same area. Modern remote sensing produces digital data, capable of being numerically manipulated within and among the spectral bands for enhancing the studied object. Consequently, understanding the spectroscopy foundations and principles can open new horizons to users by introducing a large variety of data, along with their algorithms and processing techniques, which are not able to be achieved by traditional data collection methods.

Since the early development of digital remote sensing and spectroscopy technologies, they have played a crucial role in terrestrial ecosystem science, which has gradually increased throughout the years. This is primarily due to the ability of remote sensing to provide essential biophysical variables for ecological models, such as vegetation indices (VIs), leaf area index (LAI), and land surface temperature, as described in Section 10.3. This source of valuable data appears in a large variety of spatial scales (plot, local, regional, national, continental, and global) and temporal scales (from minutes to days). In this regard, Plummer (2000) defined four approaches for linking remotely sensed data and ecological process models at a range of spatial and temporal scales: (1) using remotely sensed data to provide estimates of the variables required to drive ecological process models; (2) using remotely sensed data to test, validate, or verify the predictions of ecological process models; (3) using remotely sensed data to update or adjust ecological process model predictions; and (4) using ecological process models to understand remotely sensed data.

The chapter is aimed at reviewing the previously mentioned remote sensing capabilities that ensure advanced terrestrial ecosystem studies and applications. Note that the review is focused only on the *reflective spectrum*, this is to say, the part of the spectrum containing the VIS, NIR, and SWIR regions, and therefore the thermal and microwave instruments (e.g., radar, light detection and ranging [LiDAR]) and their respective missions are excluded. Moreover, the review mainly describes the spectral dimension of remote sensing, leaving the related image analysis techniques for another document. Section 10.2 describes the most applied past, present, and future remote sensing instruments and platforms, distinguishing between the ones with different spatial and spectral resolutions. Section 10.3 reviews the most commonly used spectral indices that are applied to broadbands and narrowbands, for vegetation, soil, and other applications. Section 10.4 covers several spectral analysis methods, including multivariate spectral analysis, multitemporal data processing, and phenological studies. Key ecological variables are tabulated in Section 10.5 with respect to their analysis means, their main spectral bands, and the preferred type of remote sensing platform. Section 10.6 lists several fundamental challenges on the way toward better integrating the remote sensing and terrestrial ecology disciplines.

10.2 Instrumentation and Platforms

By definition, remote sensing does not depend on the distance between the detected device and the object. Measurements can be conducted in the laboratory, at ground level in the field, from a drone or an unmanned aerial vehicle (UAV), aircrafts, and spacecrafts. Whatever the remote sensing device, they can be divided into four categories: point and image radiometers, and point and image spectrometers. *Radiometers* are *multispectral systems* that are characterized by up to several tens of discrete broadbands with widths of tens to hundreds of nanometers (usually >50 nm bandwidth). *Spectrometers*, however, are *hyperspectral systems* characterized by up to several hundred narrowbands, a few nanometers each, which continuously spread along the spectrum (Schaepman 2009). *Point radiometers* and *point spectrometers* integrate the incoming radiation and create a single string of wavelengths versus radiance or reflectance values. This type of instrument is usually applied in the field, sometimes with accessories such as an *integrating sphere* for separating reflectance, transmittance, and absorbance spectra or a *leaf clip* or a *contact probe* for minimizing environmental effects. *Image radiometers* and *image spectrometers* create a 2D array (spatial domain) of spectral data (third dimension) for each band. The number of bands determines the fourth dimension of a spectral cube. Hyperspectral reflectance signatures have the advantage of quantitatively assessing vegetation and soil biochemistry properties (Ustin et al. 2004; Ben-Dor et al. 2009) (Section 10.2.8).

10.2.1 Landsat Program

The *Landsat* program, a joint U.S. National Aeronautics and Space Administration (NASA) and the Geological Survey (USGS) mission, has been operated in space since 1972 till now with eight satellites belonging to four generations of sensors. The first sensor generation, named the *Multispectral Scanner* (*MSS*), onboard Landsat 1, 2, and 3, is characterized by four spectral bands in the green, red, and NIR regions, an 80 m resolution, and 18 days' revisit time. The second generation, named the *Thematic Mapper*, onboard Landsat 4 and 5, has six spectral bands in the VIS–NIR–SWIR with a 30 m resolution and one TIR band with a 90 m resolution. The temporal resolution was improved to 16 days. The third generation, named the *Enhanced Thematic Mapper Plus* (*ETM+*), onboard Landsat 7, has an additional panchromatic band with a 15 m resolution and an improved resolution of the TIR band to 60 m. Landsat 8 belongs to the fourth generation, with a payload of two sensors, the *Operational Land Imager* and the *Thermal Infrared Sensor*. The former has two additional bands in the blue and in the SWIR, while the latter has two thermal detectors. By being the longest global observation program, referenced to the same coordinate system, with repetitive acquisition of high-resolution multispectral data, the program has many scientific applications, including monitoring ecological essential variables and assessing land-cover and land-use status and changes over time for global climate research. The archived Landsat images are free (Gutman and Masek 2012).

10.2.2 SPOT Program

The *Satellite Pour l'Observation de la Terre* (*SPOT*) is a commercial moderate-spatial-resolution program for earth observation that was initiated and designed in France with some participation of Belgium and Sweden. The first satellite in the series was launched in 1986 carrying the *high-resolution visible* (*HRV*) sensor. This band setting enabled mainly vegetation monitoring and mapping. Two independent HRV sensors onboard consisted of three multispectral bands at 20 m resolution each and one panchromatic band at

10 m. The nominal revisit time of the SPOT is 26 days, and the swath width is 60 km. However, the notable advantage of the SPOT system is the ability to tilt each sensor up to ±27° off-nadir. By doing so, the system made it possible to create a digital elevation model (DEM), improving the temporal resolution (11 observing possibilities at 45° latitude and 7 at the equator) and increasing the swath width to 950 km. Since the SPOT is a commercial program that relies on advance tasking, it does not have a periodic image database. This is a drawback for change detection and time series analysis of any specific location. The first five satellites in the program were operated by the French space agency (Centre National d'Études Spatiales [CNES]). The payload of the later satellites was approved in terms of an additional SWIR band and a better spatial resolution. SPOT 4 and 5 stopped functioning in 2013 and 2015, respectively. SPOT 6 and 7 were launched in 2012 and 2014, respectively, and operated by Astrium, an aerospace manufacturer subsidiary of the European Aeronautic Defense and Space Company (EADS). Together with the two Pléiades space systems (Section 10.2.7.1), they form a constellation of four satellites, 90° apart from one another.

10.2.3 The NOAA-AVHRR Program

The *advanced very-high-resolution radiometer* (*AVHRR*), onboard the National Oceanic and Atmospheric Administration's (NOAA's) Polar-Orbiting Environmental Satellites (POES), is a broadband five-channel instrument (or six-channel, depending on the model) in the VIS, NIR, SWIR, and TIR regions. A constellation of two satellites provides at least four image acquisitions per 24 h for any location on the globe. Data are acquired in two formats: the *local area coverage* (*LAC*) with the nominal 1.1 km spatial resolution at nadir and the *global area coverage* (*GAC*) sampled and averaged onboard a satellite with the nominal 4 km spatial resolution. The primary applications of the AVHRR for the land are vegetation monitoring through *vegetation indices* (*VIs*) (Section 10.3.1), albedo (Liang 2001) (Section 10.3.8), and extracting land surface temperature (Qin et al. 2001). Despite the two drawbacks of the sensor, specifically, drifts of the satellite orbit and of the channel calibration, the main advantage of the AVHRR series for terrestrial ecology is the long-term global archived dataset of images gathered since July 1982 (Gutman and Masek 2012).

10.2.4 VEGETATION Program

The *VEGETATION* instrument, onboard SPOT 4 and 5, was aimed at daily monitoring of terrestrial vegetation cover at regional to global scales, over a long time span, for developing models of the biosphere dynamics interacting with climate models (Saint 1994). The instrument, called the HRVIR, was characterized by four spectral bands in the blue, red, NIR, and SWIR, with a 2200 km swath width and a 1.15 km pixel size at nadir.

10.2.5 MODIS Program

The *moderate-resolution imaging spectroradiometer* (*MODIS*) program comprises two similar instruments installed onboard the *Terra* and *Aqua* NASA's spacecrafts (Guenther et al. 2002). The former was launched in 1999 and the latter in 2002. Having a 2330 km swath width, the two MODIS instruments are able to observe every point on earth four times in 24 hours (morning, afternoon, and night). MODIS has 36 discrete bands in pixel sizes of 250, 500, and 1000 m, enabling a large number of land, ocean, and atmospheric applications. The environmental ones include the monitoring of large-scale biosphere dynamics in conjunction with the global carbon cycle and detecting the photosynthetic activity of land and marine plants in conjunction with greenhouse gases that are absorbed and used in plant productivity. Also worth mentioning is the ability to observe short-term events, such as dust storms, droughts, volcanic eruptions, floods, wildfires, and more. Following their policy, the NASA images are also free. Among the land archived products are surface reflectance, net and gross primary productivity, LAI and fractional photosynthetically active radiation, VIs, land surface temperature and emissivity, bidirectional reflectance distribution function (BRDF) and albedo, and snow cover. Note that unlike the AVHRR, the MODIS data are well calibrated and thus useful for time-series studies such as vegetation phenology and more.

10.2.6 Sentinel-2 Program

Sentinel-2 belongs to the European Space Agency (ESA) and the European Commission (EC) multimission program called Copernicus (formerly the GMES program) (Drusch et al. 2012). It is a polar-orbiting, multispectral high-resolution imaging mission for land monitoring, including vegetation health, land-use and land-cover status and change, inland and coastal waters, and related features. Sentinel-2 is characterized by 13 bands in the VIS, NIR, and SWIR with 10, 20, and 60 m spatial resolutions within a 290 km swath width. Sentinel-2 is a constellation of two spacecrafts. The first was launched in June 2015, and the second launch is planned for 2017. When both are in orbit, the revisit time will be 5 days under the same viewing angles. Data are free for all users.

10.2.7 High-Spatial-Resolution Satellites

10.2.7.1 Commercial Satellites

The high-spatial-resolution satellite era started in 1998 with the launch of *IKONOS*, the first commercial space system. Since then, the number of similar satellites has gradually increased, with all possessing basically the same characteristics with some variations and improvements throughout the years (Table 10.1). The multispectral bands are commonly used for a large variety

TABLE 10.1

List of the Past, Present, and Forthcoming High-Spatial-Resolution Satellites along with Their Spectral Characteristics[a]

			Bands in Spectral Regions		
			Panchromatic	VNIR	SWIR
Satellite	Country	Launch Year	Resolution (m)	No. of Bands/ Resolution (m)	No. of Bands/ Resolution (m)
Ikonos[b]	USA	1998	0.8	4 × 3.2	
QuickBird[b]	USA	2001	0.6	4 × 2.16	
Formosat 2	Taiwan	2004	2.0	4 × 8	
WorldView 1	USA	2007	0.5		
GeoEye 1	USA	2008	0.4	4 × 1.65	
RapidEye 1-5	Germany	2008		5 × 6.5	
WorldView 2	USA	2009	0.3	8 × 1.2	
TH-01	China	2010	2.0	4 × 10	
Pléiades 1	France	2011	0.5	4 × 2.0	
KompSat 2	S. Korea	2012	1.0	4 × 4.0	
Pléiades 2	France	2012	0.5	4 × 2.0	
SPOT 6	France	2012	1.5	4 × 6.0	
Deimos 2	Spain	2014	0.8	4 × 3.0	
SkySat	UK	2014	>1	3 × 2.0	
SPOT 7	France	2014	1.5	4 × 6.0	
WorldView 3	USA	2014	0.3	8 × 1.2	8 × 3.7
KompSat 3	S. Korea	2015	0.7	4 × 2.8	
WorldView 4	USA	2016	0.3	4 × 1.36	
VENμS	Israel/ France	2017		12 × 5.3	

[a] List partially provided by Abraham Tal.

[b] System does not function anymore, archive data only.

of terrestrial applications, mostly connected with vegetation, while the *panchromatic band* is used for mapping and creating a digital elevation model (DEM) created from the stereo pairs.

Among the unique capabilities of these space systems are the panchromatic band at a higher resolution than the four multispectral bands (B, G, R, and NIR). The resolution of the multispectral bands is four times bigger than that of the panchromatic one, enabling pan-sharpening (Zhang 2004). This process aimed at merging the two resolutions to create a single high-resolution color image. Since spaceborne systems operate according to commercial goals, striving to maximize the number of potential attempts in a given time period, the systems also have the ability to tilt the satellite and to acquire, typically 30°, off-nadir images. Consequently, following an image tasking, the revisit time can be as low as 1 or 2 days. The main drawback of these spaceborne systems is that, unlike the broadband ones, they lack a fixed revisit time. New images for a specific location, within a confined area and under a predefined threshold of cloud cover, can be ordered from the vendor. Consequently, the acquisition cost is relatively high and only historical images exist in the archive.

10.2.7.2 VENμS Program

Among the high-resolution satellites, the *Vegetation and Environmental New Micro Spacecraft* (*VENμS*) is an exception. VENμS is not a commercial mission but a scientific one. It is a joint venture of the Israeli Space Agency (ISA) and the French CNES. The satellite's launch is planned for late 2017. The overall aim of the VENμS scientific mission is to acquire frequent, high-resolution multispectral images of preselected sites of interest all around the world. In addition, the scientific mission is aimed at demonstrating the relevance of VENμS observation capabilities in the framework of the Copernicus program. This program represents an effort to bring environmental and security data and information providers together with users, to better understand each other, and to agree on how to make such information available to the people who need it. Copernicus will comprise a constellation of eight satellites; one of them, Sentinel-2 (Section 10.2.6), has similar and complementary spectral characteristics to the VENμS but with a coarser spatial resolution and longer revisit time.

During the VENμS mission, the satellite will fly in a near polar sun-synchronous orbit at 720 km in height and an inclination angle of 98.27°. The equator crossing time is planned to be at 10:30 AM, descending mode. The whole system can be tilted up to 30° off-nadir to enable imaging targets at up to 360 km on either side of the track and can also be tilted forward or backward to provide more flexibility in selecting scientific sites, evaluating the radiometric effects of viewing angle, etc. This orbit configuration will result in a 27 km swath, a camera resolution of 5.3 m, and the capability to observe about 120–150 preselected ground sites with a high repetitiveness of 2 days, and always with exactly the same angular acquisition conditions and minimization of directional effects. The high observing frequency is essential for detecting the dynamics of vegetation growth and of the short duration of phenological stages, as well as the rapid temporal changes of water quality.

The satellite will carry the *VENμS Super-Spectral Camera* that offers a compromise on spectral resolution to attain a high signal-to-noise ratio (SNR), despite the small ground sampled distance (GSD) of 5.3 m. This is accomplished by having 12 narrow spectral bands, referred to as "superspectral" imaging, ranging from 415 to 910 nm and including red-edge (RE) bands. The bands, varying in width from 40 nm down to 16 nm, were chosen to retrieve key information about vegetation, water, and atmosphere, along with other

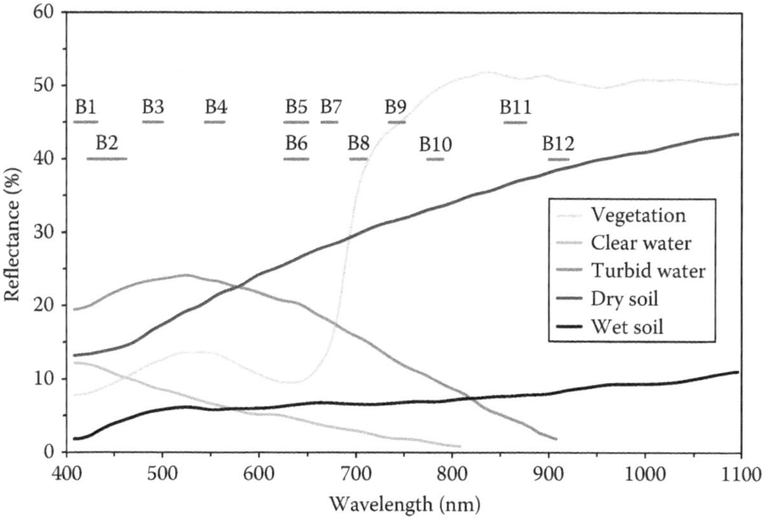

FIGURE 10.1
(See color insert.) Band settings of VENμS with respect to vegetation, soil, and water spectral signatures.

features, and are carefully located within the atmospheric absorption regions of H_2O and O_2 (Figure 10.1). One of the bands, at 620 nm, is duplicated, and both bands are positioned at the extremes of the angular field in the scan direction. The 1.5° difference in look angle between these two will allow 3D imaging that will enable the construction of a DEM of the earth surface and the assessment of cloud heights. Special attention was given to the detection of vegetation status as required by agronomists, foresters, and ecologists but the band setting could also prove useful for water quality studies in coastal areas and inland water bodies.

A chain of four product levels will be available for the end users:

- Level 0 is the basic archived product that consists of a single date and single viewing angle acquisition at a GSD of 5.3 m.
- Level 1 products are those from a single date and single viewing angle acquisition of the top of atmosphere (TOA) reflectances with a subpixel multidate registration (map projected) at a GSD of 5.3 m. A cloud mask at a coarse resolution will be associated with this product.
- Level 2 products consist of those from a single date and single viewing angle acquisition of surface reflectances after cloud masking and atmospheric correction for all spectral bands at a 10 m GSD. The algorithms used for Level 2 processing take advantage of the 2-day revisit period with constant observation angles since most

short-term variations of TOA reflectance are due to atmospheric variations. In addition, Level 2 includes a series of products derived from the reflectance data.

- Level 3 products are the same as Level 2 but consist of a 10-day composite of a single date and single angle acquisition. The aim of the Level 3 products is to reduce the data volume for users and to deliver a composite of the same variables as in Level 2 after masking the clouds based on the cloud-free pixels of the Level 2 data gathered during a short period (7–10 days) at a GSD of 10 m.

As mentioned earlier, the VENµS mission is designed to observe about 120–150 preselected ground sites. In order to select these scientific sites, an international call for proposals was released in 2014. The selection will give high priority to sites that have long-term data that can be observed by the VENµS system, to the scientific merit of the proposal, and to the limitation of the mission (coverage, storage, etc.). The selected teams will receive, within one month after acquisition, the data they requested over one or several sites, free of charge. In most cases, the products will be provided for the entire mission duration. Levels 2 and 3 and possibly Level 1 products will be available three months after the acquisitions to any scientist who does not belong to a selected team.

10.2.8 Hyperspectral Instruments

In contrast to multispectral scanners that provide information in distinct bands, hyperspectral instruments acquire data at many and contiguous spectral bands, enabling the creation of detailed spectral signatures of the detected object. Accordingly, hyperspectral data are associated with more quantitative analyses in the attempt to assess the nature of the material under study, such as chlorophyll, nitrogen, and lignin concentrations or water content in the case of vegetation biochemical properties (Goodenough et al. 2006). For assistance in this task, several spectral libraries of natural and human-made materials are provided by different agencies (e.g., http://speclab.cr.usgs.gov/spectral-lib.html; http://speclib.jpl.nasa.gov/).

As of today, most of the operational image spectrometers are onboard airborne platforms. The advantage of these systems is not only the ability to acquire data at a relatively high spatial resolution of about a meter (depending on the flight height), in addition to the high spectral resolution (of about 10 nm or less), but also the versatility to satisfy the user requirements in terms of timing and flight parameters. On the other hand, due to costs and logistics, the use of airborne systems is limited and lacks the periodic repeating nature of spaceborne systems (Asner and Heidebrecht 2003). The most common hyperspectral airborne scanners are listed in Table 10.2. Most of them cover the VIS–NIR–SWIR regions and have been used for a large variety of remote sensing applications.

TABLE 10.2

List of Hyperspectral Airborne Scanners along with Their Spectral Characteristics

Sensor	Organization (Country)	Spectral Range (nm)	No. of Bands	Spectral Resolution (nm)
DAIS 7915	DLR (Germany)	400–12,600	79	1,45
PROB-1	Earth Search Sciences Inc. (USA)	400–2,500	128	12,16
HyMap	HyVista Corp. (Australia)	450–2,500	132	15,20
CASI-1500	ITRES (Canada)	365–1,050	288	3.5
SASI	ITRES (Canada)	950–2,450	100	15
SFI	CCRS (Canada)	1208–2,445	120	10.3
AVIRIS	NASA (USA)	400–2,500	244	10
AISA DUAL	Specim (Finland)	400–2,450	498	2.9,8.5
VNIR-640	HySpex (UK)	400–1,000	128	5
Hyperspec	Headwall (USA)	400–2,500	333	5,10
CAO VSWIR	Carnegie Airborne Observatory (USA)	380–2,510	428	5
TRWIS III	TRW Space and Electronics Group (USA)	400–2,450	384	5.25,6.25

Since 2000, hyperspectral scanners have also been installed on spacecrafts. Currently, data are available from two operational hyperspectral systems, Hyperion on EO1 and Compact High-Resolution Imaging Spectrometer (CHRIS) on PROBA. *Hyperion* is a full-range instrument (375–2576 nm), consisting of 242 bands at a bandwidth of approximately 10 nm *full width at half maximum* (*FWHM*). The instrument's swath is 7.5 km, containing 256 columns (pixels) with a 30 m spatial resolution. The system was found to be suitable for retrieving geologic (mineralogical/lithological) information (Kruse et al. 2003), as well as for vegetation studies (Pignatti et al. 2009). Nevertheless, it is reported that the Hyperion data had drawbacks, such as relatively low SNR, spectral and spatial artifacts related to the array malfunction, and a significant smile effect (Dadon et al. 2010). The *Compact High Resolution Imaging Spectrometer* (*CHRIS*) acquires 13 km^2 scenes at a 17 m spatial resolution in 18 user-selected VIS and NIR wavelengths and can be reconfigured to provide 63 spectral bands at a spatial resolution of about 34 m. This system also strives to collect *bidirectional reflectance distribution function* (*BRDF*) data for a better understanding of spectral reflectances and, consequently, was designed to observe the ground at up to five different viewing angles. Houborg et al. (2015) declared that "the need for an operational [spaceborne] hyperspectral mission to produce repeatable high resolution images for all terrestrial ecosystems has not been fulfilled." Despite the current technical problems and other limitations, several hyperspectral systems are planned to be operational in the near future (Buckingham and Staenz 2008). Table 10.3

TABLE 10.3

List of the In-Orbit, Under Construction, and Forthcoming Hyperspectral Satellites along with Their Spectral Characteristics[a]

Sensor	Organization (Country)	GSD (m)	Swath at Nadir (km)	Spectral Range (nm)	Number of Bands	Spectral Resolution (nm at FWHM)	Launch Date	Status
Missions currently in orbit								
Hyperion	NASA (USA)	30	7.65	375–2576	242	10	2000	Missions currently in orbit
CHRIS	ESA (UK)	17/34	13 (nominal)	400–1050	18/63	5.6–32.9	2001	
HJ-1A	CAST (China)	100	≥50	450–950	110–128	5	2008	
HySI	ISRO (India)	506	129.5	450–950	64	~10	2008	
HICO	NASA/ONR (USA)	90	42	353–1081	128	5.7	2009	
Missions under construction								
GISAT	ISRO (India)	500	NA	NA	210	NA		Missions under construction
PRISMA	ASI (Italy)	30	30	400–2500	237	~12		
HISUI	METI (Japan)	30	15	400–2500	185	10 (VNIR) 12.5 (SWIR)		
EnMAP	DLR/GFZ (Germany)	30	30	420–2450	218	5/10 (VNIR) 10 (SWIR)		
Missions in the planning stage								
FLORIS/FLEX	ESA	300	100–150	500–780	NA	0.3–3.0		Missions in the planning stage
HYPXIM-P	CNES (France)	8	16	400–2500	>200	≤10		
HyspIRI	NASA (USA)	60	145	380–2500	>200	10		
SHALOM	ISA/ASI (Israel/Italy)	10/5	10	400–2500	200	10		

[a] Data compiled by Eyal Ben Dor.

lists the in-orbit, under construction, and forthcoming hyperspectral satellites along with their spectral characteristics.

10.3 Spectral Indices

Spectral indices are mathematical manipulations of two or more spectral bands (or spectral regions) to enhance the signal of the phenomenon under study for predicting ecosystem responses. Assuming that a multiband analysis would provide more information than a single one, reflectance values from the original input bands are converted into a single, dimensionless value. With respect to ecological research, spectral indices can be coarsely divided into vegetation, soil, and other environmental indices and further subdivided into broadband and narrowband remote sensing systems. While broadband indices are associated with multispectral systems, narrowband remote sensing is associated with hyperspectral sensors.

More than 50 different VIs are reported in the literature (Agapiou et al. 2012; Roberts et al. 2012; Stagakis et al. 2012), but only the key ones are presented in this chapter in order to demonstrate their principle in conjunction with different applications.

10.3.1 Broadband Vegetation Indices

The most familiar spectral indices are VIs that attempt to intensify plant signals for assessing their state and dynamics along with their biophysical properties. Therefore, they are the most widely used spaceborne product for monitoring the ecosystem. The broadband VIs are generally used for assessing the structural properties of the terrestrial vegetation (Roberts et al. 2012) such as LAI (Wang et al. 2005), fractional vegetation cover (Bajocco et al. 2012), aboveground biomass (Wessels et al. 2006), the fraction of absorbed photosynthetically active radiation (fAPAR) (Fensholt et al. 2004), net primary productivity (NPP) (Schloss et al. 1999), and foliar loss and damage (Vogelmann 1990).

The earliest and the most fundamental VI is the *simple ration* (*SR*), proposed by Jordan (1969):

$$SR = \frac{\rho_{NIR}}{\rho_R} \tag{10.1}$$

where ρ is the reflectance of the subscripted spectral band. The SR is based on the difference between the maximum absorption of radiation in the red band (due to the chlorophyll pigments) and the maximum reflection of radiation in the NIR band (due to the leaf cellular structure), and the fact that soil spectra, lacking these mechanisms, typically do not show such a dramatic spectral difference. Since the range of the SR values, from 0 to infinity, is considered as a drawback, the same bands were used for formulating the *normalized difference vegetation index* (NDVI) (Rouse et al. 1973; Tucker 1979):

$$NDVI = \frac{(\rho_{NIR} - \rho_R)}{(\rho_{NIR} + \rho_R)} \tag{10.2}$$

The NDVI values range from −1 to +1; thus, it is possible to distinguish between the positive values of vegetation (dense and healthy vegetation has relatively high NDVI values) and soil (which has relatively low NDVI values) and the negative values of clouds, snow, and water. Over the years, NDVI has turned out to be the most widely used VI.

Despite the extensive use of the NDVI and its various applications, several notable limitations of this index have been documented. These include (1) the sensitivity of the index to dark and/or wet soil backgrounds (Huete 1988); (2) less sensitivity ("saturation") of the index values in cases of high values of chlorophyll content, LAI, and biomass that indicate dense and multilayered canopies (Figure 10.2) (Buschmann and Nagel 1993); and (3) sensitivity to

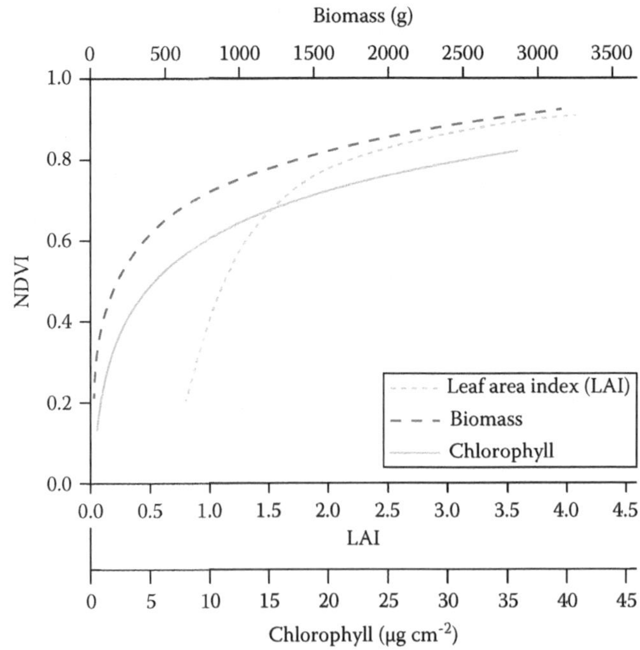

FIGURE 10.2
(See color insert.) The NDVI as a function of LAI, chlorophyll content, and biomass. Note the less sensitivity ("saturation") of the NDVI when high values of the independent variables are presented, indicating dense and multilayered canopies. (Data provided by Agustin Pimstein and Ittai Herrmann.)

atmospheric influences (Holben 1986) since aerosol increases the apparent reflectance in the red band by scattering sunlight directly to the sensor and decreases, to a lesser degree, the reflectance in the NIR by absorption of sunlight. To overcome these drawbacks, several approved VIs were proposed. Among them is the *soil-adjusted vegetation index* (*SAVI*) that is less sensitive to the soil background and approximate vegetation biophysical properties over a wide range of LAI values:

$$SAVI = \frac{\rho_{NIR} - \rho_R}{\rho_{NIR} + \rho_R + L}(1 + L) \tag{10.3}$$

where L, usually equal to 0.5, is the canopy background adjustment factor.

In order to defeat the scattering effects from aerosols suspended in the atmosphere, two VIs were proposed. The *atmospherically resistant vegetation index* (*ARVI*) (Kaufman and Tanre 1992)

$$\begin{gathered} ARVI = \frac{\rho_{NIR} - \rho_{RB}}{\rho_{NIR} + \rho_{RB}} \\ \rho_{RB} = \rho_R - \gamma(\rho_B - \rho_R) \end{gathered} \tag{10.4}$$

and the *atmospheric free vegetation index* (*AFRI*) (Karnieli et al. 2001)

$$AFRI_{SWIR3} = \frac{(\rho_{NIR} - 0.5\rho_{SWIR3})}{(\rho_{NIR} + 0.5\rho_{SWIR3})} \tag{10.5}$$

The resistance of the ARVI to atmospheric effects (in comparison to the NDVI) is accomplished by a self-correction process for the atmospheric effect on the red channel, using the difference in the radiance between the blue and the red channels to correct the radiance in the red channel. γ is equal to one. In the AFRI, the red band of the NDVI is substituted by half of the values in the SWIR3 region (2.1 μm).

Finally, the *enhanced vegetation index* (*EVI*) (Huete et al. 2002) was invented to optimize the vegetation signal with improved sensitivity in high biomass regions and improved vegetation monitoring while correcting for canopy background signals and reducing atmosphere influences.

$$EVI = \frac{\rho_{NIR} - \rho_R}{\rho_{NIR} + C_1\rho_R - C_2\rho_B + L} \times G \tag{10.6}$$

where the coefficients adopted in the EVI algorithm are $L = 1$, $C_1 = 6$, $C_2 = 7.5$, and G (gain factor) = 2.5.

10.3.2 Narrowband Vegetation Indices

As opposed to the primary use of broadband VIs, that is, assessing the structural properties of terrestrial vegetation, narrowband VIs are aimed at resolving fine spectral features along with their optimal combination for estimating pigments. Narrowband indices, retrieved from hyperspectral data, are able to detect slight modifications in leaf biochemistry and thus to quantitatively improve the identification of the material under study.

10.3.3 The Red-Edge

The *red-edge* (RE) refers to the steep increase of vegetation reflectance values from the minimum reflectance (~10%) in the red band due to chlorophyll absorption and maximum reflectance (~50%) in the NIR band due to the leaf cellular structure. This transition spectral region occurs between 700 and 800 nm. Observations show that the RE moves toward the longer wavelengths during the vegetation growth stage, while moving toward the shorter wavelengths (*blue shift*) either during the senescence stage or during vegetation stress (Figure 10.3). Bearing in mind the "saturation" weakness of the NDVI, these shifts, although stretching along only several tens of nanometers, are much more sensitive and better correlated with biophysical variables such as chlorophyll (Baret et al. 1992), nitrogen (Tarpley et al. 2000), plant stress (Smith et al. 2004), fAPAR (Vina and Gitelson 2005), vegetation fraction (Vina and Gitelson 2005), and LAI (Herrmann et al. 2011).

Since the movement of the RE is horizontal (i.e., along the wavelengths) rather than vertical (i.e., along the reflectance values), it is impossible to use standard VIs for correlating the biophysical variables with the RE shifts. Therefore, linear interpolation among several narrowbands, located along the RE, was proposed to calculate the midpoint, termed the *red-edge position* (λ_{rep}), between the red and the NIR regions (Guyot and Baret 1988):

$$\lambda_{rep} = 700 + 40\left(\frac{\left(\left(\left(\rho_{670} + \rho_{780}\right)/2\right) - \rho_{700}\right)}{\left(\rho_{740} - \rho_{700}\right)}\right) \tag{10.7}$$

This interpolation method was modified for adaptation to several spaceborne systems having at least four RE, although broad, bands, such as *MEdium-Resolution Imaging Spectrometer* (MERIS), Sentinel-2, and VENμS (Clevers et al. 2002; Herrmann et al. 2011). Since the RE position is mathematically defined as the inflection point between the positive and the negative slopes of the curve, its location can also be calculated, in the case of hyperspectral data, by the first or second derivative of the spectrum (Smith et al. 2004):

$$\rho_i' = \frac{\rho_{i+1} - \rho_{i-1}}{\lambda_{i+1} - \lambda_{i-1}} \quad 700 \leq i \geq 800 \tag{10.8}$$

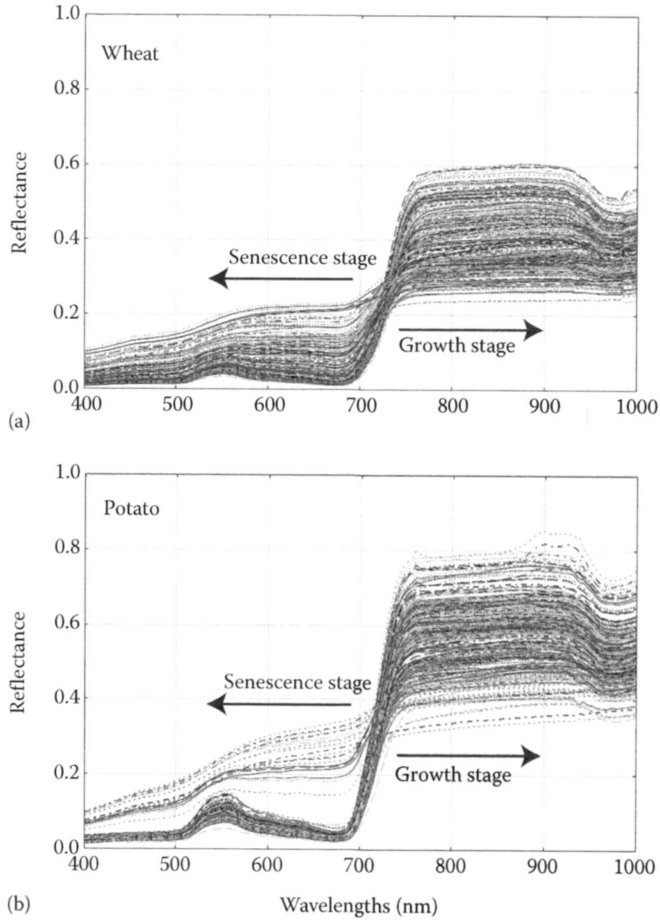

FIGURE 10.3
(See color insert.) Temporal dynamics of the RE during the phenological cycles of wheat (a) and potato (b) crops. (Data provided by Agustin Pimstein and Ittai Herrmann.)

$$\rho_i'' = \frac{\rho_{i+1}' - \rho_{i-1}'}{\lambda_{i+1} - \lambda_{i-1}} \quad 700 \le i \ge 800 \tag{10.9}$$

where i refers to a specific wavelength, typically between 700 and 800 nm. It is important to mention that the derivative method can result in two inflection points for the RE region (Zarco-Tejada et al. 2003). Notwithstanding the importance of the RE, only several multispectral spaceborne systems have appropriate bands. These are the MERIS, the Sentinel-2 (Section 10.2.6), WorldView 2,3 (Section 10.2.7.1), and the VENμS (Section 10.2.7.2).

10.3.4 Biological Pigments (Chlorophylls, Carotenoids, and Anthocyanins)

Biological pigments indicate the physiological condition of plant leaves (Gitelson et al. 2006). Among others, *chlorophylls* (green pigments) constitute the main group of pigments. They are essential for the photosynthetic activity of plants by absorbing radiation in the blue and red regions. The second group of pigments is *carotenoids* (yellow pigments) that absorb radiation in the blue region to prevent damage to the photosynthetic systems. The third group is *anthocyanins* (red pigments) that absorb radiation in the green region and protect leaves from excess light. The main pigments and the central location of their absorption features are summarized in Table 10.4.

Since each type of pigment is associated with a unique and narrow absorption feature along the VIS region, although some overlap exists, different hyperspectral indices were proposed to identify them and discriminate between them. One of the earliest was the *photochemical reflectance index* (*PRI*) (Gamon et al. 1992):

$$PRI = \frac{(\rho_{550} - \rho_{531})}{(\rho_{550} - \rho_{531})} \tag{10.10}$$

which is a normalized difference index (NDVI-like) that uses two narrow reflectance bands at 531 and 550 nm. Another one is the *chlorophyll absorption in reflectance index* (*CARI*) (Kim 1994):

$$CARI = (\rho_{700} - \rho_{670}) - 0.2(\rho_{700} - \rho_{550}) \tag{10.11}$$

This index and its following ones employed three narrowbands; the maximum depth of the chlorophyll absorption is at 670 nm in comparison to its shoulders at 550 and 700 nm, where minimum absorption occurs.

TABLE 10.4

Main Biological Pigments and the Central Location of Their Absorption Features

	In Vivo	*In Situ*
Chlorophyll-a	430, 660	450, 670
Chlorophyll-b	450, 650	
Carotenoids	445	500
β-Carotene	470	
Anthocyanins	530	550

The *modified chlorophyll absorption in reflectance index* (*MCARI*) was proposed by Daughtry et al. (2000) for assessing corn chlorophyll concentration from leaf and canopy reflectance:

$$MCARI = \left[\left(\rho_{700} - \rho_{670}\right) - 0.2\left(\rho_{700} - \rho_{550}\right)\right]\left(\frac{\rho_{700}}{\rho_{670}}\right) \tag{10.12}$$

The *transformed chlorophyll absorption in reflectance index* (TCARI) was proposed by Haboudane et al. (2002) to show that the TCARI, applied to corn, is more sensitive to chlorophyll at a lower chlorophyll leaf content:

$$TCARI = 3\left\{\left[\left(\rho_{700} - \rho_{670}\right) - 0.2\left(\rho_{700} - \rho_{550}\right)\right]\left(\frac{\rho_{700}}{\rho_{670}}\right)\right\} \tag{10.13}$$

The TCARI and the *optimized soil-adjusted vegetation index* (*OSAVI*) were combined into one index, the *TCARI/OSAVI*, in order to reduce the soil background and improve the sensitivity to chlorophyll content. The OSAVI is similar to the SAVI (Huete 1988), with an optimized parameter L (= 0.16) for improving the reduction of the soil effect on the vegetation spectra in the case of aggregated pixels (Rondeaux et al. 1996):

$$OSAVI = \frac{1.16\left(\rho_{800} - \rho_{670}\right)}{\rho_{800} + \rho_{670} + 0.16} \tag{10.14}$$

$$\frac{TCARI}{OSAVI} = \frac{3\left\{\left[\left(\rho_{700} - \rho_{670}\right) - 0.2\left(\rho_{700} - \rho_{550}\right)\right]\left(\rho_{700}/\rho_{670}\right)\right\}}{1.16\left(\rho_{800} - \rho_{670}\right)/\rho_{800} + \rho_{670} + 0.16}. \tag{10.15}$$

Numerous chlorophyll indices took advantage of the bands within the RE reflectance line (Section 10.3.3). The *normalized difference red-edge* (NDRE) (Barnes et al. 2000) has an NDVI form but replaces its bands with an RE band at 720 nm and a reference band from the NIR plateau at 790 nm:

$$NDRE = \frac{\left(\rho_{790} - \rho_{720}\right)}{\left(\rho_{790} + \rho_{720}\right)} \tag{10.16}$$

Based on the NDRE and NDVI, the *canopy chlorophyll content index* (*CCCI*) is a 2D index (Barnes et al. 2000):

$$CCCI = \frac{NDRE - NDRE_{min}}{NDRE_{max} - NDRE_{min}} \tag{10.17}$$

By scatter plotting the NDVI and NDRE, the prediction of possible $NDRE_{MIN}$ and $NDRE_{MAX}$ values is performed, and the resultant values are used to obtain the CCCI.

Other conceptual three-band models were developed and examined by Gitelson et al. (2006) and Gitelson (2012) for relating either the green band or the RE band to the main pigments. The *chlorophyll index* (*ClI*) is

$$ChlI_{green} = \rho_{NIR}\left(\frac{1}{\rho_{540-560}} - \frac{1}{\rho_{NIR}}\right) = \left(\frac{\rho_{NIR}}{\rho_{green}}\right) - 1 \quad (10.18)$$

$$ChlI_{red-edge} = \rho_{NIR}\left(\frac{1}{\rho_{690-725}} - \frac{1}{\rho_{NIR}}\right) = \left(\frac{\rho_{NIR}}{\rho_{red-edge}}\right) - 1 \quad (10.19)$$

The *carotenoids index* (*CrI*) is

$$CrI_{green} \propto \rho_{NIR}\left(\frac{1}{\rho_{510-520}} - \frac{1}{\rho_{560-570}}\right) \quad (10.20)$$

$$CrI_{red-edge} \propto \rho_{NIR}\left(\frac{1}{\rho_{510-520}} - \frac{1}{\rho_{690-710}}\right) \quad (10.21)$$

The *anthocyanins index* (*AnI*) is

$$AnI \propto \rho_{NIR}\left(\frac{1}{\rho_{530-570}} - \frac{1}{\rho_{690-710}}\right) \quad (10.22)$$

10.3.5 Nutrients (Nitrogen, Potassium, and Phosphorus)

Among the nutrients, *nitrogen* (N) is an essential element in plant growth and productivity and thus crucial to numerous ecological processes. Several absorption features are listed in the literature along the SWIR region (Curran 1989). Spectral indices derived from handheld, airborne, and spaceborne spectrometers are used for assessing N content. The majority of them are based on indirect indicators, mostly chlorophyll content, which was proven to be physiologically linked to N content. Thus, the previously listed chlorophyll indices, for example, CARI, MCARI, TCARI, and TCARI/OSAVI, have also been frequently used for assessing nitrogen.

On the other hand, since the 1510 nm is directly related to nitrogen content, SWIR-based nitrogen indices were developed and implemented. The *normalized difference nitrogen index* (*NDNI*), proposed by Serrano et al. (2002), is a log 10 transformed reflectance nitrogen index based on the absorption feature of nitrogen at 1510 nm and a reference band at 1680 nm:

$$NDNI = \frac{[\log 10(1/\rho_{1510}) - [\log 10(1/\rho_{1680})]}{[\log 10(1/\rho_{1510}) - [\log 10(1/\rho_{1680})]} \tag{10.23}$$

Herrmann et al. (2010), who explored the performances of different indices with respect to N content, found that the firm advantage of SWIR-based indices lies in their ability to predict, and in their sensitivity to, this constituent. The best index, named the *normalized ratio index* (NRI_{1510}), utilized the 1510 and 660 nm bands:

$$NRI_{1510} = \frac{(\rho_{1510} - \rho_{660})}{(\rho_{1510} + \rho_{660})} \tag{10.24}$$

The two other constituents, *potassium* (*K*) and *phosphorus* (*P*), are also important macronutrients required by plants after N. However, VIs have shown very limited success for assessing *K* and *P*. Pimstein et al. (2011) conducted an experiment in order to assess *P* and *K* in wheat plants using indices and a partial least squares-regression (PLS-R) (Section 10.4.1). The correlation coefficient (*r*) value found using a VI based on a two-wavelength index (1645 and 1715 nm) was 0.73, and the value found using a PLS-R analysis of K content was 0.88. K appears in plants as an ion K^+; therefore, it is not expected to be spectrally detected by its bonds. Pimstein et al. (2011) and Thulin et al. (2014) have shown that wavelengths that are highly related to K content mainly occur in the NIR and the SWIR since the K status is closely related to leaf structure and water regime, respectively. Studies exploring the spectral assessment of K content in *in vivo* plants have achieved less satisfactory results than studies examining the spectral assessment of K stress. P appears in plants in adenosine di- and triphosphates (ADP and ATP, respectively) and, therefore, is expected to be spectrally detected as a result of bonds with other elements. Since P affects plant development and conditions (Homolova et al. 2013), it is assumed that the material itself is not identified. Mutanga and Kumar (2007) applied a neural network (Section 10.4.1) and concluded that integrating the RE and SWIR depth of absorption features is important for P content assessment in the African savanna. Homolova et al. (2013) presented studies analyzing spectral data but did not mention studies exploring P indices; they did mention, though, that there are no pronounced P absorption features. Therefore, the exploration of early nutrient stress identification by spectral means with a high spatial resolution is suggested, along with searching for alternative methods of nutrient concentration assessment.

10.3.6 Broadband Vegetation Water Indices

Vegetation water content is another crucial variable for assessing the state and dynamics of plants with respect to water stress, drought, climate change, fire resistance, and more. *Water indices* (*WIs*) rely on the sensitivity of the SWIR bands (either 1.2 or 1.6 or 2.1 μm, denoted as SWIR1, SWIR2, and SWIR3, respectively) to leaf/canopy water content while assuming the relative stability of the NIR region to this variable. The higher the index value, the healthier the vegetation is. The *moisture stress index* (*MSI*) (Hunt and Rock 1989) is the equivalent of the SR but is based on the NIR and SWIR2 bands:

$$MSI = \frac{\rho_{NIR}}{\rho_{SWIR2}} \tag{10.25}$$

Normalized versions of the same principle have used either the SWIR1 or the SWIR2 bands. The *normalized difference infrared index* (*NDII*) was proposed by Hardisky et al. (1983):

$$NDII = \frac{(\rho_{NIR} - \rho_{SWIR2})}{(\rho_{NIR} + \rho_{SWIR2})} \tag{10.26}$$

The *normalized difference water index* (*NDWI*) was proposed by Gao (1996):

$$NDWI = \frac{(\rho_{NIR} - \rho_{SWIR1})}{(\rho_{NIR} + \rho_{SWIR1})} \tag{10.27}$$

10.3.7 Narrowband Vegetation Water Stress Indices

Hyperspectral signatures of water-stressed leaves are correlated with values of midday *leaf water potential* (ψ_l), *stomatal conductance* (*gs*), and *non-photochemical quenching* (*NPQ*) (Rapaport et al. 2015). It is shown that opposite reflectance trends at 5.3–5.5 μm and around 1.5 μm, associated with independent changes in photoprotective pigment contents and water availability, respectively, were indicative of stress-induced alterations in ψ_l, gs, and NPQ. Furthermore, combining the spectral responses at these VIs and SWIR regions yielded three normalized water balance indices that were superior to various widely used reflectance models in predicting physiological values at both the leaf and canopy levels.

The proposed *leaf water potential index* (*LWPI*) is

$$LWPI = \frac{(\rho_{1490} - \rho_{531})}{(\rho_{1490} + \rho_{531})} \tag{10.28}$$

The *stomatal conductance index* (*SCI*) is

$$SCI = \frac{(\rho_{1500} - \rho_{538})}{(\rho_{1500} + \rho_{538})} \tag{10.29}$$

And the *non-photochemical quenching index* (*NPQI*) is

$$NPQI = \frac{(\rho_{1485} - \rho_{550})}{(\rho_{1485} + \rho_{550})} \quad (10.30)$$

10.3.8 Broadband Abiotic and Other Indices

During the past few decades, it has been shown that soil spectra across the VIS–NIR–SWIR regions are characterized by significant chromophores (e.g., OH, Fe^{3+}, CO_3, and COOH) enabling the quantitative analysis of soil properties (Mathieu et al. 1998; Ben-Dor 2002; Ben-Dor et al. 2006). Consequently, broadband spectral indices have been used to characterize several soil minerals and properties. The following are several examples of soil indices. It should be noted that for detecting *soil moisture* (*SM*) the same broad SWIR bands are practically used as for assessing vegetation water content (Ben-Dor et al. 2009) (Section 10.3.6).

The *brightness index* (*BI*) quantifies the albedo over the VIS region (Escadafal and Bacha 1996; Zaady et al. 2007):

$$BI = \frac{\sqrt{\rho_B^2 + \rho_G^2 + \rho_R^2}}{3} \quad (10.31)$$

The *coloration index* (*ColI*) is sensitive to ferric oxides (Ben-Dor et al. 2006):

$$ColI = \frac{(\rho_R - \rho_G)}{(\rho_R + \rho_G)} \quad (10.32)$$

The *redness index* (*RI*) indicates hematite minerals (Ben-Dor et al. 2006):

$$RI = \frac{\rho_R^2}{(\rho_B * \rho_G^3)} \quad (10.33)$$

The *clay minerals ratio* (*CMS*) highlights hydrothermally altered rocks containing clay and alunite (Drury 1987):

$$CMR = \frac{\rho_{SWIR2}}{\rho_{SWIR3}} \quad (10.34)$$

The *ferrous minerals ratio* (*FMR*) (Drury 1987) is

$$FMR = \frac{\rho_{SWIR2}}{\rho_{NIR}} \quad (10.35)$$

The *iron oxide ratio* (*IOR*) indicates hydrothermally altered rocks that were oxidized due to iron-bearing sulfides (Drury 1987):

$$IOR = \frac{\rho_R}{\rho_B} \tag{10.36}$$

The *crust index* (*CI*) aims at distinguishing between cyanobacteria-based crusty surfaces and the exposed substrate (e.g., bare sands) (Karnieli 1997):

$$CI = \frac{(\rho_R - \rho_B)}{(\rho_R + \rho_B)} \tag{10.37}$$

The CI takes advantage of a unique spectral feature of soil biogenic crust containing cyanobacteria. The special phycobilin pigment in cyanobacteria contributes to producing a relatively higher reflectance in the blue region than does the same type of substrate without biocrusts.

The *normalized difference snow index* (*NDSI*) was developed in order to distinguish between snow and non-snow-covered areas (Salomonson and Appel 2004):

$$NDSI = \frac{(\rho_R - \rho_{SWIR2})}{(\rho_R + \rho_{SWIR2})} \tag{10.38}$$

The NDSI is based on the fact that snow reflectance is high in the VIS wavelengths and has low reflectance in the SWIR ones.

The *normalized burn ratio* (*NBR*) has been used for detecting fire scars (Miller and Quayle 2015):

$$NBR = \frac{(\rho_{NIR} - \rho_{SWIR2})}{(\rho_{NIR} + \rho_{SWIR2})} \tag{10.39}$$

This index relies on the differences the NIR and the SWIR bands have between pre- and postfire images, especially in forested landscapes.

10.3.9 Albedo

Broadband albedo is the ratio of the amount of electromagnetic radiation reflected by a surface to the amount of energy incident upon it. This is a unitless variable that is expressed as a fraction or percentage. Albedo may refer to the entire solar spectrum (0.3–4.5 μm) or merely to the visible portion

(0.4–0.7 μm). Albedo is an important earth surface variable for understanding the transfer of energy and mass from the terrestrial ecosystems to the atmosphere (Roberts et al. 2012). Liang (2004) presents regression analysis products for calculating albedo for the following sensors:

$$\alpha_{LandsatTM/ETM+} = 0.356\alpha_1 + 0.13\alpha_3 + 0.373\alpha_4 + 0.085\alpha_5 + 0.072\alpha_7 \tag{10.40}$$

$$\alpha_{AVHRR} = -0.3376\alpha_1^2 - 0.22707\alpha_2^2 + 0.7074\alpha_1\alpha_2 + 0.2915\alpha_1 + 0.5256\alpha_2 \tag{10.41}$$

$$\alpha_{MODIS} = 0.160\alpha_1 + 0.291\alpha_2 + 0.243\alpha_3 + 0.116\alpha_4 + 0.112\alpha_5 + 0.081\alpha_7 \tag{10.42}$$

$$\alpha_{VEGETATION} = 0.3512\alpha_1 + 0.1629\alpha_2 + 0.3415\alpha_3 + 0.1651\alpha_4 \tag{10.43}$$

where α refers to planetary albedo at the TOA.

10.3.10 Soil Organic Matter (Lignin, Cellulose, and Protein)

Soil organic matter (*SOM*) refers to the organic constituents in the soil. It includes undecayed plant and animal residues at various stages of decomposition, cells and tissues of soil organisms, and substances synthesized by soil organisms. It affects the soil quality through both its chemical and physical properties. In contrast to the chemical properties, and thus the spectra, of soils, rocks, and minerals that remain stable over the years, organic matter in soils varies over time (Ben Dor et al. 1997). Therefore, the SOM spectrum should be related to the soil stage, including fresh, dry, or decomposed litter, with respect to the temporal domain (Ilani et al. 2016). Several studies have suggested diagnostic absorption features for various SOM components, all within the SWIR region (Elvidge 1990; Curran et al. 1992; Ben Dor et al. 1997). Table 10.5 presents the major absorption features for *cellulose, lignin,* and *protein*.

Based on narrow spectral bands, the *cellulose absorption index* (*CAI*) was introduced by Daughtry (2001):

$$CAI = 100\left[0.5\left(\rho_{2031} + \rho_{2211}\right) - \rho_{2101}\right] \tag{10.44}$$

and the normalized difference lignin index (NDLI) was introduced by Serrano et al. (2002):

$$NDLI = \frac{\left[\log\left(1/\rho_{1754}\right) - \log\left(1/\rho_{1680}\right)\right]}{\left[\log\left(1/\rho_{1754}\right) + \log\left(1/\rho_{1680}\right)\right]} \tag{10.45}$$

TABLE 10.5
Major Absorption Features for Cellulose, Lignin, and Protein

Wavelength (nm)	SOM Component
1203	Cellulose
1358	Cellulose, lignin
1367	Cellulose, lignin
1465	Cellulose, lignin
1468	Cellulose, lignin
1582–1583	Cellulose
1726	Cellulose, lignin
1761	Cellulose, lignin
1769	Cellulose, lignin
1929–1932	Cellulose, lignin
2068	Cellulose
2111	Cellulose
2142	Lignin
2169	Lignin
2193	Protein
2331	Cellulose, lignin
2347	Cellulose, lignin
2386	Protein

Source: Ben Dor, E. et al., *Remote Sens. Environ.*, 61, 1, 1997.

10.4 Spectral Analysis Methods

10.4.1 Multivariate Analysis

As mentioned earlier, hyperspectral data, which consist of hundreds of narrow spectral bands, create almost contiguous spectral information of the detected object and enable the quantitative assessment of vegetation and soil bio-physio-chemical properties. Such detailed information cannot be provided by multispectral data. Only a few properties (e.g., chlorophyll content, SOM) are usually of interest in order to effectively determine and characterize the detected substrate. However, the multivariate nature of hyperspectral datasets is distinguished by strong multicollinearity. Often, a relatively low number of training samples (predictees) contrast with a large number of spectral bands (predictors) used by hyperspectral instruments (Atzberger et al. 2010). One of the most popular methods to conduct a regression analysis while coping with the collinearity of the hyperspectral data is the *partial least squares-regression* (*PLS-R*) technique, an advanced extension of multiple linear regression modeling. Unlike the latter, PLS-R is able to correlate collinear, noisy, and distribution-independent datasets, even when the number of predictors greatly exceeds the number of training samples (Wold et al. 2001; Hansen and Schjoerring 2003; Nguyen and Lee 2006; Thulin et al. 2014). In short, a large number of wavelengths are reduced to a smaller set of principal components, also called latent variables or factors; therefore, PLS-R may be a useful exploratory and predictive tool when applied to a hyperspectral dataset. Hansen and Schjoerring (2003) showed that the PLS-R method improved the prediction of green biomass and leaf nitrogen concentration by lowering the root mean squared error (RMSE) when compared to the best examined narrowband indices. Using PLS-R, predictive models that are better than traditional VIs were created by Pimstein et al. (2007) for dry biomass, water content, and LAI and by Pimstein et al. (2011) for predicting potassium and phosphorus contents in crop plants. Atzberger et al. (2010) demonstrated that the PLS-R method was superior to other methods, namely the stepwise multiple linear regression and principal component regression, for predicting total canopy chlorophyll content. Thulin et al. (2014) successfully used this method to construct predictive models for estimating crude protein, digestibility, lignin, and cellulose concentration in temperate pastures. Paz-Kagan et al. (2014) found suitable PLS-R models for predicting soil properties, including sand–silt–clay content, ammonium, nitrate, pH, residual water, SOM, electric conductivity, potassium, phosphorus, potential active carbon, and hydraulic conductivity. Rozenstein et al. (2015) emphasized the importance of *preprocessing transformations* (*PPTs*) prior to the PLS-R analysis, such as *Savitzky–Golay smoothing*, *autoscale*, and/or *generalized least squares weighting*, which might approve the prediction. Choosing the best performing PPTs depends on the dataset and requires testing to ascertain an effective model.

Yet PLS-R is not the only method for modeling soil and vegetation properties using hyperspectral data. Other machine learning algorithm approaches, such as *support vector machines* (*SVMs*), *artificial neural networks* (*ANN*), and *Bayesian model averaging* (*BMA*), have been used to tackle regression problems and to predict continuous variables. A review of SVM use in remote sensing is given by Mountrakis et al. (2011). Durbha et al. (2007) used SVM regression for the retrieval of LAI from a multiangle imaging spectroradiometer. Mirzaie et al. (2014) successfully used several approaches, including PLS-R and ANN, to predict vegetation water content using hyperspectral data. Zhao et al. (2013) estimated multiple foliage biochemical variables (including nitrogen, hydrogen, carbon, cellulose, lignin, chlorophyll [a or b], carotenoid, polar and nonpolar extractives, leaf mass per area, and equivalent water thickness) from 27 spectral-chemical datasets using PLS-R and BMA. This arsenal of statistical learning tools keeps expanding with time, as new methods are adapted for use with hyperspectral data.

10.4.2 Multitemporal Data Processing

Several space programs have already been in orbit for more than 30 years (Gutman and Masek 2012). These programs are the Landsat that was started in 1972, NOAA-AVHRR in 1981, and SPOT in 1986 (see Section 10.2 for details). Several other programs are new (e.g., VEGETATION and MODIS), have just been launched (e.g., Sentinel-2), or will be launched in the near future (VENµS) in an attempt to provide images for many years to come. Although not all the programs have the same spectral characteristics and/or revisit time, the long-term time series of images have considerable potential for studying terrestrial ecology in terms of the land-cover dynamics. A distinction should be made between space programs that provide (nearly) daily data (e.g., AVHRR, MODIS, VEGETATION, Sentinel-2, VENµS) and those that provide periodic images (e.g., Landsat, SPOT). The former group is capable of providing a continuous time series of short-term changes, usually applied to VIs. The temporal archive of AVHRR, sometime merged with the MODIS data, creates a long time series with statistically significant trends that may also be extended to other operated or future systems (Potter et al. 2005; Nemani et al. 2009; Karnieli et al. 2014). Several attempts have been conducted to compare NDVI datasets derived from different sensors, for example, Landsat, SPOT, AVHRR, MODIS, and VEGETATION, to construct a long-term NDVI time series (Fensholt and Sandholt 2005; Tucker et al. 2005; Brown et al. 2006; Ji et al. 2008). The latter group may provide data to conduct a change detection analysis between two or among several images, most likely to quantify interannual or intra-annual (seasonal) slow changes (e.g., Volcani et al. 2005; Karnieli et al. 2008, 2014). In this context, different change detection methods have been widely applied, such as *postclassification comparison*, *image/band/index rationing or differencing*, and *change vector analysis* (see review in Singh 1989; Mas 1999; Coppin et al. 2004; Lu et al. 2004). With respect to ecological monitoring, Willis (2015) classified the change detection applications into (1) uniform conversions between land-use and land-cover classes, (2) irregular variations such as disturbance, and (3) continuous fluctuations such as seasonal plant cycling.

10.4.3 Phenological Studies

The term *phenology* is usually defined as "the study of the timing of recurring biological phases, the cause of their timing with regard to biotic and abiotic forces, and the interaction among phases of the same or different species" (Lieth 1976). With respect to climate change in conjunction with associate alterations of the growing season, the timing of phenological events has become a significant means for assessing the environmental implications of precipitation and temperature trends in a large variety of spatial scales—from individual plant species through vegetation communities at regional, continental, and up to global scales. Traditionally, ground observations of vegetation have relied on monitoring key botanical events, namely green-up, blooming, maturity, senescence, and dormancy of specific species in a unique location such as a botanical garden (e.g., Yang and You-hao 2000). In contrast, remote sensing can upscale point and local observations and provide valuable digital data over vast areas and temporal scales at regular intervals using different environmental and climatic variables (Ganguly et al. 2010). Plant phenology modeling is a developing research field that may provide predictions of phenological cycles and their responses to climate change scenarios. In this regard, many different phenology models have been established in the last decades. These models range from simple to complex ones, depending on the species diversity, predictors, assumptions, and their spatial scales. Curnel and Oger (2006) distinguished among several types of such techniques: threshold-based, derivatives, logistic curves, moving averages, and empirical equations. These models mostly use satellite VI data in general, and NDVI in particular. Spaceborne data enable the identification of several phenological indicators, including greenness onset, maturity onset, maximum greenness, and senescence onset and offset. These indicators are used for quantifying the start, end, and length of the growing season in a specific year. Long-term repetitive data can be obtained from several spaceborne systems such as NOAA-AVHRR, MODIS, and the VEGETATION instrument (Section 10.2) that enable interannual variability in vegetation trends to be analyzed (de Jong et al. 2012).

10.4.4 Multisource Data Fusion

The *multiconcept* was proposed during the early days of modern remote sensing (Simonett et al. 1983) with the understanding that merging data from different sources would provide more useful information, denoted hereafter as *multisource*. The multiconcept includes multiband, multisensor, multiplatform, multistage, multiangle, multitemporal, multi-(spatial, spectral, and radiometric) resolution, multiphase, and multipolarization. This concept still holds today and is intensively implemented by the leading space agencies, for example, the U.S. NASA and the ESA. Of special note is NASA's constellation, called the *A-track*, which consists of several Earth-observing satellites that follow each other seconds to minutes apart along the same orbital "track." Since each sensor has different characteristics, for example, spectral bands, the constellation allows near-simultaneous observations of a wide variety of variables to aid the scientific community in advancing the knowledge of the Earth system. The EC and ESA's *Copernicus* program, based on six different spaceborne sensors (Sentinel 1–6), aims at using multisource data to get timely quality information on land, atmospheric, and oceanographic variables on a global scale. Such multisource opportunities herald a new research era for remote sensing scientists in terms of methods and algorithms that, in turn, provide new and valuable data.

Recent studies have been carried out with a new trend of attempting to conduct data fusing between different spectral bands in order to enhance the image analysis or interpretation through the new product. Fusing the high-resolution panchromatic band and the multispectral bands for the purpose of creating a single high-resolution color image was mentioned earlier (Section 10.2.7.1). Furthermore, fusions between the coarse spatial resolution and the high temporal resolution of the MODIS and the opposite spatial/temporal resolutions of the Landsat in the reflective bands were used for mapping forest disturbances (Hilker et al. 2009). The Landsat reflective bands, in conjunction with the MODIS thermal bands, were used for monitoring field-scale evapotranspiration (Kustas et al. 2003; Agam et al. 2007; Gao et al. 2012). The ability of hyperspectral systems to study the biophysical properties of vegetation and the ability of a LiDAR system to interpret the 3D vegetation structure at different levels (e.g., canopy and understory) have been cooperatively used for mapping and classifying forest areas (Mundt et al. 2006; Anderson et al. 2008; Dalponte et al. 2008). Microwave (radar) data in conjunction with reflective data were used for several applications such as extracting forest attributes (Treuhaft et al. 2004) and water areas (Hong et al. 2015). For these and for many other terrestrial ecological applications, such multisource systems have been recently installed on light aircraft, such as the *Airborne Observation Platform* operated by the National Ecological Observatory Network (Kampe et al. 2010), the *Carnegie Airborne Observatory-2* operated by the Carnegie Institution for Science (Asner et al. 2012), and the *Goddard's LiDAR, Hyperspectral and Thermal* airborne imager operated by NASA (Cook et al. 2013).

10.5 Incorporating Ecological Variables with Remote Sensing

This chapter summarizes the key ecological variables with respect to their analysis means, main spectral bands, and the types of remote sensing platforms. It shows that remote sensing, spectroscopic methods, and related spectral analyses provide essential data for a variety of ecological studies and applications. Table 10.6 aims at synthesizing all the details, demonstrating the linkage between the ecosystem variables and/or ecological applications

TABLE 10.6

Key Ecological Variables with respect to Their Analysis Means, Main Spectral Bands, and Type of Remote Sensing Platform

	Ecosystem Variable/ Ecological Application	Analysis Means	Main Spectral Bands	Remote Sensing Platform
Structural vegetation properties	LAI	Vegetation indices	VNIR	Broadband
	Fractional vegetation cover	Vegetation indices	VNIR	Broadband
	Aboveground biomass	Vegetation indices	VNIR	Broadband
	fAPAR	Vegetation indices	VNIR	Broadband
	NPP	Vegetation indices	VNIR	Broadband
	Foliar loss and damage	Vegetation indices	VNIR	Broadband
Fine spectral vegetation features	Chlorophyll content	Spectral indices	VIS, RE	Narrowband
	Carotenoid pigment	Spectral indices	VIS, RE	Narrowband
	Anthocyanin pigment	Spectral indices	VIS, RE	Narrowband
Nutrients	Nitrogen	Spectral indices	VNIR (indirect); 1510 nm (direct)	Narrowband
	Potassium (K)	Multivariate analysis		Narrowband
	Phosphorus (P)	Multivariate analysis		Narrowband
SOM	Lignin	Multivariate analysis	SWIR	Narrowband
	Cellulose	Multivariate analysis	SWIR	Narrowband
	Protein	Multivariate analysis	SWIR	Narrowband
Vegetation water content	Canopy water content	Spectral indices	NIR, SWIR	Broadband
	Leaf water potential	Spectral indices	NIR, SWIR	Narrowband
	Stomatal conductance	Spectral indices	NIR, SWIR	Narrowband
	NPQ	Spectral indices	NIR, SWIR	Narrowband

(Continued)

TABLE 10.6 (*Continued*)

Key Ecological Variables with respect to Their Analysis Means, Main Spectral Bands, and Type of Remote Sensing Platform

	Ecosystem Variable/ Ecological Application	Analysis Means	Main Spectral Bands	Remote Sensing Platform
Soil	Soil moisture	Spectral indices	NIR, SWIR	Broadband
	Brightness	Spectral indices	VIS	Broadband
	Ferric oxides	Spectral indices	VIS	Broadband
	Hematite minerals	Spectral indices	VIS	Broadband
	Clay minerals	Spectral indices	SWIR	Broadband
	Ferrous minerals	Spectral indices	SWIR	Broadband
	Soil oxidation	Spectral indices	VIS	Broadband
	Biocrusts	Spectral indices	VIS	Broadband
Albedo	Surface albedo	Spectral bands	VIS–NIR–SWIR	Broadband
Snow	Snow cover	Spectral indices	VIS, SWIR	Broadband
Fire scars	Fire scars area	Spectral indices	NIR, SWIR	Broadband

with their respective analysis means, the main applied spectral bands, and the most utilized remote system platform.

10.6 Summary

This chapter reviews remote sensing capabilities for terrestrial ecology and provides the readers with detailed information on remote sensing means, algorithms, and techniques. The document, which is written from a remote sensing perspective, presents a wide range of remote sensing analysis methods related to ecological applications. Nevertheless, despite the comprehensive and long-term link between the two disciplines, there are still fundamental challenges that must be overcome. Several steps that can be taken as derived from the chapter are listed as follows:

- The fusion and integration between different instruments that acquire data in different parts of the electromagnetic spectrum, such as high-spatial-resolution optical, hyperspectral, radar, and LiDAR systems, would initiate new methods and algorithms as well as incorporate them into ecological models in a wide range of spatial scales (Turner et al. 2004; Zhang 2010; Cook et al. 2013).
- The exploration of more diagnostic narrowbands, particularly in the RE and SWIR regions, would refine and improve the relationship between spectroscopic data and the biological, chemical, and physical properties of biotic and abiotic substrates.
- The development and implementation of advanced multivariant data analysis and image processing, such as machine learning algorithms, might improve the interpretation of hyperspectral data.
- Developments in the use of new remote sensing platforms, such as *unmanned aerial vehicles* (UAVs), commonly referred to as drones, should continue. Compared to airborne and spaceborne systems, UAVs are more affordable, inexpensive, controllable, and flexible in use and reach a very high spatial resolution (~1 m). As the radiometers and spectrometers become lighter, they can be carried onboard UAVs and utilized in plot- to local-scale terrestrial ecology applications. Drones can also be used to validate data from spaceborne missions (Panda et al. 2016).
- Remote sensing and physical models can be combined for terrestrial ecosystem applications since such models make a crucial contribution to the accurate determination of ecosystem properties and to the derivation of ecosystem information products (Goodenough et al. 2006; Baret 2016).

Acknowledgments

The author thanks Prof. E. Ben Dor, Dr. A. Pimstein, Dr. I. Herrmann, Dr. O. Rozenstein, and Mr. A. Tal for their fruitful contributions. The manuscript is based on research that was partially funded by the European Union's FP7 programme under grant agreements no. 262060 and no. 312690, and by the European Union's Horizon 2020 Research and Innovation Programme under grant agreements no. 641762 and no. 654359.

References

Agam, N., Kustas, W.P., Anderson, M.C., Li, F., and Neale, C.M.U. (2007). A vegetation index based technique for spatial sharpening of thermal imagery. *Remote Sensing of Environment*, **107**, 545–558.

Agapiou, A., Hadjimitsis, D.G., and Alexakis, D.D. (2012). Evaluation of broadband and narrowband vegetation indices for the identification of archaeological crop marks. *Remote Sensing*, **4**, 3892–3919.

Anderson, J.E., Plourde, L.C., Martin, M.E., Braswell, B.H., Smith, M.-L., Dubayah, R.O., Hofton, M.A., and Blair, J.B. (2008). Integrating waveform lidar with hyperspectral imagery for inventory of a northern temperate forest. *Remote Sensing of Environment*, **112**, 1856–1870.

Asner, G.P. and Heidebrecht, K.B. (2003). Imaging spectroscopy for desertification studies: Comparing AVIRIS and EO-1 Hyperion in Argentina drylands. *IEEE Transactions on Geoscience and Remote Sensing*, **41**, 1283–1296.

Asner, G.P., Knapp, D.E., Boardman, J., Green, R.O., Kennedy-Bowdoin, T., Eastwood, M., Martin, R.E., Anderson, C., and Field, C.B. (2012). Carnegie airborne observatory-2: Increasing science data dimensionality via high-fidelity multi-sensor fusion. *Remote Sensing of Environment*, **124**, 454–465.

Atzberger, C., Guerif, M., Baret, F., and Werner, W. (2010). Comparative analysis of three chemometric techniques for the spectroradiometric assessment of canopy chlorophyll content in winter wheat. *Computers and Electronics in Agriculture*, **73**, 165–173.

Bajocco, S., De Angelis, A., and Salvati, L. (2012). A satellite-based green index as a proxy for vegetation cover quality in a Mediterranean region. *Ecological Indicators*, **23**, 578–587.

Baret, F. (2016). Canopy biophysical variables retrieval from the inversion of reflectance models. In P.S. Thenkabail (Ed.), *Land Resources Monitoring, Modeling, and Mapping with Remote Sensing* (pp. 23–46). Boca Raton, FL: CRC Press, Taylor & Francis Group.

Baret, F., Jacquemoud, S., Guyot, G., and Leprieur, C. (1992). Modeled analysis of the biophysical nature of spectral shifts and comparison with information-content of broad bands. *Remote Sensing of Environment*, **41**, 133–142.

Barnes, E.M., Clarke, T.R., Richards, S.E., Colaizzi, P.D., Haberland, J., Kostrzewski, M., Waller, P. et al. (2000). Coincident detection of crop water stress, nitrogen status and canopy density using ground-based multispectral data. In P.C. Robert, R.H. Rust, and W.E. Larson (Eds.), *Proceedings of the Fifth International Conference on Precision Agriculture* (pp. 1–15). Bloomington, MN: ASA-CSSA-SSSA.

Ben-Dor, E. (2002). Quantitative remote sensing of soil properties. In D.L. Sparks (Ed.), *Advances in Agronomy*, Vol. 75 (pp. 173–243). San Diego, CA: Elsevier–American Press.

Ben-Dor, E., Chabrillat, S., Dematte, J.A.M., Taylor, G.R., Hill, J., Whiting, M.L., and Sommer, S. (2009). Using imaging spectroscopy to study soil properties. *Remote Sensing of Environment*, **113**, S38–S55.

Ben Dor, E., Inbar, Y., and Chen, Y. (1997). The reflectance spectra of organic matter in the visible near-infrared and short wave infrared region (400–2500 nm) during a controlled decomposition process. *Remote Sensing of Environment*, **61**, 1–15.

Ben-Dor, E., Levin, N., Singer, A., Karnieli, A., Braun, O., and Kidron, G.J. (2006). Quantitative mapping of the soil rubification process on sand dunes using an airborne hyperspectral sensor. *Geoderma*, **131**, 1–21.

Brown, M.E., Pinzon, J.E., Didan, K., Morisette, J.T., and Tucker, C.J. (2006). Evaluation of the consistency of long-term NDVI time series derived from AVHRR, SPOT-Vegetation, SeaWiFS, MODIS, and Landsat ETM+ sensors. *IEEE Transactions on Geoscience and Remote Sensing*, **44**, 1787–1793.

Buckingham, R. and Staenz, K. (2008). Review of current and planned civilian space hyperspectral sensors for EO. *Canadian Journal of Remote Sensing*, **34**, S187–S197.

Buschmann, C. and Nagel, E. (1993). In vivo spectroscopy and internal optics of leaves as basis for remote-sensing of vegetation. *International Journal of Remote Sensing*, **14**, 711–722.

Clevers, J., De Jong, S.M., Epema, G.F., Van der Meer, F.D., Bakker, W.H., Skidmore, A.K., and Scholte, K.H. (2002). Derivation of the red edge index using the MERIS standard band setting. *International Journal of Remote Sensing*, **23**, 3169–3184.

Cook, B.D., Corp, L.A., Nelson, R.F., Middleton, E.M., Morton, D.C., McCorkel, J.T., Masek, J.G., Ranson, K.J., Ly, V., and Montesano, P.M. (2013). NASA Goddard's LiDAR, Hyperspectral and Thermal (G-LiHT) airborne imager. *Remote Sensing*, **5**, 4045–4066.

Coppin, P., Jonckheere, I., Nackaerts, K., Muys, B., and Lambin, E. (2004). Digital change detection methods in ecosystem monitoring: A review. *International Journal of Remote Sensing*, **25**, 1565–1596.

Curnel, Y. and Oger, R. (2006). Agrophenology indicators from remote sensing: State of the art. In B. Baruth, A. Royer, and G. Genovese (Eds.), *Remote Sensing Support to Crop Yield Forecast and Area Estimates* (pp. 31–38). Stresa, Italy: ISPRS.

Curran, P.J. (1989). Remote-sensing of foliar chemistry. *Remote Sensing of Environment*, **30**, 271–278.

Curran, P.J., Dungan, J.L., Macler, B.A., Plummer, S.E., and Peterson, D.L. (1992). Reflectance spectroscopy of fresh whole leaves for the estimation of chemical concentration. *Remote Sensing of Environment*, **39**, 153–166.

Dadon, A., Ben-Dor, E., and Karnieli, A. (2010). Use of derivative calculations and minimum noise fraction transform for detecting and correcting the spectral curvature effect (smile) in Hyperion images. *IEEE Transactions on Geoscience and Remote Sensing*, **48**, 2603–2612.

Dalponte, M., Bruzzone, L., and Gianelle, D. (2008). Fusion of hyperspectral and LIDAR remote sensing data for classification of complex forest areas. *IEEE Transactions on Geoscience and Remote Sensing*, **46**, 1416–1427.

Daughtry, C.S.T. (2001). Discriminating crop residues from soil by shortwave infrared reflectance. *Agronomy Journal*, **93**, 125–131.

Daughtry, C.S.T., Walthall, C.L., Kim, M.S., de Colstoun, E.B., and McMurtrey, J.E. (2000). Estimating corn leaf chlorophyll concentration from leaf and canopy reflectance. *Remote Sensing of Environment*, **74**, 229–239.

de Jong, R., Verbesselt, J., Schaepman, M.E., and de Bruin, S. (2012). Trend changes in global greening and browning: Contribution of short-term trends to longer-term change. *Global Change Biology*, **18**, 642–655.

Drury, S. (1987). *Image Interpretation in Geology*. London, U.K.: Allen & Unwin.

Drusch, M., Del Bello, U., Carlier, S., Colin, O., Fernandez, V., Gascon, F., Hoersch, B. et al. (2012). Sentinel-2: ESA's optical high-resolution mission for GMES operational services. *Remote Sensing of Environment*, **120**, 25–36.

Durbha, S.S., King, R.L., and Younan, N.H. (2007). Support vector machines regression for retrieval of leaf area index from multiangle imaging spectroradiometer. *Remote Sensing of Environment*, **107**, 348–361.

Elvidge, C.D. (1990). Visible and near-infrared reflectance characteristics of dry plant materials. *International Journal of Remote Sensing*, **11**, 1775–1795.

Escadafal, R. and Bacha, S. (1996). Strategy for the dynamic study of desertification. In R. Escadafal, M.A. Mulders, and L. Thiombiano (Eds.), *Monitoring Soils in the Environment with Remote Sensing and GIS* (pp. 19–34). Ouagadougou, Burkino Faso: Orstom Editions.

Fensholt, R. and Sandholt, I. (2005). Evaluation of MODIS and NOAA AVHRR vegetation indices with in situ measurements in a semi-arid environment. *International Journal of Remote Sensing*, **26**, 2561–2594.

Fensholt, R., Sandholt, I., and Rasmussen, M.S. (2004). Evaluation of MODIS LAI, fAPAR and the relation between fAPAR and NDVI in a semi-arid environment using in situ measurements. *Remote Sensing of Environment*, **91**, 490–507.

Gamon, J.A., Penuelas, J., and Field, C.B. (1992). A narrow-waveband spectral index that tracks diurnal changes in photosynthetic efficiency. *Remote Sensing of Environment*, **41**, 35–44.

Ganguly, S., Friedl, M.A., Tan, B., Zhang, X., and Verma, M. (2010). Land surface phenology from MODIS: Characterization of the Collection 5 global land cover dynamics product. *Remote Sensing of Environment*, **114**, 1805–1816.

Gao, B.C. (1996). NDWI—A normalized difference water index for remote sensing of vegetation liquid water from space. *Remote Sensing of Environment*, **58**, 257–266.

Gao, F., Kustas, W.P., and Anderson, M.C. (2012). A data mining approach for sharpening thermal satellite imagery over land. *Remote Sensing*, **4**, 3287–3319.

Gitelson, A.A. (2012). Nondestructive estimation of foliar pigment (chlorophylls, carotenoids, and anthocyanins) contents: Evaluating a semianalytical three-band model. In P.S. Thenkabail, J.G. Lyon, and A. Huete (Eds.), *Hyperspectral Remote Sensing of Vegetation* (pp. 141–165). Boca Raton, FL: CRC Press.

Gitelson, A.A., Keydan, G.P., and Merzlyak, M.N. (2006). Three-band model for non-invasive estimation of chlorophyll, carotenoids, and anthocyanin contents in higher plant leaves. *Geophysical Research Letters*, article no. L11402, **33**.

Goodenough, D.G., Li, J.Y., Asner, G.P., Schaepman, M.E., Ustin, S.L., and Dyk, A. (2006). Combining hyperspectral remote sensing and physical modeling for applications in land ecosystems. In *2006 IEEE International Geoscience and Remote Sensing Symposium*, Denver, CO, Vols. 1–8 (pp. 2000–2004).

Guenther, B., Xiong, X., Salomonson, V.V., Barnes, W.L., and Young, J. (2002). On-orbit performance of the earth observing system moderate resolution imaging spectroradiometer; first year of data. *Remote Sensing of Environment*, **83**, 16–30.

Gutman, G. and Masek, J.G. (2012). Long-term time series of the Earth's land-surface observations from space. *International Journal of Remote Sensing*, **33**, 4700–4719.

Guyot, G. and Baret, F. (1988). Utilisation de la haute resolution spectrale pour suivre l'etat des couverts vegetaux. In *Fourth International Colloquium Spectral Signatures of Objects in Remote Sensing*, Aussois, France (pp. 279–286).

Haboudane, D., Miller, J.R., Tremblay, N., Zarco-Tejada, P.J., and Dextraze, L. (2002). Integrated narrow-band vegetation indices for prediction of crop chlorophyll content for application to precision agriculture. *Remote Sensing of Environment*, **81**, 416–426.

Hansen, P.M. and Schjoerring, J.K. (2003). Reflectance measurement of canopy biomass and nitrogen status in wheat crops using normalized difference vegetation indices and partial least squares regression. *Remote Sensing of Environment*, **86**, 542–553.

Hardisky, M.A., Klemas, V., and Smart, R.M. (1983). The influence of soil-salinity, growth form, and leaf moisture on the spectral radiance of spartina-alterniflora canopies. *Photogrammetric Engineering and Remote Sensing*, **49**, 77–83.

Herrmann, I., Karnieli, A., Bonfil, D.J., Cohen, Y., and Alchanatis, V. (2010). SWIR-based spectral indices for assessing nitrogen content in potato fields. *International Journal of Remote Sensing*, **31**, 5127–5143.

Herrmann, I., Pimstein, A., Karnieli, A., Cohen, Y., Alchanatis, V., and Bonfil, D.J. (2011). LAI assessment of wheat and potato crops by VENμS and Sentinel-2 bands. *Remote Sensing of Environment*, **115**, 2141–2151.

Hilker, T., Wulder, M.A., Coops, N.C., Linke, J., McDermid, G., Masek, J.G., Gao, F., and White, J.C. (2009). A new data fusion model for high spatial- and temporal-resolution mapping of forest disturbance based on Landsat and MODIS. *Remote Sensing of Environment*, **113**, 1613–1627.

Holben, B.N. (1986). Characteristics of maximum-value composite images from temporal AVHRR data. *International Journal of Remote Sensing*, **7**, 1417–1434.

Homolova, L., Maenovsky, Z., Clevers, J.G.P.W., Garcia-Santos, G., and Schaeprnan, M.E. (2013). Review of optical-based remote sensing for plant trait mapping. *Ecological Complexity*, **15**, 1–16.

Hong, S., Jang, H., Kim, N., and Sohn, H.-G. (2015). Water area extraction using RADARSAT SAR imagery combined with landsat imagery and terrain information. *Sensors*, **15**, 6652–6667.

Houborg, R., Fisher, J.B., and Skidmore, A.K. (2015). Advances in remote sensing of vegetation function and traits. *International Journal of Applied Earth Observation and Geoinformation*, **43**, 1–6.

Huete, A., Didan, K., Miura, T., Rodriguez, E.P., Gao, X., and Ferreira, L.G. (2002). Overview of the radiometric and biophysical performance of the MODIS vegetation indices. *Remote Sensing of Environment*, **83**, 195–213.

Huete, A.R. (1988). A soil-adjusted vegetation index (SAVI). *Remote Sensing of Environment*, **25**, 295–309.

Hunt, E.R. and Rock, B.N. (1989). Detection of changes in leaf water-content using near-infrared and middle-infrared reflectances. *Remote Sensing of Environment*, **30**, 43–54.

Ilani, T., Herrmann, I., Karnieli, A., and Arye, G. (2016). Characterization of the biosolids composting process by hyperspectral analysis. *Waste Management (New York, N.Y.)*, **48**, 106–114.

Ji, L., Gallo, K., Eidenshink, J.C., and Dwyer, J. (2008). Agreement evaluation of AVHRR and MODIS 16-day composite NDVI data sets. *International Journal of Remote Sensing*, **29**, 4839–4861.

Jordan, C.F. (1969). Derivation of leaf-area index from quality of light on forest floor. *Ecology*, **50**, 663–666.

Kampe, T.U., Johnson, B.R., Kuester, M., and Keller, M. (2010). NEON: The first continental-scale ecological observatory with airborne remote sensing of vegetation canopy biochemistry and structure. *Journal of Applied Remote Sensing*, **4**, article no. 043510.

Karnieli, A. (1997). Development and implementation of spectral crust index over dune sands. *International Journal of Remote Sensing*, **18**, 1207–1220.

Karnieli, A., Gilad, U., Ponzet, M., Svoray, T., Mirzadinov, R., and Fedorina, O. (2008). Assessing land-cover change and degradation in the Central Asian deserts using satellite image processing and geostatistical methods. *Journal of Arid Environments*, **72**, 2093–2105.

Karnieli, A., Kaufman, Y.J., Remer, L., and Wald, A. (2001). AFRI—Aerosol free vegetation index. *Remote Sensing of Environment*, **77**, 10–21.

Karnieli, A., Qin, Z., Wu, B., Panov, N., and Yan, F. (2014). Spatio-temporal dynamics of land-use and land-cover in the Mu Us Sandy Land, China, using the change vector analysis technique. *Remote Sensing*, **6**, 9316–9339.

Kaufman, Y.J., and Tanre, D. (1992). Atmospherically Resistant Vegetation Index (ARVI) for EOS-MODIS. *IEEE Transactions on Geoscience and Remote Sensing*, **30**, 261–270.

Kim, M.S. (1994). The use of narrow spectral bands for improving remote sensing estimation of fractionally absorbed photosynthetically active radiation (fAPAR). *Department of Geography*. College Park, MD: University of Maryland.

Kruse, F.A., Boardman, J.W., and Huntington, J.F. (2003). Comparison of airborne hyperspectral data and EO-1 Hyperion for mineral mapping. *IEEE Transactions on Geoscience and Remote Sensing*, **41**, 1388–1400.

Kustas, W.P., Norman, J.M., Anderson, M.C., and French, A.N. (2003). Estimating subpixel surface temperatures and energy fluxes from the vegetation index-radiometric temperature relationship. *Remote Sensing of Environment*, **85**, 429–440.

Liang, S. (2004). *Quntative Remote Sensing of Land Surfaces*. Hoboken, NJ: John Wiley & Sons, Inc.

Liang, S.L. (2001). Narrowband to broadband conversions of land surface albedo I algorithms. *Remote Sensing of Environment*, **76**, 213–238.

Lieth, H.H. (1976). Contributions to phenology seasonality research. *International Journal of Biometeorology*, **20**, 197–199.

Lu, D., Mausel, P., Brondizio, E., and Moran, E. (2004). Change detection techniques. *International Journal of Remote Sensing*, **25**, 2365–2407.

Mas, J.F. (1999). Monitoring land-cover changes: A comparison of change detection techniques. *International Journal of Remote Sensing*, **20**, 139–152.

Mathieu, R., Pouget, M., Cervelle, B., and Escadafal, R. (1998). Relationships between satellite-based radiometric indices simulated using laboratory reflectance data and typic soil color of an arid environment. *Remote Sensing of Environment*, **66**, 17–28.

Miller, J.D. and Quayle, B. (2015). Calibration and validation of immediate post-fire satellite-derived data to three severity metrics. *Fire Ecology*, **11**, 12–30.

Mirzaie, M., Darvishzadeh, R., Shakiba, A., Matkan, A.A., Atzberger, C., and Skidmore, A. (2014). Comparative analysis of different uni- and multi-variate methods for estimation of vegetation water content using hyper-spectral measurements (vol 26, pg 1, 2014). *International Journal of Applied Earth Observation and Geoinformation*, **28**, 260.

Mountrakis, G., Im, J., and Ogole, C. (2011). Support vector machines in remote sensing: A review. *ISPRS Journal of Photogrammetry and Remote Sensing*, **66**, 247–259.

Mundt, J.T., Streutker, D.R., and Glenn, N.F. (2006). Mapping sagebrush distribution using fusion of hyperspectral and lidar classifications. *Photogrammetric Engineering and Remote Sensing*, **72**, 47–54.

Mutanga, O. and Kumar, L. (2007). Estimating and mapping grass phosphorus concentration in an African savanna using hyperspectral image data. *International Journal of Remote Sensing*, **28**, 4897–4911.

Nemani, R., Hashimoto, H., Votava, P., Melton, F., Wang, W., Michaelis, A., Mutch, L., Milesi, C., Hiatt, S., and White, M. (2009). Monitoring and forecasting ecosystem dynamics using the Terrestrial Observation and Prediction System (TOPS). *Remote Sensing of Environment*, **113**, 1497–1509.

Nguyen, H.T. and Lee, B.W. (2006). Assessment of rice leaf growth and nitrogen status by hyperspectral canopy reflectance and partial least square regression. *European Journal of Agronomy*, **24**, 349–356.

Panda, S.S., Rao, M.N., Thenkabail, P.S., and Fitzerald, J.E. (2016). Satellites and sensors from different eras and their characteristics. In P.S. Thenkabail (Ed.), *Remotely Sensed Data Characterization, Classification, and Accuracies*. Boca Raton, FL: CRC Press, Taylor & Francis Group.

Paz-Kagan, T., Shachak, M., Zaady, E., and Karnieli, A. (2014). A spectral soil quality index (SSQI) for characterizing soil function in areas of changed land use. *Geoderma*, **230**, 171–184.

Pignatti, S., Cavalli, R.M., Cuomo, V., Fusilli, L., Pascucci, S., Poscolieri, M., and Santini, F. (2009). Evaluating Hyperion capability for land cover mapping in a fragmented ecosystem: Pollino National Park, Italy. *Remote Sensing of Environment*, **113**, 622–634.

Pimstein, A., Karnieli, A., Bansal, S.K., and Bonfil, D.J. (2011). Exploring remotely sensed technologies for monitoring wheat potassium and phosphorus using field spectroscopy. *Field Crops Research*, **121**, 125–135.

Pimstein, A., Karnieli, A., and Bonfil, D.J. (2007). Wheat and maize monitoring based on ground spectral measurements and multivariate data analysis. *Journal of Applied Remote Sensing*, **1**, 013530.

Plummer, S.E. (2000). Perspectives on combining ecological process models and remotely sensed data. *Ecological Modelling*, **129**, 169–186.

Potter, C., Tan, P.N., Kumar, V., Kucharik, C., Klooster, S., Genovese, V., Cohen, W., and Healey, S. (2005). Recent history of large-scale ecosystem disturbances in North America derived from the AVHRR satellite record. *Ecosystems*, **8**, 808–824.

Qin, Z.H., Dall'Olmo, G., Karnieli, A., and Berliner, P. (2001). Derivation of split window algorithm and its sensitivity analysis for retrieving land surface temperature from NOAA-advanced very high resolution radiometer data. *Journal of Geophysical Research: Atmospheres*, **106**, 22655–22670.

Rapaport, T., Hochberg, U., Shoshany, M., Karnieli, A., and Rachmilevitch, S. (2015). Combining leaf physiology, hyperspectral imaging and partial least squares-regression (PLS-R) for grapevine water status assessment. *ISPRS Journal of Photogrammetry and Remote Sensing*, **109**, 88–97.

Roberts, A.D., Roth, L.K., and Perroy, L.R. (2012). Hyperspectral Vegetation indices. In P.S. Thenkabail, J.G. Lyon, and A. Huete (Eds.), *Hyperspectral Remote Sensing of Vegetation* (pp. 309–328). Boca Raton, FL: CRC Press.

Rondeaux, G., Steven, M., and Baret, F. (1996). Optimization of soil-adjusted vegetation indices. *Remote Sensing of Environment*, **55**, 95–107.

Rouse, J.W., Haas, R.H., Schell, J.A., and Deering, D.W. (1973). Monitoring vegetation systems in the Great Plains with ERTS. In S.C. Freden, E.P. Mercant, and M.A. Becker (Eds.), *Third Earth Resources Technology Satellite-1 Symposium* (pp. 309–317). Washington, DC: NASA.

Rozenstein, O., Paz-Kagan, T., Salbach, C., and Karnieli, A. (2015). Comparing the effect of preprocessing transformations on methods of land-use classification derived from spectral soil measurements. *IEEE Journal of Selected Topics in Applied Earth Observations and Remote Sensing*, **8**, 2393–2404.

Saint, G. (1994). *"VEGETATION" Onboard SPOT 4 Mission Specification*. Ispra, Italy: European Commission Directorate-General Joint Research Centre Institute for Environment and Sustainability.

Salomonson, V.V. and Appel, I. (2004). Estimating fractional snow cover from MODIS using the normalized difference snow index. *Remote Sensing of Environment*, **89**, 351–360.

Schaepman, M.E. (2009). Imaging spectrometers. In T.A. Warner, M. Duane Nellis, and G.M. Foody (Eds.), *The SAGE Handbook of Remote Sensing* (pp. 166–178). London, U.K.: SAGE Publications Ltd.

Schloss, A.L., Kicklighter, D.W., Kaduk, J., Wittenberg, U., and Participants Potsdam, N.P.P.M.I. (1999). Comparing global models of terrestrial net primary productivity (NPP): Comparison of NPP to climate and the Normalized Difference Vegetation Index (NDVI). *Global Change Biology*, **5**, 25–34.

Serrano, L., Penuelas, J., and Ustin, S.L. (2002). Remote sensing of nitrogen and lignin in Mediterranean vegetation from AVIRIS data: Decomposing biochemical from structural signals. *Remote Sensing of Environment*, **81**, 355–364.

Simonett, D.S., Green, R.G., Reeves, G., Estes, J.E., Bertke, S.E., and Sailer, C.T. (1983). The development and principles of remote sensing. In R.N. Colwell (Ed.), *Manual of Remote Sensing*. Falls Church, VA: American Society of Photogrammetry.

Singh, A. (1989). Digital change detection techniques using remotely-sensed data. *International Journal of Remote Sensing*, **10**, 989–1003.

Smith, K.L., Steven, M.D., and Colls, J.J. (2004). Use of hyperspectral derivative ratios in the red-edge region to identify plant stress responses to gas leaks. *Remote Sensing of Environment*, **92**, 207–217.

Stagakis, S., Gonzalez-Dugo, V., Cid, P., Guillen-Climent, M.L., and Zarco-Tejada, P.J. (2012). Monitoring water stress and fruit quality in an orange orchard under regulated deficit irrigation using narrow-band structural and physiological remote sensing indices. *ISPRS Journal of Photogrammetry and Remote Sensing*, **71**, 47–61.

Tarpley, L., Reddy, K.R., and Sassenrath-Cole, G.F. (2000). Reflectance indices with precision and accuracy in predicting cotton leaf nitrogen concentration. *Crop Science*, **40**, 1814–1819.

Thulin, S., Hill, M., Held, A., Jones, S., and Woodgate, P. (2014). Predicting levels of crude protein, digestibility, lignin and cellulose in temperate pastures using hyperspectral image data. *American Journal of Plant Sciences*, **5**, 997–1019.

Treuhaft, R.N., Law, B.E., and Asner, G.P. (2004). Forest attributes from radar interferometric structure and its fusion with optical remote sensing. *Bioscience*, **54**, 561–571.

Tucker, C.J. (1979). Red and photographic infrared linear combinations for monitoring vegetation. *Remote Sensing of Environment*, **8**, 127–150.

Tucker, C.J., Pinzon, J.E., Brown, M.E., Slayback, D.A., Pak, E.W., Mahoney, R., Vermote, E.F., and El Saleous, N. (2005). An extended AVHRR 8-km NDVI dataset compatible with MODIS and SPOT vegetation NDVI data. *International Journal of Remote Sensing*, **26**, 4485–4498.

Turner, D.P., Ollinger, S.V., and Kimball, J.S. (2004). Integrating remote sensing and ecosystem process models for landscape- to regional-scale analysis of the carbon cycle. *Bioscience*, **54**, 573–584.

Ustin, S.L., Roberts, D.A., Gamon, J.A., Asner, G.P., and Green, R.O. (2004). Using imaging spectroscopy to study ecosystem processes and properties. *Bioscience*, **54**, 523–534.

Vina, A. and Gitelson, A.A. (2005). New developments in the remote estimation of the fraction of absorbed photosynthetically active radiation in crops. *Geophysical Research Letters*, **32**, L17403.

Vogelmann, J.E. (1990). Comparison between 2 vegetation indexes for measuring different types of forest damage in the north-eastern United States. *International Journal of Remote Sensing*, **11**, 2281–2297.

Volcani, A., Karnieli, A., and Svoray, T. (2005). The use of remote sensing and GIS for spatio-temporal analysis of the physiological state of a semi-arid forest with respect to drought years. *Forest Ecology and Management*, **215**, 239–250.

Wang, Q., Adiku, S., Tenhunen, J., and Granier, A. (2005). On the relationship of NDVI with leaf area index in a deciduous forest site. *Remote Sensing of Environment*, **94**, 244–255.

Wessels, K.J., Prince, S.D., Zambatis, N., Macfadyen, S., Frost, P.E., and Van Zyl, D. (2006). Relationship between herbaceous biomass and 1-km(2) Advanced Very High Resolution Radiometer (AVHRR) NDVI in Kruger National Park, South Africa. *International Journal of Remote Sensing*, **27**, 951–973.

Willis, K.S. (2015). Remote sensing change detection for ecological monitoring in United States protected areas. *Biological Conservation*, **182**, 233–242.

Wold, S., Sjostrom, M., and Eriksson, L. (2001). PLS-regression: A basic tool of chemometrics. *Chemometrics and Intelligent Laboratory Systems*, **58**, 109–130.

Yang, Z.-h. and You-hao, E. (2000). A phenology research on the main xylophyte in arid desert area: A example on cultivated plants of Minqin Desert Botanical Garden. *Xibei Zhiwu Xuebao*, **20**, 1102–1109.

Zaady, E., Karnieli, A., and Shachak, M. (2007). Applying a field spectroscopy technique for assessing successional trends of biological soil crusts in a semi-arid environment. *Journal of Arid Environments*, **70**, 463–477.

Zarco-Tejada, P.J., Pushnik, J.C., Dobrowski, S., and Ustin, S.L. (2003). Steady-state chlorophyll a fluorescence detection from canopy derivative reflectance and double-peak red-edge effects. *Remote Sensing of Environment*, **84**, 283–294.

Zhang, J. (2010). Multi-source remote sensing data fusion: Status and trends. *International Journal of Image and Data Fusion*, **1**, 5–24.

Zhang, Y. (2004). Understanding image fusion. *Photogrammetric Engineering and Remote Sensing*, **70**, 657–661.

Zhao, K., Valle, D., Popescu, S., Zhang, X., and Mallick, B. (2013). Hyperspectral remote sensing of plant biochemistry using Bayesian model averaging with variable and band selection. *Remote Sensing of Environment*, **132**, 102–119.

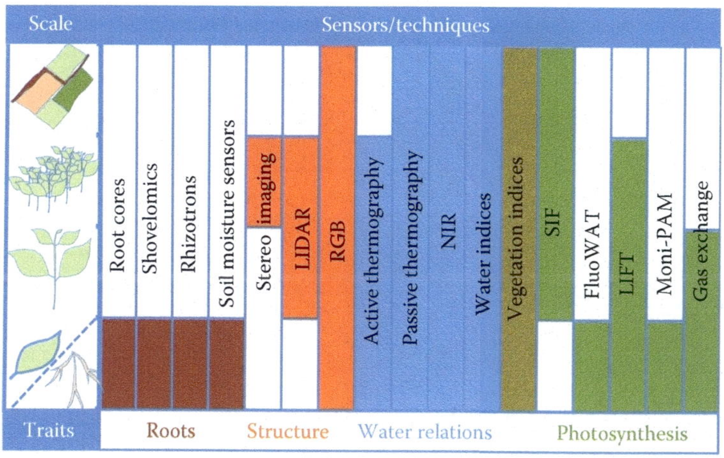

FIGURE 3.2
Scheme of the sensors and techniques to characterize plant phenotypic traits for roots, structure, water relations, and photosynthesis, where the colored part of the column indicates the scale they can be employed for. These are (1) roots mostly by excavation, (2) individual leaves, for example, by clip on devices, (3) capturing whole plants or measuring organs within canopy, (4) on top of canopy mostly by mobile platforms, and (5) field from flying platforms and satellite.

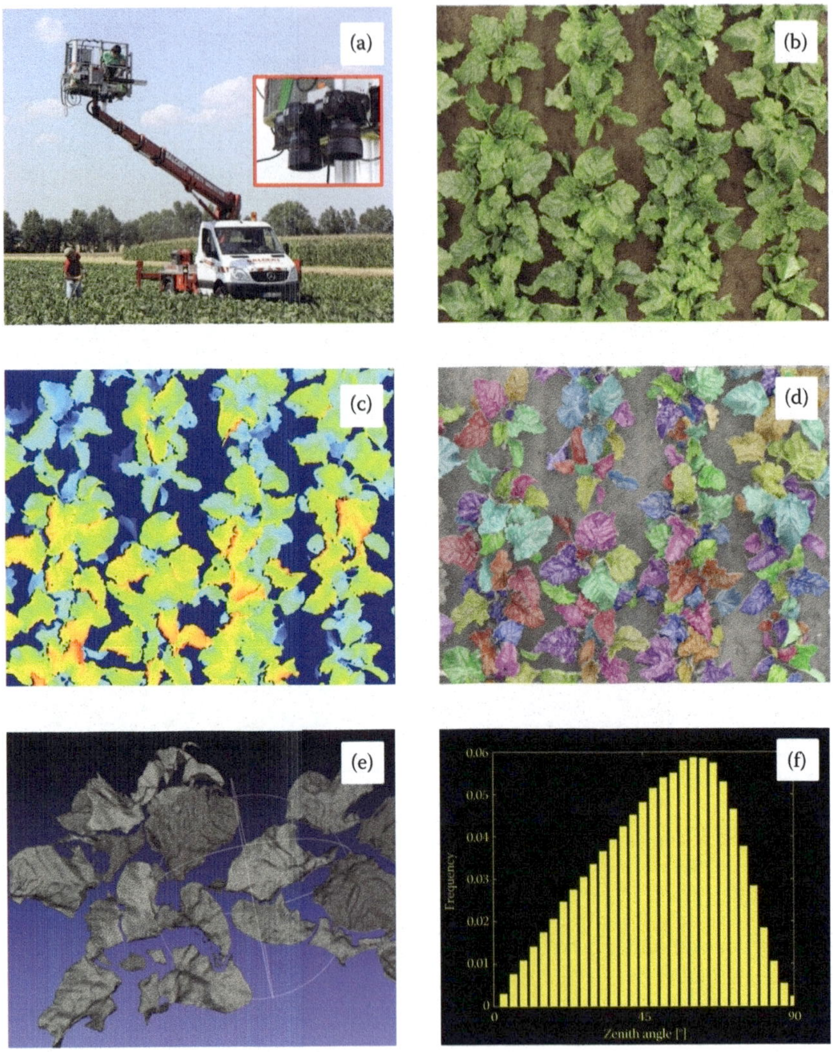

FIGURE 3.3
Estimation of the leaf angle distribution from a 3D reconstruction of a sugar beet canopy: (a) The stereo camera setup (see insert) has been mounted on a cherrypicker 3.5 m top of canopy; (b) shows a region of interest with ~6 plants in the original left camera RGB; (c) filtered depth (disparity) map of (b) with pixel colors indicating the object depth; (d) single leaf segmentation, which is used for further individual leaf surface modeling; (e) leaf surface models are used to derive different leaf traits like leaf area, canopy area, or leaf angle distributions; (f) the leaf angle distribution of the zenith angle of the previous reconstruction; the zenith angle ranges from 0° (flat leaf surface) to 90° (errected surface) and can derived locally or for the complete leaf.

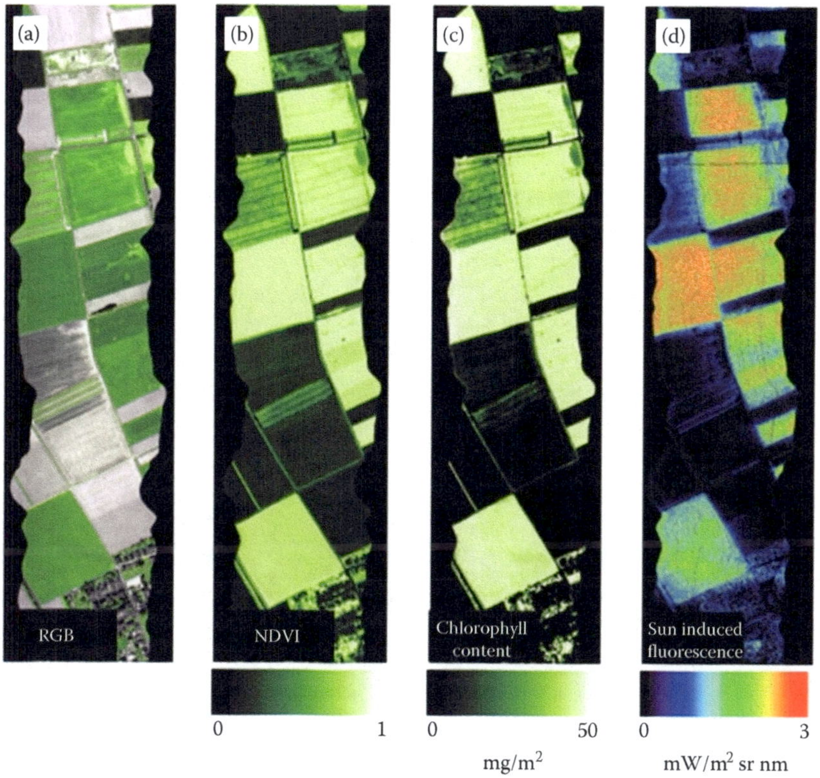

FIGURE 3.4
Airborne maps of different vegetation products and sun-induced fluorescence (SIF) from an agricultural area in Western Germany. The flight line was recorded on August 23, 2012, from 600 m height, at 13:50 local time (UTC + 2 h), which was approximately 1 h after solar noon. (a) For the pseudo-RGB image, reflectance bands at 696, 708, and 677 nm are used for the red, green, and blue channel of the image. (b) The normalized difference vegetation index (NDVI) was calculated as NDVI = NDVI =R758 − R670R758 + R670. (c) Leaf chlorophyll content was calculated using the Gaussian process regression method (Verrelst et al. 2012, 2013). (d) Sun-induced fluorescence (F760) was calculated using the 3FLD method with an empirical correction of effective transmittance in the relevant wavebands. Dominant green vegetation in this area at the time of observation were sugar beet and corn on one side and grassland on the other side. The two vegetation types were in contrasting periods of their seasonal development. Sugar beet and corn had a dense fully mature canopy in contrast grassland had reached senescence. A visual evaluation already indicates that fluorescence shows a different pattern compared to other vegetation variables. This indicated the added value of the sun-induced fluorescence signal that provides a new window into photosynthetic functioning that potentially can be mapped on large areas.

FIGURE 6.2
View of SPRUCE experiment infrastructure with (a) exterior view of experimental chamber, (b) interior view of experimental chamber, and (c) aerial view of the S1 Bog site. (Pictures a and c are Oak Ridge National Laboratory Images from: http://mnspruce.ornl.gov. Image b is a PHENOCAM network SPRUCE image from http://phenocam.sr.unh.edu/webcam/gallery/.)

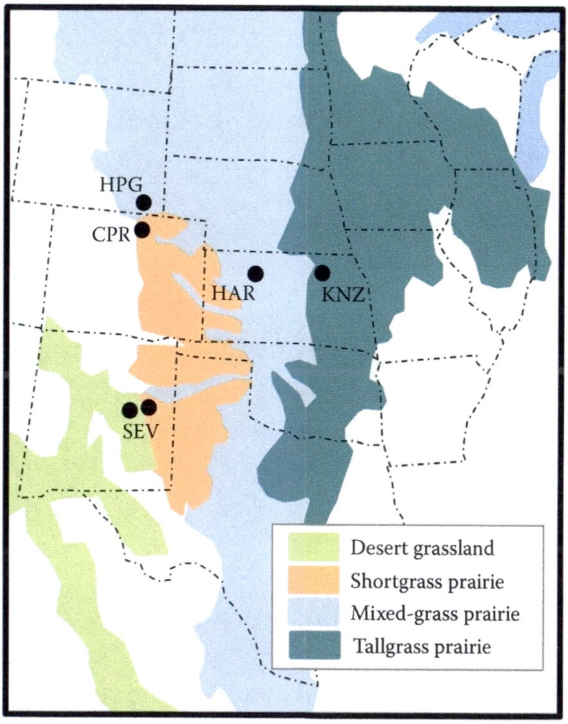

FIGURE 6.3
Map of the Midwest United States showing locations of the six experimental sites of the EDGE project. SEV, Sevilleta National Wildlife Refuge, NM; CPR, Central Plains Experimental range; HPG, High Plains Grassland Research Center; HAR, Hays Agricultural Research Center; KNZ, Konza Prairie Biological Center.

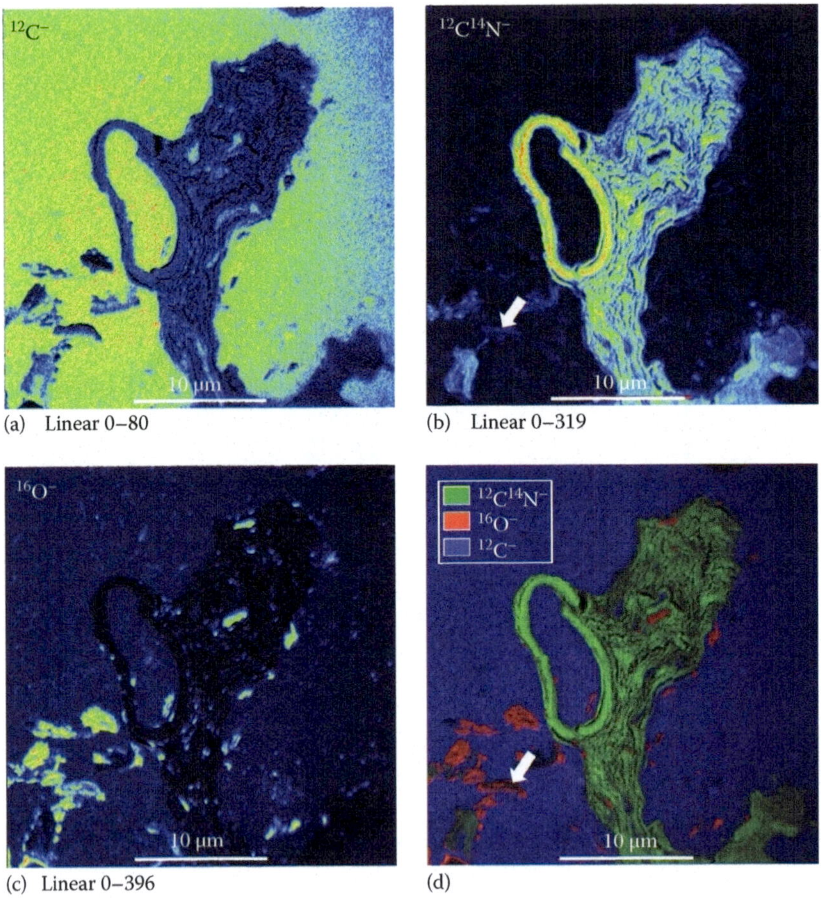

FIGURE 8.1
NanoSIMS images of the distribution of $^{12}C^-$ (a), $^{12}C^{14}N^-$ (b), $^{16}O^-$ (c), and a composite image of the three secondary ions (d) derived from a resin-embedded section of a particulate organic matter–rich permafrost soil structure (Typic Aquiturbel, Barrow, Alaska). The distinctive different distributions between secondary ion counts of $^{12}C^-$ and $^{12}C^{14}N^-$ (d) indicate the good spatial differentiation between the used epoxy resin (a, Araldite 502) and the soil organic matter (b). The $^{16}O^-$ shows the distribution of mineral particles; the arrow indicates mineral-associated organic matter. (From Mueller, C.W., TU Munich, Germany, unpublished data.)

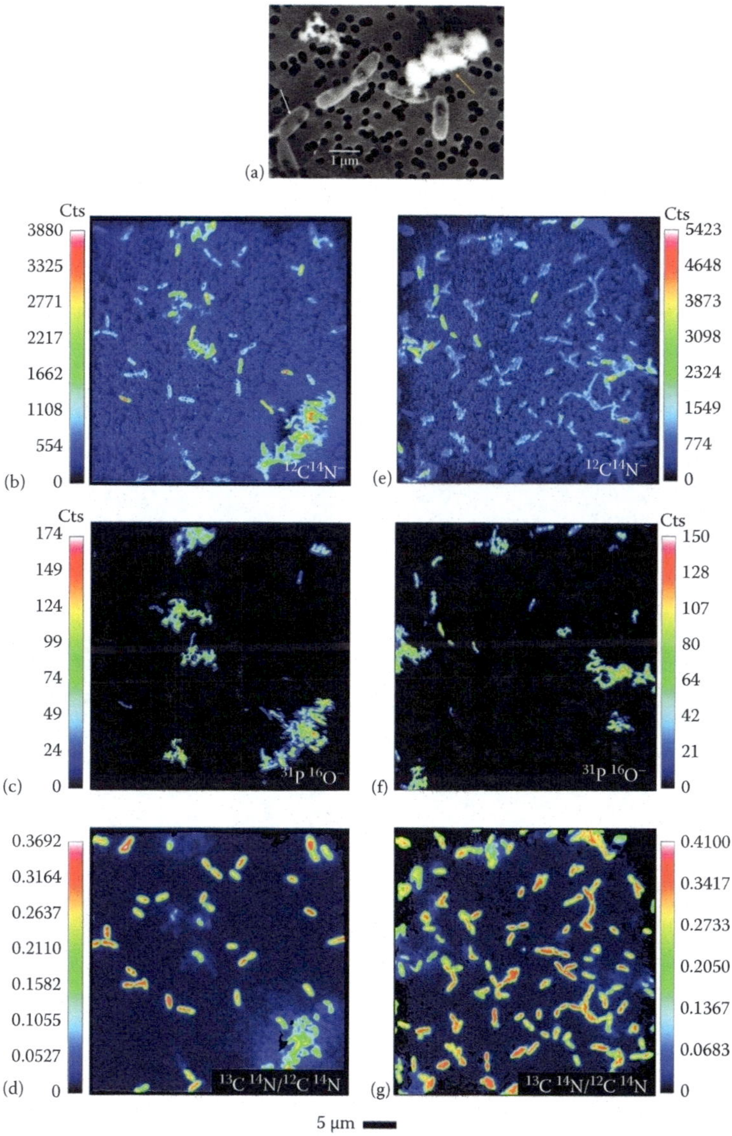

FIGURE 8.2
Acidovorax sp. strain BoFeN1 is inducing Fe mineralization (here Fe-phosphate, image a, orange arrow) as a consequence of Fe oxidation in the presence of large amount of dissolved phosphate in the environment. A cell culture was exposed to the mineralization media for 4 days before being transferred to a media with ^{13}C-labeled acetate and no dissolved phosphate. Cells were then harvested after 1 day (NanoSIMS images b–d) or 4 days (e–g) to visualize the evolution of the bacterial colony. CN images (b, e) show the cells location, PO images (c, f) the precipitated minerals. The ^{13}C-enrichment (d, g) reveals the assimilation of acetate by living cells; this is a marker of active bacteria (e.g., living cells). Cells showing significant mineralization are free of ^{13}C-enrichment, showing that mineralization induces cell death. (Modified from Miot, J. et al., *Front. Microbiol.*, 6, 2015.)

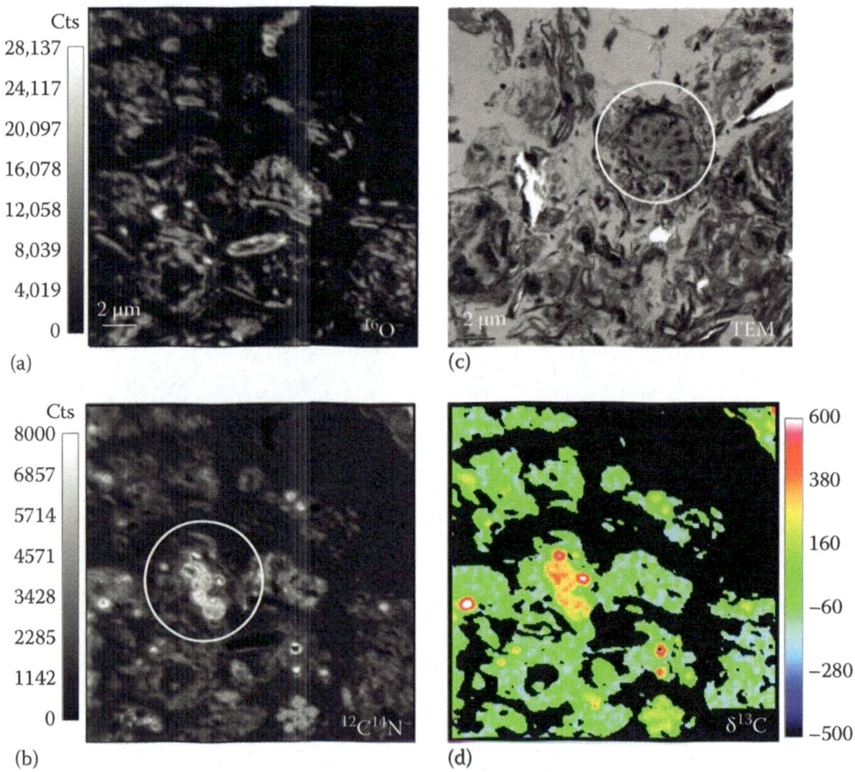

FIGURE 8.3
NanoSIMS and TEM (top right) images of an earthworm cast recovered after a 24-week labeling experiment. ^{16}O image (a) is showing minerals, the $^{12}C^{14}N$ image (b) the organic matter. Thanks to the TEM image (c), we can identify the typical morphology for bacterial colonies and we observe that bacteria have trapped C from the labeled litter (shown by the $\delta^{13}C$ image (d)). (Modified from Vidal, A. et al., *Soil Biol. Biochem.*, 93, 8, 2016.)

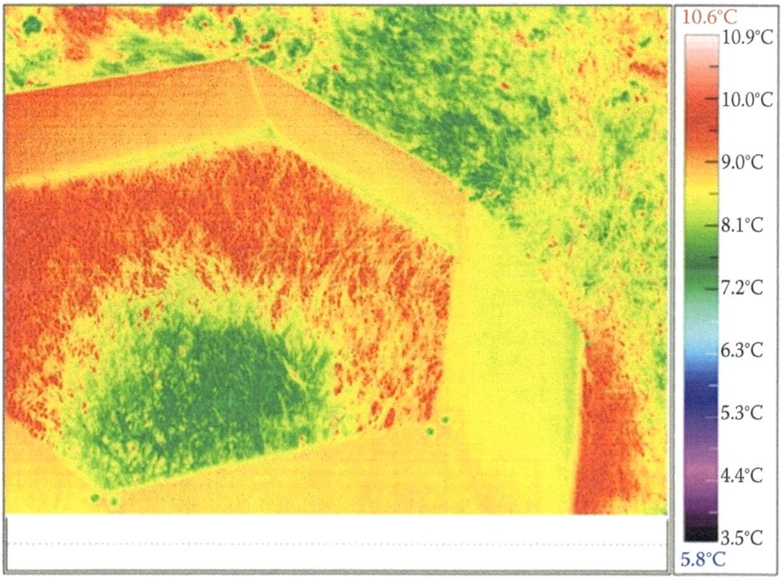

FIGURE 9.1
Infrared image of a passive OTC covering a lawn, taken on a sunny day. (Copyright H.J. De Boeck.) Note the significant differences in surface temperature within the chamber.

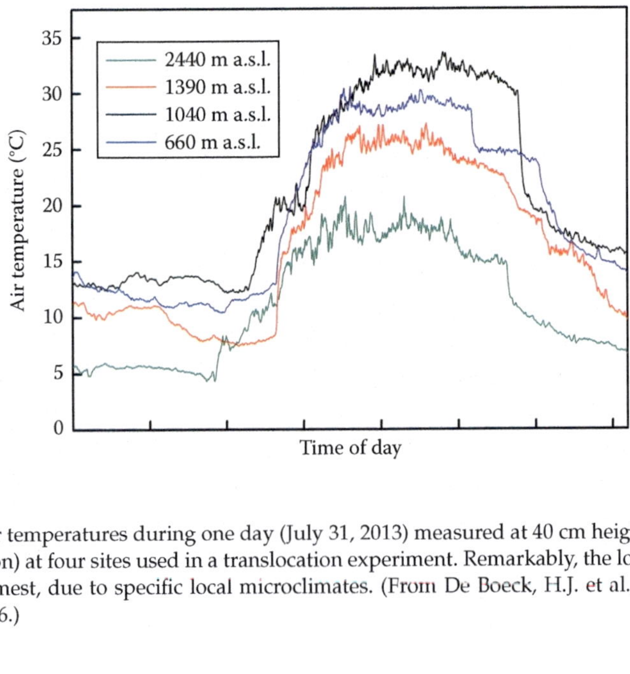

FIGURE 9.2
Course of air temperatures during one day (July 31, 2013) measured at 40 cm height (just above the vegetation) at four sites used in a translocation experiment. Remarkably, the lowest site was not the warmest, due to specific local microclimates. (From De Boeck, H.J. et al., *New Phytol.*, 209, 531, 2016.)

FIGURE 9.3
FORHOT site in Iceland, where an earthquake in 2008 led to parts of the forest and grassland being exposed to marked increases in soil temperature. A gradient with plots warmed with 1°C–20°C at 10 cm depth was established here. (Copyright H.J. De Boeck.)

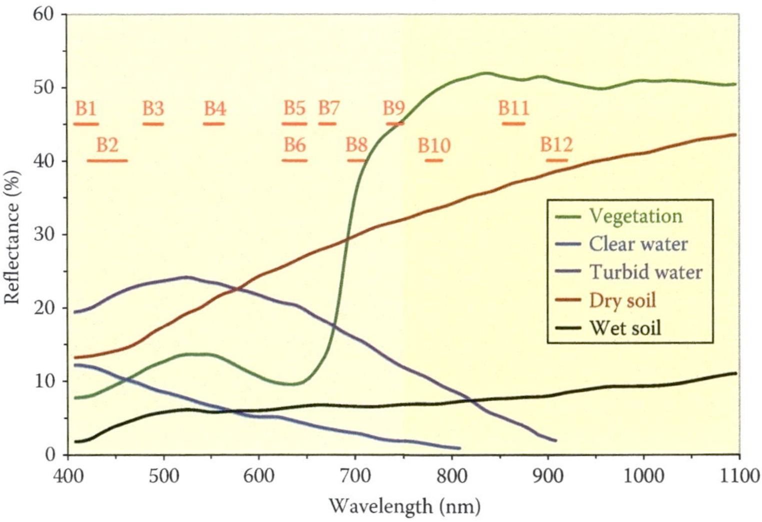

FIGURE 10.1
Band settings of VENμS with respect to vegetation, soil, and water spectral signatures.

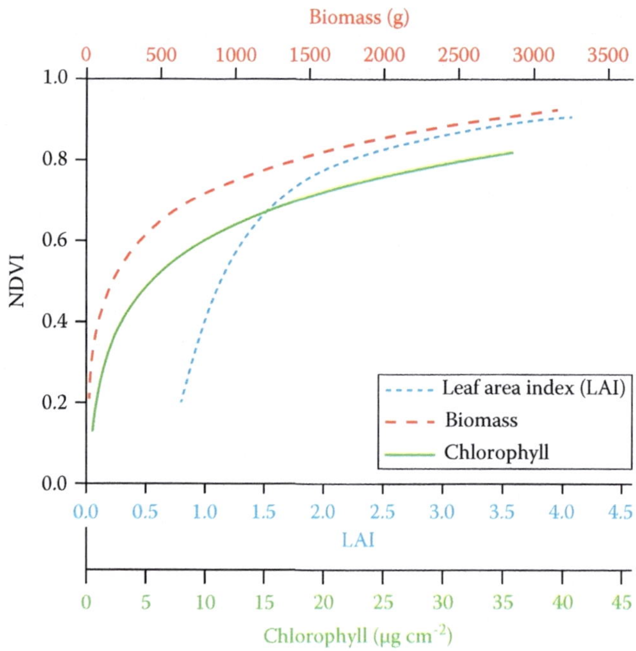

FIGURE 10.2
The NDVI as a function of LAI, chlorophyll content, and biomass. Note the less sensitivity ("saturation") of the NDVI when high values of the independent variables are presented, indicating dense and multilayered canopies. (Data provided by Agustin Pimstein and Ittai Herrmann.)

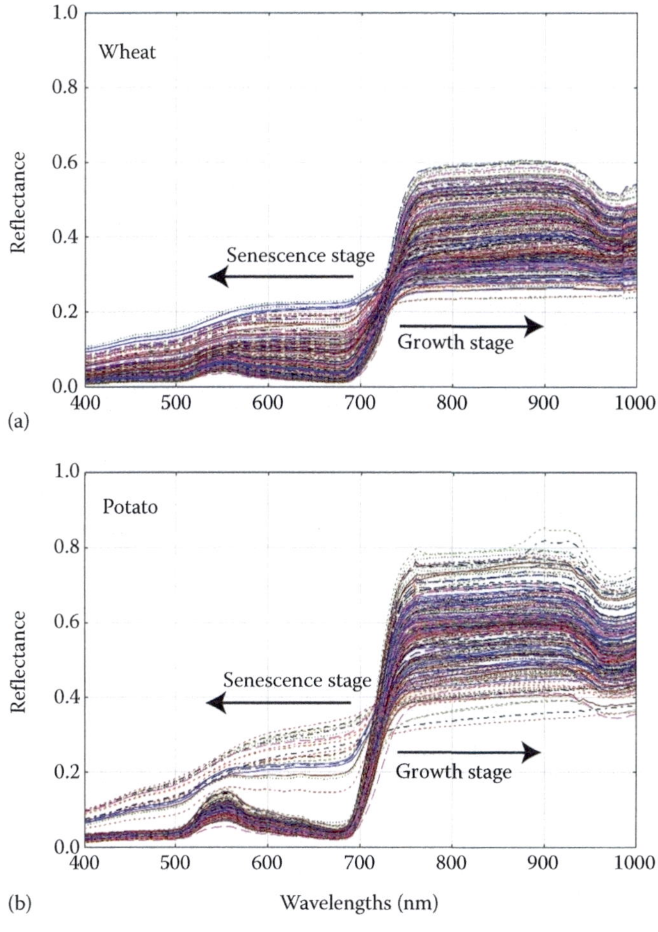

FIGURE 10.3
Temporal dynamics of the RE during the phenological cycles of wheat (a) and potato (b) crops. (Data provided by Agustin Pimstein and Ittai Herrmann.)

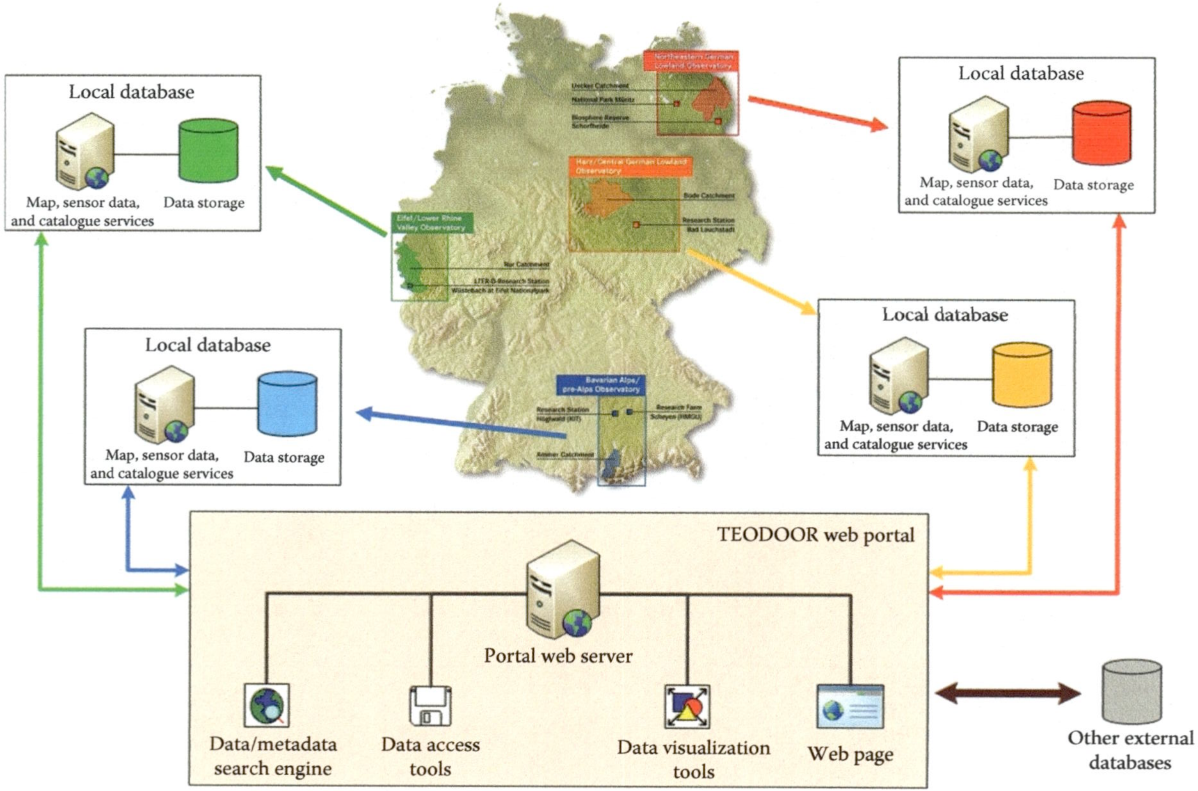

FIGURE 11.2
The design of the TERENO infrastructure for the distribution of the data.

FIGURE 14.1
Location and protection status of the PAs considered in ECOPOTENTIAL and European biogeographic regions.

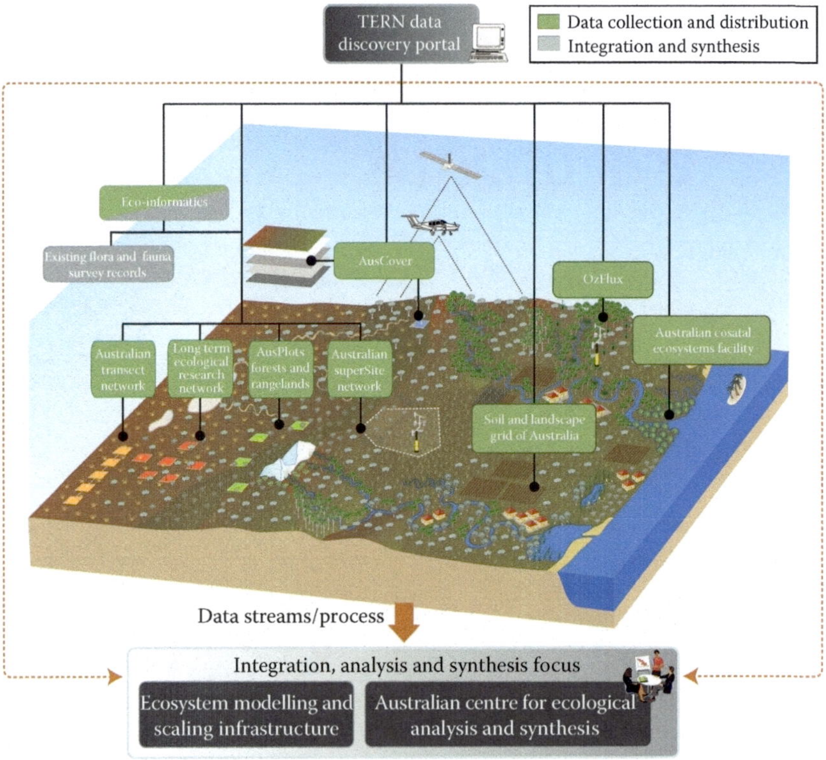

FIGURE 16.1
TERN's infrastructure for ecosystem science: How TERN fits together.

11

A Blueprint for a Distributed Terrestrial Ecosystem Research Infrastructure

Heye Bogena, Harrie-Jan Hendricks Franssen, Carsten Montzka, and Harry Vereecken

CONTENTS

Abstract

Although there is growing awareness that continuous observation of the various terrestrial subsystems (i.e., atmosphere, hydrosphere, pedosphere, biosphere) is essential in improving our understanding of the complex influences of global change on terrestrial ecosystems, existing networks of long-term research infrastructures tend to address only specific issues (e.g., carbon balance). In this chapter, we advocate the need for long-term, distributed terrestrial ecosystem research infrastructures (TERIs) for a new level of fully integrated, multidisciplinary global change research. TERI networks would offer significant potential to address a wide range of challenging environmental problems. Knowledge gained through these networks will be critical in understanding, detecting, and forecasting changes in terrestrial ecosystems that affect important ecological services upon which society depends. Such a network of integrated observation platforms will provide information necessary for societies to adapt to broad-scale changes such as those associated with land use, demographic, and climate change.

In this chapter, we discuss the blueprint for a network of TERIs. Based on the example of the TERENO infrastructure, we present the complexities and challenges confronting the design and implementation of ecological monitoring networks.

Keywords: Terrestrial ecological monitoring network, Remote sensing, Scaling, Modeling

11.1 Introduction

During the last century, mankind impacted fundamental environmental and physical processes that sustain life on earth in an unprecedented way (McCarthy et al. 2001). We are now starting to observe broad-scale ecological impacts of global climate change, including changes in the timing of ecological processes (Westerling et al. 2006). In particular, polar and high-altitude regions are already showing the highest rate of warming connected to climate change (UNEP and WMO 2011). The increase in global average surface temperature by 0.85°C ± 0.2°C is expected to further increase by 1.5°C–4.8°C until 2100 (IPCC 2013). Through modifications of the earth's surface and global biogeochemical cycles, productive ecosystems have lost to desertification processes, reduced availability and reliability of clean and abundant sources of water and food, increased risks to natural hazards such as floods and fires, and increased disease and exposure to harmful chemicals (Bellamy et al. 2005; Young and Harris, 2005). The specific human influences and drivers of these observed changes are complex and multiscaled in nature (MEA 2005).

There is an increased recognition that ecological research infrastructures, providing long-term information on the development of states and fluxes of the various terrestrial subsystems (i.e., atmosphere, hydrosphere, pedosphere, biosphere), are essential in improving our understanding of these complex relationships between humans and environment and in developing policies and strategies to reduce impacts from anthropogenic change, to develop adaptation strategies and to secure a more sustainable future (Allen et al. 2014, Balvanera et al. 2013; Fraser et al. 2013; IPCC 2013; Peters et al. 2014). However, the existing networks of research infrastructures tend to address specific issues often related to specific subsystems within ecosystems and, as a result, represent different geographies, temporal and spatial scales, and attributes of the environment (CENR 1997). We also lack coordinated monitoring programs that provide integrated, standardized, and quality-checked data across multiple ecological systems at various temporal and spatial scales in order to make studies and comparison more robust.

Moreover, a critical need exists to synthesize biophysical, ecological, social, and economic information to increase our understanding of the significance of interactions among processes that impact upon ecological systems over broader time and space than currently possible. New developments in science and technology provide new opportunities for collecting and organizing data that could greatly expand our monitoring capabilities and, hence, develop new understandings.

In this chapter, we present the challenge of designing and implementing long-term distributed terrestrial ecosystem research infrastructures (TERIs) for a new level of fully integrated, multidisciplinary global change research. We further describe how such infrastructures in combination with remote sensing and prognostic modeling efforts can be used to provide early warning and forecasting of potentially irreversible conditions and trends in processes that maintain important ecological services.

The remainder of this chapter is organized as follows. First, we will address the current challenges in ecosystem research and the need for distributed TERIs. We then describe the TERI implementation and design, which is followed by a presentation of complementary remote monitoring and scale transfer using modeling approaches. To illustrate the TERI concept, we use the existing national TERI (TERENO) as an example. Finally, we present concluding remarks and an outlook.

11.2 Challenges in Terrestrial Ecosystem Research

Up-to-date research networks have mainly focused on specific subsystems, such as land surface, various parts of the atmosphere and the subsurface, and natural ecosystems, without considering their different buffer capacities (e.g., soil nutrient buffer capacity, flood retention zones) and process time scales (Bogena et al. 2006). Although there is a growing awareness of subsurface terrestrial systems (e.g., vadose zone, groundwater aquifers) being affected by global change and the direct impact of human activity, less attention has been paid to important subsurface characteristics and processes (e.g., observed decrease in soil organic matter, expected changes in the soil moisture regime, evapotranspiration, groundwater recharge, and runoff generation). For instance, subsurface lateral flow is widely considered to be an important hydrological flux that is poorly understood and difficult to measure and quantify (Vereecken et al. 2015). In the framework of TERI, high-frequency isotope analysis and hydrogeophysical methods (e.g., multireceiver electromagnetic induction, ground-penetrating radar) could be used to better determine this important hydrologic process from the field to the catchment scale. In addition to the impact of global change, terrestrial systems are strongly influenced by local and direct human activities causing dramatic changes in the terrestrial system functions, which are not yet monitored comprehensively (e.g., changes in land use, deforestation, overfertilization, opencast mining, groundwater withdrawal).

In our view, the increasing anthropogenic impact on the terrestrial systems and their associated changes require a multiscaled, consistent, and longer-term utilization of environmental research from local sites to regional scales with corresponding personnel and financial expenditure. Here we emphasize the importance of regional scale because at this scale, interactions between land surface and atmosphere and land management consequences (e.g., political decisions on bioenergy of fertilizer taxation) can be investigated. On the other hand, comprehensive experimental studies on multisubsystem interactions and scale dependencies are needed to better understand the response of terrestrial systems to changing environmental conditions (e.g., Bogena et al. 2015; Burt et al. 2008; Katul et al. 2012). Numerous experiments (e.g., summarized in Giardina and Ryan 2000) have been conducted in order to evaluate the potential future effects of climate change on soil carbon dynamics. However, these experiments rarely continue for more than a few years and thus never provide information on the response of the large, slower pools that will dominate feedback from the soil to the atmosphere over time scales of decades or more (Powlson 2005). The effects of climate change are much more complex than just an increase of a single parameter, since compensation and acclimatization processes may occur (Davidson et al. 2000). Thus, in order to increase the ecosystem process understanding, that is, in terms of multisystem interactions and scale dependencies, the continued multiscale fully integrated observation of terrestrial states, fluxes, and properties by TERI needs to be combined with more detailed, long-term experimental studies.

11.3 Implementation and Design

As elaborated in previous chapters, the implementation of long-term TERIs is urgently needed to enable a better study of the terrestrial system in order to provide guidance in how (1) to sustain the ability of terrestrial systems to adapt to environmental change as well as their resilience, (2) to maintain ecosystem services, and (3) to provide scientifically justified solutions to land use conflicts. This is underpinned by the strategy paper of the DFG on "long-term perspectives and infrastructure in terrestrial research in Germany—a systemic approach" (DFG, 2013). A terrestrial system in this context is defined as a system consisting of the subsurface environment, the land surface including the biosphere, the lower atmosphere, and human impact on the different scientific areas of study. Since this system is organized along a hierarchy of spatiotemporal scales and structures ranging from the local scale to the regional scale, the TERI concept needs to follow a multiscale approach consisting of a set of spatially distributed, site-based terrestrial observatories. In order to characterize terrestrial-atmospheric feedback and human-use gradients, the spatial scale of each terrestrial observatory (which contributes toward the overall network) should range between 1,000 and 10,000 km^2 including several intensively instrumented super test sites (field or headwater catchment scale). These super test sites need to be representative, that is, they should cover the main types of land use, soil, geology, and topography found in the region of the terrestrial observatory to enable upscaling of local observations (see Chapter 4 for more details on scaling issues). In addition, super test sites can host manipulation experiments on ecosystem response to changing environmental conditions and can be used for testing new observation techniques (e.g., Bogena et al. 2013). Since hydrologic processes exert a fundamental control on aquatic and terrestrial metabolism and nutrient cycling, catchments represent an ideal fundamental unit of the TERI network (Bogena et al. 2006), providing the construct to close or constrain the water, energy, and mass balances (Jensen and Illangasekare 2011). By combining individual terrestrial observatories into the overall TERI network, continental- to global-scale questions between anthropogenic and natural factors can begin to be addressed in a systematic fashion, similar to existing distributed research infrastructures that focus on specific science disciplines, for example, Integrated Carbon Observation System (ICOS) that aims for a European-scale carbon balance.

Natural sites that have minimal human impact should be a core design element of a TERI network as reference sites. A "near" natural site such as state owned (as in Siberia, Russia), wildlands of the Bureau of Land Management (as in the United States), restricted access zones (as in Brazil), private ownership of reserves (as in Costa Rica), natural heritage sites, can provide a baseline understanding of how a least-human-impacted one would behave and allow for robust comparisons with sites of varying degrees of land management and human-impacted systems. These types of comparative analyses will enable spatially integrative understanding of the long-term environmental changes due to human impact. Moreover, a comparative design that includes natural areas will also provide the context to socioeconomic research questions. Albeit the integration of ecosystem science and socioeconomic questions is still nascent (Burkhard et al. 2010) and has not yet been implemented in existing distributed TERIs.

Another requirement for the development of TERI network is the availability of appropriate measurement platform and sensing techniques. The TERI instruments need to be designed for robust, long-term, consistent measurements of important terrestrial fluxes of mass and energy and state variables, for example, soil moisture, and the factors that control these exchanges between the terrestrial subsystems. The observational strategies need to be designed in a way to address the expected changes in the processes at relevant time and space scales. For example, near-real-time measurements by fixed installed instruments enable to assess rapid, short-term process rates such as the effects of diurnal patterns (e.g., evapotranspiration) and fast stochastic changes in system states and fluxes (e.g., runoff discharge, effects of fire, rain pulse events). On the other hand, the measurement design needs to be able to also detect slower processes that occur over seasons and years (e.g., soil compaction, droughts). Finally, TERI measurement capabilities must also have the flexibility to add additional capabilities to address the ability to discover future unforeseen processes and address important processes in the future that are not known today.

A primary challenge is to determine the key process (among disciplines) and the relevant time and space scales by which they occur and to design measurement systems that can effectively quantify these time and space scales. For example, in hydrology, one faces the challenge of large spatial and temporal variability of soil moisture from the soil-aggregate-to-plot-to-field and stand-to-landscape scales (Vereecken et al. 2014). Factors controlling soil moisture include atmospheric forcing, topography, soil properties, vegetation, and human impacts (e.g., irrigation, groundwater withdrawal), which all interact in complex, nonlinear ways. One solution is to deploy a sensor network consisting of a multitude of small sensor nodes embedded in the environment that can quantify spatially heterogeneous phenomena, for example, temperature or soil moisture fields, with high temporal and spatial resolution (Bogena et al. 2010). Such data enable the determination of covariance structures of autocorrelated drivers to assist in upscaling and downscaling of state variables in time and space (Loescher et al. 2007). Slower changes in system states might best be observed at broader scales than local scales, that is, field, stand, ecosystem scales. Such would be the example of measuring the below-canopy microclimate of many ecosystems within the landscape. The physical structure of these systems does not change rapidly and can be assessed on a campaign basis through the use of wireless ad hoc sensor networks (Inagaki et al. 2010). Such ad hoc sensor networks are extremely flexible and can be used temporarily to assess spatial heterogeneity of ecosystem states in areas that are not part of the "fixed" sensor network, thus increasing spatial coverage of TERI. In addition, these ad hoc networks can be deployed in a chronosequence design to substitute time for space. The associated requirement would be for these mobile platforms to be interoperable and the data should have the same or similar uncertainties as the fixed long-term measurements. Moreover, these ad hoc sensor networks can operate periodically on a regular or event-driven basis, thus allowing to capture other nonlinear behaviors in the landscape that otherwise could not be measured, that is, heat islands, shifting agriculture, flooding, wildfires. Due to the large amount of information gathered by such instrumentations, dataflow needs to be accompanied by high-performance computational systems for data acquisition, storage, and processing (see Chapter 4).

In order to link in situ information from local monitoring systems to the regional, continental, or global scale, the TERI design needs to be linked with a hierarchy of process-based models. Local-scale information such as parameter fields can be obtained from a combination of direct measurements (Weihermueller et al. 2009), pedotransfer functions that enable predicting key soil properties and parameters from easily measurable properties and state variables (e.g., Qu et al. 2015; Vereecken et al. 2010). In return, these scale-dependent process-level measurements can be used to drive and validate process-based field-scale models (e.g., Fang et al. 2015). Another approach is to use data assimilation techniques at the next larger scale that updates both states and parameters of the terrestrial systems using local-scale information. To continue to evaluate the efficacy of these model approaches, long-term consistent measures of fluxes of energy, water, and mass (e.g., evapotranspiration, infiltration, groundwater flow, runoff discharge) need to be measured (e.g., selected test sites) in order to further validate these models over years and decades and to test new theories. Being able to incorporate land management with more traditional biophysical models is an active area of research, particularly within a scaling framework. Again, using hydrology as an example, catchment models like WASIM-ETH (e.g., Montzka et al. 2011) or TerrSysMP (Kurtz et al. 2016) are able to reproduce management activities like reservoir use, groundwater withdrawal, and irrigation and place them in a context by which they can be scaled further to the regional or continental scale. Alongside with the mandate to scale to larger areas comes the need to integrate data from a variety of remote sensing sources (see Chapter 3 for more details). A nice example that implements the concept of hierarchical scales and scale transformation is the collaborative research center "patterns in soil–plant atmosphere systems" that uses the Eifel/Lower Rhine observatory of the TERENO initiative (Chapter 5); this is a major research thrust to develop novel approaches in upscaling and use of data assimilation approaches (Simmer et al. 2015).

An important set toward the development of a TERI network is the selection of appropriate sites. A promising approach to identify TERI sites from the continental to the global scale is the Budyko framework (Budyko 1974), which describes an empirical global relationship between the evaporative index and climatic dryness index. The Budyko framework has been used to assess the sensitivity of river discharge to climatic change (Donohue et al. 2011; Renner et al. 2012) and to analyze climate and vegetation controls on the surface water balance and evapotranspiration (Williams et al. 2012). The Budyko framework has also been used to describe and subdivide terrestrial systems with respect to energy- and water-limited systems (van der Velde et al. 2014). In addition, further considerations with respect to dominant vegetation and land management should be included to cover all important ecosystems (e.g., tropical rainforest, temperate rainforest, tropical wetlands, desert, managed temperate forest). In addition, more vulnerable and less vulnerable sites, for example, to climate change, can be compared in order to better understand scaling of ecosystem stability. Ecosystem stability in this sense can range from regeneration via resilience (returning to a previous state in a short term) to constancy to persistence. In Section 4 and 5, we describe how remote sensing techniques and model base scale transfer can support the upscaling of local information obtained by the TERI observations systems.

11.4 Complementary Earth Observation

Earth observation (EO) describes the collection of information about planet Earth's physical, chemical, and biological systems via remote sensing technologies supplemented by earth surveying techniques, encompassing the collection, analysis, and presentation of data. Here we review EO techniques because TERI field measurements are typically used for accurate estimation of biogeophysical parameters at the local scale but are not able to provide area-wide information (Kampe et al. 2010). EO sensors are able to return environmental quantities in order to anchor and scale ecosystem properties and support the in situ instrumentation of TERI to monitor the temporal development of spatial patterns (Lausch et al. 2015; Müller et al. 2014; Rudolph et al. 2015) in a sound observatory concept. On the other hand, the in situ observation of TERI can significantly contribute to the validation of EO data products and to the calibration of retrieval models needed to transfer raw EO measurements into environmental variables (Montzka et al. 2013).

Over the last two decades, we have seen great advances in the development and deployment of remote sensors for EO (CEOS 2015). These include a wide range of biophysical quantities, including temperature profiles, soil characteristics, geology, landforms, elevation, chemical quality of surface waters, and vegetation characteristics (e.g., Eisele et al. 2012). Moreover, advances in computing and digital communication have dramatically increased our ability to process, store, analyze, and integrate remote sensing data. Remote sensors are typically designed to monitor specific environmental variables.

In the following, we present an overview of important remote sensing techniques that are relevant for TERI. In most cases, remote sensors operate in the visible (VIS, 400–750 nm), near-infrared (NIR, 750–1,400 nm), short-wave infrared (SWIR, 1,400–3,000 nm), mid-wave infrared (3,000–6,000 nm), thermal infrared (6,000–15,000 nm), and microwave (MW, 1 mm to 1 m) regions of the electromagnetic spectrum. However, these sensors just record electromagnetic waves, which need to be transferred to the respective environmental variables by retrieval models. There are retrieval models with different complexities, from statistical regression to physically based models where additional parameters are needed (e.g., for more details, see

Verrelst et al. 2015). The challenge is not only to develop a remote sensing system that is able to observe the record (reflectance, emissivity) in adequate spatial and temporal resolution but also to relate the raw data to quantitative environmental properties.

Vegetation monitoring is typically performed in the VIS and NIR bands. Here, multispectral systems such as Landsat (Irons et al. 2012), ASTER (Yamaguchi et al. 1998), RapidEye (Krischke et al. 2000), and Sentinel-2 (Drusch et al. 2012) have been used extensively to characterize land cover status and change. This includes not only the provision of spectral indices but also the derivation of higher-level data such as leaf area index (LAI) (Ali et al. 2015; Delegido et al. 2011; Ganguly et al. 2012; Qu et al. 2015) and physiognomic land cover maps (Homer et al. 2015) for further application in environmental models. For the purpose of long-term monitoring in terrestrial observatories, the Landsat missions provide valuable information dating from now back to 1972. The current Landsat 8 satellite had been launched in 2013, and plans for the launch of the successor Landsat 9 in 2023 already existed. Similar to the in situ component of TERI, the long-term observational strategy is inherent to the Landsat science program, making it an excellent partner to identify environmental changes over several years and decades. However, in high-biomass ecosystems, passive optical sensors typically fail to accurately estimate LAI and aboveground biomass. Light detection and ranging (LIDAR) has been shown to provide valuable information about the vegetation structure. Linking both the biophysical quantities and structure are key design attributes of many environmental research infrastructures and campaign-style research activities.

Higher-resolution VIS and NIR data (<5 m) can be obtained by spaceborne systems such as IKONOS and QuickBird that permit the analysis of more detailed land-surface characteristics, including urban environments and impervious surfaces (Boyle et al. 2014). However, to detect finer-scale environmental characteristics especially in VIS and NIR, high-resolution applications are more and more delivered by unmanned air vehicles (UAVs) due to their high flexibility and availability at reduced costs (Colomina and Molina 2014). With this variety of platforms, EO is able to provide information from local to global scale, where the implementation of scaling strategies gains more and more importance (see also Section 11.5).

Although UAVs are able to increase the spatial resolution of multispectral systems, hyperspectral sensors are able to better resolve the electromagnetic spectrum with hundreds of narrow spectral channels (spectral resolution) from VIS to SWIR at moderate spatial resolution (Govender et al. 2007). Currently, the only operational spaceborne imaging spectrometer covering the full spectral range from 4 to 25 nm is EO-1 Hyperion (Pearlman et al. 2003). With this kind of information, for example, different vegetation species can be separated with high accuracy or the health of forests and crops can be estimated. Moreover, hyperspectral systems enable the remote investigation of soil characteristics, such as the soil chemical composition. Specific absorption features deliver information on the iron oxide, organic carbon, or clay fraction, which correlate well with in situ measurements of soil properties (Babaeian et al. 2015; Lagacherie et al. 2012). Currently, for environmental observatories, airborne sensors provide higher spatial resolution at a better signal-to-noise ratio. Available sensors are AISA Eagle/Hawk (Lausch et al. 2013; Feilhauer et al. 2014) and HySpex (Möckel et al. 2014). The new spaceborne Environmental Mapping and Analysis Program is currently built for launch in 2018 (Guanter et al. 2015). Hyperspectral sensors in combination with LIDARs are the key elements of several already institutionalized remote sensing observatories, such as the Carnegie Airborne Observatory (Asner et al. 2012) and the Airborne Observation Platform (AOP) of the National Ecological Observatory Network (NEON, Kampe et al. 2010). In addition to ground measurements, AOP data at submeter to meter spatial scales will help to bridge scales from organism and stand scales to the scale of satellite-based remote sensing.

Although typical hyperspectral sensors record the spectra in bands of 3–25 nm, a special group of hyperspectral remote sensors started recently to operate with ≤1 nm spectral resolution. The reason for this new development is the existence of solar-induced fluorescence. Approximately 82% of the absorbed light in a leaf is used for carbon assimilation while the remaining part is lost as heat and dissipated as emissions of chlorophyll fluorescence. Fluorescence, the radiant flux emitted, is therefore the most direct measurable variable of photosynthetic efficiency (van der Tol et al. 2014). It is obvious that this information is of high value for terrestrial observatories. An airborne remote sensor is the HyPlant system (Rossini et al. 2015), where a typical hyperspectral imager is combined with a fluorescence imager, a special module that acquires data at high spectral resolution (0.25 nm) in the spectral region of the two oxygen absorption bands (670–780 nm) and is dedicated to measure the vegetation fluorescence signal. Recently, with FLEX, the European Space Agency selected this concept to be established on a spaceborne platform to globally record chlorophyll fluorescence on a $300 \times 300\ m^2$ resolution.

Active microwave data (radar) open a large variety of applications by investigating environmental parameters such as terrain elevation change (Montagh et al. 2013) or vegetation structure (Reigber et al. 2000). A widely used further application is the estimation of near-surface soil moisture at the subcatchment to regional scale, exploiting the different dielectric properties of dry soil and liquid water (Kornelsen et al. 2013). Due to the active transmission, the strength of the radar signal is distinctively higher than for passive microwave sensors, which depend on the weak natural emission of radiation from the earth. Therefore, passive MW systems on spaceborne platforms are typically very coarse (>36 × 36 km), so that for terrestrial observatories airborne systems are preferred (Hasan et al. 2014). New generations of radar instruments will become available within the next years, for example, the U.S.–Indian NISAR mission to be launched in 2020 or the German Tandem-L mission (Moreira et al. 2015) for a planned launch in 2024.

Where in some disciplines the forward simulation of the remote sensor record is far advanced, other disciplines still make use of ground-based manual measurements using remotely sensed parameter maps for regionalization and pattern recognition (e.g., Lagacherie et al. 2012). Here, the grand challenge is to develop and employ fully physical forward models describing the radiative transfer processes from the investigated variable to the sensor record.

11.5 Scaling

Upscaling of water, energy, and matter fluxes is challenging as the relevant processes that govern exchange processes between the land and the atmosphere act on very different time scales. This can be nicely illustrated for processes controlling the terrestrial carbon cycle. For example, whereas variability in photosynthetic activity and associated carbon uptake act on very short time scales related to variations in photon flux density (among others), the slow decay of recalcitrant carbon pools acts on time scales of years to decades. Many of these processes are parameterized with semiempirical equations including parameters that are difficult to determine or not easily available (e.g., Weihermueller et al. 2013). Parameter estimation plays therefore an essential role in improving estimates of ecosystem carbon balances and in future projections of the land carbon source or sink term.

For example, Post et al. (2016a) estimated eight sensitive ecosystem parameters of the Community Land Model (CLM) version 4.5 (Oleson et al. 2013) with the help of 1-year-long time series of net ecosystem exchange (NEE), measured by eddy covariance stations in Germany and France. The parameters were estimated by the $DREAM_{ZS}$ algorithm (Laloy and Vrugt 2012; Ter Braak and Vrugt 2008), a Markov chain Monte Carlo–based algorithm, which is particularly CPU efficient. The estimated ecosystem parameters were evaluated with an additional year of NEE data at the same sites. The evaluation was made in terms of the root-mean-square error of 30-minute NEE measurements, the yearly accumulated NEE sum, the mean daily NEE cycle, and the mean annual NEE cycle. The results showed that the estimated parameters reproduced the measured NEE data in the evaluation period better than the default CLM parameters. The estimated ecosystem parameters were also evaluated at other sites, situated ~500 km away from the estimation sites in eastern Germany, with 1 year of measured time series of NEE data. The estimated ecosystem parameters were evaluated at sites that had the same plant functional type (PFT) as the estimation site. Furthermore, at these evaluation sites, it was found that the estimated ecosystem parameters improved the reproduction of the measured NEE data compared to the default ecosystem parameters. Only for grassland sites, no improvement was found.

Finally, the estimated ecosystem parameters were assigned to the different corresponding PFTs of the Rur catchment in western Germany and a CLM simulation was made for the complete catchment (Post et al. 2016b). Moreover, a CLM simulation with the default ecosystem parameters was carried out. The simulated LAI values were compared with the remotely sensed LAI values by RapidEye. The difference between simulated LAI values and measured LAI values was considerably smaller if the estimated ecosystem parameters were used as input.

This study illustrated the potential of upscaling carbon fluxes by measurements of NEE fluxes at single EC towers. It can be expected that the results further improve for longer NEE time series and the inclusion of other data sources like biomass or LAI. Nevertheless, the study also pointed to the fact that although the characterization of the net carbon balance improved with the estimated ecosystem parameters, the optimized parameters also could mask model structural errors. In the context of a European ecosystem infrastructure, it is expected that upscaling would be improved if at existing sites NEE, respiration, biomass, and LAI would be determined simultaneously. A further extension of the network, covering different types of agricultural crops, would also be an advantage.

In order to monitor the processes controlling matter fluxes in soil–plant–atmosphere systems, the described hydrological and carbon monitoring concepts have to be extended by adding other relevant subsystems (e.g., atmosphere) and by a socioeconomic analysis. This may include the identification of socioeconomic drivers that control the intensification and/or deintensification processes (e.g., demographic and technological change) and the feedback to the ecological system (e.g., on the state of soil, water, and air quality) as well as the analysis of ecological impacts of land use changes across and within sectors (e.g., from agricultural to industrial types or from chemical to energy production).

11.6 TERENO Observatory Network in Germany

In the following, we present the TERENO initiative in Germany as an example for a distributed TERI network. The general aim of TERENO is to conduct integrated and long-term observation studies of climate change and global change impacts on terrestrial ecosystems across Germany (Bogena et al. 2012; Zacharias et al. 2011). TERENO combines observations with comprehensive larger-scale experiments and integrated modeling to increase our understanding of the functioning of terrestrial systems and the complex interactions and feedback mechanisms among their different subsystems. Such a combination is needed since the monitoring of terrestrial systems alone is not sufficient to understand the relevant processes of environmental

FIGURE 11.1
Schematic view of a typical TERENO observatory platform, including measuring systems for the determination of regional precipitation fields using (1) weather radars and (2) precipitation gauging networks (Diederich et al. 2015), (3) micrometeorological eddy covariance towers (Mauder et al. 2013), (4) sensor networks (Rosenbaum et al. 2012), (5) Runoff monitoring systems (Stockinger et al. 2014), (6) weighable lysimeter systems (Hannes et al. 2015), (7) ground-based and airborne remote sensing platforms, (8) airborne campaigns (Hasan et al. 2014), (9) satellite-borne data (Rötzer et al. 2014), (10) geoarchiving systems (Kienel et al. (2013), and (11) tree growth monitoring systems (Simard et al. 2014).

changes, due to the interrelation of many influencing factors. Within TERENO, four terrestrial observatories were selected because they represent typical landscapes in Germany and other central European countries, which are predicted to be highly vulnerable to the effects of global and climate change. Furthermore, the four terrestrial observatories within these regions can be expected to most appropriately exhibit the dominant terrestrial processes and the different roles of groundwater, surface water, soils, and their links to the atmospheric boundary layer. All of these selected regions either are already affected by climate change or will probably react sensitively in the foreseeable future (e.g., Pfeifer et al. 2015; Vautard et al. 2014). Figure 11.1 gives a schematic overview of the typical instrumentation of a terrestrial observatory as realized in the TERENO network. Mobile measurement platforms are used both for the monitoring of dynamic processes at the local scale and the determination of their spatial patterns at the regional scale.

The TERENO infrastructure is characterized by a high degree of automation and operationalization, for example, for the near-real-time derivation of value-added information products based on in situ data (Borg et al. 2014; Sorg and Kunkel 2015) or on EO data (Missling et al. 2014). During instrumentation of the four terrestrial observatories, local data infrastructures

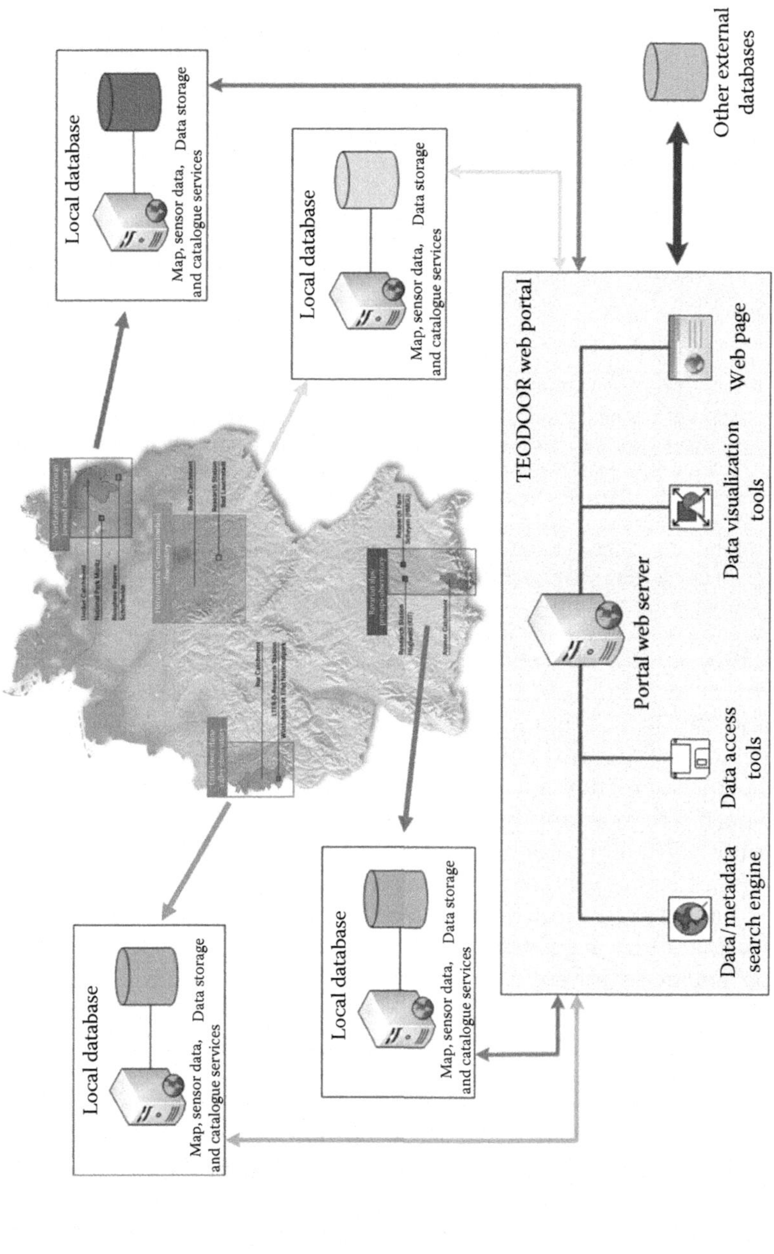

FIGURE 11.2
(See color insert.) The design of the TERENO infrastructure for the distribution of the data.

were implemented (Kunkel et al. 2013). The central data portal TEODOOR facilitates the online provision of decentralized TERENO data (Figure 11.2). It is hosted by Forschungszentrum Jülich and can be accessed via https://teodoor.icg.kfa-juelich.de/. TEODOOR uses common standards for the metadata description (e.g., Open Geospatial Consortium [http://www.opengeospatial.org/ogc], Consortium of Universities for the Advancement of Hydrologic Science [https://www.cuahsi.org/#sthash.FcOhNZWf.dpuf]) of data sets based on the INSPIRE directive (2007/2/EC) for spatial data infrastructures (http://inspire.jrc.ec.europa.eu) allowing for a search throughout the entire database. Such standard protocols for accessing environmental data sets are used to guarantee compatibility to the related individual data infrastructures of the TERENO partners. The TEODOOR portal allows versatile community access to data sources. Data governance and data stewardship programs and data architecture and data management programs are much more effective if they are supported by a directive concerning the data management policy. A data policy statement (TERENO, 2015), required for data processing and data exchange, was developed in a common approach by all TERENO partners. A main aspect of the data policy was the definition of the data ownership (intellectual property rights) and data access rights concerning the directives of funding organizations differentiated by types of digital resources, their process status, the data creator, and the data source. As a rule, all data are freely accessible within the TERENO community and accessible also to the public as soon as at least a first quality check was performed on the data and no other usage restrictions are existent, for example, due to ongoing PhD studies or external copyright issues.

11.7 Conclusion

In this chapter, we discussed the challenge in ecosystem research to embrace all subsystems of the terrestrial system in order to better understand the impact of climate and land use change on terrestrial ecosystems. We suggest that a new, multiscale fully integrated TERI network is needed, which capitalizes on the existing experience and knowledge gained from existing networks (e.g., Long-Term Ecological Research, ICOS, NEON) and the recently established critical zone observatories. We advocate that the TERENO example could serve as a blueprint for a distributed TERI as outlined earlier. Such an infrastructure could provide answers to key questions related to ecosystems and terrestrial research from the continental to the global scale, for example, how ecosystems are changing or adapting to global change stressors, identification of the determinants of ecosystem resilience, and threshold interactions resulting in system shifts. In addition, a TERI network can give answers on how to support systems that are more resilient to global change effects and to develop adaptive measures that are needed to sustain ecosystem services.

The ecological science community would strongly benefit from the establishment of a TERI network exploiting the full potential of novel measurement technologies allowing to better falsify hypotheses, to identify underlying organizational principles, to correct biased remote sensing data, and to gain increased insights into the nature of these biases and at the same time reduce the uncertainty in knowledge about states, fluxes, and parameters of the terrestrial system. Linking these observations with remote sensing, large-scale models and data assimilation approaches offer the perspective of predicting the terrestrial water cycle at regional to continental scales or even to global scale.

References

Ali, M., C. Montzka, A. Stadler, G. Menz, F. Thonfeld, and H. Vereecken (2015) Estimation and validation of RapidEye-based time series of leaf area index for winter wheat in the Rur catchment (Germany). *Remote Sensing*, 7(3), 2808–2831.

Allen, D.C., Cardinale, B.J., and Wynn-Thompson, T. (2014) Toward a better integration of ecological principles into ecogeoscience research. *BioScience*, 64, 444–454.

Anderson, J., Martin, M.E., Smith, M-L, Dubayah, R.O., Hofton, M.A., Hyde, P., Peterson, B.E., Blair, J.B., and Knox, R.G. (2006) The use of waveform LIDAR to measure northern temperate mixed conifer and deciduous forest structure in New Hampshire. *Remote Sensing of Environment*, 105, 248–261.

Anselmi, S., Chiesi, M., Giannini, M., Manes, F., and Maselli, F. (2004) Estimation of Mediterranean forest transpiration and photosynthesis through the use of an ecosystem simulation model driven by remotely sensed data. *Global Ecology and Biogeography*, 13, 371–380.

Asner, G.P., Knapp, D.E., Boardman, J., Green, R.O., Kennedy-Bowdoin, T., Eastwood, M. et al. (2012) Carnegie airborne observatory-2: Increasing science data dimensionality via high-fidelity multi-sensor fusion. *Remote Sensing of Environment*, 124, 454–465, doi:10.1016/j.rse.2012.06.012.

Asner, G., Scurlock, J.M.O., and Hicke, J.A. (2003) Global synthesis of leaf area index observations: Implications of ecological and remote sensing studies. *Global Ecology and Biogeography*, 12, 191–205.

Baatz, R., Bogena, H., Hendricks Franssen, H.-J., Huisman, J.A., Wei, Q., Montzka, C., and Vereecken, H. (2014) Calibration of a catchment scale cosmic-ray soil moisture network: A comparison of three different methods. *Journal of Hydrology*, 516, 231–244.

Babaeian, E., Homaee, M., Montzka, C., Vereecken, H., and Norouzi, A.A. (2015) Towards estimating soil hydraulic properties by hyperspectral remote sensing. *Vadose Zone Journal*, 14(3). doi:10.2136/vzj2014.07.0080.

Balvanera, P., Siddique, I., Dee, L., Paquette, A., Isbell, F., Gonzalez, A. et al. (2013) Linking biodiversity and ecosystem services: Current uncertainties and the necessary next steps. *BioScience*, 64, 49–57.

Bartuszevige, A.M., Gorchov, D.L., and Raab, L. (2006) The relative importance of landscape and community features in the invasion of an exotic shrub in a fragmented landscape. *Ecography*, 29, 213–222.

Binenko, V.I., Donchenko, V.K., Andreev, V.L., and Ivanov R.V. (2001) Some results and perspectives of using unmanned air vehicles for ecological monitoring tasks. *Ecological Chemistry*, 10, 21–31.

Bogena, H.R., Bol, R., Borchard, N., Brüggemann, N., Diekkrüger, B., Drüe, C. et al. (2015) A terrestrial observatory approach for the integrated investigation of the effects of deforestation on water, energy, and matter fluxes. *Science China: Earth Sciences*, 58(1), 61–75, doi: 10.1007/s11430-014-4911-7.

Bogena, H.R., Herbst, M., Huisman, J.A., Rosenbaum, U., Weuthen, A., and Vereecken, H. (2010) Potential of wireless sensor networks for measuring soil water content variability. *Vadose Zone Journal*, 9(4), 1002–1013, doi:10.2136/vzj2009.0173.

Bogena, H.R., Huisman, J.A., Baatz, R., Hendricks Franssen, H.-J., and Vereecken, H. (2013) Accuracy of the cosmic-ray soil water content probe in humid forest ecosystems: The worst case scenario. *Water Resources Research*, 49(9), 5778–5791, doi: 10.1002/wrcr.20463.

Bogena, H., Kunkel, R., Krüger, E., Zacharias, S., Pütz, T., Schwank, M. et al. (2012) TERENO – Long-term monitoring network for terrestrial research. *Hydrologie und Wasserbewirtschaftung*, 56(3), 138–143.

Bogena, H., Schulz, K., and Vereecken, H. (2006) Towards a network of observatories in terrestrial environmental research. *Advances in Geosciences*, 9, 109–114.

Bohlman, S.A., Adams, J.B., Smith, M.O., and Peterson, D.L. (1998) Seasonal foliage changes in the eastern Amazon Basin detected from Landsat Thematic Mapper satellite images. *Biotropica*, 30, 376–391.

Borg, E., Schiller, C., Daedelow, H., Fichtelmann, B., Jahncke, D., Renke, F., Tamm, H-P., and Asche, H. (2014) Automated generation of value-added products for the validation of remote sensing information based on in-situ data. In Murgante, B. et al. eds., ICCSA 2014, Guimarães, Portugal, June 30–July 3, 2014, *Proceedings Part I*. Vol. 8579 LNCS, Springer, pp. 393–407.

Boyle, S.A., Kennedy, C.M., Torres, J., Colman, K., Pérez-Estigarribia, P.E., and de la Sancha, N.U. (2014) High-resolution satellite imagery is an important yet underutilized resource in conservation biology. *PLoS ONE*, 9(1), e86908.

Burkhard, B., Petrosillo, I., and Costanza, R. (2010) Ecosystem services—Bridging ecology, economy and social sciences. *Ecological Complexity*, 7(3), 257–259.

Burt, T.P., Howden, N.J.K., Worrall, F., and Whelan, M.J. (2008) Importance of long-term monitoring for detecting environmental change: Lessons from a lowland river in South East England. *Biogeosciences*, 5, 1529–1535.

CENR (Committee on Environment and Natural Resources) (1997) Integrating the Nation's environmental monitoring and research networks and program: A proposed framework. *Subcommittee on Ecological Systems*, Washington, DC, 82pp.

CEOS (Committee on Earth Observation Satellites) (2015): Satellite Earth Observations in support of Climate Information Challenges. *The CEOS Earth Observation Handbook, Special 2015 COP21 Edition Special*, 37pp.

Clark, J.S., Carpenter, S.R., Barber, M., Collins, S., Dobson, A., Foley, J.A. et al. (2001) Ecological forecasts: An emerging imperative. *Science*, 293, 657–660.

Clevers, J.G.P.W., Schaepman, M.E., Mücher, C.A., de Wit, A.J.W., Zurita-Milla, R., and Bartholomeus, H.M. (2007) Using MERIS on Envisat for land cover mapping in the Netherlands. *International Journal of Remote Sensing*, 28, 637–652.

Colomina, I. and Molina, P. (2014) Unmanned aerial systems for photogrammetry and remote sensing: A review. *ISPRS Journal of Photogrammetry and Remote Sensing*, 92, 79–97, doi:10.1016/j.isprsjprs.2014.02.013.

Davidson, E.A., Trumbore, S.E., and Amundson, R. (2000) Soil warming and organic carbon content. *Nature*, 408, 789–790.

Delegido, J., Verrelst, J., Alonso, L., and Moreno, J. (2011) Evaluation of sentinel-2 red-edge bands for empirical estimation of green LAI and chlorophyll. *Sensors*, 11(7), 7063–7081.

DFG (Deutsche Forschungsgemeinschaft) (2013) Long-term perspectives and infrastructure in terrestrial research in Germany—A systemic approach. DFG-Strategy paper, Bonn, Germany, 29 p. (in German).

Diederich, M., Ryzhkov, A., Simmer, C., Zhang, P., and Troemel, S. (2015) Use of specific attenuation for rainfall measurement at X-band radar wavelengths part I: Radar calibration and partial beam blockage estimation. *Journal of Hydrometeorology*, 16, 487–502.

Diehl, R.H., Larkin, R.P., and Black, J.E. (2003) Radar observations of bird migration over the Great Lakes. *The Auk*, 120, 278–290.

Drusch, M., Del Bello, U., Carlier, S., Colin, O., Fernandez, V., Gascon, F. et al. (2012) Sentinel-2: ESA's optical high-resolution mission for GMES operational services. *Remote Sensing of Environment*, 120, 25–36, doi:10.1016/j.rse.2011.11.026.

Eisele, A., Lau, I., Hewson, R., Carter, D., Wheaton, B., Ong, C., Cudahy, J., Chabrillat, S., and Kaufmann, H. (2012) Applicability of the thermal infrared spectral region for the prediction of soil properties across semi-arid agricultural landscapes. *Remote Sensing*, 4(11), 3265–3286.

Fan, W., Randolph, J.C., and Ehman, J.L. (1998) Regional estimation of nitrogen mineralization in forest ecosystems using geographic information systems. *Ecological Applications*, 8, 734–747.

Feilhauer, H., Dahlke, C., Doktor, D., Lausch, A., Schmidtlein, S., Schulz, G., and Stenzel, S. (2014) Mapping the local variability of Natura 2000 habitats with remote sensing. *Applied Vegetation Science*, 17, 765–779.

Fraser, L.H., Henry, H.A.L., Carlyle, C.N., White, S.R., Beierkuhnlein, C., Cahill, Jr. J.F. et al. (2013) Coordinated distributed experiments: An emerging tool for testing global hypotheses in ecology and environmental science. *Frontiers in Ecology and the Environment*, 11, 147–155.

Ganguly, S., Nemani, R., Zhang, G., Hashimoto, H., Milesi, C., Michaelis, A. et al. (2012) Generating global leaf area index from landsat: Algorithm formulation and demonstration. *Remote Sensing of Environment*, 122, 185–202.

GEOSS (2005) The Global Earth Observation System of Systems (GEOSS) 10-year implementation plan. 11 pp. http://earthobservations.org/

Giardina, C.P. and Ryan, M.G. (2000) Evidence that decomposition rates of organic carbon in mineral soil do not vary with temperature. *Nature*, 404, 858–861.

Govender, M., Chetty, K., and Bulcock, H. (2007) A review of hyperspectral remote sensing and its application in vegetation and water resource studies. *Water SA*, 33, 145–151.

Guanter, L., Kaufmann, H., Segl, K., Förster, S.,Rogass, C., Chabrillat, S. et al. (2015) The EnMAP spaceborne imaging spectroscopy mission for earth observation. *Remote Sensing*, 7(7), 8830–8857.

Haines-Young, R. and Weber, J-L. (2006) *Land Accounts for Europe 1990–2000: Towards Integrated Land and Ecosystem Accounting*. European Environment Agency, Copenhagen, Denmark, ISBN 92-9167-888-0.

Hannes, M., Wollschläger, U., Schrader, F., Durner, W., Gebler, S., Pütz, T., Fank, J., von Unold, G., and Vogel, H.-J. (2015) High-resolution estimation of the water balance components from high-precision lysimeters. *Hydrology and Earth System Science*, 19, 3405–3418.

Hasan S., Montzka, C., Rüdiger, C., Ali, M., Bogena, H., and Vereecken, H. (2014) Soil moisture retrieval from airborne L-band passive microwave using high resolution multispectral data. *Journal of Photogrammetry and Remote Sensing*, 91, 59–71.

Homer, C.G., Dewitz, J.A., Yang, L., Jin, S., Danielson, P., Xian, G., Coulston, J., Herold, N.D., Wickham, J.D., and Megown, K. (2015) Completion of the 2011 National Land Cover Database for the conterminous United States—Representing a decade of land cover change information. *Journal of Photogrammetry and Remote Sensing*, 81(5), 345–354.

Huisman, J.A., Sperl, C., Bouten, W., and Verstraten, J.M. (2001) Soil water content measurements at different scales: accuracy of time domain reflectometry and ground-penetrating radar. *Journal of Hydrology*, 245, 48–58.

Inagaki, M.N., Fukatsu, T., Hirafuji, M., and Nachit, M.M. (2011) Monitoring soil moisture under wheat growth through a wireless sensor network in dry conditions. *Journal of Environmental Science and Engineering*, 5, 428–431.

IPCC (Intergovernmental Panel on Climate Change) (2013) Contribution of Working Group I to the Fifth Assessment Report of the Intergovernmental Panel on Climate Change. In: Stocker, T.F., D. Qin, G.-K. Plattner, M. Tignor, S.K. Allen, J. Boschung, A. Nauels, Y. Xia, V. Bex and P.M. Midgley (eds.), *Climate Change 2013: The Physical Science Basis*, 1535 pp. Cambridge University Press, Cambridge, United Kingdom and New York, NY.

Irons, J.R., Dwyer, J.L., and Barsi, J.A. (2012) The next Landsat satellite: The Landsat data continuity mission. *Remote Sensing of Environment*, 122, 11–21, doi:10.1016/j.rse.2011.08.026.

Jagdhuber, T., Hajnsek, I., and Papathanassiou, K. (2015) An iterative generalized hybrid decomposition for soil moisture retrieval under vegetation cover using fully polarimetric SAR. *IEEE Journal of Selected Topics in Applied Earth Observations and Remote Sensing*, 8(8), 3911–3922.

Jensen, K.H. and Illangasekareb, T.H. (2011) HOBE: A hydrological observatory. *Vadose Zone Journal*, 10(1), 1–7.

Jones, K.B., Bogena, H., Vereecken, H. and Weltzin, J.F. (2010) Design and importance of multi-tiered ecological monitoring networks. In Mueller F. et al. (eds.), *Long-Term Ecological Research: Between Theory and Application*, pp. 355–374, Springer, Dordrecht, the Netherlands.

Jones, K.B., Nash, M.S., Wade, T.G., Walker, J., Neale, A.C., Müller, F. et al. (2008) Cross-European landscape analyses: Illustrative examples using existing spatial data. In Petrosillo, I. et al. (eds.), *Use of Landscape Sciences for the Assessment of Environment Security*, pp. 258–316, Springer, Dordrecht, the Netherlands.

Jongman, R.H.G., Bunce, R.G.H., Metzger, M.J., Mucher, C.A., Howard, D.C., and Mateus, V.L. (2006) Objectives and applications of a statistical environmental stratification of Europe. *Landscape Ecology*, 21, 409–419.

Kampe, T.U., Johnson, B.R., Kuester, M., and Keller, M. (2010) NEON: The first continental-scale ecological observatory with airborne remote sensing of vegetation canopy biochemistry and structure. *Journal of Applied Remote Sensing*, 4(1), 043510, doi: 10.1117/1.3361375.

Kastens, J.H., Price, K.P., Kastens, D.L., and Martinko, E.A. (2001) Forecasting preharvest crop yields using time series analysis of AVHRR NDVI composite imagery. *Proceedings, Annual Convention. American Society of Photogrammetric Engineering and Remote Sensing*, St. Louis, MO, April 23–27, 2001.

Katul, G.G., Oren, R., Manzoni, S., Higgins, C., and Parlange, M.B. (2012) Evapotranspiration: A process driving mass transport and energy exchange in the soil–plant-atmosphere-climate system. *Reviews of Geophysics*, 50, 1–25.

Kerr, J.T., Kharouba, H.M., and Currie, D.J. (2007) The macroecological contribution to global change. *Science*, 316, 1581–1584.

Kienel, U., Dulski, P., Ott, F., Lorenz, S., and Brauer, A. (2013) Recently induced anoxia leading to the preservation of seasonal laminae in two NE-German lakes. *Journal of Paleolimnology*, 50(4), 535–544.

Kornelsen, K.C. and Coulibaly, P. (2013) Advances in soil moisture retrieval from synthetic aperture radar and hydrological applications. *Journal of Hydrology*, 476, 460–489.

Korres, W., Reichenau, T.G., Fiener, P., Koyama, C.N., Bogena, H.R., Cornelissen, T. et al. (2015) Spatio-temporal soil moisture patterns—A meta-analysis using plot to catchment scale data. *Journal of Hydrology*, 520, 934–946.

Krajewski, W.F., Anderson, M.C., Eichinger, W.E., Entekhabi, D., Hornbuckle, B.K., Houser, P.R. et al. (2006) A remote sensing observatory for hydrologic sciences: A genesis for scaling to continental hydrology. *Water Resources and Research*, 42, W07301.

Krischke, M., Niemeyer, W., and Scherer, S. (2000) RapidEye – Satellite based geo-information system. *Acta Astronaut*, 46, 307–312. doi: 10.1016/s0094-5765(99)00219-2.

Kunkel, R., Sorg, J., Eckardt, R., Kolditz, O., Rink, K., and Vereecken, H. (2013) TEODOOR: A distributed geodata infrastructure for terrestrial observation data. *Environmental Earth Sciences*, 69, 507–521.

Kurt, W., He, G., Kollet, S., Maxwell, R., Vereecken, H., and Hendricks Franssen, H.J. (2016) TerrSysMP-PDAF (version 1.0): A modular high performance data assimilation framework for an integrated land surface-subsurface model. *Conditionally accepted for publication in Geoscientific Model Development*.

Lagacherie, P., Bailly, J.S., Monestiez, P., and Gomez, C. (2012) Using scattered hyperspectral imagery data to map the soil properties of a region. *European Journal of Soil Science*, 63, 110–119.

Laloy, E. and Vrugt, J.A. (2012) High-dimensional posterior exploration of hydrologic models using multiple-try DREAM(ZS) and high-performance computing. *Water Resources Research*, 48, doi:10.1029/2011WR010608.

Lausch, A., Blaschke, T., Haase, D., Herzog, F., Syrbe, R.-U., Tischendorf, L., and Walz, U. (2015) Understanding and quantifying landscape structure—A review on relevant process characteristics, data models and landscape metrics. *Ecological Modelling*, 295, 31–41.

Lausch, A., Zacharias, S., Dierke, C., Pause, M., Kühn, I., Doktor, D., Dietrich, P., and Werban, U. (2013) Analysis of vegetation and soil pattern using hyperspectral remote sensing, EMI and gamma ray measurements. *Vadose Zone Journal*, 12, 4, doi:10.2136/vzj2012.0217.

Loescher, H.W., Jacobs, J., Wendroth, O., Robinson, D.A., Poulos, G.S., McGuire, K., Reed, P., Mohanty, B., and Krajewski, W. (2007) Enhancing water cycle measurements for future hydrologic research. *Bulletin of the American Meteorological Society*, 88, 669–676, doi: 101175/BAMS-88-5-669.

Loveland, T.R., Sohl, T.L., Stehman, S.V., Gallant, A.L., Sayler, K.L., and Napton, D.E. (2002) A strategy for estimating the rates of recent United States land-cover changes. *Photogrammetric Engineering and Remote Sensing*, 68, 1091–1099.

Lupo, F., Linderman, M. Vanacker, V. Bartholome, E., and Lambin, E.F. (2007) Categorization of land-cover change processes based on phenological indicators extracted from time series of vegetation index data. *International Journal of Remote Sensing*, 11–12, 2469–2483.

Mauder, M., Cuntz, M., Drüe, C., Graf, A., Rebmann, C., Schmid, H.P., Schmidt, M., and Steinbrecher, R. (2013) A strategy for quality and uncertainty assessment of long-term eddy-covariance measurements. *Agricultural and Forest Meteorology*, 169, 122–135.

McCarthy, J.J., Canziani, O.F., Leary, N.A., Dokken, D.J., and White, K.S. (2001) *Climate Change Impacts, Adaptation, and Vulnerability. IPCC Third Assessment Report*, Cambridge University Press, Cambridge, UK.

Millennium Ecosystem Assessment (MEA). (2005) *Ecosystems and Human Well-Being: Synthesis*. Island Press, Washington, DC.

Möckel, T., Dalmayne, J., Prentice, H.C., Eklundh, L., Purschke, O., Schmidtlein, S. and Hall, K. (2014) Classification of grassland successional stages using airborne hyperspectral imagery. *Remote Sensing*, 6, 7732–7761.

Montzka, C., Bogena, H.R., Weihermueller, L., Jonard, F., Bouzinac, C., Kainulainen, J. et al. (2013) Brightness temperature and soil moisture validation at different scales during the SMOS validation campaign in the Rur and Erft catchments, Germany. *IEEE Transactions on Geoscience and Remote Sensing*, 51(3), 1728–1743.

Montzka C., Pauwels, V.R.N., Hendricks Franssen, H.-J., Han, X., and Vereecken, H. (2012) Multivariate and multiscale data assimilation in terrestrial systems: A review. *Sensors*, 12(12), 16291–16333, doi:10.3390/s121216291.

Moreira, A., Krieger, G., Hajnsek, I., Papathanassiou, K., Younis, M., Lopez-Dekker, P. et al. (2015) Tandem-L: A highly innovative bistatic SAR mission for global observation of dynamic processes on the earth's surface. *IEEE Geoscience and Remote Sensing Magazine (GRSM)*, 3(2), 8–23.

Motagh, M., Wetzel, H.-U., Roessner, S., and Kaufmann, H. (2013) A TerraSAR-X InSAR study of landslides in southern Kyrgyzstan, Central Asia. *Remote Sensing Letters*, 4(7), 657–666.

Müller, B., Bernhardt, M., and Schulz, K. (2014) Identification of catchment functional units by time series of thermal remote sensing images. *Hydrology and Earth System Sciences* 18(12), 5345–5359.

Oleson, K., Lawrence, D.M., Bonan, G.B., Drewniak, B., Huang, M., Koven, C.D. et al. (2013) Technical description of version 4.5 of the Community Land Model (CLM). *NCAR Technical Notes*, Boulder, CO, pp. 420.

Ollinger, S.V., Aber, J.D., and Federer, C.A. (1998) Estimating regional forest productivity and water yield using an ecosystem model linked to GIS. *Landscape Ecology*, 13, 323–334.

Pearlman, J.S., Barry, P.S., Segal, C.C., Shepanski, J., Beiso, D. and Carman, S.L. (2003) Hyperion, a space-based imaging spectrometer. *IEEE Transactions on Geoscience Remote Sensing*, 41(6), 1160–1173.

Peters, D.P.C., Loescher, H.W., SanClements, M.D., and Havstad, K.M. (2014) Taking the pulse of a continent: Role of observatories and long-term research networks to fill critical knowledge gaps. *Ecosphere*, 5(3), 1–23.

Pfeifer, S., Bülow, K., Gobiet, A., Hänsler, A., Mudelsee, M., Otto, J., Rechid, D., Teichmannand, C., and Jacob, D. (2015) Robustness of projected changes in seasonal and extreme precipitation from different ensembles of regional climate change simulations for Germany. *Atmosphere*, 6(5), 677–698.

Post, H., Hendricks Franssen, H.J., Han, X., Baatz, R., Montzka, C., Schmidt, M., and Vereecken, H. (2016b) Upscaling of net carbon fluxes from the plot scale to the catchment scale: Evaluation with NEE and LAI data. *Submitted to Biogeosciences*.

Post, H., Vrugt, J.A., Fox, A., Kumbhar, P., Vereecken, H., and Hendricks Franssen, H.J. (2016a) Estimation of Community Land model parameters with DREAM(ZS) for an improved assessment of net carbon fluxes at European sites. *Submitted to Journal of Geophysical Research-Biogeosciences*.

Powlson, D. (2005) Will soil amplify climate change? *Nature*, 433, 204–205.

Qu, Y., Han, W., and Ma, M. (2015) Retrieval of a temporal high-resolution leaf area index (LAI) by combining MODIS LAI and ASTER reflectance data. *Remote Sensing*, 7, 195–210.

Ramsey III, E., Rangoonwala, A., and Ehrlich, R. (2005) Mapping the invasive species, Chinese Tallow with EO1 satellite Hyperion hyperspectral image data and relating tallow percent occurrences to a classified Landsat Thematic Mapper landcover map. *International Journal of Remote Sensing*, 26,1637–1657.

Reigber, A. and Moreira, A. (2000) First demonstration of airborne SAR tomography using multibaseline L-band data. *IEEE Transactions on Geoscience and Remote Sensing*, 38, 44–46.

Rosenbaum, U., Bogena, H.R., Herbst, M., Huisman, J.A., Peterson, T.J., Weuthen, A., Western, A., and Vereecken, H. (2012) Seasonal and event dynamics of spatial soil moisture patterns at the small catchment scale. *Water Resources Research*, 48(10), W10544.

Rossini, M., Nedbal, L., Guanter, L., Ac, A., Alonso, L., Burkart, A. et al. (2015) Red and far red Sun-induced chlorophyll fluorescence as a measure of plant photosynthesis. *Geophysial Research Letters*, 42(6), 1632–1639.

Rötzer, K., Montzka, C., Bogena, H., Wagner, W., Kidd, R., and Vereecken, H. (2014) Catchment scale validation of SMOS and ASCAT soil moisture products using hydrological modelling and temporal stability analysis. *Journal of Hydrology*, 519, 934–946.

Rudolph, S., van der Kruk, J., von Hebel, C., Ali, M., Herbst, M., Montzka, C., Pätzold, S., Robinson, D.A., Vereecken, H., and Weihermüller, L. (2015) Linking satellite derived LAI patterns with subsoil heterogeneity using large-scale ground-based electromagnetic induction measurements. *Geoderma*, 241–242, 262–271.

Running, S.W., Nemani, R.R., Heinsch, F.A., Zhao, M., Reeves, M., and Hashimoto, H. (2004) A continuous satellite-derived measure of global terrestrial primary production. *BioScience*, 54, 547–560.

Sawaya, K.E., Olmanson, L.G., Heinert, N.J., Brezonik, P.L., and Bauer, M.E. (2003) Extending satellite remote sensing to local scales: Land and water resource monitoring using high-resolution imagery. *Remote Sensing of Environment*, 88, 144–156.

Scholes, R.J., Walters, M., Turak, E., Saarenmaa, H., Heip, C.H.R., Tuama, E.O. et al. (2012) Building a global observing system for biodiversity. *Current Opinion in Environmental Sustainability*, 4, 1–8.

Simard, S., Blume, T., Heidbüchel, I., Heinrich, I., Dreibrodt, J., Güntner, A., and Helle, G. (2015) Interactions and feedbacks of a temperate lake ecosystem in NE Germany. *Geophysical Research Abstracts*, 17, EGU2015-10912.

Simmer, C., Thiele-Eich, I., Masbou, M., Amelung, W., Bogena, H., Crewell, S. et al. (2015) Monitoring and modeling the terrestrial system from pores to catchments – the transregional collaborative research center on patterns in the soil-vegetation-atmosphere system. *Bulletin of the American Meteorological Society*, 96, 1765–1787.

Šimůnek, J., van Genuchten, M.Th., and Šejna, M. (2008) Development and applications of the HYDRUS and STANMOD software packages and related codes. *Vadose Zone Journal*, 7, 587–600.

Smith, R.A., Alexander, R.B., and Schwarz, G.E. (2003) Natural background concentrations of nutrients in streams and rivers of the conterminous United States. *Environmental Science and Technology*, 37, 3039–3047.

Sorg, J. and Kunkel, R. (2015) Conception and implementation of an OGC-compliant sensor observation service for a standardized access to raster data. *ISPRS International Journal of Geo-Information*, 2015(4), 1076–1096.

Stockinger, M., Bogena, H., Lücke, A., Diekkrüger, B., Weiler, M., and Vereecken, H. (2014) Seasonal soil moisture patterns control transit time distributions in a forested headwater catchment. *Water Resources and Research*, 50(6), 5270–5289.

Ter Braak, C.J.F. and Vrugt, J.A. (2008) Differential Evolution Markov Chain with snooker updater and fewer chains. *Statistics and Computing*, 18, 435–446. doi:10.1007/s11222-008-9104-9.

TERENO (2015) TERENO Data policy. http://teodoor.icg.kfa-juelich.de/downloads/TERENO Data policy.pdf/ (accessed 20 Nov 2015).

Treuhaft, R.N., Asner, G.P., Law, B.E., and Van Tuyl, S. (2002) Forest leaf area density profiles from the quantitative fusion of radar and hyperspectral data. *Journal of Geophysical Research-Atmospheres*, 107(D21), 4568–4578.

UNEP and WMO (2011) *Integrated Assessment of Black Carbon and Tropospheric Ozone, Summary for Decision Makers*, http://www.unep.org/dewa/Portals/67/pdf/BlackCarbon_SDM.pdf (accessed 4 Feb 2016).

van der Tol, C., Verhoef, W., Timmermans, J., Verhoef, A., and Su, Z. (2009) An integrated model of soil-canopy spectral radiances, photosynthesis, fluorescence, temperature and energy balance. *Biogeosciences*, 6, 3109–3129.

Vautard, R., Gobiet, A., Sobolowski, S., Kjellstroem, E., Stegehuis, A., Watkiss, P. et al. (2014) The European climate under a 2°C global warming. *Environmental Research Letter*, 9, doi:10.1088/1748-9326/9/3/034006.

Vereecken, H., Huisman, J.A., Hendricks Franssen, H.J., Brüggemann, N., Bogena, H.R., Kollet, S., Javaux, M., van der Kruk, J., and Vanderborght, J. (2015) Soil hydrology: Recent methodological advances, challenges, and perspectives. *Water Resources Research*, 51, 2616–2633.

Vereecken, H., Huisman, J.A., Pachepsky, Y., Montzka, C., van der Kruk, J., Bogena, H., Weihermueller, L., Herbst, M., Martinez, G., and Vanderborght, J. (2014) On the spatio-temporal dynamics of soil water content at the field scale. *Journal of Hydrology*, 516, 76–96.

Verrelst, J., Camps-Valls, G., Munoz-Mari, J., Rivera, J.P., Veroustraete, F., Clevers, J.G.P.W., and Moreno, J. (2015) Optical remote sensing and the retrieval of terrestrial vegetation bio-geophysical properties—A review. *ISPRS Journal of Photogrammetry and Remote Sensing*, 108, 273–290. doi:10.1016/j.isprsjprs.2015.05.005.

Walker J.P., Houser, P.R., and Willgoose, G.R. (2004) Active microwave remote sensing for soil moisture measurement: a field evaluation using ERS-2. *Hydrologic Processes*, 1811, 1975–1997.

Weihermueller, L., Graf, A., Herbst, M., and Vereecken, H. (2013) Simple pedotransfer functions to initialize reactive carbon pools of the RothC model. *European Journal of Soil Science*, 64(5), 567–575.

Weihermueller, L., Huisman, J.A., Graf, A., Herbst, M., and Vereecken, H. (2009) Multistep outflow experiments for the simultaneous determination of soil physical and carbon dioxide production parameters. *Vadose Zone Journal*, 8, 772–782.

Westerling, A.L., Hidalgo, H.G., Cayan, D.R., and Swetnam, T.W. (2006) Warming and earlier spring increases western U.S. forest wildlife activity. *Science*, 313, 940–943.

Yamaguchi, Y., Kahle, A.B., Tsu, H., Kawakami, T., and Pniel, M. (1998) Overview of Advanced Spaceborne Thermal Emission and Reflection Radiometer (ASTER). *IEEE Transactions on Geoscience and Remote Sensing*, 36, 1062–1071. doi:10.1109/36.700991.

Young, S.S. and Harris, R. (2005) Changing patterns of global-scale vegetation photosynthesis, 1982–1999. *International Journal of Remote Sensing*, 26, 4537–4563.

Zacharias, S., Bogena, H., Samaniego, L., Mauder, M., Fuß, R., Pütz, T., Frenzel, M. et al. (2011) A network of terrestrial environmental observatories in Germany. *Vadose Zone Journal*, 10(3), 955–973.

Section IV

Data Management and Access

12

Computational Challenges in Global Environmental Research Infrastructures

Paul Martin, Yin Chen, Alex Hardisty, Keith Jeffery, and Zhiming Zhao

CONTENTS

Abstract

Environmental science research is increasingly dependent on the collection and analysis of large volumes of data gathered via wide-scale deployments of sensors and other observation sources. Meanwhile, researchers are being called upon to address global societal challenges that are inextricably tied to the stability of our native ecosystems. These challenges are intrinsically interdisciplinary in nature, forcing scientists to collaborate across traditional disciplinary boundaries. The role of research infrastructure in this context is to support researchers in their interactions with a host of different data sources and analytical tools, as well as with each other, but no single environmental research infrastructure can hope to fully encompass the entire research ecosystem that has arisen to support the study of environmental science. The challenge therefore is for new environmental research infrastructures to exhibit sufficient technical interoperability between the different services they offer so as to permit researchers to freely and effectively interact with the full range of research assets potentially available to them, allowing them to collaborate and conduct innovative interdisciplinary research regardless of the particular research community to which they belong. Realising this ideal however requires a broad understanding of the fundamental commonalities of environmental science research infrastructure services as well as the development and wide adoption of common foundational services. It also requires a pragmatic bridging between the different standards and controlled vocabularies currently in use or preparation by different scientific communities, a process that can be expedited by the use of a standard reference model and the use of a formal framework for semantically linking similar concepts in different contexts.

12.1 Introduction

Environmental science addresses both our understanding of the physical world and our relationship with that world. As humanity continues to expand and place increasing pressure on rapidly diminishing natural resources, a number of societal challenges have arisen that can only be addressed by innovative new science that combines knowledge and expertise drawn from across all environmental domains – whether they focus on the earth, oceans, atmosphere or biosphere. These challenges include managing and adapting to climate change, overpopulation, food security and disaster prevention and relief and maintaining biodiversity in fragile but vital ecosystems.

One characteristic shared by all of these societal challenges is that they concern large, interconnected systems of considerable complexity. Carbon emissions in one corner of the world can affect the global climate, changing rainfall patterns in another part of the world that then has an impact on crop yields. The loss of species biodiversity causes a dramatic change in local ecosystems, resulting in mass die-offs and gradual desertification of wide regions. The shrinking of glaciers changes the temperature and composition of oceans, affecting their interaction with the atmosphere. Natural disasters (e.g. rising sea levels, earthquakes or tsunamis) cause immediate chaos and death but also affect trade networks and thus the global economy, which can lead to resource shortages and geopolitical instability or even change how people interact with the land as they are forced to change livelihood, feeding additional ecological impacts. Forecasting or even simply extrapolating the potential consequences of these scenarios requires interdisciplinary, data-intensive science – science that interleaves theory, models and data in order to describe a combination of complex, closely related systems that in the past have had to be considered, at least to some degree, in isolation.

It is only recently that computational resources and data collection facilities have existed to permit large-scale analysis and integration of significant volumes of data from multiple sources in real time. Nevertheless, the environmental systems under study remain extremely complex, and the interactions across different environmental systems are still not fully understood. However, the problem is not simply one of scientific understanding but also of how to support the practical integration of data and methods needed to develop this understanding. One distinction between the environmental sciences and some of the other data-rich fields such as astronomy and high-energy physics is the diversity and spread of data sources, as well as the variety of forms that data can take. Whereas in those other fields the number of data sources is often (relatively) few and concentrated around large-scale research facilities, albeit with extremely rich data yields, the environmental sciences typically have a greater number of different data sources, many (but not all) yielding modest quantities of data individually, but huge amounts in aggregate. The challenge then becomes how best to integrate these data and to provide the infrastructure necessary to do so. Environmental science has long been based on the collection and analysis of empirical data, but the quantities and scope of data being gathered via dedicated instruments and observations now outstrip the capacity of classical research methods. It is therefore necessary to support the development of a range of tools and services to be made available to researchers who wish to explore new sources of data. Scientists are also being challenged to collaborate on a global scale across traditional domain boundaries to discover and interact with data from many different, sometimes unfamiliar, research contexts. It is also necessary therefore to provide new research environments that support cross-disciplinary collaboration, allowing researchers to share new approaches to data analysis and integration, and take advantage of their peers' expertise and technical knowledge.

Environmental science research infrastructures aggregate technical infrastructure with standardised practices for data handling and experimentation in order to support a particular range of environmental science research activities. This technical infrastructure often already exists, at least partially, in the form of large-scale sensor and observer networks with both dedicated and leased data curation facilities, data dissemination and analysis tools and other research assets but is (initially at least) typically configured and managed independently, forcing researchers to learn the best approach to exploiting each asset individually, provided that they are fortunate enough to become aware of their existence in the first place. The challenge then for a new research infrastructure is to build an integrated system with coherent data management services and practices, in order to present a simple way for researchers in the field to interact with a particular body of related resources. Examples of research infrastructures in Europe that strive to address this challenge include LifeWatch for biodiversity (LifeWatch 2016), the European Plate Observing System (EPOS) for solid earth science (EPOS 2016), the Integrated Carbon Observation System (ICOS) for carbon science (ICOS 2016) and the European Multidisciplinary Seafloor and water-column Observatory for ocean/marine science (EMSO 2016). However, to fully address global environmental challenges, it is important that all research activities be well integrated in order to enable data-intensive system-level science (Foster and Kesselman 2006). This requires having common policies, protocols and standards in order to realise optimal coordination, harmonisation, and integration of data, applications and other services shared between research infrastructures. However, the complex nature of environmental science often results in the development of isolated environmental research infrastructures that meet only the immediate requirements and needs of a specific research community, with very limited interoperability of data, data access mechanisms and data processing tools. Interoperability is key to streamlining the process of interdisciplinary research – ensuring that common standards and interfaces are used as widely as possible allowing data sets, tools and services to be composed in innovative and unexpected ways with the minimum of additional engineering. Such technological inter-compatibility encourages inventive research by removing unnecessary technical barriers. This is why forums such as ENVRI (ENVRIplus 2016) and the Research Data Alliance (RDA 2016) have been established to identify and support certain common operations shared by different research infrastructures in different domains. By accelerating the construction of standardised solutions for technical problems common to the research infrastructure, the hope is that this will simplify the process of developing common policies for data management across researcher communities and so promote a wider adoption of standards that enhance technical interoperability between different infrastructure initiatives.

The integration of standards and best practices requires however a strong formal understanding of the architecture, norms and processes of research infrastructure at various levels: social, physical and technical. The focus of this chapter is primarily on computing infrastructure – data archives, online services, networks, etc. The construction of a standard model of computational research infrastructure – a *reference model* – can provide a shared taxonomy of concepts by which to understand different aspects of such infrastructure and also provide a common basis for understanding the different standards, specifications and schemas currently used to describe data, services, processes and policies relating to environmental science research. It is much easier to understand the requirements of developing infrastructure, to discuss experiences and problems and to identify existing solutions that can be adapted to new contexts if the architects of research infrastructure have a common reference point and use the same jargon. Similarly, controlled vocabulary for technical specification is also essential if we wish to describe and implement interoperable services, being fundamental to addressing the heterogeneity of data from different contexts and providing a basis for both correct automated interpretation of data and automated translation of data between different contexts and formats. Therefore, we need to consider how to construct a *semantic linking framework* that can guide the process of integrating different controlled vocabularies and translating between them where necessary. This means having access to tried and tested techniques and tools for translating descriptions made in one specification language into another, whether automatically or by expert analysis (or more likely, both). This chapter discusses some of the essential characterisation of research infrastructures from the technological and computational perspective. It identifies some of the key technical services needed to realise technological interoperability between different research infrastructures; these services all raise challenges in their implementation, the solutions to which often determine how easily they can interact with other services, especially when exchanging data, or how easily they can be transplanted into different infrastructures. It describes how the construction of a standard reference model for environmental science research infrastructures might assist current and future infrastructure developments. It also describes how the development of a semantic linking framework can be used to enhance interoperability by bridging the semantic gap between the many different vocabularies used in different scientific disciplines to characterise environmental data and processes, and we discuss how such a framework can also help the development of the interoperable services identified earlier.

12.2 Characterising Research Infrastructures

Modern scientists interact with a host of resources in order to do their work, including instruments, databases, analytical tools and simulation platforms. Regardless of the different methods that they might apply, their research efforts can be thought of in terms of a series of interactions between different actors and resources. The role of research infrastructures then is to support researchers in the conduct of their research by materially supporting a subset of these interactions.

'Research infrastructure' is thus a term that can be used in a broad range of contexts – technically speaking, a research infrastructure can be thought of here as simply a deployment of technologies or practices that support a set of research activities conducted by a group of researchers. At a more practical level, however, it would appear that the research infrastructure, in the information technology context at least, commonly refers to the technical integration of large-scale data collection with data curation and data processing facilities behind a unified service interface (e.g. a single data portal for accessing data sets). They exist to organise the facilities and technologies needed to provide researchers with the means to interact with a particular collection of data, tools and services and to strengthen the community that exists around it.

From the societal perspective, research infrastructures augment or extend the primarily extra-personal research networks that exist among researchers, laboratories and other organisations, typically by connecting researchers to useful computational services or data. Our concern here is mainly with *computational* infrastructure – infrastructure that provides computational tools and services and associated informatics that, for example, support the discovery of online data sets, the execution of data mining processes and the transfer of data over electronic networks. Such infrastructure comes in many forms: fundamental 'e-infrastructure' for computation, storage and networking, domain-specific infrastructure providing services and tools of interest to specific scientific disciplines and virtual research environments for improving coordination and collaboration among researchers. However, all such technological infrastructure exists in the context of pre-existing social structures that form the basis for the very collaborations that lead to development of technological infrastructure in the first place. Smith (2011) shows an increase in global collaboration and networking in research since the turn of the last century and makes the case for further enhancing transnational cooperation. Wagner (2009) argues that the scientific world is now best characterised by self-organising networks of researchers who collaborate 'not because they are told to but because they want to'. These social research networks are often invisible to policymakers, with their impact being not always formally recognised but nonetheless representing a critical informal structure for global research and knowledge sharing. With that in mind, it is foolish to ignore (or worse interfere with) existing social networks when identifying the research interactions that a research infrastructure might enable.

Likewise, certain research actions can be automated, but certain other actions can only currently be conducted with human expertise – for example, the orchestration and deployment of data analyses on computational hardware are inherently amenable to automation, but the expert selection of specific statistical analyses can generally only be accomplished through the guidance and intuition of a trained scientist – indicating that there is an inherent limit to the extent to which technology can be integrated into research practice. Nevertheless, to the extent that technology *can* be integrated, it behoves us to consider how best to synthesise computational research infrastructure to best support research communities.

There are a number of different architectural 'configurations' of environmental science research infrastructures, depending on the particular kinds of research activities that constitute their primary objectives or mission. However, there are common elements found with varying degrees of emphasis in most infrastructures. In Chen et al. (2013a), it is posited that environmental science research infrastructures can be functionally

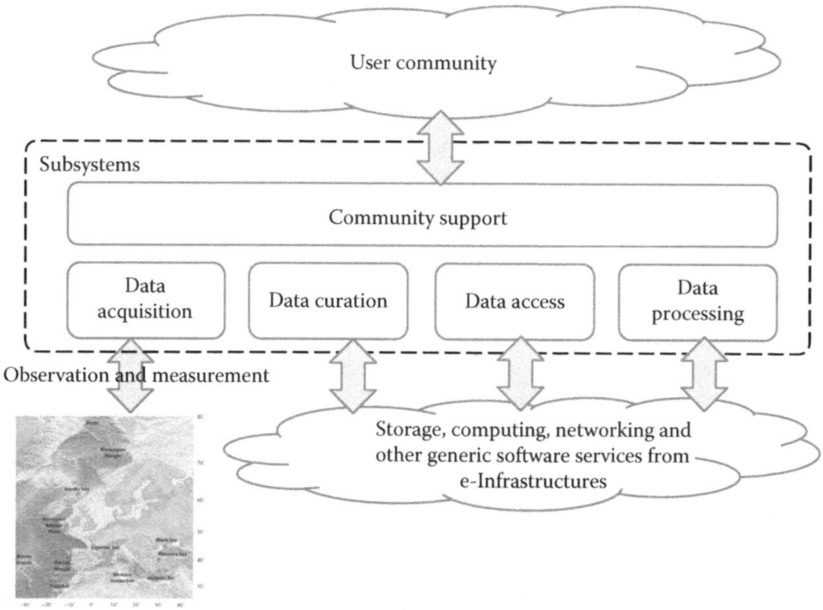

FIGURE 12.1
The main subsystems within a typical environmental science research infrastructure.

decomposed into five distinct subsystems of *data acquisition, data curation, data access, data processing* and *community support*. The relationship between the five subsystems and the broader user community, the underlying technical resources enlisted by the infrastructure and the accumulation of observations and measurements in the field or in laboratories are illustrated in Figure 12.1. Chen et al. (2013a) go on to make a distinction between *large-scale observatory systems*, which focus on how to acquire and curate data from a specific collection of instruments or other observation sources and then on how best to allow research communities to access that data, and *comprehensive integration infrastructures*, which focus on providing unified platforms for data processing and community support, often on behalf of a number of observatory systems within a specific domain. This distinction embodies two different (but overlapping) perspectives on research infrastructure: infrastructure that is constructed around *instruments* (sources of scientific observations and measurements) and infrastructure that is constructed around *services* (platforms for data discovery, analysis and integration). The challenges for the former include how to handle the real-time ingestion of data from the instrument network (i.e. how to provide sufficient network capacity and ensure that data are quality checked and processed as quickly as possible), the packaging and its curation of acquired data and how to provide timely access to those data and any derived products to a chosen user community. The challenges of the latter focus on how to provide access to computing resources and code, how to facilitate more complex experimental workflows and how to help users disseminate results both formally (e.g. in academic publications) and among their peers (for collaboration purposes) in a manner that supports verifiability and reproducibility. Both models are concerned with provenance (where data came from and what was done with them on the way), availability (short and long term) and discoverability (data and services exist to be used, and that requires researchers to know about them).

Focusing on large-scale observatory systems, some focus on providing dedicated support for a 'single' (albeit often physically distributed) instrument, very similar to examples in astronomy like the low-frequency array for radio astronomy (LOFAR 2016) or in high-energy physics like the Large Hadron Collider (CERN 2016). The principal challenge for these infrastructures is to handle the influx of data from the main instrument and to pass it on (after quality checking) to its target community. For example, the European Incoherent Scatter Scientific Association provides a three-dimensional incoherent scatter research radar to study the upper atmosphere and near-Earth space (EISCAT 2016). The infrastructure itself consists of the antenna arrays, the signal processing system, the network and the data distribution system. The beam-formed sample data, together with data from the interferometry system and some high-volume data from other supporting instruments, are streamed to a large ring buffer designed to hold several days' worth of data, after which the data will be overwritten. The ring buffer serves to store raw data long enough to allow it to be interpreted and processed. The interpreted data can then be transferred to a permanent data archive. Simultaneously, a second copy of the incoherent scatter data is separately passed through default signal processing in order to produce preliminary data sets needed for real-time experiments. The infrastructure must be able to cope with a data ingestion rate of at least 18 Gb/s per site, preferably scaling up to somewhere between 50 and 100 Gb/s per site. Besides the issue of managing data ingestion, which can only really be resolved by investing in building dedicated transport routes for data from the instrument to the main ring buffer, the primary challenge for such an infrastructure is how best to store and make available the data collected, especially for data that have a limited lifetime or are likely to be heavily accessed. This may require the regular transport of data to a secondary data centre, which can provide additional curation services such as making catalogues available for querying and search.

Other large-scale observatory systems are more highly distributed, in the sense that they acquire data from a highly dispersed set of sources and often require intermediary data centres simply to handle the local data acquisition and minimise network congestion. The challenge is much the same as for single-instrument infrastructures, but there is the additional difficulty of handling multiple sites and determining where it is most practical to process the raw data. For example, Euro-Argo (2016) is the European contribution to Argo, a global ocean observing system comprised of a large network of robot floats distributed across the world's oceans. Robot floats serve to monitor heat, salt transport, ocean circulation and the ability of the ocean to absorb excess carbon dioxide from the atmosphere. Euro-Argo is an infrastructure that supports an array of approximately 800 floats providing enhanced coverage of oceanic conditions in European seas and providing researchers access to quality-controlled data via client services such as offered by the European earth observation programme for global environmental monitoring Copernicus (2016). Data collection in Euro-Argo is based on periodic communication with its robot float network. Every 10 days, a float dives 2000 m and then rises to the surface to transmit data by a satellite link. More than 200 such transmission cycles can be performed during the float's 4-year lifespan. In this case, the processing workload is staggered. Data assembly (basically the packaging of raw information into useful, self-describing data sets) is performed at designated centres, which receive data from satellite operators and perform automatic quality control. Quality-controlled data sets are passed onwards and made available to selected researchers somewhere between 24 and 48 h of original transmission, with general availability within 6–12 months of transmission. Data are also delivered to other regional sites, which perform more comprehensive and specialised analysis, including integration with other data sources.

An example of a 'comprehensive integration infrastructure' is LifeWatch (2016). The difficulty for comprehensive integration infrastructures is identifying what facilities are available (fundamentally determined by the experimental sites contributing resources to the infrastructure) and defining a suite of common functions or services to be supported across all facilities – for example, supporting the execution of data analyses via a specific scripting language, supporting the browsing of data libraries via a specific protocol or supporting the composition of executable workflows for doing multiple tasks in sequence or parallel. The main objective of LifeWatch is to put in place a unified analytical platform for new and existing biodiversity data, a task which requires a great deal of effort to standardise technologies and practices before even considering how to deal with the intrinsic heterogeneity of data. The emphasis is on a distributed network of services providing secure access across multiple organisations and providing relevant analytical and modelling tools to research collaborations. Just as large-scale observatory systems need to address how to provide a single view on disparate data sets, these comprehensive integration infrastructures must provide a single view on computational tools, the availability of which is often dependent on a number of independent sites working 'beneath' the infrastructure.

The distinction between large-scale observatory systems and comprehensive integration platforms is imprecise – many infrastructures combine aspects of both in practice. In the case of the EPOS (2016), which is intended to provide a unified research infrastructure for the solid earth sciences in Europe, the research infrastructure is essentially a platform for providing a unified set of core services by which to interact with a confederation of more narrowly focused, large-scale observatory systems, in the form of existing national and international data centres and experimental facilities. EPOS encompasses a range of different data products – in seismology, for example, a number of data centres provide quality control and archival for real-time seismic waveform data collected from more than 500 broadband stations in Europe. EPOS is intended to build upon existing data acquisition and curation facilities by adding a new layer of interoperability between constituent infrastructures and presenting a standard interface to researchers – essentially trying to realise many of the goals of interoperable infrastructure within a single federated structure.

For comprehensive integration infrastructures, there is usually an existing set of large-scale observatory systems upon which additional research infrastructure is being constructed. Typically, these observatory systems are organisationally independent, participating voluntarily in the greater infrastructure in accordance with some agreed set of policies. It is therefore useful to distinguish between 'integrated' infrastructures, where all core resources are administered under one central authority, and infrastructures founded on collaboration among different authorities.

A common scenario where multiple organisations cooperate to build a research infrastructure is where key elements of the core infrastructure's operations are delegated to specialised sub-infrastructures, such as delegating long-term data archival to a specialist data centre. Another scenario is where dedicated data or a computational infrastructure is enlisted to handle more immediate data curation and access, including services for cataloguing and data identification (and therefore citation resolution). For example, EUDAT (2016) is an initiative concerned with the integration of data sources for research. It provides a number of key data management services to research communities, essentially allowing research infrastructures to delegate some data curation and access functions to EUDAT-affiliated institutions. A number of current research infrastructures (such as EPOS) are looking closely at how to use data infrastructure such as EUDAT to manage some of their long-term data needs.

A number of infrastructures demonstrate a tendency to separate operations by sub-discipline, especially where this reflects existing standards of practice. The ICOS (2016) is a research infrastructure specialised in quantifying and understanding greenhouse gas fluxes; its objective is to harmonise the measurement of greenhouse gases across sites in Europe and increase the availability and accessibility of the data being collected. ICOS divides its operations by 'themes' of atmosphere, ecosystem and ocean, meaning that it could be thought of as an example of a federated infrastructure. Each theme has its own dedicated thematic centre, which acts as a coordinating site and common data centre for the network of observation sites contributing to the theme. EPOS, again, does a similar division between seismology, volcanology and several other specific solid earth science domains. This pragmatic decision to create a federated infrastructure allows independent segments of their respective research communities to plan and develop their activities and infrastructure based on their related but distinct needs and ensures that infrastructure development is not unduly held back by issues and conflicts that can be localised to certain components of the overall architecture. On the other hand, this does mean that the need for standardisation of services and interfaces is particularly important in order to ensure interoperability within the infrastructure and to present a unified research infrastructure to the outside world.

Research infrastructures do not exist in isolation, rather they are an aggregation of services and networks, both technical and social, that augment the activities of researchers. Many self-identified research infrastructures overlap in function and resources with other sources of infrastructure to the extent that many data centres and research sites actively contribute to many different initiatives, each of which might present themselves as a 'research infrastructure'. For example, the Global Biodiversity Information Facility (GBIF 2016) is a distributed open data infrastructure for accessing species-level data about global biodiversity. Many infrastructures contribute directly to GBIF, and GBIF, in turn, contributes to infrastructures like LifeWatch. What is important then is not 'which initiatives do and do not constitute research infrastructures', but how these different initiatives cooperate materially to support the ambitions of researchers, regardless of the organisational umbrellas under which particular resources belong.

As such, when we speak of 'interoperable' research infrastructure, what we are really interested in is the system of services that all of these different initiatives and facilities collectively provide and the technical barriers that might prevent them from being easily exploited by a wider community. These barriers might be privileged access, poor documentation or customisability or an inability to selectively retrieve specific results, but one of the most pernicious barriers is a lack of standardisation – that certain tools and services in one context simply do not work in the same way as equivalent tools and services in a different context and as such require the would-be user to have to learn a new set of protocols and adapt their working practices in order to take advantage of them. Identifying commonalities between different research infrastructures and developing services according to a standard model where possible or at least establishing intermediary brokers that can translate between different contexts would significantly improve the technical interoperability of infrastructure.

12.3 Requirements for Interoperable Data Services

There are a number of research activities that a research infrastructure can support, the most important of which involve the curation of scientific data and tools. It is important to ensure both their long-term availability and immediate accessibility, as well as to provide a full accounting of their provenance (where they came from) and use (both internally in order to evaluate their importance, and externally to foster trust in their validity as research assets). New environmental science research infrastructures need to provide a number of services to their respective communities if they want to maximise their usefulness, including but not limited to the following:

Resource discovery. Every research asset, whether they be data sets, code, documentation, instruments or tools, should be catalogued, and the contents of these catalogues should be made available to researchers and be searchable via a variety of means.

Data harmonisation. Data are recorded in different formats in different levels of detail and are then catalogued in different ways. One of the roles of research infrastructure is to harmonise how data are collected and provide a unified model of interaction with those data.

Data preservation. Data stored should be kept in good condition, ensuring that the data are made accessible over a longer term. This entails not only maintaining backups of the data but also maintaining documentation regarding the use and purpose of the data and ensuring that the tools needed to interact with the data remain themselves available.

Provenance recording. In order for researchers to correctly evaluate the relevance of different data to their research, it is necessary to be able to answer questions about where data have come from, how they were gathered, and what has been done to them (e.g. in terms of quality control and error correction).

Workflow composition. There exist a variety of models, algorithms, code and services for processing data that a researcher might employ. Every computational experiment has a workflow that can be partially or wholly automated by research infrastructure. The use of automated workflows also assists in the reproduction of experiments, a key problem in verifying the results of peers.

Computational task deployment. Where the infrastructure is used for processing, it is necessary to optimise the use of limited computing resources (including supercomputers, network bandwidth and temporary storage) for the tasks assigned. This requires information about both the application needs and the execution environments available.

Publication of research. Publishing research outputs (whether in the form of papers, documentation, models, code or data sets) must be done in a way that permits citation, supports corroboration of experimental results and is persistent. The research infrastructure can support this by hosting essential assets in a stable environment.

Accounting. It is important to know how data and tools are used and by whom, making sure that the researchers who benefit from access to resources are actually using them and ensuring that the benefits are recognised by the institutions ultimately providing the resources as well as by the funding agencies that authorise funding for those institutions. Authenticating who is using an infrastructure and where providing authorisation for use of restricted resources is necessary is also important in many cases, though this has to be balanced against the desire for open access to data for any interest in pursuing research questions.

To be considered 'interoperable', these services should adhere to common standards wherever applicable and should present their functionality using standard interfaces that allow researchers to interact with them directly or as part of a more complex task workflow.

In the following sections, we consider in more detail some of the services that need to be developed and their requirements, focusing on the *provision* of research assets (curation, cataloguing and provenance), the *generation* of research outputs (processing, workflow and optimisation), and the *publication* of research outputs (identification, discovery and citation).

12.3.1 Curation, Cataloguing and Provenance

One of the most fundamental responsibilities conferred to research infrastructure is the reliable long-term storage and preservation of accumulated research data, whether that be raw data extracted from instruments or observed in the field (or laboratory) or interpreted data obtained via analysis and processing. What distinguishes a modern research infrastructure from a simple data archive is how it makes that data accessible – an infrastructure should provide as clear a path as possible from the sources of data to scientists and their experiments and provide a set of companion services that provide useful functions on data. This entails not merely storing the data but also providing simple but effective mechanisms by which investigators can search data (and indeed other research assets) using various catalogues and then retrieve those data to feed into computational processes.

Curation services therefore exist to support the curation of data at all points in the data life cycle. They provide the means to ingest data gathered from the field or the laboratory into a curation framework. The integration of data curation facilities with data acquisition networks results in continuous (or at least frequent) ingestion of new data over time that must be initially processed, quality checked and stored. Most environmental science research infrastructures have a 'staggered' curation sub-infrastructure, consisting, for example, of an initial 'buffer' of a certain capacity for newly acquired data, followed by a regional data centre for 'chunking' data into discrete data sets (if necessary), quality processing and initial metadata annotation, followed by storage at a central facility for the medium to long term. Some experimental activities (e.g. volcanic monitoring) require access to new data almost immediately, necessitating the creation of dedicated pipelines for acquiring data as soon as they are available.

Once acquired, environmental data should be packaged into data sets based on geography, time, specific events or other criteria. These data sets should be annotated with characteristic information, including a unique identifier and a record of how the data were obtained. Data sets should be stored safely in some repository and preferably replicated either to improve access (e.g. provide multiple redundant locations from which to retrieve data) or to ensure long-term availability (e.g. store a copy of the data in an archive, with offline records produced on long-life media such as tape). Replica sites can be managed by the core research infrastructure or can be farmed out to designated dedicated data infrastructure on the research infrastructure's behalf. It is also important to institute policies for handling changes in storage and data formats over the long term, whether by preserving certain retrieval technologies or regularly migrating data to new models so that resources remain accessible far into the future.

The provenance of curated artefacts is also important. Researchers need to know the source of data, the methods used to acquire them, the quality processes that they have been subjected to and so forth. This information inspires trust in the infrastructure, ensures a degree of accountability and promotes reproducibility of experiments. Provenance services should record the evolution of data by tracking each operation processed – such services have to be further developed, harmonised and integrated into existing and future research infrastructures. It is necessary to carefully consider how to integrate better provenance tracking within existing services and workflows however and what tools are needed to work with the resulting provenance information. Provenance services need to trace the entire research data life cycle from acquisition through curation through processing. In a federated infrastructure, provenance recording also helps to correctly attribute the efforts of different participating institutions, which is important for their own reporting processes.

As already alluded to, data are not the only important asset maintained by research infrastructures. Code, documentation, instruments, tools, processes and other useful assets that contribute to the research process also need curation. The formulation of policies for preserving and ensuring access to those assets can be as fundamental a concern of research infrastructure development programmes as the curation of scientific data sets. Many of them have the same curation requirements as pure data – the need for persistent identifiers, rich metadata descriptions and a high level of availability. The notion of 'research objects' – packages of information necessary for the reproducibility of research – is becoming especially important (Bechhofer et al. 2010).

Quality control and annotation can be considered essential curation services but have differing requirements. Quality control focuses on identifying errors or gaps in data and on flagging those errors and gaps or even repairing them using extrapolation or default values where these do not unduly damage the integrity of the data (this is most common for continuous data streams rather than discrete measurements). Quality control is generally part of the standard workflow for fully ingesting newly acquired data and is often performed in near real time. Annotation focuses on enriching the metadata associated with a data set and allowing researchers and other agents to make observations about the data and communicate those observations to other users of the data. The process of quality control may lead to annotation of the data, but annotation in general can be performed at any time while the data in question is under curation, at any point in its life cycle after ingestion.

In a sense, interdisciplinary research begins with the drawing together of data from different sources. Interoperable data cataloguing allows for the discovery, access, retrieval and integration of data from multiple infrastructures, making it a key component in an interdisciplinary research environment. To support interoperability, it is imperative to ensure that data sets and other research assets are adequately prepared for use by various services. This entails associating substantive metadata, including provenance records, with every asset – for environmental science, rich metadata for geospatial semantic annotation is particularly important. In order to ensure that these metadata exist, scientists and technicians should be supported by a range of flexible services for automatic curation and semantic annotation so as to reduce the burden of producing metadata and thus increase the likelihood that good metadata practices are upheld. To realise this, curation services need to be developed that consider all of the different 'levels' of data (from raw to various degrees of interpretation, derivation and integration) and should comply with relevant standards.

Meanwhile, a common data provenance service standard can provide data tracing services for data evolutions across different infrastructures. Standardised interfaces for querying, accessing and integrating provenance data can then be realised. In practice, linking all infrastructures to a single provenance service is not feasible, so instead the use of standards for provenance collection is applied in each infrastructure's dedicated provenance architecture so as to allow a distributed provenance network to emerge that can (in principle at least) be treated as a single unified service. Standardised interfaces for querying, accessing and integrating provenance data should be realised. Some degree of semantic linking is necessary to harmonise the key components and standards used for provenance and querying.

Being a provider of research assets is one of the primary roles of a technical research infrastructure, but the key purpose of providing such assets is to allow researchers to use those assets to achieve some research output, such as by analysing data sets gathered from sensors. Often, however, the data sets provided by the research infrastructure are difficult to process either because of their large size or because the processing necessary is challenging to configure and execute. Thus, another important role adopted by many infrastructures is to provide facilities for computation close to the data itself (rather than requiring all data to be transferred to a researcher's personal machine first) and access to preconfigured processing services (which can be trusted to produce accurate, high-precision results). This is our next concern.

12.3.2 Processing and Optimisation

Environmental system-level science increasingly relies on large volumes of heterogeneous data as produced by various research infrastructures. Data processing services can make it significantly easier for scientists to aggregate data from multiple sources and to conduct a range of experiments and analyses upon those data, when those services are sufficiently well designed and accessible. In principle, researchers can always retrieve data from data centres and perform any analysis they wish on that data using their own private facilities. In practice, however, this poses a number of difficulties. For example, computation can be prohibitively expensive. Many analyses, especially deep analyses of large, codependent data sets, outstrip the capacity of desktop/laptop computers. High-performance (HPC) or high-throughput computation (HTC) is not universally available, essentially locking out researchers whose sponsoring institutions have not had the foresight to invest in such facilities. Alternatively, the facilities that are available may be overstretched, with time on them very limited. If certain processes are deemed valuable to the community at large, it should be made possible to acquire additional computational capacity within the auspices of a research infrastructure. Another problem is that data movement itself can be prohibitive, whether due to the size of data sets or limited bandwidth for network transfers. Doing comprehensive analysis and data mining on large data sets requires computational facilities and data to be brought together; traditionally, this entails bringing the data to the computer. With the (many) large data sets now being made available, the simple act of downloading all the available data needed to conduct a particular data-intensive process may by itself be hugely time consuming. Having computational facilities at the data centres, and scheduling processes there, removes the need to transport the data anywhere else – and research infrastructure initiatives can provide a framework for putting such facilities in place.

The environmental sciences are producers of 'big data' – data that come in the form of a large number of varied data sets, many of which are themselves very large or are generated very rapidly. In many cases, these data are dispersed in small scattered data sets, which are updated frequently (with periods in minutes or even seconds). Parallelisation of computing tasks is often necessary to handle that update frequency and to ensure that core data analyses and experiments can be performed on schedule, keeping up with the arrival rates of new data. In other cases, the data arrive more slowly, with periods of hours or days, but to fully analyse and integrate all relevant data sets still requires extensive cross-correlation of data elements. This can also benefit from parallelisation.

In many cases, researchers have to configure their own workflows. Writing code, preparing tools and composing processes to realise a complete experimental pipeline requires time and considerable technical expertise. While many researchers are indeed very technically capable, this is still effort that is being diverted away from fundamental research and exploration of data. Some researchers will indeed be comfortable with and wish to configure their own workflows while exploring new methods. For many others, however, support for common or fundamental tasks provided as a service by an infrastructure would greatly increase the efficiency of a research community – more so if there is support for the composition of tasks to create more complex workflows, along with the ability to share and reuse those workflows.

Moreover, there is a significant replication of common tasks. There are many standard processes that researchers in various scientific disciplines like to apply to certain kinds of data. If these processes are applied to a data set within the scope of a specific research infrastructure, then the results of those processes can be shared with the community at large by the same infrastructure, avoiding a lot of unnecessary repetition of computation. It is sometimes difficult however to trust in the reliability of results produced by others, particularly if you are staking your own research on them. Processes conducted in private, isolated from oversight, may be subject to unknown flaws that cast doubt on the results then produced. Even should no indication of error be present, a scientist who wishes to use the results of some analysis on a data set for their own research has to decide whether to put their trust in those prior results or to repeat the analysis in their own environment. The citation of data, tools and methods in research addresses this concern to a degree, but research infrastructures can also address this concern in the trusted research environments that they offer to researchers. Computations performed within the auspices of a trusted research infrastructure can be annotated with metadata describing the provenance of the results, including pointers to data sources and to the specific methods and tools used to perform the computation, allowing investigators to make better judgements about the quality (and trustworthiness) of derived data sets.

Data processing services should make it easier for investigators to aggregate data from multiple sources and then perform systematic analysis on those data. Of increasing interest is how to support the entire life cycle of computational experimentation by allowing researchers to take full advantage of the underlying e-infrastructure, that is, the computers and networks available for working with experimental data, available to them. Specific data processing services are often (but not always) domain specific. However, generic mechanisms and languages exist for enhancing the usability and integration of processing elements to support interdisciplinary system-level science. It is highly desirable that any provision of a data processing facility to deal with the requirements of research infrastructures reuses (to the greatest extent possible) tools already developed. In addition, the extensibility of processing services is of paramount importance – new algorithms, models and techniques need to be brought into any framework very easily to achieve significant impact. Workflow composition services focus on the engineering and technological aspects of managing entire life cycles of computing tasks and application workflows for the efficient utilisation of underlying computational infrastructure. In particular, the service should enable scientists to enrich the data processing environment by easily injecting new algorithms to be also reused by others. There are a number of different workflow management systems designed for scientific computing, such as Pegasus (Deelman et al. 2005), Taverna (Oinn et al. 2004) and Kepler (Ludäscher et al. 2006), though few have been integrated specifically into public research infrastructure. The use of 'big data' analysis tools based on MapReduce and similar formalisms can also augment experimentation if used effectively; however, the automation of experimental configuration remains difficult.

Flexible monitoring and diagnosis services for data processing allow researchers to verify that their experiments are operating as intended and engender trust in the system. By evaluating the characteristic experiments that researchers want to conduct and developing common services, different possible avenues of optimisation can be identified. Much of this optimisation will be bespoke – custom solutions for specific problems. However, there is also potential for generic optimisation, performed in advance or during runtime. Such generic performance optimisation focuses on mechanisms for making decisions about the deployment and orchestration of resources, services, data sources and potential execution infrastructures so as to increase the overall efficacy of the whole system, allowing agents to schedule the execution of environmental big data applications more efficiently. Service-level agreements and modelling the infrastructure-level quality of service can augment this, allowing expert systems to make decisions on resources, services, data sources and potential execution infrastructures and to then schedule the execution. Such services can extend existing optimisation mechanisms for resources and provide an effective control model for applications at runtime. A semantic linking framework can support generic decision procedures at service, infrastructure and network levels and provide effective mapping between application-level quality attributes onto infrastructure-level quality of service attributes of computing, storage and network. We consider some of the benefits of such a framework later in this chapter.

Given the generation of research outputs, it is very important that these outputs can be published in a manner that allows fellow researchers to verify, replicate and build upon them in order to further increase the body of available knowledge. It is also necessary that the raw assets used in the production of research outputs be made not only as accessible to researchers as possible but also 'publishable', in the sense that interested parties can identify and refer back to them without needing particular knowledge of the research infrastructures which happen to provide them.

12.3.3 Identification, Discovery and Citation

Research infrastructures support the activities of researchers by providing data, tools and services. However, these assets are only useful if researchers are aware of their existence and find them sufficiently accessible. The capability to discover research assets is a significant problem in a global research context. The proliferation of research infrastructure presents a range of opportunities to the agile researcher, but these opportunities cannot be realised unless researchers are both made aware of the kind of assets available to them and are able to effectively seek them out on their own initiative. The role of discovery services in research infrastructure is to provide the tools needed by researchers to pull information about useful research assets on demand. To a lesser extent, notification services can also be used to push information to the researcher where it is deemed worthwhile to do so.

The ability to cite sources is fundamental to research. Statements can be verified, prior experiments can be replicated, and credit can be properly attributed. Increasingly it has become important to cite data, models (often in the form of code) and tools as well as prior research publications – this can be attributed to the massive increase in data volumes and the increasing complexity of data analysis, which has led to a state of affairs where, without the ability to retrieve the exact same data sets and analytical models and tools actually used in the research, there is little-to-no basis by which interested parties can actually validate the research of their peers and hold them accountable for their conclusions.

The discovery and citation of data and other assets relies on the ability to unambiguously identify objects. At the most fundamental level, this entails being able to describe the data to an extent that an agent familiar with the data can retrieve them on request. Given the vast quantities of data sets being handled, and the desire to automate basic curation functions, data sets are generally given their own unique name or identifier that can be used to recall the data on demand within a given context (such as a specific data centre or archive). Generally, the 'names' of data have limited scope, only applicable within a single institutional context – when making data available publicly, it becomes important to try to ensure that it can be referred to using a genuinely unique identifier so that conflicts with other similarly named data are avoided. Associating the object with another, more widely used namespace (such as used for URLs on the web) can help with this. If the 'domain' of an identifier is unknown, then the identifier by itself may be insufficient for retrieving the data – the use of a globally unique identifier associated with a resolution service (which acts on behalf of a range of different data-carrying institutions) can assist with this. The European Persistent Identifier Consortium (ePIC 2016) is an example of an initiative that provides such identifiers as well as identifier resolution. Another system, used for scientific publications in particular, is the digital object identifier (DOI) system (DOI 2016), with which can be associated various metadata.

The principal role played by research infrastructures in the context of resource identification is simply to be the community-preferred place to find those resources. If the community knows that a given infrastructure maintains all the important research products in a given research sphere, then it can be used as the default portal for discovery of those products.

The ability to refer to data and the artefacts that allow for the manipulation of those data by citing their respective identifiers allows colleagues (and other agents) to retrieve research assets for themselves and provides a means to attribute those assets to the infrastructures and institutions responsible for making them available. Environmental research infrastructures integrate a large number of observational and experimental sites, administered by a variety of different institutions that are responsible for the operation, funding and maintenance of the different sites. It is often extremely important to these institutions that the research outputs produced using their resources are correctly attributed to them, as much for political and financial reasons as for scientific prestige. Thus, any open access policy for data held by an infrastructure needs to acknowledge the source of the data and those responsible for making them available. It is also important that data providers are able to track the usage of their data, both to prove their importance and to refine their own understanding of how the data are used.

Optimisation of identification and citation models and technologies will be necessary because of the need to handle a truly vast number of different data objects – in a future where data, concepts, instruments and services are all citable, we can assume that there will be a corresponding explosion of persistent identifiers. The need to be able to efficiently and reliably resolve these identifiers and to direct investigators to the correct information artefacts is likely to become increasingly pressing. To identify resource use across federated infrastructures or between interoperable infrastructures, it is desirable to implement common policy models for describing persistent identifiers for certain classes of data object, which can then be used to publish and cite data used in research. Several services for data identification (e.g. via DOIs) already exist, but there are still questions as to how best to apply them to the scientific process – for example, should different persistent identifiers be supplied for different versions of the same data set? Does a continuous data stream merit a direct identifier or only the chunked output sets? Should raw data be given identifiers if in many cases they will rarely be accessed or will be discarded after a few months, or should focus be given mainly to commonly accessed derivative data sets? How should data generated during modelling or simulation be treated, given that it may (or may not) be more efficient to simply rerun the original process (which itself should also be citable)? At what point are researchers overwhelmed by a glut of persistent identifiers, and what is the role of data curators in what data should be permanently identified and what is designated 'limited access'?

In practice, any significant efforts to harmonise data citation require collaboration with existing academic publishers, who wield considerable influence on current community behaviours and the effectiveness of citation mechanisms (which are currently focused on research paper citation but increasingly involve generic citation mechanisms such as DOIs).

An interoperable data identification and citation service should aim to adhere to common policy models for using persistent identifiers for publishing and citing data and should use existing technologies where possible. It should furthermore be operated in close cooperation with existing initiatives like RDA.

Resolving many of the issues associated with discovery, identification and citation requires common agreements among a range of different stakeholders, and many of the issues regarding the provision of research assets or the generation of research outputs also bear influence, because the internal composition of resources in a research infrastructure (whether technological or otherwise) determines what is often the 'simplest' or least invasive approach to, for example, assigning persistent identifiers to individual data sets, instruments and other assets. As such, the adoption of standard architectures, taxonomies and other tools for describing the research infrastructure can play a role in establishing a fundamental orthodoxy that makes many of these issues easier to resolve, and it is this that we now address.

12.4 Building Interoperable Computational Infrastructures for Research

Interoperable tools which use standard APIs and can be used together in different configurations can make a huge impact on interdisciplinary research if made available to researchers. Interdisciplinary research after all relies on the integration of research processes founded in different research disciplines. If one accepts that the use of research infrastructure services is increasingly vital to expand the horizon of current innovation, then it is necessarily entailed that the integration of the experimental processes that are supported by those services can only happen if those services can be made to interact. Such interaction can be manually mediated by the efforts of technicians and (often) junior researchers, but this approach is both time consuming and rarely generalisable to anything beyond the specific technologies being worked with. In essence, the time and intellectual capital of researchers are being increasingly diverted towards solving technical problems, rather than to genuine research.

It is unlikely that there will ever be a single unified research infrastructure for all aspects of science, yet the challenges humanity faces require the ability to cross conventional scientific boundaries with a minimum of friction. Efforts are underway to consolidate within specific disciplines or areas of interest, to reduce the fragmentation of specific scientific communities. Nevertheless, we still need to accept that the needs of certain communities (as well as certain political realities) will always result in a degree of independence and technological drift. As such, we still need to be able to efficiently build bridges between different research infrastructures where the potential for interdisciplinary research exists. This requires a toolkit of interoperable data standards, protocols and service specifications that can be used to build interoperability layers that must be inserted between technically distinct infrastructures. Such interoperability layers are needed to streamline the interaction between data and services of different origins, automating where possible the establishment of pathways for interdisciplinary research or at least simplifying the task of creating translation tools for combining specific services and data products.

The design, construction and maintenance of effective research infrastructure pose political, economic and technical challenges (Womersley 2010) – especially for primarily academic or research-oriented institutions – but these challenges are shared widely. In practice, considerable knowledge already exists regarding a range of issues typical to research infrastructure development. In that regard, it should be feasible to pool expertise already present in infrastructure projects in order to both share solutions and prevent the same mistakes from constantly recurring. To do so however, there needs to be common forums for discussion and standard frames of reference (in terms of language and common understanding) by which to relate past experiences to new initiatives. In that regard, the foundation of interoperable architectures for research infrastructure must be a common model for research infrastructure that infrastructure developers and system architects can refer to. To improve cooperation and interoperability between infrastructure projects, attempts have been made to produce such a reference model for environmental science research infrastructures. The principal goals of such a model should be to capture high-level characteristics of operations common to environmental science research infrastructures and to establish a lexicon for describing the parts and composition of such infrastructures to be used by research communities in future infrastructure development efforts.

Any archetypical model of environmental research infrastructure should not however exist in ignorance of the multitude of standards, protocols and policies already established for many of the operations of research infrastructure. Nor should it be ignorant of the current practices of existing infrastructure. In order to shape both the design and validation of a reference model, there should be a framework by which the concepts defined by the model can be related in terms of their semantics to relevant concepts articulated by different specifications, for example, to link metadata concepts to the description of information flow in the life cycle of a curated data set. Such a semantic linking framework can then also be used to produce mappings between different controlled vocabularies (e.g. metadata standards, service descriptions, data schemas) that are needed to realise interoperability between different infrastructure services.

Given a well-defined reference model and semantic linking framework then, it only remains to consider how new services that fulfil the needs described in the previous section can be defined and deployed on suitable e-infrastructure. A demonstration of how e-infrastructures cut across the different key services needed by interoperable infrastructure is described in Section 12.3 (Figure 12.2). The reference model and the semantic linking model inform the architecture design, which draws upon and informs the construction of all services. This architecture takes the requirements of the

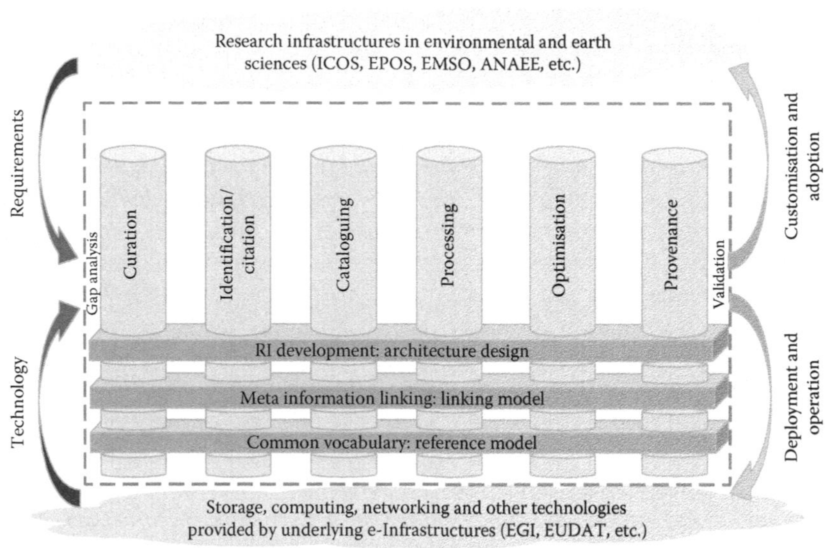

FIGURE 12.2
The intersection of key services with cross-cutting actions in environmental science research infrastructure construction.

research infrastructure initiatives and the technologies provided by existing 'e-infrastructure' providers (being providers of storage, computational power and networking on demand) and guides the development of new services on top of those technologies and the adoption of those services by the research community.

The following sections argue for the rigorous modelling of infrastructure and consider some of the issues that must be faced in the course of such modelling.

12.4.1 Reference Modelling

Existing interoperability solutions mainly focus on specific levels of interaction: between infrastructures (Ngan et al. 2011), between middleware (Blair and Grace 2012) and between workflows (Zhao et al. 2006). Interoperation is typically achieved via iterative steps: building adapters or connectors between two infrastructures and then deriving new service layer models for standardisation via community efforts. Such iterations can continuously promote the evolution of standards for infrastructures (and particularly those service layers) but will not completely solve all interoperability problems as long as the diversity between infrastructures remains great and there still exist missing links between standards (Riedel et al. 2009). Providing interoperability solutions only at a specific layer without a global view of the entire technology stack hampers the convergence of service layers. White et al. (2012) argued that an interoperability reference model is needed to complement the model of the application and infrastructure. This argument can be extended to the design of environmental science research infrastructures.

A reference model provides a framework for communicating complex concepts in precise terms and a methodology for describing and rationalising the design and development of an instance of the modelled artefact. In recent years, the construction of a reference model for environmental research infrastructures has been seen as essential for developing the research field globally into one that can coherently address the interdisciplinary challenges facing the earth and society. This realisation has been partly driven by the rapid proliferation of new research infrastructures and the recurring problems that arise in their development and use.

Having a reference model to refer to during the development of a research infrastructure confers a number of benefits. For one, a reference model provides a common vocabulary for key concepts, helping a community to share and discuss ideas more efficiently and precisely. A reference model also helps a community converge on a single common vision by providing a means to clearly express it. A standard model can allow a proposed resource, service or technology to be evaluated in the context of the larger proposed infrastructure, making it easier to spot omitted functionality or violations of standard practice, and the exercise of fitting existing infrastructures into a standard model can make it easier to identify existing solutions to recurring problems. Finally, a reference model helps to identify points at which interoperability has to occur, thus pinpointing the standards and protocols that might be applicable.

An example of a reference model developed for a specific research infrastructure is that of the LifeWatch Reference Model (Hernandez-Ernst et al. 2010), which provides guidelines for specification and implementation of the LifeWatch infrastructure. The LifeWatch Reference Model is built upon the ORCHESTRA Reference Model (Usländer 2007), an architectural framework for distributed processing and geospatial computing, which is itself founded on the Reference Model for Open Distributed Processing (RM-ODP) (Linington et al. 2011). The approach taken for the LifeWatch model was generalised for environmental research infrastructures in (Chen et al. 2013b, Zhao et al. 2015a) – this model decomposes 'research infrastructure' based on the five different viewpoints prescribed for distributed systems by RM-ODP (also known as ISO/IEC 10746). However, the model cannot be deemed to be complete – not all viewpoints prescribed by ODP are addressed, and the validation of the model against real infrastructures is lacking. There is still a need for a general reference model for environmental research infrastructure that encompasses the full scope of issues described earlier.

Nevertheless, the use of standards such as ODP that deconstruct complex systems by viewpoint seems to have merit. The fundamental idea that we can break down complexity by focusing on certain specific concerns in one context, with the presumption that any absent information will be present in another viewpoint, is appealing. It borrows from the idea of blueprints in construction and mechanical engineering. ODP in particular considers five viewpoints: *enterprise* (the interaction between agents in the system), *information* (the evolution and handling of information during execution of the system), *computation* (the decomposition and distribution of logical functionality in the system), *engineering* (the mapping of logical to physical resources as well as the data channels that exist between physical resources) and *technology* (the technologies and standards used by the system). In principle, there are other possible decompositions that could be used instead; however, the fundamental idea (that of decomposition of complex systems by viewpoint) underpins one facet of semantic linking, as we describe in the following.

12.4.2 Semantic Linking

Interdisciplinary experimentation requires integration of data and methods from different scientific disciplines. These data and methods are increasingly being provided as part of dedicated research infrastructure. Different research communities have different working practices and use different technical standards to model data and processes, so technical incompatibilities often exist between data sets, tools and services deployed within different infrastructures. Composing an experimental workflow across research infrastructures often requires bespoke engineering to allow the different components to correctly interact with one another. Interdisciplinary data-intensive research therefore requires an understanding of all the workflow components (including data) involved in the activity – essentially the semantics (and pragmatics) of the different components needs to be understood before they can be made to interoperate. A formalised, standard vocabulary for shared concepts and processes can be used to define this understanding more precisely and communicate it to others – such as provided by a reference model as described earlier. A generic, globally operational ontology that describes all aspects of research and computational infrastructures applicable in all contexts however is infeasible (not to mention cumbersome) to develop. Instead, the construction of interfaces for interoperability often depends on the ability to translate from one local controlled vocabulary to another, essentially ensuring that the inputs provided to various processes and services and the outputs extracted from those processes and services adhere to the expected formats, regardless of the actual provenance of those inputs and outputs. Providing a translation component between two different contexts can be (and often has to be) done manually, especially if the tolerance for translation error is low. Nevertheless, if the vocabularies used in both the source and the target contexts have been formally defined, it is at least possible to define a mapping between vocabularies that can then be used by a generic broker to manage the translation programmatically, rather than relying on custom brokers for every pairwise combination of foreign components.

The proliferation of semantic annotation of components, data and services, for example, in the form of Linked Open Data (Bizer et al. 2009), is founded on the principle that some of the burden of finding associations between disparate data sets and services should be taken off the shoulders of researchers (whose awareness of available data sets and tools may be unavoidably narrow) and taken on by the same discovery services that provide access to research assets in the first place by virtue of making it possible to automatically infer correspondences between the metadata attached to those assets. Semantic annotation provides the basis for semantic linking, the activity of providing translations between different concept spaces so as to allow agents to reason about scientific (and other) data between and within specific semantic models.

Semantic linking is often investigated in the context of ontology matching, mapping or alignment (Ehrig 2007). The key task is to compare similarity between entities from different semantic models and measure the similarity distances at different layers: the *data* layer, comparing data values and objects; the *ontology* layer, comparing the labels and concepts of entities; and the *context* layer, comparing semantic entities with inclusion of application contexts.

Linking pairs of information models via semantic linking, allowing for incremental improvements to the interoperability of infrastructure components, may prove more pragmatic than waiting for a universal ontology to describe 'everything'. However, a complete pairwise bridging of all information models used by all potentially interoperable research infrastructures is no more practical than the development of that one universal ontology, and moreover is unnecessary. Just as it is often necessary to use intermediate brokering for flexible service composition, it is necessary to use an intermediary concept model for semantic linking, reducing the number of required mappings between pairs of ontologies and their internal concepts (Martin et al. 2015).

The role of a semantic linking framework is simply to formalise the methodology for establishing semantic correspondences more efficiently, by allowing concepts in data models, specifications and other controlled vocabularies and languages to be linked via generic concepts defined by a core reference model, instead of being linked to each other on a pairwise basis. By having a generic reference model for environmental science research infrastructure and associating as many existing standards for data, services and technologies to the concepts defined by the reference model as possible, it becomes easier to compare and indirectly link the standards themselves to one another, using the core reference model as a 'concept exchange'. This also serves the purpose of validating the reference model, as a complete model for environmental science research infrastructures can legitimately be expected to be able to describe most, if not all, of the concepts specified in any standard used by researchers in their experiments. Moreover, when the reference model takes the multi-viewpoint approach described in the previous section, there are additional benefits. Different standards and ontologies focus on different aspects of research and technology, so, in principle, there should be a natural fit between most models and a specific viewpoint (though models that straddle multiple viewpoints do exist). The internal correspondences between concepts in different viewpoints can therefore provide a means to find associations between models addressing different viewpoints via those correspondences, allowing the construction of a network of (indirect) concept relations. The coverage of a set of standards (e.g. as used collectively by a research infrastructure to describe all aspects of its operation) can be evaluated by how completely it maps to the set of concepts described across all viewpoints of the reference model.

The actual process of semantic linking between two concept ontologies (including between a dedicated vocabulary and a reference model) can be manual, automated or a mix of both. In any case, it involves several iterations of the following steps:

Preprocessing of features by looking at a small set of excerpts from the overall ontology specification to describe a specific entity

Definition of the search space in the ontology for candidate alignment

Computation of the similarity between two entities from different ontologies

Aggregation of the different similarity results of each entity pair, depending on the algorithms used

Derivation of the final linking between entities using different interpretation mechanisms, including the analysis of human experts

Semantically linking information models from different environmental research infrastructures remains difficult however, even ignoring additional complications regarding, for example, multilingual research. Information resources (e.g. data sets, documents and descriptions) from different infrastructures often do not share common vocabularies due to their individual idiosyncrasies coupled with the different contexts these information sources address. Moreover, the diversity of metadata standards used by different infrastructures (and in particular their potential evolution, extension or adjustment to address specific needs) makes it costly to sustain and use any semantic linking model. To alleviate the cost, an effective linking model needs to focus on the interoperability gaps between research infrastructures, where semantic linking is most useful. It should also identify which viewpoints that specific concepts and concept models address, and in doing so identify which practical aspects of infrastructure design are influenced by those concepts.

12.4.3 Deploying New Services on e-Infrastructure

Research infrastructure should either provide services that directly support key research activities or interoperate with existing facilities already in use by research communities. In principle, these services should be the following:

Accessible. The assets of a research infrastructure should be made as available as possible to a wide range of users, and any services should be made as simple as possible in order to allow users to integrate them quickly into their own working practices.

Accountable. All resources should be properly annotated with appropriate metadata generated at all stages in an experimental workflow, so as to allow the provenance of research results to be traced from the end product back to the source.

Translatable. A certain degree of semantic mapping may be required to bridge the operational gap between the different knowledge organising systems required by different scientific and technical domains, but tools and resources need to be formally documented in order to make this possible.

Adaptable. Available resources change and user demands fluctuate; core research infrastructure services must be elastic and fault tolerant and provide programmatic interfaces for ad hoc service composition.

Open yet secure: Although most research data are open, there is a need to protect more sensitive data, protect the privacy of researchers, attribute credit to individuals and organisations, embargo new research prior to publication and preserve authority and accountability constraints when transferring data between different technical and political domains.

Based on the demands collected from each domain – whether currently represented by a research infrastructure or not – requirements of research infrastructures and their individual current solutions should be characterised with consideration for underlying common technologies and engineering challenges. Common operations (covering general and overarching activities) are characterised in several iterative steps involving research communities, infrastructure developers and technology providers. A reference model developed for constructing research infrastructures can be applied in the design and implementation of cross-infrastructure common services as well. Such an approach is used to reduce risk; the risk of developing new services is ameliorated by pooling resources and drawing upon the expertise of a broader technical community. It is also used to maximise utilisation of e-infrastructures. There exist a number of initiatives that provide technological infrastructure, generally based on 'grid' or 'cloud' computing, which are intended to provide or host services for public research. Deploying new services on these e-infrastructures reduces the need to invest individually on new computational infrastructure and makes efficient use of prior public investments. Sharing responsibility and effort for the development of common services does not simply reduce risk but also promotes the cross-pollination of ideas that leads to different infrastructure initiatives solving recurring problems in the same way and using one another's results to their mutual advantage, which maximises interoperability – the simplest way to ensure that the resources provided by different infrastructures interoperate is by using the same standards and technologies in the first place. Moreover, even for quite different data sets, processes and tools, if the means to interoperate with a standard service is developed internally by an infrastructure development, then it is simpler for semantic links to be developed with other artefacts in other infrastructures that likewise have had an interoperation interface developed with the shared service.

Data-intensive approaches allow researchers to define assumptions, extract evidence and validate theories based on large quantities of observations, measurements, documents and other forms of data collected from a variety of possible sources. These approaches can only be effectively enabled, however, in the presence of a supporting 'virtual research environment', a kind of integrated desktop for common services acting as a virtual laboratory for researchers. Such virtual research environments should not only provide the necessary tools for searching, accessing and integrating data and software to realise the many different workflows that constitute scientists' research activities but should ideally also provide tools for enabling collaboration. Such environments must be underpinned by research infrastructure, essentially acting as the unified interface for all research activities.

Common services can be deployed in generic data infrastructures provided by publicly funded project such as EUDAT (2016) and organisation such as EGI Foundation (EGI 2016), which can then operate them on behalf of specific research infrastructures. This approach aligns with current trends in the provision of computational infrastructure, especially grid- or cloud-based (e.g. EGI) and data-centric projects (e.g. EUDAT) (Jeffery et al. 2015), and the developments being proposed (and in some cases implemented) under the umbrella of community initiatives such as RDA. Similarly, infrastructures operated by commercial organisations can also be exploited.

Traditionally, research infrastructure has been built around data centres hosting data gathered by specific deployments of scientific instrumentation, or data gathered by field researchers. Most research infrastructure projects concentrate on the integration of multiple data centres behind a common service interface or the standardisation of processes and data products in order to increase *internal* interoperability within a research community. However, the development of truly interoperable research infrastructure requires support for complex application workflows that can be made available to any and all researchers under many different contexts (Mork et al. 2015). The process of brainstorming, planning and implementing data-intensive experiments has to be accomplished without having to acquire privileged access to limited resources. To realise this ideal of rapid innovation requires autonomous deployment and configuration of resources on demand, which is only feasible by enlisting scalable virtualised architecture such as that provided by the cloud (e.g. via some intelligent workbench such as described in Zhao et al. (2015b,c)) and being able to optimise the movement and processing of data at a low level (and indeed the movement of code), for example, by making use of programmable networks (Koulouzis et al. 2016).

Deployed services are ultimately validated only by how they are used by the research communities. Infrastructure development projects should define representative study cases by which to evaluate the utility of any prototype. Such study cases should be selected to preferably involve as broad a range of research interests as possible, across traditional disciplinary boundaries, and have a clear impact, such as the study of the mechanisms of carbon sequestration in the biosphere (Sedjo and Sohngen 2012). Any study case analysis should be articulated using a reference model, and any existing applicable standards (e.g. applying to the types of data sets typically used in the scenario) should be linked to relevant model concepts. Well-defined success criteria for this study case should be assigned and tracked alongside validation. The resources available or required on e-infrastructures should be identified, from network connectivity to data storage and processing capabilities. Data delivery to stakeholders should be optimised, and specific first test actions should be taken into account.

This process of fundamental modelling, creating a common conceptual vocabulary and understanding, semantic linking of applicable specifications and standards and deployment of common operations to generic e-infrastructure is key to the construction of interoperable research infrastructure – with the ultimate goal of maximising the interoperability at infrastructure, service and application levels.

12.5 Conclusions

There is a duty of care that we have assumed for our world. To preserve our environment or to adapt it to our needs without unintended (and possibly disastrous) consequences, we need to be able to analyse and understand the hugely complex environmental systems that determine the state of our planet in so many different ways. These systems transcend the conventional boundaries of modern scientific disciplines, and so they demand extensive interdisciplinary collaboration by researchers of many different specialities using data drawn from a variety of sources. Such interdisciplinary collaboration requires interoperability of technology and information – the integration of disparate experimental methodologies necessitates an equivalent integration of data and processing across different operational contexts. To facilitate such interoperability, research infrastructures need to be constructed to be both extensible and flexible.

An extensible research infrastructure is needed to handle the integration of new experimental sites, new services and new data sources. It is tempting to define the research infrastructure as beginning and ending with the integration of a specific set of facilities within a single organisational umbrella. In truth, the range of research assets, experimental sites and data available to a research community changes continuously. Moreover, while there are reasons for individual institutions to retain some degree of identity in a network of research infrastructures (e.g. for proper attribution), the purpose of facilitating interoperation is to remove technical boundaries that might limit scientific experimentation. Thus, the ability to integrate the research assets of neighbouring infrastructures in order to present an open, unified research infrastructure to users is essential.

The notion of a 'flexible' research infrastructure is simply a reinforcement of the notion of an extensible infrastructure. Flexibility is required to take full advantage of available resources (including computational, storage and network resources), to provide redundancy (and therefore increased reliability) and to manage changes in infrastructure topology (e.g. whether that be due to redeployments of sensor networks, reconfigurations of services or the setting up of temporary field laboratories for research). One way by which flexibility of a research infrastructure can be realised is via the adoption of grid- or cloud-specific e-infrastructure to provide selected services (such as for data storage or processing). Grid computing has been used by academic projects for over a decade to provide access to HPC and storage facilities. Cloud computing, which inherits many of the ideas of the grid, provides elastic virtualised generic infrastructure for hosting a range of services with a minimum of prior planning and configuration. It can be used to provide agile on-demand experimental facilities for researchers without intensive prior negotiation, and as a technology, it benefits from widespread industrial support and investment.

There is an increasing interest on the part of environmental science research communities in exploiting HPC and HTC) as part of generic (i.e. not installed for specific purposes) e-infrastructure. Building an open framework for data processing requires the integration of resources from many different infrastructures of different types. This includes not only the core research infrastructures but also other generic 'data' or computing infrastructures that provide specific support services for data and processing. Access to such data and computing infrastructures is often moderated by umbrella projects and organisations such as EUDAT, EGI and the Partnership for Advanced Computing in Europe (PRACE 2016), which allow for unified brokerage of resources of various kinds from different physical sites. An example of the use of such e-infrastructure for data processing is to analyse and predict the spread of infectious diseases. Mosquito-borne infections resulting in diseases like West Nile fever, chikungunya, dengue, Usutu and Sindbis have (re-)emerged in Europe during the recent decades, the result of globalisation and climate change granting new opportunities for pathogens to colonise or re-establish themselves in new areas. Statistical correlation approaches such as species distribution modelling are invaluable methods for predicting disease outbreaks. A portal provided by LifeWatch is used to provide high-quality biological data for mosquito species, while the Biodiversity Virtual e-Laboratory (Hardisty 2016) is used to access relevant environmental information and provide a series of ecological modelling algorithms. Finally, HTC resources provided by EGI are used to model a number of different climate scenarios for many different disease-carrying species.

To comprehend and plan complex, extensible and flexible research infrastructure requires a well-defined model of both the infrastructure and the context in which the infrastructure exists (in terms of community, engineering, standards and of course the research process itself). The various components and the different concerns of stakeholders can all be mapped out and represented by a multi-viewpoint model in a way that allows developers to identify common operations, recurring issues and gaps in their planning or implementation efforts. Such a reference model is also invaluable for disseminating the results to others and to generalise best practice and technology selection to be applicable to a number of different research infrastructures that might be defined using the same core model. If the model can be formally specified in a machine-readable format, then it can also be used to support semantic linking and other automated activities.

A semantic linking framework provides a pragmatic means to support interoperability between data and services from different research infrastructures by guiding the construction of semantic mappings between different controlled vocabularies: metadata models, service specification standards, operational policies, etc. Such semantic mappings allow for analysis of the coverage of different models and specifications but also facilitate the practical translation of data from one context to another. This potentially permits the construction of interoperability services between different operational environments, allowing (for instance) the construction of multi-infrastructure workflows by which researchers can conduct experiments using the resources and other assets made available by different e-infrastructures. In the absence of a unified research environment, such semantic linking may prove necessary for encouraging a much needed interdisciplinary research.

It is important for environmental science research infrastructures to embrace open data administration policies so as to provide additional support to researchers and facilitate the contribution of research to governmental policy. Fundamentally, all of the advantages of semantic interoperability described so far can only be realised if as broad a church as possible of researchers from a range of institutions (e.g. universities, national research centres and industries) have access to the data and services provided by the research infrastructure. Despite the challenges, however, the prognosis for future environmental science research infrastructure is good. There is an increasing level of collaboration between different infrastructure development initiatives and an increasing availability of dedicated computational infrastructure for generic data curation and processing, which can be adapted to the needs of research communities. An increasing recognition of the importance of data modelling, especially for cataloguing and tracing data provenance, will lead to the introduction of better standards (and better adoption of those standards). In turn, given the support of semantic linking and other metadata management methodologies, this should allow the production of more unified service interfaces and greater interoperability, thus encouraging more system-level science to address the global environmental challenges that have motivated the construction of advanced environmental science research infrastructure in the first case.

Acknowledgments

This work was supported by the European Union's Horizon 2020 research and innovation programme under grant agreements No. 654182 (ENVRI[PLUS] project), No. 676247 (VRE4EIC project) and No. 643963 (SWITCH project).

References

Bechhofer, S., D. De Roure, M. Gamble, C. Goble, and I. Buchan. 2010. Research objects: Towards exchange and reuse of digital knowledge. In *First International Workshop on The Future of The Web for Collaborative Science (FWCS2010)*, Raleigh, NC: Nature Proceedings.

Bizer, C., T. Heath, and T. Berners-Lee. 2009. Linked data—The story so far. *International Journal on Semantic Web and Information Systems* 5(3):1–22.

Blair, G. and P. Grace. 2012. Emergent middleware: Tackling the interoperability problem. *IEEE Internet Computing* 1: 78–82.

CERN. 2016. The Large Hadron Collider | CERN. http://home.cern/topics/large-hadron-collider (accessed March 8, 2016).

Chen, Y., A. Hardisty, A. Preece et al. 2013a. Analysis of common requirements for environmental science research infrastructures. In *International Symposium on Grids and Clouds*, Taipei, Taiwan: Proceedings of Science.

Chen, Y., P. Martin, B. Magagna et al. 2013b. A common reference model for environmental science research infrastructures. In *Proceedings of the 27th Conference on Environmental Informatics—Informatics for Environmental Protection, Sustainable Development and Risk Management*, eds. B. Page, A. Fleischer, and J. Gobel, pp. 665–673. Hamburg, Germany: Shaker Verlag.

Copernicus. 2016. Copernicus. http://www.copernicus.eu/ (accessed March 8, 2016).

Deelman, E., G. Singh, M.-H. Su et al. 2005. Pegasus: A framework for mapping complex scientific workflows onto distributed systems. *Scientific Programming* 13(3):219–237.

DOI. 2016. Digital observation identifier system. https://www.doi.org/ (accessed March 8, 2016).

EGI. 2016. EGI site. https://www.egi.eu/ (accessed March 8, 2016).

Ehrig, M. 2007. *Ontology Alignment: Bridging the Semantic Gap*. New York: Springer-Verlag.

EISCAT. 2016. EISCAT_3D | The next generation incoherent scatter radar system. https://eiscat3d.se/ (accessed March 8, 2016).

EMSO. 2016. EMSO—European multidisciplinary seafloor and water-column observatory—Home. http://www.emso-eu.org/ (accessed March 8, 2016).

ENVRIplus. 2016. ENVRIplus—RIs providing shared solutions for science and society. http://www.envriplus.eu/ (accessed March 8, 2016).

ePIC. 2016. European persistent identifier consortium. http://www.pidconsortium.eu/ (accessed March 8, 2016).

EPOS. 2016. EPOS. https://www.epos-ip.org/ (accessed March 8, 2016).

EUDAT. 2016. EUDAT—Research data services, expertise & technology solutions. http://eudat.eu/ (accessed March 8, 2016).

Euro-Argo. 2016. http://www.euro-argo.eu/ (accessed March 8, 2016).

Foster, I. and C. Kesselman. 2006. Scaling system-level science: Scientific exploration and IT implications. *Computer* 39(11):31–39.

GBIF. 2016. Free and open access to biodiversity data | GBIF.org. http://www.gbif.org/ (accessed March 8, 2016).

Hardisty, A., F. Bacall, N. Beard et al. 2016. BioVeL: A virtual laboratory for data analysis and modelling in biodiversity science and ecology. *BMC Ecology*. doi: 10.1186/s12898-016-0103-y.

Hernandez-Ernst, V., A. Poigné, J. Giddy et al. 2010. LifeWatch deliverable 5.1.3: Technical construction plan—Reference Model. http://orca.cf.ac.uk/56502/ (accessed March 8, 2016).

ICOS. 2016. Unraveling Earth's greenhouse gas balance with measurements | ICOS EU. https://www.icos-ri.eu/ (accessed March 8, 2016).

Jeffery, K., G. Kousiouris, D. Kyriazis et al. 2015. Challenges emerging from future cloud application scenarios. *Procedia Computer Science* 68:227–237.

Koulouzis, S., A. Belloum, M.T. Bubak, Z. Zhao, M. Živković, and C. de Laat. 2016. SDN-aware federation of distributed data. *Future Generation Computer Systems* 56:64–76.

LifeWatch. 2016. LifeWatch ERIC official portal. http://www.lifewatch.eu/ (accessed March 8, 2016).

Linington, P.F., Z. Milosevic, A. Tanaka, and A. Vallecillo. 2011. *Building Enterprise Systems with ODP: An Introduction to Open Distributed Processing*. Chapman & Hall/CRC Press. ISBN 9781439866252.

LOFAR. 2016. LOFAR | LOFAR. http://www.lofar.org/ (accessed March 8, 2016).

Ludäscher, B., I. Altintas, C. Berkley et al. 2006. Scientific workflow management and the Kepler system. *Concurrency and Computation: Practice and Experience* 18(10):1039–1065.

Martin, P., P. Grosso, B. Magagna et al. 2015. Open information linking for environmental research infrastructures. *2015 IEEE 11th International Conference on e-Science*, pp. 513–520. IEEE.

Mork, R., P. Martin, and Z. Zhao. 2015. Contemporary challenges for data-intensive scientific workflow management systems. *Proceedings of the 10th Workshop on Workflows in Support of Large-Scale Science*. Austin, TX: ACM.

Ngan, L., Y. Feng, S. Rho, and R. Kanagasabai. 2011. Enabling interoperability across heterogeneous semantic web services with OWL-S based mediation. *2011 IEEE Asia-Pacific Services Computing Conference*, pp. 471–476. Jeju Island, South Koren: IEEE.

Oinn, T., M. Addis, J. Ferris et al. 2004. Taverna: A tool for the composition and enactment of bioinformatics workflows. *Bioinformatics* 20(17):3045–3054.

PRACE. 2016. PRACE research infrastructure. http://www.prace-ri.eu/ (accessed March 8, 2016).

RDA. 2016. RDA | research data sharing without barriers. https://rd-alliance.org/ (accessed March 8, 2016).

Riedel, M., E. Laure, T. Soddermann et al. 2009. Interoperation of world-wide production e-Science infrastructures. *Concurrency and Computation: Practice and Experience* 21(8):961–990.

Sedjo, R. and B. Sohngen. 2012. Carbon sequestration in forests and soils. *Annual Review of Resource Economics* 4(1):127–144.

Smith, C.L. ed. 2011. *Knowledge, Networks and Nations: Global Scientific Collaboration in the 21st Century*. London, UK: The Royal Society. https://royalsociety.org/topics-policy/projects/knowledge-networks-nations/report/ (accessed March 8, 2016).

Usländer, T. ed. 2007. Reference model for the ORCHESTRA architecture (RM-OA) V2 (Rev 2.1). Orchestra project. http://portal.opengeospatial.org/files/?artifact_id=20300 (accessed March 10, 2016).

Wagner, C.S. 2009. *The New Invisible College: Science for Development*. Washington, DC: Brookings Institution Press.

White, L., N. Wilde, T. Reichherzer et al. 2012. Understanding interoperable systems: Challenges for the maintenance of SOA applications. *2012 45th Hawaii International Conference on System Science*, pp. 2199–2206. Maui, HI: IEEE.

Womersley, J. ed. 2010. Cost control and management issues of global research infrastructures. European Commission, Directorate-General for Research. Luxembourg: Publications Office of the European Union. https://ec.europa.eu/research/infrastructures/pdf/cost_control.pdf (accessed March 10, 2016).

Zhao, Z., S. Booms, A. Belloum, C. De Laat, and B. Hertzberger. 2006. VLE-WFBus: A scientific workflow bus for multi e-science domains. *Second IEEE International Conference on e-Science and Grid Computing*. Amsterdam, the Netherlands: IEEE.

Zhao, Z., P. Martin, P. Grosso et al. 2015a. Reference model guided system design and implementation for interoperable environmental research infrastructures. *2015 IEEE 11th International Conference on eScience*, pp. 551–556. Munich, Germany: IEEE.

Zhao, Z., P. Martin, J. Wang et al. 2015b. Developing and operating time critical applications in clouds: The state of the art and the SWITCH approach. *Procedia Computer Science* 68:17–28.

Zhao, Z., A. Taal, A. Jones et al. 2015c. A software workbench for interactive, time critical and highly self-adaptive cloud applications (SWITCH). *2015 15th IEEE/ACM International Symposium on Cluster, Cloud and Grid Computing (CCGrid)*, pp. 1181–1184. Shenzhen, China: IEEE.

13

ÆKOS: Next-Generation Online Data and Information Infrastructure for the Ecological Science Community

David J. Turner, Anita K. Smyth, Craig M. Walker, and Andrew J. Lowe

CONTENTS

Abstract

Ecological data are inherently complex, covering a diverse range of context-dependent concepts that create challenges for secondary users both in terms of interpretation and integration. Interpretation is particularly challenging when different collection protocols, measurement standards and classification systems are in use, and in many cases are not described in enough detail to be reproducible. Integration can also be challenging because many data collection activities are small scale and rely on bespoke data management practices. Furthermore, the multitude of different ways that ecological system observations are made leads to difficulties with aligning similar but not synonymous concepts, a challenge further exacerbated by the lack of sufficient context.

The Advanced Ecological Knowledge and Observation System (*ÆKOS*) utilizes a flexible knowledge representation approach, which allows us to integrate data into a common information model. All ecological data are stored and exposed to users at a site level allowing them to interact directly with the data. This is different from most other repositories that employ a data set storage and metadata search paradigm, whereby the data remain essentially opaque to the user until downloaded.

ÆKOS also places significant emphasis on the provision of detailed contextual information to foster reproducibility. In particular, information about sampling design, data collection, measurement protocols and classification systems employed are all provided to enable researchers to interpret the underlying data and make an informed assessment of the potential utility and appropriateness of the data.

Although only recently developed, the value of the approach *ÆKOS* uses is being recognized as a leading global platform for supporting excellent science, reproducible reuse and scientific reward. The system is available online at http://www.aekos.org.au.

13.1 Introduction

Data drive scientific discovery and the sharing of research results through scientific articles, the currency of mainstream science (Hanson et al. 2011). The increasing availability of data and analytical resources is transforming the way that ecology as a discipline interacts with large data sets and is driving new insights in the field (White et al. 2015). Data can be repurposed in many beneficial ways, for example, for predictive modelling as well as in the subsequent testing and validation (Ferrier 2012). While data sharing is not new, the open data revolution is enabling data to be shared more widely and well beyond the peer networks of original collectors. Not only is this leading to a data deluge (Bell et al. 2009), but the increasing anonymity with which data are shared creates challenges for knowledge preservation as well as driving discussion of what constitutes 'appropriate' reuse. As a result, there is a need for increasingly sophisticated data management systems to protect the integrity of the data, improve the efficiency with which it is used and credit original data authors (McKiernan et al. 2016).

In many cases, repurposing data involves the creation of 'new' data sets (secondary data), which represent aggregates of data from multiple sources that have been carefully assembled and curated (often for a purpose different to that intended by the original collector). Data assembly and curation requires the user to have a solid understanding of the underlying concepts represented by the data set as well as the technical skills to be able to manipulate and transform the data itself. Infrastructure that facilitates the integration of data and the transfer of important contextual knowledge will further help facilitate knowledgeable reuse.

Openly publishing data are increasingly mandated and a growing number of platforms are now available to allow researchers to fulfil these obligations. Nevertheless, many researchers have expressed concern that the current approaches do not provide adequate safeguards against inappropriate or unethical behaviour (Lindenmayer and Likens 2013, Lindenmayer et al. 2015). In some cases, these concerns are such that the researchers are prepared to boycott scientific journals that require depositing of data in open repositories as part of paper publishing (e.g. Mills et al. 2015). This creates challenges for those responsible for data infrastructure if they are to truly meet the needs of their user community. In the case mentioned previously, it is clear that researchers collecting the data require that it be represented in a way that not only reflects its inherent characteristics but also guides its proper as well as ethical reuse (*sensu* Duke and Porter 2013, box 2) in a similar way to that afforded to the reuse of knowledge from scientific papers. Thus, in encouraging proper use of open (publicly accessible) ecological research data, it is essential that data publishers provide appropriate hard and informative soft infrastructure. Hard infrastructure (e.g. storage, discovery, access, web services) that meet expectations of primary (data creators) and secondary data users is already recognized by most repositories as being critical to successful uptake. It is equally important though that educational soft infrastructure describing the principles and practice of open research data are also front and centre of data services in order that users are guided on the proper use of others' data.

In this chapter, we take a researcher-centric view outlining many of the challenges facing those who wish to build suitable data infrastructure for the ecosystem science community. We offer solutions to many of these challenges and describe our own infrastructure offering – *ÆKOS* – a next-generation infrastructure for primary and secondary ecosystem data users.

The Advanced Ecological Knowledge and Observation System (*ÆKOS*) was built to address the challenges facing 'Open Data' in ecology today. *ÆKOS* is designed to bring together different types of heterogeneous data, optimize its usage and help realize its enormous reuse potential. Ecological data can be deposited, stored and published in ways that not only maximizes the proper and ethical reuse of data but also increases research efficiencies and opens the door for data publication reward systems. The system supports the discovery of 'site level' ecological survey data that can be integrated across multiple, disparate data sets. To enhance reproducibility, that is, the appropriate repurposing of data based on a consistent understanding of the nuances of original data (a key premise of the scientific method), *ÆKOS* also provides 'quality assured' descriptions of the data along with associated collection methods and other pertinent contextual information. Ensuring reproducibility helps minimize data misuse – a key concern for many ecologists who are wary about open data publication.

The Terrestrial Ecosystem Research Network (*TERN*), an Australian enterprise that establishes research infrastructure and scientific networks, has developed *ÆKOS*. TERN enables sustained, long-term collection, storage and sharing of ecosystem data to meet terrestrial ecosystem research and natural resource management needs (Thurgate et al. 2016).

13.1.1 Publishing Ecological Data to Support Intelligible Reuse

Science practices encourage the publication of results, together with associated contextual information. Such practices aid interpretation of data, support reproducibility and also build confidence and hence trust by enabling independent verification of the results (Bechhofer et al. 2013, Kepes et al. 2014). However, if ecologists are to effectively reuse others' data, they need to be able to understand as unambiguously as possible descriptions of the methods, data measurements, analyses (for derived data) and the meaning of terms used in those descriptions. In this respect, many repositories utilize metadata as the basis of providing this context. The question here is whether current offerings are adequate to enable intelligible reuse of ecological data.

Information and terminology need to be forensically described to a point that it becomes reproducible, once this is true, researchers will be able to consistently repurpose data. Nevertheless, not all descriptions are equal and the degree to which published data are independently understandable is highly variable (Figure 13.1). In actuality, there is a continuum of degrees of systematic bias due to the quality and completeness of metadata that ranges from low (biased or incompletely verifiable), to high (unbiased or highly verifiable). It is generally only unbiased data that tend to hold up to scientific scrutiny and utility over time (Peng 2009). The challenge when publishing data is therefore to minimize the level of systematic bias in order to maximize reproducibility. Not only is this likely to involve improvements to hard and soft infrastructure, but it also involves greater engagement with data creators.

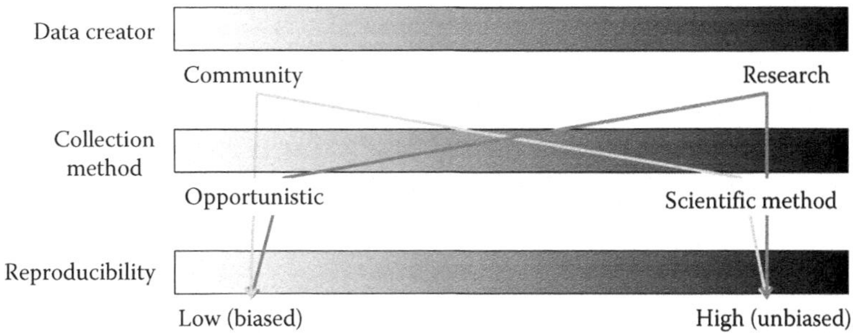

FIGURE 13.1
Data reproducibility for reuse is a continuum ranging from low to high and is related to the scientific method used to collect the data which is influenced by the scientific experience and skills of data creators. Low reproducibility implies the data were collected using opportunistic (haphazard) collection methods and therefore may be biased spatially and temporally (e.g. data collected repeatedly along roads in a single season). High reproducibility implies the data were collected using a scientific survey method at randomly selected places and times and therefore are unbiased and tend to hold up to scientific scrutiny.

The adoption and rate of data publishing in ecology (and across the environmental sciences) has been slower than for other disciplines. This is despite the plethora of online ecological data repositories (Costello et al. 2013; Hampton et al. 2013; Reichman et al. 2011; Soranno et al. 2014), implying that there are underlying challenges that need to be resolved. During our work to develop *ÆKOS*, consultations with the Australian ecosystem science community* (unpublished user requirements of researchers) and publications by the International Council for Science† have indicated a strong pull for platforms and services that align closely with science practice to balance the technology and operational push for infrastructure (Bach et al. 2012). In addition, 10 features currently not provided by online ecological data platforms were identified (Table 13.1).

These requirements were organized into three main groups. The first of which (1–3) is related to the depth of information contained within the repository. In this respect, secondary users were keen to be able to get hold of the raw site level data and not just summary or aggregates. Furthermore, they expected data to be sufficiently described as to be reproducible. Users specifically requested that data be accompanied by rich descriptions of concepts measured as well as the methods used for collection. Information regarding provenance was also considered important for the user to understand how data have been managed as well as details of any curation activities undertaken.

* http://ecosystemscienceplan.org.au/Events-pg26776.html#Workshop reports.
† International Council for Science.

TABLE 13.1

Essential Requirements for Online Ecological Data Services Identified by the Terrestrial Ecosystem Research Network's Ecological Science Community

1. *Primary (raw) site-based scientific data.* The data should include high-quality primary data collected at the site level using scientific methods.
2. *Reproducible and verifiable data.* Knowledge about the data, meaning of terms, data attributes, measurement and the scientific methods use to collect the data must be unambiguously described and appropriately catalogued to allow discovery, access and assessment, providing good precision and recall for searches.
3. *Data provenance.* The provenance of data and its temporal stability (versioning of updated, curated and derivative works) must be tracked to match data with original citations in published articles (data-article publishing).
4. *Single web tool for ecologists.* All ecological data should be available through a single web tool which supports data archiving, discovery and open access, with the option to collaborate with data authors.
5. *Seamless user experience.* A 'shop front' for the platform is required that enables users to efficiently discover and visualize the information content of potentially suitable data products and then extract these in a format that lends itself to further processing.
6. *Data relationships.* Visuals of the relationships among the observations and different methods to assist data reproducibility.
7. *Persistence.* Metadata and unique identifiers should persist beyond the life of the data.
8. *Licensing.* Practical data licensing and conditions of use.
9. *Scientific rewards.* Data citations should be described to enable quick scholarly credit and legal attribution.
10. Digital object identifier (DOI). Data citations should include a unique hyperlinked identifier such as a DOI and should facilitate access to the data themselves.

The second group of requirements (4–7) described how users wished to interact with online repositories. Ideally, they wanted a unified shop front that seamlessly fit in with their science practice. Also important was the need to be able to visualize relationships between different data entities and have persistent identifiers for all artefacts.

Data creators lodging data in repositories introduced a third group of requirements (8–10) intended to help protect their investment. In this respect, it was important that data were accompanied by clear right of access statements, including details of the license and any requirements for attribution. Beyond this, users were also looking for a return on their investment via scientific rewards, including the ability to track data citations.

The challenges associated with building a repository that fulfils the requirements mentioned earlier are many and varied, involving a combination of both hard and soft infrastructures. We will continue to describe and unpack these issues and some of their suggested solutions in the following sections.

13.1.2 Making Heterogeneous Ecological Data Reuseable

Ecology is the scientific study of relationships between organisms and their environment and the interactions between organisms (Attiwil and Wilson 2003). It is studied at various levels of organization (e.g. individuals of a species, populations of a species, communities of species populations, ecosystems of a region, biomes and biospheres) and ecologists alter their frame of reference at each level. Under this broad banner sits a diverse range of sub-disciplines and a diverse suite of researchers and institutions generating the associated data (including environmental survey data). Each of these research groups collects ecological data for different and sometimes overlapping purposes creating many challenges for data management and integration.

Data collected by ecologists take many forms ranging from simple opportunistic observation through to elaborate and highly structured sampling programmes. Generally, the latter employ some form of a site-based paradigm, whereby the ecologist undertakes their research at one or more suitable study areas. There are exceptions and these are alluded to later in the chapter. Within these study areas, more formal sampling units are generally set up as study sites. Sampling is undertaken within and across study sites for the study area(s). Within sites, there can be nested sampling (e.g. subsampling and sub-subsampling units). Whether at the site level or more granular nesting levels, data are collected using a range of field sampling methods (including plots, quadrats, transects, trap lines and arrays, and other systematic collection methods). Observations may relate to genetics, behaviour, populations or multispecies studies relative to landscape features. Measurements may be or quantitative or (commonly in ecology) qualitative.

Data are generally collected and analyzed with a clear goal in mind; sometimes this relates to a form of inventory or surveillance program; other times the goal is to answer a question or hypothesis related to long-term monitoring or experimental projects. Typically, the structure and content of data is closely related to both the question and the method of origination (e.g. analytical process, field observation protocol, remote sensing). The result is that ecological observations tend to be a product of where, when and how observations are made and are thus considered context dependent. The large number of potential observations (e.g. tens of thousands) combined with variations in the approach to record observations (e.g. different methods, automated versus manual data capture) as well as subsequent differences in the chosen measurement standards (e.g. diameter at breast height – 500 mm, 1.3 m) gives rise to complexity in ecological data and is why we describe it as heterogeneous in nature. A good understanding of the nuances in the data is therefore necessary for correct interpretation of the underlying data (data context).

Heterogeneity in ecological data manifests at many levels. Basically, studies may focus on a myriad of different organisms or assemblages, observing or measuring different facets of each. Beyond this, the design may be based on one or more underlying conceptual frameworks (Mills et al. 2015), which potentially leads to different scales of observation (both spatial and temporal) and site stratification (e.g. by ecosystem type, disturbance type). Design also affects the number and location and types of sites, sampling methods (e.g. plots, quadrats, animal trap arrays) and the length of sampling period and associated sampling effort.

Heterogeneity also exists between projects through differences in the types of field/laboratory methods (and variants therein) used to collect data. While field protocols tend to follow conventional approaches, they are often nevertheless refined on a project basis to cater for local design. Variations may also exist across and within an individual study in order to accommodate site-specific characteristics (e.g. soil cores not undertaken at a site due to a rocky substrate). For long-term studies, heterogeneity may occur through time for several reasons including the reshaping of original science questions, new approaches replacing obsolete ones, resource setbacks and unforeseen climatic events (Lindenmayer and Likens 2013, Michener et al. 1997, 2011). Beyond project design, heterogeneity also exists in more nuanced ways such as how the data are handled over the duration of the project and later on by any new custodian in the future. When taken as a whole, facilitating reuse of ecological data is challenging owing to the data's diverse and heterogeneous nature and because of the need to provide significant contextual description in order to make it reproducible. This situation is often further exacerbated as ecologists interact with data in different ways and this affects hard and soft infrastructure design.

Ecologists tend to be *both* data creators and primary data users and collect observational and experimental data that are conventionally reported in a way that is replicable by subsequent investigators (Atici et al. 2013). Being data creators facilitates their subsequent role as primary data users because they have a first-hand understanding of the collection protocols and the nuances associated with the study, aiding them to interpret the data and findings. This close coupling between creators and primary users often results in a lot of important contextual knowledge being handled implicitly within the study, that is to say it is never formally recorded. While this is generally not immediately problematic to the primary user, it creates several challenges from the perspective of publication as well as subsequent reuse. In particular, if the goal is to publish data in a form that makes it reproducible and hence reusable for a specific purpose, the challenge becomes how to capture sufficient contextual knowledge and convey it in a way that facilitates a clear and consistent interpretation by secondary users who have no previous knowledge of the data.

From a data management perspective, most primary users (including data creators from now on) make some effort to try and retain the data in case they find a new use for it. Nevertheless, management practices are generally not sophisticated (Hampton et al. 2013) meaning that data are often inadvertently discarded or lose relevance either through accident or because the original investigation moves on to new questions (Michener et al. 1997). Publication of the data set will help avoid this fate; the challenge of course is to prepare and describe it to a level that makes it reproducible.

13.1.2.1 Challenges Associated with Publication

The challenges associated with ecological data publication and reuse can be viewed from three different perspectives. First, the secondary user is the individual or team that will make use of and thus repurpose the data. Second, it is the perspective of the data creator, whose role is to supply the data for publication. Finally, is the infrastructure builder's perspective. It is necessary to overcome the challenges associated with all three perspectives to build a truly effective data publication solution.

13.1.2.1.1 Challenges from the Secondary User's Perspective

A secondary user needs to acquire, understand and be able to manipulate data to suit their purpose. Many challenges are associated with these tasks, and it has been reported that ecologists relying on other people's data for research can spend well over half of their effort obtaining, collating and verifying the reproducibility of the data (Zilinski et al. 2014).

In terms of discovery, ecological data are currently widely dispersed and this in conjunction with the large amount of heterogeneity creates a challenge for secondary users. This is because only a subset of the data is likely to be useful for a particular purpose, and these gems need to be found within the growing deluge. Users would have traditionally been limited to accessing data via sharing through peer networks, and, if only a relatively small number of data sets were available, a researcher could collaborate with data creators and review each one and assess its suitability for their purpose. This approach may be effective at small scales but is impractical as the volume of data grows. In the latter case, the secondary user needs to maximize research efficiencies and to access to more sophisticated tools that allow them to quickly filter down to the most suitable, candidate data sets.

Ecological data heterogeneity also creates challenges for data comprehension, as it cannot be assumed that any data were collected in a certain way or are appropriate for a certain purpose without clearly reviewing the design and protocols used to collect the data as well as any subsequent post processing. This knowledge provides critical insights into how the data can be appropriately re-purposed (Lindenmayer and Gibbons 2012).

In many cases when re-purposing data, a researcher will draw together data from multiple data sets. Here too, any heterogeneity between the data sets is likely to create challenges for the secondary user. In particular, different data collection methods or measurement standards may result in observations that are subtly different on a conceptual level. The extent to which this is important will vary depending on the magnitude of the difference and also the intended purpose of use. Nevertheless, the only way for the researcher to make an informed decision is through careful consideration of the underlying context (Krebs 2012).

Models and other classification systems are commonly used in studies, and not only do these need to be understood by secondary users, but they also create an integration challenge where data are to be drawn from multiple data sets. To illustrate this, consider a simple classification where one researcher has classified trees according to height as short, average or tall, while a second researcher independently and additionally uses very short and very tall. Clearly, the secondary data user needs access to the definitions of each of these categories (i.e. how tall is an average-sized tree) in order to correctly interpret them. The conundrum however arises when they discover that the categories between the two classifications only partially overlap. Depending on the intended use of the data, it may be appropriate to aggregate these concepts or otherwise treat them in a way to make them useable. Either way, this is a decision for the end user and can only be made when sufficient context is provided.

13.1.2.1.2 Challenges from the Primary User's Perspective

From the point of view of primary users, there are two key challenges that need to be overcome in order to facilitate data publication. The first relates to the amount of effort associated with the production of appropriate data publication artefacts (e.g. field journals, lab notebooks), which at the time of *ÆKOS'* development was poorly sponsored, often because publication then was done as an afterthought. The second key challenge relates to appropriate credit being given to the primary user and concerns relating to potential data misuse (Mills et al. 2015).

In terms of effort associated with publication, the nature of many data collection programmes is such that they are small, targeted for a purpose, and often analyzed by the same individuals who undertake the observations. Hence, much of the focus in data management is on the observation data itself and preparing it for use, rather than documenting the surrounding processes and knowledge, which are 'known' or implicit to the initial user base. Consequently, generating good-quality artefacts to support data publication places a significant burden on the primary users who frequently have to author the materials from scratch and who often feel that it is actually the secondary users that are the direct beneficiaries of their investment (Lindenmayer and Likens 2013). The data itself are also often structured in a way that makes them amenable to the primary user's needs rather than in a form or structure required for publishing in a repository. This creates an additional workload for the primary user who then has to 'clean up' their data for publication. Clearly, if the effort associated with publication is too onerous, then either it won't get done or, in cases where mandated, the standards produced often fail to meet a reproducible level.

Effective publication to support reuse requires good contextual descriptions, which are often available as a mixture of recorded information and colloquial knowledge. While many existing metadata schemes (such as ecological metadata language [EML]) provide a mechanism for data publishers to richly describe their data, our experience talking to users is that the tooling is not set up in a format that is intuitive to the domain. Furthermore, many users view that what guidance is available is quite technical in nature (written from an IT specialist perspective) and therefore not immediately informative to ecologists. This intensifies the challenges associated with providing adequate contextual description when combined with the perceived burden associated with data publication.

The challenges associated with publishing good descriptive artefacts are further exacerbated where self-service models are used. Under such scenarios, creating consistent publication artefacts and assuring quality of those products is difficult both because few levers exist to control and influence what is produced. Also, there are many parties to influence or govern. The clear exception here is the emergence of data papers (e.g. *Nature Scientific Data*), which undergo a similar peer review process to that of scientific publications. Such publications demonstrate that with the imposition of higher standards accompanied by clear guidelines and feedback to authors, the challenges associated with making data reproducible are not insurmountable.

13.1.2.1.3 *Challenges from the Infrastructure Builder's Perspective*

The role of the data repository is to aggregate data sets from multiple collectors and to store these and present them to secondary users in a format that supports intelligible reuse. The challenges can be grouped according to a number of key interrelated functions of the system, which include storage, integration, enrichment, ingestion, interaction and extraction.

Several challenges exist with respect to storage, the most difficult of which pertain to the underlying information structure of the data. This is because ecological data contain a mixture of classification, description and quantitative measurement. The heterogeneous nature of the data also means there is often limited overlap across sources as well as a lot of ambiguity regarding how to handle similar but not identical concepts. The result is that the data can be considered both sparse and semi-structured (such as opportunistic observations) from a data management perspective. These characteristics make it not immediately amenable to management using traditional IT approaches, that is, it is difficult to create a unified infrastructure that manages heterogeneous data structures and formats, yet provides consistent functionality and user experience. This challenge is further exacerbated by the fact that ecology is still rapidly evolving meaning that the nature and type of data being collected frequently changes. This further complicates attempts to build the unified infrastructure, as the underlying information model needs to constantly evolve to handle new data sets that don't fit the existing structure. Changing the model then has further implications for existing data within such a system.

As a consequence, a fragmented set of systems and tools has come into existence, which tends to distribute rather than centralize access to data. This means that individual data management environments generally adopt one of the following approaches (which are generally in contrast to what the users actually want):

- Choosing to store only a narrow 'standard' subset of information
- Becoming highly tailored to a particular project/local need
- Focussing on the storage of products derived from the original data such as spatial layers

We often see variations on these types of solutions manifest at multiple levels from local research group repositories through to major data aggregators. While such approaches are often pragmatic (from the builder's perspective), they do not solve the underlying challenges presented by the heterogeneous nature of the data nor do they suit the ecosystem science community. Instead, these approaches shift the problem on to the community that invariably ends up investing a significant amount of effort (re)assembling the data into a suitable format.

In order to improve reusability, Mills et al. (2015) proposed that data should be brought together (federated) at a site level and integrated on common entities and infrastructure elements. While this clearly intensifies the challenges associated with data storage, having a consistent structure across data sets greatly simplifies subsequent data use by reducing the amount on manipulation required and facilitating understanding.

The challenges associated with integrating data from multiple sources are a consequence of the underlying heterogeneity of ecological data and can be grouped into five classes as proposed by Bach et al. (2012). Using their meanings, the following can be considered:

- Structural heterogeneity refers to the issue when information can be represented multiple ways within- and cross-data models.
- Syntactic heterogeneity is when multiple descriptions are used for the same value among databases (e.g. biodiversity or biota for a collection of plants, animals and microorganisms).
- Semantic heterogeneity refers to the issue of differences in meaning, interpretation and usage of data within and across databases.

- Technical heterogeneity refers to the issue of physically exchanging data among independently designed databases with different data formats.
- Data model heterogeneity refers to the issue of databases using different data models.

While these challenges are clearly interrelated, the first three have a direct impact on any attempt to integrate data from an information modelling perspective. In contrast, the latter two are more important from an infrastructure perspective and will be further discussed as part of data ingestion.

Structural, syntactic and semantic heterogeneity are all examples of the types of nuances exhibited by ecological data and help reflect the type and depth of contextual knowledge required in order to make a published data set reproducible. In terms of data integration, a user would generally require access to this knowledge in order to map the various concepts to a common basis. From a repository perspective, the challenge is therefore to create tools that allow for this concept mapping to happen transparently and efficiently. Importantly, every effort must be made to ensure that the underlying meanings are not distorted in the process, as this would in itself constitute a form of data entropy (Michener et al. 1997). It must also be acknowledged that the underlying information model may itself need to evolve over time. As such, a key challenge in this space is to ensure that the information model (and associated tooling) can be restructured to best fit the available data rather than distorting the data by forcing it to fit a sub-optimal model.

Before integration can occur, data sets need to have a way of being deposited into the repository along with any additional pertinent information, a process we refer to as data ingestion. In this respect, challenges associated with both technical and data model heterogeneities are important. The net result of this is that individual data sets can be expected to be stored in numerous different ways. The resulting challenge is therefore to develop a flexible approach for connecting to and coupling with foreign data systems. For one-off data transfers, this could be achieved using relatively ad hoc processes; however, many data sets are likely to be actively curated meaning that data transfer processes (once established) should be largely automated to allow for periodic refreshing of the data.

Given that a repository needs to store more than just the data, in the sense that essential contextual information needs to be stored alongside it to ensure it is independently understandable, it is necessary that any ingestion process take this into account. This raises additional challenges primarily because (as we described earlier) primary data users don't often explicitly describe all of the pertinent information as part of their practice. Even where this is the case, it will often be necessary to access the information in different formats – which is necessary in order to undertake data enrichment.

Data enrichment refers to the process of attaching important contextual information onto the data. The traditional approach is for repositories to provide metadata, normally as a separate document that travels along with the data set and is no doubt a beneficial step towards reproducibility, especially if such a document supplies sufficient contextual detail for a secondary user to understand and interpret it.

We have already discussed the challenges associated with acquiring suitably rich context from data creators as well as those associated with its interpretation from a secondary user's perspective. The remaining challenges relate to how best to represent the information to maximize the efficiency of the system. While this may seem like a 'value add', it is actually growing in importance with the volume of data available as part of the deluge. The result is that users are now requiring more sophisticated ways of being able to interact with data in order to quickly assess reproducibility. As an example, a user that is interested in soil observations may want to review any data sets that contained a soil component and not just those where it was the primary focus. The problem in this respect is that the key contextual information may be buried within a much larger document (and given less importance by the author), making it more difficult to find what they need therefore leading them to discard potentially useful data. The challenge of this is to find a way to store and represent contextual information at a level of granularity that best meets the needs of the user community. How this is subsequently presented to the users is an interface challenge.

The strength of a repository that overcomes the challenges previously mentioned can only be realized if it is matched with a user interface that ties the various components together in a way that meets the users' needs and expectations. Repositories therefore need to ensure that the resulting 'portal' fits in with the user's workflow and is intuitive to use. While this may seem self-evident, a lot of data management systems are designed from a database and/or information technology perspective (Bach et al. 2012). In contrast, a lot of 'typical' users of ecological data frequently lack these high-end skills. This creates a number of infrastructure challenges as the volume and structure of the data clearly lend themselves to the need to undertake complex operations and thus would be ideally suited to sophisticated query and data manipulation tools. Given that users are likely to use such systems infrequently, it is unrealistic to believe that they will learn these tools (and remember them between visits). The challenge is therefore to still enable complex operations but do so via a simple and intuitive interface.

Similar challenges also exist in the extraction space. For example, when working on their own systems, users often store data in a structure that mimics their intended analytical process (and often in spreadsheets) rather than forms that lend themselves to efficient curation and re-processing. As a result, they are often unfamiliar with the tools necessary to manipulate data within databases. Thus, many users will need to be able to download the data in simple formats if they are to be able to use it. Of course, this is not that straightforward as the data are generally structured in a way that better lends itself to either a graph-oriented or relational model.

When examined in this way, it becomes apparent that there are many challenges that need to be overcome in order to better support ecological data publication with a focus on intelligible reuse. We present in the following text our approach to address many of these challenges.

13.2 The *ÆKOS* Approach to Support the Intelligible Reuse of Ecological Data

13.2.1 Solving the Business and Information Challenges

ÆKOS was designed as a repository to support the publication of ecological data. In this respect, we decided to focus our attention on rich plot-based data on the basis that this was a recognized gap in Australia and despite it being clear that it was arguably the most challenging space to achieve progress in. To limit the scope, we also chose to largely ignore several other types of ecological data (at least from the initial prototype). First, biodiversity data (species by location observations) were deemed out of scope because this type of data is already aggregated in a national repository through the Atlas of Living Australia (ALA), a member node of GBIF. Second, spatial and gridded data are also already available via several thematic national government and research repositories. This type of data is also suited to a different style of infrastructure, so that combining them in a single system seemed to be counterproductive. Similar arguments were also used to exclude time-series sensor data. In all cases, this allowed us to better focus our solution and at the same time avoids duplication of effort with these other initiatives.

To build *ÆKOS*, we adopted an adaptive strategy, which was necessary at the time as we did not have a complete understanding of the scope of the problem or how the resulting implementation would look. The overall approach was to identify design requirements based on the needs of the user community. To this end, we established user reference groups and additionally solicited feedback through a range of other channels (including questionnaires, feedback buttons on the portal, product demonstrations to research groups). Implementation and feedback was an iterative process and as requirements became clearer, so did the challenges described in the previous sections. Several innovative approaches were prototyped and tested by the end users with the most promising design elements incorporated into the emergent design. Taking this approach minimized the risk of failure, meaning that we avoided unproven technologies, as there was a risk they wouldn't scale to production levels or alternatively would not be supported in future. We also chose flexible directions that kept as many options open as possible.

We adopted several additional fundamental principles consistent with addressing the challenges associated with publication of reproducible ecological data with the overall goal of thereby facilitating its reuse. First, given the complex and context-sensitive nature of the data, publication was considered more of a knowledge transfer challenge than simply a data transfer challenge. While we expand upon this in more detail later, this fundamentally means that the data and important contextual information need to be coupled and thus considered together as complimentary elements of knowledge. With this in mind, the second principle was to then present all data and information as fully as possible because every user will have different needs of the data set. While we can obviously determine some of what would be considered important knowledge, we cannot predict every specific use case. Similarly, it was also considered important not to change any of the underlying data or information and instead preserve an exact copy of what was received from data creators. Thus, the third principle was that any manipulation of the data needed to happen in a way that was reproducible and hence transparent to the user community. The actual mechanism used to do this is described in the following text. Finally, in order to maximize usability, the tools we built needed to easily fit with users' scientific practice, focusing on generating efficiencies and benefits for them rather than expecting them to adapt to the system.

13.2.2 Opting for a Centralized Service

Many of the challenges associated with publishing ecological data to a standard that renders it independently understandable have been described earlier. To best address these varied challenges, a decision was made to use a centralized model whereby key data management activities such as data processing and conceptual modelling would be undertaken by the TERN's Eco-informatics team. Centralization enabled quality controls over metadata to be implemented without the influence and governance challenges that would be present in a distributed model. It also ensured a level of quality assurance, enhanced consistency and uniform publication, which together addressed several of the challenges faced by secondary users through improvements to overall usability. Critically, it enabled a coordinated approach to dealing with the myriad of challenges associated with heterogeneity. More information on how this was achieved is described in the next section.

By centralizing the knowledge transfer of the data, the role of data creators is reduced to that of subject matter experts maximizing their value to the process and minimizing their additional workload. Knowledge transfer specialists manage all other technical steps associated with publication internally. Complex skills required for information processing therefore become a shared expertise, developed and applied in a uniform way. The approach enables data creators to remain involved in the process while minimizing the overall burden which is important given they are typically poorly sponsored for this work and also are unlikely to have appropriate capacity given that informatics lies across the boundary of several sub-disciplines. A published example of where this has been used effectively is for TERN's AusPlots program, which collects ecological plot data across Australia using standard protocols. By transferring the data publication role to *ÆKOS*, they are able to better focus on building sophisticated data capture and curation tools (Tokmakoff et al. 2016).

13.2.3 Implementing Dynamic Infrastructure

We stated earlier that using a centralized approach provides the opportunity to handle the challenges associated with heterogeneous data in a coordinated way. The key to accomplishing this rests in the use of dynamic infrastructure comprised of five key components that together optimize the importation and management of knowledge (data coupled with contextual information) within the system: (1) knowledge transfer tools, (2) the information model, (3) data enrichment, (4) knowledge representation model, and (5) the *ÆKOS FIXER* language.

13.2.3.1 Knowledge Transfer Tools

The highly heterogeneous nature of source data requires that the mechanisms used to access and import them need to be highly flexible. This is accomplished through a combination of tools and the use of knowledge transfer specialists. Information transfer relies heavily on templates that are designed to capture important contextual details in a standardized way. Templates exist for methods as well as overall project design and are generally populated by the knowledge transfer specialists in consultation with data creators to ensure that key information is captured uniformly.

Data transfer is undertaken using a software tool that is designed to run on the data creator's infrastructure and can be configured to take a regular 'snapshot' which is then compressed and encrypted and sent via the internet to the *ÆKOS* servers. The use of snapshots allows data updates to be transferred in an entirely automated way. Given that the data model and storage infrastructure varies, the role of the data transfer specialist is to undertake the initial configuration in consultation with the data creator. A plain text scripting language is used so as to be transparent to the information technology administrators in order for them to be able to assess and be confident that the software is not undertaking any malicious activities. This is particularly important in cases where the software is installed on large institutional systems such as government agencies. The underlying principle of the data transfer tool is that it captures whatever data are fed to it and simply transfers it. This model makes the tool robust to any changes to data structure that the custodian may inadvertently make. The result is that if changes to the data structure are to cause problems then the process is most likely to break down once on *ÆKOS* infrastructure where it can be more readily diagnosed and addressed. Similarly, when modifications are required to the *ÆKOS* information model they do not require the data transfer tools to be updated.

13.2.3.2 Information Model

At the heart of *ÆKOS* is an information model, which includes a formal ontology and vocabulary. The ontological basis of the model provides a platform to support data integration from multiple and disparate sources (Madin et al. 2008) and enables further integration with other data collection and archival platforms. This model also defines all of the concepts within the system, relationships among terms (e.g. synonyms, preferred terms) and any other rules or constraints for their use within the system.

The information model represents a common basis for the representation of all knowledge within the system and all incoming data are mapped into this form. As new data sets are added, the existing model can be reviewed and extended or modified as necessary, enabling it to evolve over time. The model also contains necessary processing and handling instructions for the system itself. As an example, quantitative measurements are always displayed with a value and corresponding unit to remove any ambiguity (i.e. 10 m).

13.2.3.3 Data Enrichment

The complex, nuanced nature of ecological data means that effort is required to describe and interpret the context of observations. With this in mind, we see ecological data as part of a broader information landscape, all levels of which need to be described in order to transfer sufficient knowledge to understand the data. *ÆKOS* employs several approaches to provide adequate enrichment of the data (Figure 13.2).

In general, we found that breaking information down into smaller fragments improved the speed of comprehension because users can quickly identify immediately pertinent details. Attaching this information as closely as was practical to the relevant observation data further improved user experience. For example, if the user is examining a particular aspect of the data, then it helps if they can directly access the relevant protocol associated with its derivation. To achieve this, pertinent contextual information is directly attached to the data to form an observation unit that more completely describes the observed entity and its originating process (Figure 13.3).

Related data are connected together and annotated with semantics so as to link various observations associated with a project into an observation set. Linkages can be further extended to draw in details concerning related knowledge. A good example here is where an observation is recorded against

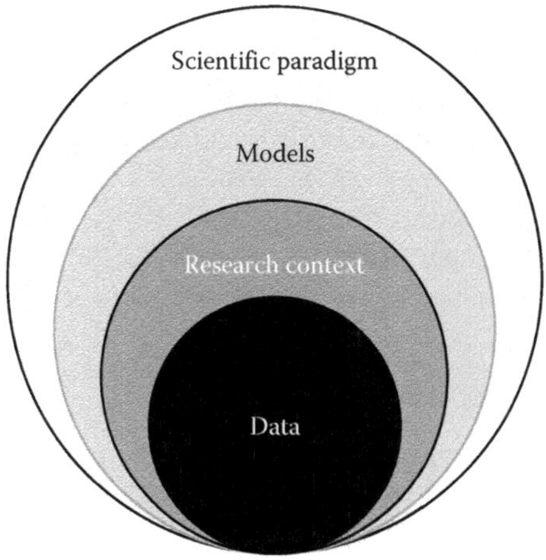

FIGURE 13.2
Levels of knowledge for enrichment of data in *ÆKOS* include data (entity, attributes, values), research context such as the sampling methods, models of information such as classification systems like taxonomy and the overarching scientific paradigm of data creators.

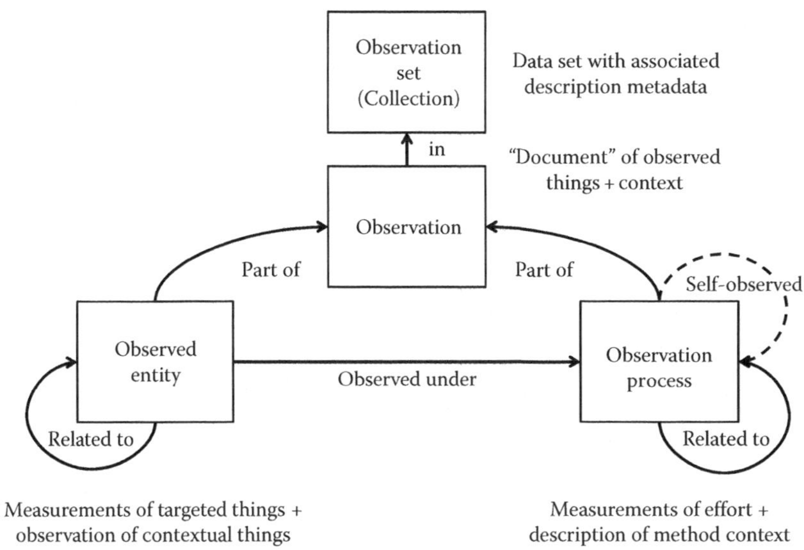

FIGURE 13.3
ÆKOS adopts the notion that context should be embedded with and not separated from observation. We therefore tie the entity being observed with the process that led to the observation. This is important as many observations in ecology are context specific; therefore, embedding processes in this way ensures that users have access to the knowledge during any examination or interpretation.

some form of classification, in which case links can be drawn to the associated values and underlying definitions.

Higher-level information such as that associated with sampling design, measurement protocols and overall project objectives are handled using description templates because information of this type is primarily targeted at human readers and not for machine processing. The use of standard templates allows readers to learn document structure in order to quickly find the required information.

13.2.3.4 Data Representation

The *ÆKOS* information model was developed in order to provide a common basis for the representation of observation data from different sources and is implemented within the system using flexible knowledge representation. Our approach to this problem is similar to publishing a set of books discussing overlapping topics, written by different authors in different languages. Using this analogy, our job is to translate the set of disparate books into a common language and style to make a monograph series (Figure 13.4a and b). Data can thus be represented in a structure that more closely matches what was actually observed or measured rather than a more abstract form such as data tables (Figure 13.4c).

The system uses a graph-oriented storage approach to represent a study site, which is a familiar concept to ecologists. Under this model, observations made at a single location (i.e. a plot) as part of a program are grouped together providing an intuitive level of organization. Focussing on the study site is in direct contrast to most other online repositories that use the data set as the basic unit and that generally store data in a format that is 'opaque' to the user until such time as it is downloaded. The benefit of the *ÆKOS* information system approach is that users are able to search for and interact with information on individual study plots and even directly view individual data records (Figure 13.4c).

Flexible knowledge representation also facilitates integration of data where appropriate. The challenge of course is that many concepts in ecology may appear similar but are not identical due to differences in the underlying processes that led to their origin. As an example, consider an observation that records the diameter of a tree. Given that most tree trunks are not perfect cylinders, it is obvious that the recorded diameter will depend on the height of the measurement as well as how the diameter was determined (i.e. maximum, minimum, average and others). Thus, while two different researchers may set out to measure fundamentally the same characteristic, if they apply different measurement standards, then the resulting concepts will be similar but not synonymous.

Whether the distinction between two similar concepts is important needs to be determined on a case-by-case basis and will depend on how the data are intended to be used. As such, the decision needs to be made by the

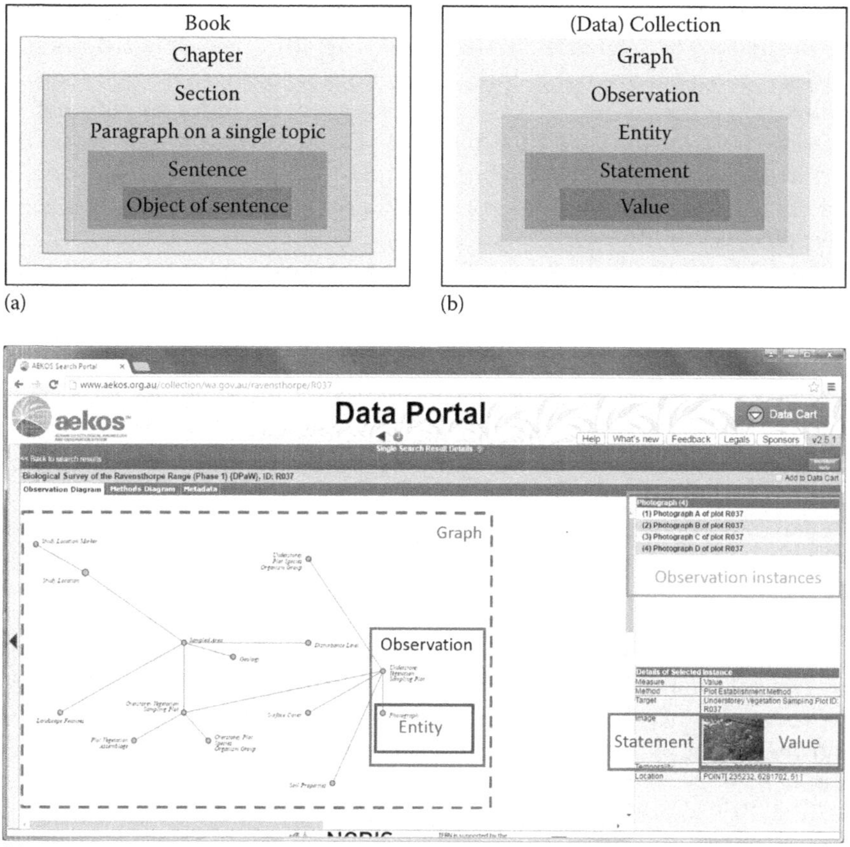

FIGURE 13.4
ÆKOS knowledge representation is similar to that of a book which is comprised of a large number of sentences arranged in an orderly way into higher-level constructs including paragraphs sections and chapters (a). The key structures in *ÆKOS* are statements that provide context to a value and that pertain to entities (generally the target of one or more discrete observations). These in turn are arranged into observations which reflect the ecological concept of a sample (b). These relationships are then represented graphically within the *ÆKOS* portal (c), allowing users to view plot-level information in a way that matches the way they collect primary data.

secondary user and not *a priori* by the system. Interpreting the concepts behind the data is often challenging for secondary data users because the concepts are frequently underspecified or imprecisely described. The result is that it is difficult to determine if there is a difference and if so what the magnitude of this difference is and whether it requires special handling.

In implementing *ÆKOS*, we align concepts as closely as possible but do not merge them unless they were truly synonymous. By then providing rich context to the observations, users are now able to recognize and assess any nuances present within the data. An exception to this rule is where we are trying to facilitate data discovery. In this case, small differences between concepts are not helpful and tend to complicate search requests. To overcome this issue, *ÆKOS* employs a separate index model that targets key traits of the data (that users are likely to want to search against) and deliberately reduces variability. While the resulting concepts are not strictly faithful representations of the underlying data, they vastly improve their discoverability.

13.2.3.5 The ÆKOS FIXER *Language (Instruction, Transform and Enrichment)*

To make data ingestion efficient, we opted to use a scripting language to manage the full process. All necessary instructions were stored together in a single text readable format and then executed with a series of simple commands. Instead of adopting a generic language, we decided to build a domain-specific language which has been dubbed the FIXER (Federated Ingestion X[Trans]fer and Enrichment Ruleset). The customized language meant that the structure and terminology employed could be optimized for the task at hand. Importantly, it also meant that the language could be designed in a way that was more intuitive to ecological domain data experts rather than requiring specialist IT training. As a result, the full data ingestion process can be undertaken by knowledge transfer specialists and treated in an efficient and consistent way to ensure a higher-quality product.

The *FIXER* handles all aspects of data ingestion. This includes definition of the information model, instructions to map a new data source to the common framework and all enrichment. As a result, re-running scripts is straightforward. By editing the scripts, the system can easily accommodate changes to both the model and originating data. The use of the scripts also aids quality assurance because all aspects of the ingestion process are recorded in a transparent and reproducible way (Figure 13.5).

13.2.4 Facilitating Reuse via the Data Portal

13.2.4.1 Discovery: Data, Metadata and Methods

A key challenge here is that ecological data are currently widely dispersed, meaning it is housed in a multitude of separate storage systems, the majority of which are not actually discoverable and/or accessible via the internet (Reichman et al. 2011). One of the drivers behind our initiative was to partner with data creators and primary data users (and initially those with large data holdings), who had limited or no online data presence, and work with them to publish their data via *ÆKOS*. By aggregating data into a single system, users can search a large number of data holdings via a single

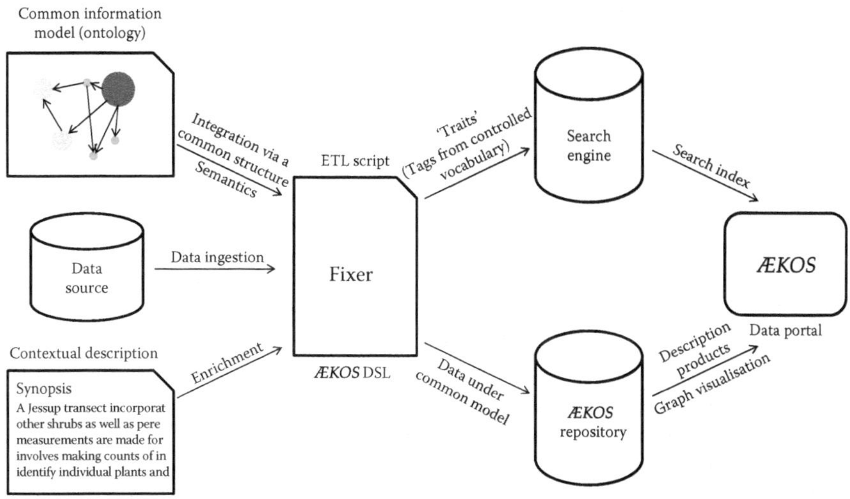

FIGURE 13.5
ÆKOS integrates data by mapping from a source to our common information model and attaching enriched contextual knowledge as close to individual data points as practical. To make the process transparent and reproducible, a specialized scripting language (FIXER) is used to manage the process. The output of this is an enriched data artefact that is stored in the ÆKOS repository at a study location level. A secondary artefact squashes out excess variability and feeds into the indexing model to facilitate powerful searches.

interface. *ÆKOS* is clearly not unique in this respect and there are a growing number of online data repositories. Enabling users to interact with larger, well-organized repositories is an improvement on past practice. Having a range of sophisticated domain-level repositories represents a good bridge between smaller isolated data stores and a single monolithic system which would be impractical to maintain. Furthermore, by setting *ÆKOS* up to on-publish rich metadata, initiatives such as *DataOne* can harvest these and provide users with a convenient entry point across multiple systems.

ÆKOS employs a range of features to assist users to make data discovery efficient. We see discovery as a combination of search and initial review. In this respect, search is essentially a process whereby users set criteria, which can be used as a basis for filtering results.

In the most basic sense, a lot of repositories work by storing a metadata document for each data product. The search engine scans these for 'key terms' as specified by the user and matches are returned as the results for review. Although straightforward, this approach has several disadvantages. First, records are only matched if the user specifies the exact (or very similar) keywords or terms used in the metadata. If they understand the process, the user can partially compensate for this by repeating the search with different related terms, but this is not particularly efficient and the user never really knows if they have identified all potentially useful records.

ÆKOS in contrast uses its rich information model and employs controlled vocabularies and links concepts through semantic expansion, meaning that the system can automatically match synonyms and related terms. Each term is also explicitly defined to the user so as to remove any ambiguity as to what is being matched – an important requirement for the ecological science community.

ÆKOS also provides metadata to users as well as on publishing to other systems. The system does not adhere to any particular standard but instead exploits the rich information model to capture raw concepts. These can then be assembled by the system into a structure that conforms to a specific standard where required. Maintaining a rich model gives us the flexibility to convert the data holding into standard formats where required and at the same time not constraining them when capturing and storing knowledge. Currently, ÆKOS supports EML and Registry Interchange Format (RIF-CS), but we could easily map to other relevant standards.

A related challenge with free text searches is that a single term may have a different meaning depending on the context, leading systems to falsely match and return large numbers of irrelevant records. *ÆKOS* addresses this problem through the use of search facets that allow users to specify the context of the term. For example, a species search for a kangaroo will return the results pertaining to the iconic Australian animal, whereas the same term used in a geographic context will return different results (e.g. Kangaroo Island). Each facet in *ÆKOS* is then associated with a different search interface (called a 'picker') and is optimized for information of that type. Rather than simply relying on the content of metadata, *ÆKOS* extends its search capabilities into the actual data. This is possible because data within the system is integrated rather than being stored as opaque objects (e.g. zipped up data tables).

When combined, the previously mentioned approaches enable users to more precisely identify data of interest and minimize false matches. Nevertheless, depending on the number of criteria defined in a search, a large number of records may still be returned, requiring the user to invest time examining each. *ÆKOS* incorporates several features to make this process more efficient. Search results are presented in a standard format and incorporate a number of descriptive elements designed to convey useful information. The goal is to provide key details about the record in a way that can be rapidly assessed by the user.

13.2.4.2 Assessment of Reproducibility

If users are able to gain a detailed understanding of how data have been collected and subsequently manipulated, then they are in a good position to determine if it is suitable. As such, we see this assessment process fundamentally as a knowledge transfer or comprehension challenge. *ÆKOS* approaches this issue by providing initial information during the discovery phase. The goal here is to allow users to make a rapid assessment and differentiate potentially useful results from irrelevant results. Presenting information in a way that enables rapid assessment also greatly assists with overcoming many of the challenges associated with the data deluge. Once the user has a manageable list of products, they can target individual results in order to get a more detailed understanding. Rapid comprehension is facilitated by breaking information up into smaller units and presenting this to the user using different perspectives. In each case, key summary information is presented first allowing the user to make a judgement as to whether the data appear promising and therefore whether they want to invest further time in a more detailed assessment.

13.3 Summary and Next Steps

The challenges associated with the publication of ecological data are varied and in many cases can be traced to the heterogeneity of the data and in turn to the way that the discipline has traditionally operated. Here, we have attempted to outline many of the more problematic challenges that have been identified and in particular those relating to the interplay between heterogeneity and the context dependency of the data in building *ÆKOS*. The result is that significant information is required to make ecological data independently understandable, which invariably places a burden on data creators who don't perceive clear benefits for their investment. Without this investment, secondary users struggle to fully comprehend the nuances of the data, which either renders it unusable or, of more concern, leaves data open to misinterpretation and inappropriate reuse. From a technology perspective, the highly variable, sparse and semi-structured nature of ecological data also challenges many information technology storage paradigms.

Throughout the development of *ÆKOS*, we have taken a user-centric approach to understanding the data publication problem and have attempted to identify and solve many of the challenges facing the domain. *ÆKOS* is about bringing together high-quality ecological data and publishing it in a manner that facilitates intelligible reuse. The system has now been live for two years and is continuing to evolve as we work with the community to iteratively identify and address improvements.

ÆKOS adopts a study site–centric view of ecological data, a feature that sets it apart from many other repositories. Integrating data and focussing at the site level represent a new paradigm for data storage systems and is an outcome of our approach to information modelling, which focuses on enabling intelligible reuse. Thus, the focus on sites really reflects a manifestation of the underlying information model, which requires us to treat knowledge in a way that, as closely as practical, mimics the way the observations have been made and thus how they will be perceived by practitioners. *ÆKOS* is therefore able to store and represent knowledge in a granular way that approximates reality in ecological science practice rather than a more abstract data storage format, which greatly facilitates comprehension by secondary users through the use of an intuitive presentation format.

Comprehension is also improved by addressing some of the heterogeneity within the data. This is accomplished by presenting knowledge in an integrated way against a common model, but importantly, without distorting the original concepts. Furthermore, data are enriched and then bound with detailed contextual information to create units of knowledge that are then linked to related observations and project descriptors to enable the data set to become independently understandable from any reference point.

To date, *ÆKOS* has focused on solving the challenges associated with acquiring data sets from data creators, storing them in an integrated way in an online repository and presenting them to secondary users in a manner that supports discovery and reproducibility. While there are clearly opportunities to further improve these aspects, a key limitation of the system relates to data extraction.

Currently, users are able to extract data in two formats, the first utilizes a rich relational format suitable for importation into standard databases. The second generates a biodiversity perspective, which is essentially a cut-down version of the data that only includes details on the site and species observations via standard Darwin Core. As such, these formats represent two extremes, the former being far too complex (hundreds of tables) for most users while the latter is oversimplified (flat file consisting of a dozen columns) losing much of the richness of the data. We are currently working on an extraction wizard that will allow users to filter out concepts that are not of interest to them as it is rare that secondary users will require all of the data associated with a particular study. The wizard will also include the ability to combine concepts into fewer relational tables as well as support formats such as RDF.

We are also looking to support more advanced users through machine interfaces and via direct linkages to software packages such as R. It is anticipated that these types of interfaces would facilitate greater interoperability with other data systems and future virtual laboratories.

Acknowledgments

We thank Drs. T. Clancy and N. Thurgate for providing very helpful comments on earlier drafts and the editors for the invitation to contribute to this book. An anonymous reviewer also provided detailed comments and advice that resulted in an overall improvement in the quality of the manuscript. TERN is supported by the Australian Government through the National Collaborative Research Infrastructure Strategy (NCRIS).

References

Atici, L., S.W. Kansa, J. Lev-Tov, and E.C. Kansa. 2013. Other people's data: A demonstration of the imperative of publishing primary data. *Journal of Archaeological Method and Theory* 20 (4):663–681. doi:10.1007/s10816-012-9132-9.

Attiwil, P. and B. Wilson. 2003. Ecology in Australia. In *Ecology: an Australian Perspective*, edited by P. Attiwil and B. Wilson, pp. 1–12. New York: Oxford University Press.

Bach, K., D. Schäfer, N. Enke, B. Seeger, B. Gemeinholzer, and J. Bendix. 2012. A comparative evaluation of technical solutions for long-term data repositories in integrative biodiversity research. *Ecological Informatics* 11: 16–24. doi:10.1016/j.ecoinf.2011.11.008.

Bechhofer, S., I. Buchan, D. De Roure, P. Missier, J. Ainsworth, J. Bhagat, P. Couch et al. 2013. Why linked data is not enough for scientists. *Future Generation Computer Systems-the International Journal of Grid Computing and Escience* 29 (2): 599–611. doi:10.1016/j.future.2011.08.004.

Bell, G., T. Hey, and A. Szalay. 2009. Beyond the data deluge. *Science* 323 (5919):1297–1298. doi: 10.1126/science.1170411.

Costello, M.J., W.K. Michener, M. Gahegan, Z-Q. Zhang, and P.E. Bourne. 2013. Biodiversity data should be published, cited, and peer reviewed. *Trends in Ecology & Evolution* 28 (8):454–461. doi:10.1016/j.tree.2013.05.002.

Duke, C.S. and J.H. Porter. 2013. The ethics of data sharing and reuse in biology. *BioScience* 63 (6):483–489. doi:10.1525/bio.2013.63.6.10.

Ferrier, S. 2012. Big-picture assessment of biodiversity change: Scaling up monitoring without selling out on scientific rigour. In *Biodiversity Monitoring in Australia*, edited by D. Lindenmayer and P. Gibbons. Collingwood, Victoria, Australia: CSIORO Publishing.

Hampton, S.E., C.A. Strasser, and J.J. Tewksbury. 2013. Growing pains for ecology in the twenty-first century. *BioScience* 63 (2):69–71. doi: 10.1525/bio.2013.63.2.2.

Hampton, S.E., C.A. Strasser, J.J. Tewksbury, W.K. Gram, A.E. Budden, A.L. Batcheller, C.S. Duke, and J.H. Porter. 2013. Big data and the future of ecology. *Frontiers in Ecology and the Environment* 11 (3): 156–162. doi:10.1890/120103.

Hanson, B., A. Sugden, and B. Alberts. 2011. Making data maximally available. *Science* 331 (6018): 649. doi:10.1126/science.1203354.

Kepes, S., A.A. Bennett, and M.A. McDaniel. 2014. Evidence-based management and the trustworthiness of our cumulative scientific knowledge: Implications for teaching, research, and practice. *Academy of Management Learning & Education* 13 (3): 446–466. doi:10.5465/amle.2013.0193.

Krebs, C.J. 2012. Biodiversity monitoring in Canada's Yukon: The community ecological program. In *Biodiversity Monitoring in Australia*, edited by D. Lindenmayer and P. Gibbons, pp. 151–157. Collingwood, Victoria, Australia: CSIRO Publishing.

Lindenmayer, D. and P. Gibbons. 2012. Can we make biodiversity monitoring happen in Australia? Moving beyond "It's the thought that counts". In *Biodiversity Monitoring in Australia*, edited by D. Lindenmayer and P. Gibbons, pp. 193–202. Collingwood, Victoria, Australia: CSIRO.

Lindenmayer, D. and G.E. Likens. 2013. Benchmarking open access science against good science. *Bulletin of the Ecological Society of America* 94 (4): 338–340. doi:10.1890/0012-9623-94.4.338.

Lindenmayer, D.B., E.L. Burns, P. Tennant, C.R. Dickman, P.T. Green, D.A. Keith, D.J. Metcalfe et al. 2015. Contemplating the future: Acting now on long-term monitoring to answer 2050's questions. *Austral Ecology* 40 (3): 213–224. doi:10.1111/aec.12207.

Madin, J.S., S. Bowers, M.P. Schildhauer, and M.B. Jones. 2008. Advancing ecological research with ontologies. *Trends in Ecology & Evolution* 23 (3): 159–168. doi:10.1016/j.tree.2007.11.007.

McKiernan, E.C., P.E. Bourne, C.T. Brown, S. Buck, A. Kenall, J. Lin, D. McDougall, B.A. Nosek, K. Ram, C.K. Soderberg, J.R. Spies, K.Thaney, A. Updegrove, K. H. Woo, and T. Yarkoni. 2016. How open science helps researchers succeed. *eLife* 5: e16800. doi:10.7554/eLife.16800.

Michener, W.K., J.W. Brunt, J.J. Helly, T.B. Kirchner, and S.G. Stafford. 1997. Nongeospatial metadata for the ecological sciences. *Ecological Applications* 7 (1): 330–342. doi:10.2307/2269427.

Michener, W K., J. Porter, M. Servilla, and K. Vanderbilt. 2011. Long term ecological research and information management. *Ecological Informatics* 6 (1): 13–24. doi: 10.1016/lecoinf.2010.11.005.

Mills, J.A., C. Teplitsky, B. Arroyo, A. Charmantier, P.H. Becker, T.R. Birkhead, P. Bize et al. 2015. Archiving primary data: Solutions for long-term studies. *Trends in Ecology & Evolution* 30 (10): 581–589. doi:10.1016/j.tree.2015.07.006.

Peng, R.D. 2009. Reproducible research and biostatistics. *Biostatistics* 10 (3): 405–408. doi:10.1093/biostatistics/kxp014.

Reichman, O.J., M.B. Jones, and M.P. Schildhauer. 2011. Challenges and opportunities of open data in ecology. *Science* 331 (6018): 703–705. doi:10.1126/science.1197962.

Soranno, P.A., K.S. Cheruvelil, K.C. Elliott, and G.M. Montgomery. 2014. It's good to share: Why environmental scientists' ethics are out of date. *BioScience* 65 (1): 69–73. doi:10.1093/biosci/biu169.

Tokmakoff, A., B. Sparrow, D. Turner, and A. Lowe. 2016. AusPlots Rangelands field data collection and publication: Infrastructure for ecological monitoring. *Future Generation Computer Systems* 56: 537–549. doi:10.1016/j.future.2015.08.016.

White, R.L., A.E. Sutton, R. Salguero-Gómez, T.C. Bray, H. Campbell, E. Cieraad, N. Geekiyanage et al. 2015. The next generation of action ecology: Novel approaches towards global ecological research. *Ecosphere* 6 (8):art134. doi:10.1890/ES14-00485.1.

Zilinski, L.D., D.A. Scherer, D.M. Bullock, D. Horton, and C.E. Matthews. 2014. Evolution of data creation, management, publication, and curation in the research process. *Transportation Research Record* (2414):9–19. doi:10.3141/2414-02.

14

Comprehensive and Coordinated Approach of GEOSS to Ecosystem Challenges

Antonello Provenzale and Stefano Nativi

CONTENTS

Abstract

Terrestrial and marine ecosystems provide essential goods and services to human societies. In the last several decades, however, anthropogenic pressures are causing serious threats to ecosystem integrity, functions, and processes. In turn, ecosystem degradation and loss of ecosystem services can seriously affect human well-being and climate processes on local and regional scales, potentially amplifying the negative effects of global change. Knowledge-based conservation, management, and restoration policies are thus urgently needed in order to ensure the delivery of ecosystem benefits in the face of increasing anthropogenic pressures. The Group on Earth Observations Global Ecosystem Initiative (GEO ECO) uses newly available monitoring methodologies that combine approaches in geosciences and biosciences, remotely sensed data, and in situ observations to provide open-access information on the state and ongoing changes of ecosystems and ecosystem services. The best use is made of existing and future earth observations and field monitoring data, complemented by appropriate interpretation tools and data services, as well as ecosystem models able to use these data. The GEO ECO combines new activities developed in the framework of the European H2020 ECOPOTENTIAL and Satellite-Based Wetland Observation Service projects, with continuing global ecosystem mapping activities carried forward from the former GEO ecosystems task from the first decade of GEO and Global Earth Observation System of Systems (GEOSS), in collaboration with the U.S. Geological Survey, Esri, and a number of international ecosystems experts. In these endeavors, all data, scientific results, models, and information will be made accessible and available through a system of cloud-based open platforms implementing virtual laboratories. Such platforms will be a major contribution to the GEOSS common infrastructure, reinforcing the GEOSS Data Collection of Open Resources for Everyone and in harmony with the Open Data global vision.

Keywords: Global ecosystem monitoring, Ecosystem services and processes, Geosphere–biosphere interactions, Open data

14.1 Introduction

Terrestrial and marine ecosystems provide essential goods and services to human societies (Daily 1997). Over the past decades, however, anthropogenic pressure has caused serious threat to ecosystem integrity, functions, and processes that potentially lead to habitat degradation; creation of uncertainty related to anthropogenically made ecosystems, that is, "novel ecosystems"; and the increased risk of ecosystem collapse (Hobbs et al. 2009,

Keith et al. 2013). Ecosystem degradation and loss of services can seriously affect human well-being and economies and potentially amplify the negative effects of global change on local and regional scales (http://www.unep.org/ maweb/en/Framework.aspx).

In addition to scientific discovery, knowledge-based conservation, management, and restoration policies are urgently needed in order to continue to provide and improve ecosystem benefits in the face of these increasing anthropogenic pressures. In order to do this, effective monitoring of the state and trends in ecosystem conditions and services are also needed (MAES 2014, Tallis et al. 2012, Spalding et al. 2013). New monitoring methodologies are now available that combine approaches in geo- and bioscience, remote sensed data, and in situ data (CBD 2014). Till now, however, the use of remote sensing results in the assessment of ecosystem state and changes has been limited by the lack of specific data products that could be easily used for decision-making, that is, natural resource managers, planners, and field personnel (e.g., in natural parks). To achieve maximum benefit from the existing and incoming (new) earth observation (EO) data and other in situ monitoring data, appropriate interpretation tools, data services, and ecosystem models should be developed and used in combination for stakeholders and users outside the research community. Hence, knowledge must come from the codevelopment of data, its information, and data analytics, with the relevant stakeholders to identify the required research outputs and support the use of new data and tools (Stocker 2015). Finally, synergies must be sought among international and European initiatives, projects, and networks, in order to avoid duplication of work and make best use of the available time and funding resources. In the following sections, we address how such challenges are addressed in the framework of the Group on Earth Observations (GEO) and of the large European project "ECOPOTENTIAL." In this project, data, information, and knowledge are managed by applying the DIKW (data, information, knowledge, wisdom) model (Zins 2007), which is about understanding and connecting: information is an added-value product generated by understanding data and working out relations among them and with physical and/or social phenomena, while understanding information and working out valuable patterns generate knowledge.

14.2 GEO View

The GEO was established in 2005 as a voluntary partnership of governments and organizations that envisions "a future wherein decisions and actions for the benefit of humankind are informed by coordinated, comprehensive, and sustained Earth observations and information." To meet this goal, GEO is creating the Global Earth Observation System of Systems (GEOSS) that will link EO resources and data worldwide across multiple societal benefit areas

(SBAs) and make those resources and data available for better informed decision-making and scientific understanding. There were originally nine SBAs: one such SBA was devoted to the global monitoring of ecosystems, and others to agriculture, water, climate, and biodiversity (during 2005–2015). Owing to the scientific and thematic overlap and interactions among the ecosystem and biodiversity SBAs, from 2016 onward, the ecosystem and biodiversity SBAs have been joined into a single biodiversity and ecosystem sustainability SBA. Other changes from the 2005–2015 SBAs are the establishment of disaster resilience, energy and mineral resources management, food security and sustainable agriculture, infrastructure and transportation management, public health surveillance, sustainable urban development, and water resources management SBAs. The choice of these areas emerged from several discussions and meetings between scientists and stakeholders from developed and developing countries that took place in 2015 and identify the most important areas for which data and knowledge access are crucial that the previous group of SBAs did not specifically address. Moreover, partners, particularly those in developing countries, wish to include more practical applications in the SBA structure. In parallel to this general SBA structure, and in order to provide an efficient and sustainable structure for the dissemination of knowledge and results, from 2016, GEO will be organized around a set of community activities, flagships, and initiatives (see www.earthobservations.org), which must be supported by funded projects over a time scale of a few years. GEO community activities intend to codevelop the support system needed for information management, data distribution and sharing; GEO flagships represent an operational stage for the distribution of data and information on a specific issue, while GEO initiatives are activities at a preoperational stage leading to institutional integration and collection of the relevant data and knowledge.

One of such global initiatives is the GEO ECO, the GEO Global Ecosystem Initiative. The goal of this initiative is to collect data, information, and knowledge (including model outputs and model codes themselves) on specific ecosystem types, on their state and ongoing changes, and make them available through the GEOSS system of portals. Special emphasis is given to remote sensing data and products to provide spatially extended information on ecosystems. GEO ECO will work at two different levels: (1) a global mapping of ecosystem characteristics (to be discussed in the following text) and (2) specific information on relevant study sites and ecosystems (usually obtained from protected areas [PAs] where a large body of measurements is often available). In the early stage of development, special focus will be on mountain ecosystems, arid/semiarid environments, wetlands, coastal and marine ecosystems, followed by extensions to other ecosystem types.

14.2.1 Mountain Ecosystems

Mountain ecosystems, rich in endemic and endangered species, are directly linked to downstream regions through ecosystem goods and services, including benefits to watersheds, slope stability, discharge regulation, food and energy production, recreational services and options for tourism. Mountain ecosystems are "sentinels of change" as they are highly sensitive to the impacts of modifications associated with climate and/or land-use change. In addition, pilot sites in mountains integrate a spectrum of altitudinal zones and ecosystems in one PA. The role of mountain regions has been acknowledged at the UN Conference Rio + 20. The spatial heterogeneity of mountains exhibits methodological challenges for EOs (cloudiness, shade, etc.), making these areas excellent training grounds for the development of robust approaches. In Europe, mountainous PAs of international value exist in all climatic zones and latitudes. ECOPOTENTIAL will make use of this "natural experiment."

14.2.2 Arid Ecosystems (Including Semiarid)

Arid ecosystems (including semiarid) exhibit unique pathways of ecosystem function and specialized ecosystem services and represent life under extreme conditions. Such water-limited ecosystems can be vulnerable to the current impacts associated with global change. According to climatic projections, large areas in southern Europe are exposed to the risk of facing significantly drier conditions, and collapse of previous ecosystem functioning can occur as a consequence of increased climatic variability. Here especially, uncertainties are high about future ecosystem behavior. In water-limited ecosystems, temporal variability must in particular be addressed by EO and field data. Hence, we see a strong contribution of these sites to improving the monitoring of temporal dynamics in drylands, a biome that is home to some 2.3 billion people worldwide (http://web.undp.org/).

14.2.3 Coastal/Marine Ecosystems

Coastal/marine ecosystems are "an integrated and essential component of the Earth's ecosystem and are critical to sustaining it" (Rio + 20 outcome document The Future We Want, 2012). Rio + 20 also noted that the health of oceans and marine biodiversity are negatively affected by the impact of human activities, leading to a loss of biodiversity, decreased abundance of species, damage to habitats, and loss of ecological functions and ultimately ecosystem services [14]. Countering these threats is only possible through sustained monitoring and development of indicators to inform policy makers and coastal/marine managers. Coastal areas are transition zones between ecosystems that are of extreme importance for biodiversity and for the exchange, migration, and refuge of species with complex habitat requirements. In consequence, we see this category of ecosystems and pilot sites as a representative for approaches that focus on capturing the mobility of organisms within and between ecosystems.

For such ecosystems, available data and metadata will be collected and made available when possible, models will be developed and implemented, and information will be collected on specific websites, and portals linked with the GEO portals.

The GEO activities are based on voluntary contributions and no direct funding from GEO is available for the initiatives and flagships. Thus, the support should come from externally and independently funded projects. The European Union is supporting GEO through a set of calls that explicitly indicate that the selected projects should be conducted in the framework of—and provide support to—the GEO activities. In 2015, two H2020 projects focused on the use of EOs (both remote sensing and in situ) for the assessment of ecosystem services were selected by the reviewers of the European community and funded (Call SC5-16-2014). The two projects are ECOPOTENTIAL, a large consortium focused on the assessment of mountain environments, drylands, transitional coastal lagoons, and large marine ecosystems (LMEs), which include >25 European and non-European PAs of international relevance, and Satellite-Based Wetland Observation Service (SWOS), a smaller project focused on developing an operational, remote sensing–based, wetland observation service in support of international conventions, regulations, and policy frameworks. ECOPOTENTIAL includes a programmatic emphasis on macrosystems ecology, cross-scale interactions, and coupled geosphere–biosphere processes, and it has the goal of building a GEO Ecosystems Community of Practice, that is, a community of stakeholders that has codeveloped the analytics and best practices, and users trained in the use of the GEOSS system of portals.

In addition to the H2020 activities, global ecosystem mapping work (which began at the inception of GEO and GEOSS) is being carried forward and included in GEO ECO. In partial satisfaction of the former GEO ecosystems activity to "map standardized, robust, and practical global ecosystems for terrestrial, freshwater, and marine environments," a new global terrestrial ecosystems map was produced in collaboration with the U.S. Geological Survey (USGS: Roger Sayre and coworkers), Esri, and a number of international ecosystems experts. This new global ecological land units (ELUs) product is a first-of-its-kind, globally comprehensive, high-resolution, and data-derived characterization (http://rmgsc.cr.usgs.gov./ecosystems/pubs.shtml). While the global terrestrial ecosystems map is now completed, the global marine and global freshwater ecosystem maps are still outstanding. A major collaboration is now underway to produce a first-of-its-kind, 3D global ecological marine unit (EMUs) map in fashion to the ELUs. The global EMUs map will be developed as a short-term (1–2 years) foundational activity of the GEO ECO initiative, and an analog global ecological freshwater unit (EFU) map will be advanced as a longer-term (2–3 years) GEO ECO activity. It is now to the research and user community to get engaged in exploiting such wealth of information, and GEO ECO is ready to help along these lines by organizing training courses and workshops on the use of EO data in ecosystem mapping and characterization.

The scope, timeframe, significance, and resourcing of the two European projects and the two global ecosystem mapping projects (marine and freshwater) are consistent with the nature of the concept and process for developing new GEO initiatives in its second phase of development (2016–2025). Significant H2020 funding for these two European projects and significant in-kind support available for the global ecosystem mapping efforts from Esri are promising indicators of the commitment and likelihood of success for these activities in GEO ECO.

Based on these existing perspectives and results, the GEO ECO initiative intends to build upon available results on a regional or continental scale (e.g., from the ECOPOTENTIAL project) and extend them to a global scale, identifying PAs of international relevance where the same methodology used in ECOPOTENTIAL can be applied. Parallel to this, GEO ECO intends to support the efforts of extending and improving the ELU, EMU, and EFU maps currently in development.

14.3 The European H2020 ECOPOTENTIAL Project

To characterize the current state and the ongoing and expected changes in biodiversity, ecosystem functions, processes, and services, the transdisciplinary European H2020 ECOPOTENTIAL project adopts a blend of EO data and services, in situ monitoring and measurements, and modeling development, including the estimate of cross-scale interactions and future change scenarios (standard IPCC RCP scenarios coupled with available land-use change scenarios).

To tackle the scientific goals of ECOPOTENTIAL and other GEO efforts, we face several challenges:

A first issue concerns the large amount of available ecosystem data (the "Big Data" problem, see Heffernan et al. 2014), from both remote sensing devices and in situ monitoring activities, which should be used and interpreted. Such data come in diverse formats and provide different information and should be harmonized and made available through efficient and easy-to-use platforms that can serve expert users (scientists, technical staff), the stakeholder and policymaker communities, and concerned citizens. ECOPOTENTIAL will develop a suite of data access systems and virtual laboratories for this purpose, building on previous and current projects (e.g., FP7 GEOWOW, H2020 ConnectinGEO) and on the GEO/GEOSS brokering approach (see Section 14.4.4).

A second challenge involves how to address the emerging notions of cross-scale interactions and macrosystem ecology, which explicitly recognizes the presence of multiple and potentially contrasting spatial and temporal scales of ecosystem processes and their drivers. And as such, calls for the need to determine on which scales the information of process rates and their drivers are important to inform other scales, that is, upscaling and downscaling (Heffernan et al. 2014, Soranno and Schimel 2014). ECOPOTENTIAL will develop conceptual approaches to address macrosystem ecology and foster European (and global) research that will provide the context on transfer of information at one scale to different regional to continental scales.

The risk of losing essential ecosystem services and benefits is not just a scientific topic but has profound implications on the society and on economy (MAES Working Group 2014). For this reason, it is essential to develop solid science-policy interfaces to transfer scientific and technological knowledge into citizen information and policy strategies. ECOPOTENTIAL will build upon these needs and construct a system of combined policy options and of capacity building/outreach/dissemination activities at different levels, with the goal of defining and communicating the best strategies and actions for the improvement of ecosystem benefits and the definition of the needs of future PAs.

The ECOPOTENTIAL projects' activities and pilot actions target a set of internationally recognized PAs in Europe, European territories, and beyond; these include mountain, arid and semiarid, and coastal and marine ecosystems (Figure 14.1). These PAs such are chosen because they are exposed to a variety of pressures, which can change their ecosystem state (Marris 2011, Potts et al. 2014, and Chapter 10). ECOPOTENTIAL sites cover most European biogeographical regions and represent UNESCO world natural heritage sites, biosphere reserves, national parks, and important Natura 2000 sites. Additionally, two LMEs in the Mediterranean and the Caribbean are included. Many of the sites considered here exhibit a variety of protection categories. Finally, many PAs are also directly or indirectly linked to European Long-Term Ecological Research (LTER) sites (http://www.lter-europe.net/).

The variety of PAs included in the project, their different protection status, and the spectrum of involved ecosystems and biogeographical regions are a major strength of ECOPOTENTIAL. The project profits from the enormous range of ecological contexts that occur at the European scale and are aiming to create widely applicable knowledge and approaches that can be implemented in the future beyond the specific PAs of this project. It is impossible to realize such an ambition on national scales. This setting allows for addressing questions related to the status and ongoing changes of ecosystem

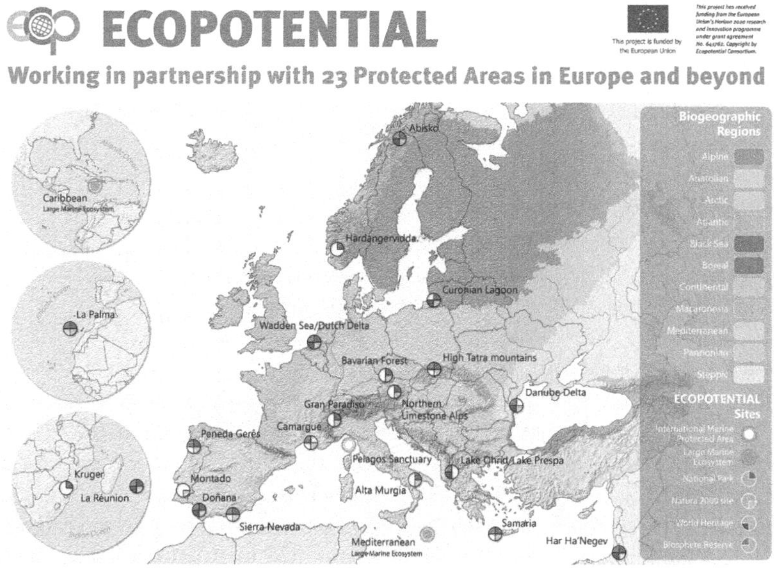

FIGURE 14.1
(**See color insert.**) Location and protection status of the PAs considered in ECOPOTENTIAL and European biogeographic regions.

functions and services in varied environmental conditions and for different protection levels. Based on this, the project will be able to define the needs of future PAs. A major strength of ECOPOTENTIAL is its focus on essential biodiversity variables (EBVs) and other essential variables (EVs) able to synthesize the required information (Pereira et al. 2013).

Building on the knowledge gained in individual PAs, the ECOPOTENTIAL project addresses cross-scale ecological interactions and landscape–ecosystem dynamics on regional to continental scales (Heffernan et al. 2014, Soranno et al. 2014). In this way, ECOPOTENTIAL will provide specific examples of the upscaling of local concepts to pan-European conditions, blending modeling efforts at the scale of individual PAs with large-scale ecosystem and climate models and providing a framework to assess changes in ecosystem services on the continental scale. ECOPOTENTIAL addresses the whole spectrum of ecosystem-related services, by (1) developing ecosystem data services, with special emphasis on Copernicus services, (2) implementing model output services to distribute the results of the modeling activities, and (3) estimating current and future ecosystem services and benefits, combining ecosystem functions (supply) with beneficiaries' needs (demand).

In ECOPOTENTIAL, all data, model results, and acquired knowledge will be made available on common and open platforms, coherent with the

GEOSS data sharing principles and fully interoperable with the GEOSS common infrastructure (GCI). ECOPOTENTIAL will be conducted in the context of the implementation of the Copernicus EO Component and in synergy with the European Space Agency (ESA) Climate Change Initiative. The project activities will contribute to Copernicus and non-Copernicus contexts for ecosystems and will create an Ecosystem Data Service for Copernicus (ECOPERNICUS), a new open-access geospatial data/products retrieval portal and web coverage service using a dedicated online server, which will enable users (from the scientific and management communities) to recover needed information on ecosystem state and changes.

ECOPOTENTIAL is developing strict links with the stakeholder, PA management, and policy-making communities by activating a permanent stakeholder consultancy group and involving PA managers, stakeholders, nature conservation associations, economic sectors, and concerned citizen groups into the definition and discussion of the crucial social and economic needs of local populations. In synthesis, ECOPOTENTIAL addresses the data-to-information-to-decision-making process for ecosystem services, using the interdisciplinary approach of earth system and natural sciences and building on biogeographical developments and the coupled dynamics of climate–landscape–ecosystem processes.

The main reason why the ECOPOTENTIAL project was established is that a large amount of remote sensing (mainly satellite) observations and in situ measurements (here collectively called earth observations) is now available, but till now the use of these data in the development of knowledge-based conservation and management policies has been rather limited. Thus, ECOPOTENTIAL's main goal is to develop a suite of EO products that can help scientists, users, policy makers and stakeholders in the study and management of ecosystems and ecosystem services. To do so, ECOPOTENTIAL opted for focusing its activities mainly in PAs of international relevance, both because these are the "family jewels" of the European environment and host important and endangered ecosystems and because a large quantity of in situ data is available from PAs, thanks to the extensive research activities taking place in those sites.

In summary, ECOPOTENTIAL's most important objectives are as follows:

- Make extensive use of EO data in combination with in situ monitoring. Data from existing archives and new missions (in particular, Sentinel sensors and VENµS) and from ground-based monitoring networks such as those of the LTER Network, the GEO Global Biodiversity Observation Network (GEO BON), the Global Biodiversity Information Facility (GBIF), and the Ocean Biogeographic Information System (OBIS) will be used and integrated to provide a picture of the state and changes in biodiversity, natural capital, and ecosystem services provided by key marine and terrestrial PAs.

- Create an ecosystem data service related to the Copernicus space component (ECOPERNICUS), providing a significant contribution to the research requirement for the Copernicus operational services and allowing simplified access to EO data and products and to ecosystem models' outputs.
- Create a corpus of innovative, field-tested, peer-reviewed, and documented monitoring methodologies to define the ecological status of current and future PAs, based on EO and in situ data.
- Develop a conceptual framework guiding the integration of data, models, and scenarios toward a new vision of ecosystem structure, change, and services. Refine the concept of EBVs, essential ocean variables, and essential climate variables, and define an overarching EV approach in combination with new essential water variables, essential geo variables, and essential social and environmental variables. Link existing and new EVs to EO data and in situ monitoring data.
- Develop new ecosystem models able to make the best use of EO and monitoring data, enhancing our knowledge on ecosystem nonlinearity, complexity, and uncertainties and predicting ecosystem changes in key PAs. The models developed in ECOPOTENTIAL will be able to assimilate EO and monitoring data using a suite of advanced mathematical techniques.
- Address the issues related to cross-scale interactions and landscape–ecosystem dynamics, including biological, geomorphological, climatic, social, and economic connections and emergent properties across scales. Special emphasis is devoted to the propagation of uncertainty across scales, carefully evaluating the different sources of uncertainty and the possibilities to reduce them.
- Quantify ecosystem services, taking into account social demand. Develop a knowledge-based decision support toolbox for management and ecosystem services, based on improved access to PA data and models, and implement an information system to assess the status of terrestrial and marine ecosystem services and resource benefits.
- Develop a list of requirements of future PAs. The scale mismatch between climate models and ecosystem dynamics will be addressed, implementing upscaling and downscaling procedures to obtain a European-wide survey of endangered ecosystems at different spatial and temporal scales.

- Make data, scientific results, models, and information accessible and available through a cloud-based open platform implementing virtual laboratories. This is the ECOPOTENTIAL Virtual Laboratory Platform. The platform will be a major contribution to the GCI, reinforcing the GEOSS Data Collection of Open Resources for Everyone (Data-CORE). By the end of the project, new prototype products and ecosystem services, based on improved access (notably via GEOSS) and long-term storage of ecosystem EO data and information in existing PAs, will be made available.
- Address, through participation of all players in scientific, social, and political decisions, key issues in ecosystem management. A permanent stakeholder consultancy group will be created, which will be kept active after the end of the project creating a GEO Ecosystem Community of Practice. This group will contribute to GEO BON and to GEO Blue Planet and interact with LTER-Europe, the Global Ocean Observing System (GOOS), Biology and Ecosystems (including EuroGOOS and MONGOOS), and OBIS.
- Improve evidence-based environmental policy making, enhancing administrative efficiency and contributing to transparent decision-making. The ECOPOTENTIAL results will directly benefit resource managers and active community stakeholders of selected terrestrial/coastal ecosystems and LMEs. Tested, evidence-based environmental policy making and administrative efficiency will be realized, as well as improved evidence-based environmental policy making and political decisions.
- Develop efficient capacity building at all levels, through training courses and user take-up of EO and monitoring data services and ecosystem models. Citizen science activities in local PA sites will actively involve rangers, wardens, and local citizens in collecting data about natural resources. Development of innovative/interactive outreach tools and dissemination activities will have strong focus on visual components, and networking will be designed in close collaboration with the users.
- Involve small and medium enterprises (SMEs) on EO and monitoring data recovery and services, capacity building, outreach and citizen science, development of virtual laboratories, and apps for the access to geoinformation to ensure commercial uptake and long-term continuation of activities. Involve the expert knowledge of large private companies/foundations/organizations, on ecosystem services and monitoring, fully integrating their contribution.
- Develop a strong European activity within the GEO ecosystem tasks and GEO global initiatives and support/implement the general vision endorsed by GEO/GEOSS.

14.4 GEOSS Information System across Scientific Domains

A central part of GEO's mission is to build the GEOSS information system*: a set of coordinated, independent EO, information, and processing systems that interact and provide access to diverse information for a broad range of users in both public and private sectors. GEOSS links these systems to strengthen the monitoring of the state of the Earth. It facilitates the sharing of environmental data and information collected from the large array of observing systems contributed by countries and organizations within GEO. Furthermore, GEOSS ensures that these data are accessible, of identified quality and provenance, and interoperable to support the development of tools and the delivery of information services. Thus, GEOSS information system increases our understanding of Earth processes and enhances predictive capabilities that underpin sound decision-making: it provides access to data, information, and knowledge to a wide variety of users.

GEOSS information system is intended to support human decision-making and action; therefore, it must be viewed as a work system where humans and machines perform processes and activities using resources to produce specific products and services for customers (i.e., GEOSS Users). In summary, the GEOSS information system is a combination of hardware, software, infrastructure, and expert people organized to facilitate planning, control, coordination, and decision-making in GEO. To build the GEO information system, the GEOSS program applies a "system of systems" (SoS) approach: it consists of developing a central GCI that, proactively, links together existing and planned information and processing systems around the world and supports the need for the development of new systems where gaps currently exist.

14.4.1 GEOSS Information System Community

In the past 10 years, a GEOSS community emerged including four main stakeholders, following a supplier–consumer pattern: (1) information and processing resource providers, (2) GCI component providers, (3) EO application developers, and (4) GEOSS end users (Figure 14.2). The GEOSS information supply system is comprised of the GEOSS resource providers and the GCI providers (i.e., ESA, USGS, CNR, IEEE), while the GEOSS consumers are the GEOSS application developers (e.g., downstream services SMEs) and the GEOSS end users (e.g., data scientists, global changes researchers, teachers, practitioners, decision makers, and citizens).

GEO has devoted considerable efforts to building the GEOSS information system that has made EOs discoverable and in part accessible (see the next

* http://www.earthobservations.org/geoss.php.

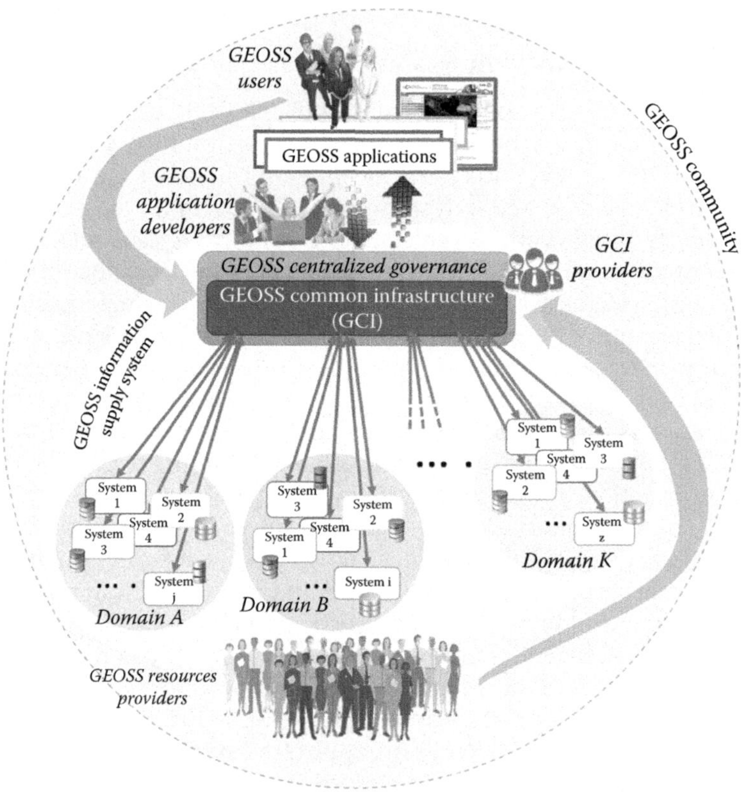

FIGURE 14.2
The GEOSS community.

section on GCI). This information system comprises a large set of resources providers:

Observation (data) systems: These include ground-, air-, water-, and space-based sensors, field surveys, and citizen observatories. GEO works to coordinate the planning, sustainability, and operation of these systems, aiming to maximize their added-value and use.

Information and processing systems: These include hardware and software tools needed for handling, processing, and delivering data from the observation systems to provide information, knowledge, services and products.

To continue leveraging these successes through 2025 and increase EO accessibility and (re)use, GEO will evolve GEOSS and the GCI to meet current and emerging needs by

Extending the user audience to decision makers and the general public

Placing additional focus on the accessibility and usability of EO resources to improve our scientific understanding of the Earth processes and enhance our predictive capabilities that underpin sound decision-making

Providing a service framework to engage partners and user communities in evolving the current infrastructure to enable collaborative tools for cocreation of products and services suitable for effective exploitation by user communities

Evolving the current SoS component-based architecture with an open-systems platform that is flexible, sustainable, and reliable for data access, integration and use, and the delivery of knowledge-based products and services

14.4.2 New GEOSS Information System Strategic Goals and Implementation Plan

GEO is constructing GEOSS based on cycles of 10-year implementation plans—the first plan covered the last 10 years, from 2005 to 2015, while the new period has just started and will end in 2025. Each implementation plan defines a vision statement for GEOSS, its purpose and scope, expected benefits, and a set of targeted "societal benefit areas."

The SBAs recognized by the new GEOSS implementation plan are (GEO 2016) (1) biodiversity and ecosystem sustainability, (2) disaster resilience, (3) energy and mineral resources management, (4) food security and sustainable agriculture, (5) infrastructure and transportation management, (6) public health surveillance, (7) sustainable urban development, and (8) water resources management.

The new plan builds on the following actions:

To implement the "GEOSS data sharing principles": GEO recognizes that the societal benefits arising from EOs can only be fully achieved through the sharing of data, information, knowledge, products, and services. GEO has therefore promoted fundamental principles for data sharing, expanding the trend toward open data worldwide. Thus, as it embarks on its second decade, GEO now aims to implement the following GEOSS data sharing principles:

Data, metadata, and products will be shared as open data by default, by making them available as part of the GEOSS Data-CORE without charge or restrictions on reuse, subject to the conditions of registration and attribution when the data are reused.

Where international instruments, national policies, or legislation preclude the sharing of data as open data, data should be made available with minimal restrictions on use and at no more than the cost of reproduction and distribution.

All shared data, products, and metadata will be made available with minimum time delay.

To implement the "GEOSS data management principles": To further maximize the value and benefits arising from EO data, GEO will continue to work with partners to promote the use of data management principles,* which are based on discoverability, accessibility, usability, preservation, and curation. These principles address the need for common standards and interoperability arrangements. This will ensure that data and information of different origin and type are comparable and compatible, facilitating their integration into models and the development of applications to derive decision support tools.

Stakeholders' engagement: A key tenet of GEO's vision is that EO data should serve societal needs. The value of these data is fully realized when it is transformed into useable knowledge and information to address societal needs. GEO will therefore convene key stakeholders across the provider–user spectrum to codesign a process to systematically identify and document EO needs for addressing specific problems within the scope of the SBAs. Recognized stakeholders include

United Nations agencies, treaties, and conventions

GEO members and participation organizations

Communities of practice

Private sector

GEO engages with stakeholder communities and acts as a broker, connecting users, data providers, engineers, scientists, and other relevant experts to create solutions to global challenges that transcend both national and disciplinary boundaries. Drawing on these partners, which encompass government departments, the private sector, civil society, and academia, GEO will implement a set of core functions essential for the realization of its strategic objectives (GEO 2016). The present strategic plan for implementing GEO considers the following core functions:

Identifying user needs and addressing gaps in the information chain

Sustaining foundational observations and data

Fostering partnerships and mobilizing resources

Advancing GEOSS and best practice in data management and sharing

Implementing sustained global and regional services

Cultivating awareness, building capacity, and promoting innovation

---

* https://www.earthobservations.org/documents/dswg/201504_data_management_principles_long_final.pdf.

These actions are implemented through four instrument types:

1. *GEO community activities*: They allow stakeholders to cooperate flexibly in a bottom-up fashion and with a low initiation cost.
2. *GEO initiatives*: They allow members and participating organizations to coordinate their actions and contributions toward a common objective within an agreed, yet flexible framework.
3. *GEO flagships*: They allow members and participating organizations with a policy-relevant mandate to spin up a dedicated operational service serving common needs and/or well-defined user groups.
4. *GEO foundational tasks*: They allow GEO to implement selected, often enabling, tasks to achieve the core functions' objectives. For instance, two foundational tasks are devoted to operate and evolve the GCI, while a dedicated task is building the GEOSS knowledge base.

To achieve its strategic objectives, GEOSS needs to be interdisciplinary, drawing on natural, economic, and social sciences and work with stakeholders to answer research questions and communicate recommendations for implementing solutions. Thus, for issues such as climate change, food security, water availability, and ecosystem services sustainability, it is increasingly recognized that we need to develop an integrated multidisciplinary approach to advance our understanding of the complex relationships between environmental and social phenomena (GEO 2009).

14.4.3 GEOSS Interoperability Needs and Implementation Approach

Berners-Lee declared in 2010 "the year open data went global." Since then, hundreds of nations, regions, and cities across the world have launched their own open data initiatives. This open data global movement is characterized by a philosophy and a set of practices of making the data collected by government agencies freely available to the public. Valuable examples are the USGS decision to adopt a policy (begun in 2008) of free and open access to Landsat data and the more recent European Union and ESA commitment to provide open and free access to the Copernicus (i.e., Sentinel satellites) data. In the framework of the Digital Agenda for Europe initiative, the European Union has been working for Open Data portals* to facilitate access to and reuse of public sector information. Open data portals are web-based interfaces designed to make it easier to find reusable information. Like library catalogs, they contain metadata records of datasets published for reuse, that is, mostly relating to information in the form of raw, numerical data and not to textual documents. Analogously,

* http://ec.europa.eu/digital-agenda/open-data-portals.

the Unites States launched the "data.gov"* portal to find data, tools, and resources to conduct research, develop web and mobile applications, and design data visualizations.

The open data global movement creates many opportunities for science to address climate changes' challenges developing an effectively integrated multidisciplinary approach. However, globally shared data need to be harnessed by a new breed of data infrastructures that are based not only on the interoperability of data systems for a specific domain area, but also on the interoperability of multiple disciplines in the physical and social sciences, engineering, and humanities (GEO 2007). For disciplinary and domain applications, systems interoperability largely deals with the adoption of agreed technologies, standards, specifications, and interfaces with a disciplinary/domain services protocol or means of information exchange, if available (GEO 2007). A domain infrastructure requires to be able to address domain resources (or components), achieving interoperability for observations and data models, service interfaces, processing schemes, terms, etc. According to a study of the European Commission (2006), an infrastructure interoperability encompasses at least three overarching and different aspects:

1. Semantics, which ensures that exchanged information is understandable and usable by any application or user involved.
2. Technology, which concerns the technical issues of linking up computer and information systems, the definition of open interfaces, data formats, and protocols.
3. Organization, which deals with modeling organizational processes, aligning information architectures with organizational goals, and helping these processes to cooperate. This category can also include important interoperability challenges, such as data policy, legal, cultural, and people harmonization.

However, multidisciplinary efforts make more complex demands on the type of systems and arrangements needed to support cross-domain activities (GEO 2007). Interconnecting existing disciplinary systems has traditionally introduced limitations to their autonomy and scope. Because different disciplines may have different approaches to data and modeling and different vocabularies (these may be called cultural aspects) and even different interface protocols, bridging across disciplines is a more complex challenge. Thus, interoperability among diverse disciplinary and domain systems must be pursued adopting more flexible and sustainable approaches, the GEOSS brokering approach (see the next paragraph), to introduce such flexibility and evolvability (GEO 2009, 2015). Brokering philosophy is formulated to

* https://www.data.gov./.

handle such differences without limiting the autonomy and without putting a significant investment burden on existing disciplinary systems (Nativi et al. 2012, Vaccari et al. 2012).

14.4.4 Brokering Approach

The notion of "system of systems" (SoS), and the related "system of systems engineering" process, emerged in many fields of applications (De Laurentis 2009) to address the common problem of integrating many independent, autonomous systems, frequently of large dimensions, in order to satisfy a global goal while keeping them autonomous (Karcanias and Essami 2010). SoS can be usefully described as large-scale integrated systems that are heterogeneous and consist of subsystems that are independently operable on their own but are networked together for a common goal (Jamshidi 2005).

The GEOSS mission is to build a global SoS across multiple domains: it is conceived to leverage hundreds of existing and heterogeneous enterprise systems, which belong to different domains and contribute to address the eight SBAs (or great challenges) and the 11 Communities of Practices recognized by GEO, that is, air quality, biodiversity, carbon, coastal zone, cryosphere, energy, forests, geohazards, global agricultural monitoring, health and environment, and water cycle.

In such an ecosystem of domain infrastructures, multidisciplinary interoperability has been traditionally pursued on a one-to-one basis or by asking the stakeholders (i.e., both GEOSS users and resource providers) to be able to utilize the plethora of interoperability standards (both international and community based) characterizing the different disciplinary systems. Clearly, this has represented a high entry barrier for developing cross-disciplinary science and applications (Nativi et al. 2011, 2013). For this reason, a new solution was proposed first by a European FP7 project (Vaccari et al. 2012, Santoro et al. 2010, De Laurentis 2009) and then by a US-NSF initiative (Karcanias and Essami 2010), namely, the brokering approach.

The brokering approach follows these principles to make existing infrastructures and data systems interoperable, in a SoS framework (Nativi et al. 2012):

- To keep the existing capacities as autonomous as possible by interconnecting and mediating between standard-based and non-standard-based capacities
- To supplement, without supplanting, the individual systems' mandates and governance arrangements
- To assure a low entry barrier for both the resource providers and the end users
- To be flexible enough so as to accommodate the existing systems as well as future ones

To build in an incremental fashion upon the existing infrastructures (information systems) and incorporate heterogeneous resources by introducing distribution and mediation functionalities

To specify interoperability arrangements focusing on the modularity of interdisciplinary concepts rather than just on the technical interoperability of systems

14.4.5 The GEOSS Common Infrastructure

Realizing a SoS, GEOSS is composed of contributed supply systems, ranging from systems collecting primary data, to systems concerned with the creation and distribution of information products (Nativi et al. 2015). Although all GEOSS systems continue to operate within their own mandates and will evolve, GEOSS systems can leverage each other so that the overall GEOSS becomes much more than the sum of its component systems (GEO 2007). This was achieved by implementing a digital infrastructure that coordinates access to these systems, interconnecting and harmonizing their data, applications, models, and products: the GEOSS common infrastructure (GEO 2007).

Through the GCI, GEOSS resources, including EO data (satellite, airborne, in situ, models), information services, standards, and best practices, can be searched, discovered, and accessed by scientists, policy leaders, decision makers, and those who develop and provide information services across the entire spectrum of users (GEO 2005). The GCI includes three main components: (1) the GEOSS web portal, (2) the GEO discovery and access broker (DAB), and (3) the component and service registry (CSR) (Figure 14.3).

14.4.6 The GEOSS Web Portal

The GEOSS web portal (aka GEOSS portal) offers a single web-based user interface to discover and access the comprehensive data produced by the GEOSS community. A new version of the portal is under development. Developed by the ESA, this new portal is going to make it easier and faster to integrate diverse datasets, identify relevant data and portals of contributing systems, access models and other decision-support tools, and preview data before downloading. The portal is powered by the DAB developed by the National Research Council of Italy (CNR) (GEO 2005).

14.4.6.1 The DAB

Any request received by the GEOSS web portal is forwarded to the DAB, which connects user requests to an ever-increasing number of databases and information systems around the world, that is, the GEOSS resources

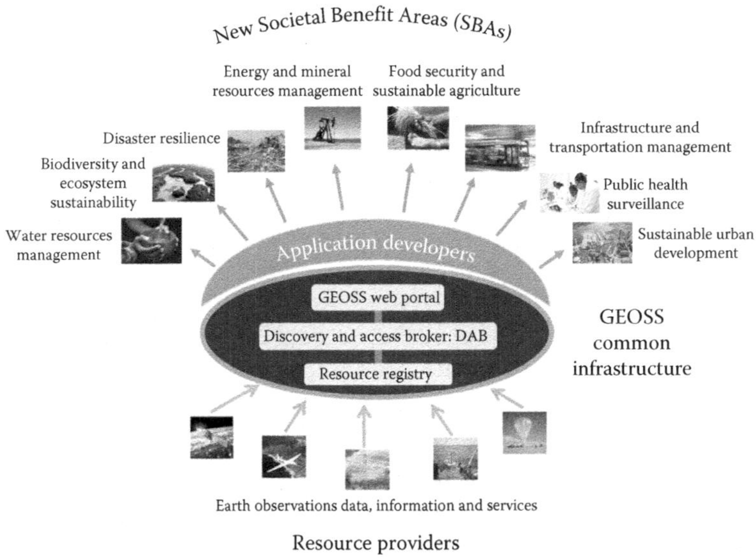

FIGURE 14.3
The GEOSS common infrastructure.

supply system provided by the SoS enterprise systems. DAB applies the brokering principles to interconnect the many enterprise systems constituting GEOSS, that is, the global SoS managed by GEO. Through the DAB services, GCI relaxes the requirement for implementing a common data model and exchange protocol, providing the necessary mediation and transformation functionalities in a transparent way to the SoS components (Nativi et al. 2015). The DAB supports more than 50 well-used and standard protocols, commonly implemented by the GEOSS data and information and service suppliers to share their resources, harmonizing them to provide a unique and consistent response to the GEOSS user requests.

The DAB exposes a set of well-used standard internet interfaces and high-level JavaScript APIs* enabling the developers' stakeholders to implement applications and sophisticated downstream services for the end users. The APIs implement discoverability, accessibility, and simple transformation (i.e., data encoding transformation, coordinate reference systems mapping, data subsetting, and data resolution change) functionalities.

14.4.6.2 The CSR

This tool enables both data and information providers to register their resources and also to share their knowledge and experiences of standard

* http://www.geodab.net/#!apis/ch644.

and interoperability and best practices in all fields of EO (GEO 2005). CSR is developed and managed by USGS.

In the first decade of GEOSS, the GCI has played a crucial role to build the GEOSS information system (see Figure 14.1). The next section will discuss the current contribution of the GCI to the GEOSS information system structure and assets, and the Big Data challenges it had to face.

14.4.7 GCI Contribution to the GEOSS Information System

The main role played by the GCI is to realize the GEOSS information system and noticeably the GCI components' interconnections (Figure 14.3). Presently, more than 145 enterprise systems (spanning from a simple database to complex digital federated infrastructures) are brokered by the GCI. They are the actual GEOSS resource providers and share about 1.3 million of datasets for more than 200 million of discoverable and potentially accessible elements, that is, single files. GEO DAB makes use of the services provided by the CSR and a pool of semantic engines, aligned and managed by the EC-JRC (Joint Research Centre of the European Commission in Ispra). Besides the GEOSS web portal, other (community) portals and applications access the GCI, through the DAB, to discover and access the GEOSS resources.

Recently, GEO decided to engage the private sector considering no-profit bodies (e.g., NGOs) and private companies. GCI has been pursuing interoperability tests with some private systems, in addition to the already public brokered ones (Figure 14.4).

14.4.8 GEOSS Information System and the Big Data Challenges

Big Data topic immediately emerges when considering large and heterogeneous EO systems. The GEOSS information system (and the GCI) goals pose challenges along all the Big Data dimensionalities (Nativi et al. 2015). Each Big Data challenge (commonly known as 'V' axes: volume, variety, velocity, veracity, visualization) required the GEOSS and the GCI to devise and operate ad hoc solutions and strategies (Nativi et al. 2015). This may be considered as the third important GCI evolution, while the second one is the adoption of the brokering approach.

Presently, the GEOSS information system adopts a fully brokering approach implementation, building on cloud computing technology: GCI (i.e., its three main components) moved in the Cloud, realizing a public cloud-based software ecosystem that characterizes the present GEOSS information systems. Community application developers can join this software ecosystem by using the GCI/DAB cloud-based APIs to develop new applications and community-driven portals.

For example, the presently operational DAB configuration takes advantage of the following cloud-specific elements (Nativi et al. 2015):

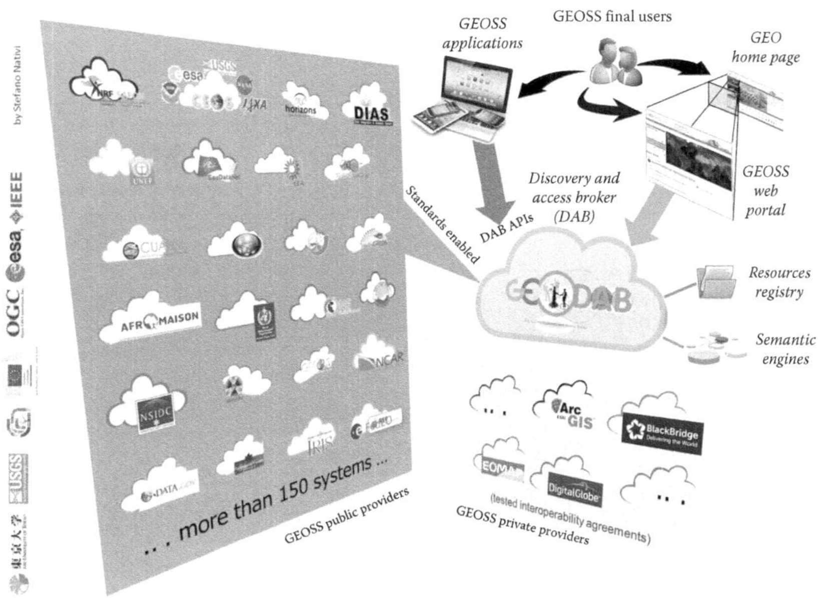

FIGURE 14.4
GCI contribution to the GEOSS information system.

Routing service: This is a Domain Name System service, specifically tailored to be used in cloud environments, that is, the cloud routing service can use cloud-specific functionality to optimize traditional routing functionalities (e.g., route requests to healthy machines).

Load balancer: Provides external client applications with a unique entry point and routes each request to the machine with the lowest workload.

Instance: A virtual machine provisioned by the cloud provider.

Clone instance cluster: A set of instances, every instance in the cluster is assumed to be running the same application with the same configuration (they are clones).

Auto-scaling cluster: An instance cluster that is able to add/remove instances on-the-fly according to a set of scaling rules.

Cluster: A set of instances, every instance can be running different applications with different configurations.

The main solutions and strategies, adopted by the GEOSS information system (and GCI), for addressing Big Data challenges were analyzed by Nativi et al. (2015); in particular, Table 14.1 provides an executive summary of them.

TABLE 14.1

GEOSS Information Systems and GCI Strategies and Solutions to Address Big Data Challenges

Big Data Challenges		Solutions Adopted to Address the Challenges
Volume	*Discovery challenges* High number of catalogs, inventory, listing services to be brokered Large number of metadata records Large number of users' discovery requests	Reduce the number of matching results, by supporting advanced constraints in addition to the more traditional "what," "where," and "when." Design and apply a ranking metrics and related paging strategy. Support distributed queries, along with the harvesting approach, to reduce the number of large metadata records to be stored and managed by the DAB. Use of load balancing and auto-scaling clusters to support a large number of queries.
	Access challenges High number of data services to be brokered Large amount of datasets Big data volume Large number of users' access requests	Use of server-side transformation functionalities to limit downloaded data. Supplement missing transformation functionalities (not supported by data servers). Support data caching and map tiling. Use of load balancing and auto-scaling clusters.
Variety	*Discovery challenges* Support of highly heterogeneous metadata models and discovery service interfaces Publication of the set of metadata models and discovery interfaces implemented by GEOSS users' applications Long-term data access sustainability in a multidisciplinary environment	Introduction of a brokering tier dedicated to mediation of service interfaces and metadata models harmonization in a transparent way for both users and data providers. Design and implementation of a brokering semantic and metadata model used. Extensible architecture of brokering to support new service interfaces and metadata models.

(Continued)

TABLE 14.1 (*Continued*)

GEOSS Information Systems and GCI Strategies and Solutions to Address Big Data Challenges

Big Data Challenges		Solutions Adopted to Address the Challenges
	Access challenges Support of highly heterogeneous data models, encoding formats, and access service interfaces Publication of the set of data models, encoding format, and access interfaces implemented by GEOSS users' applications Long-term data access sustainability in a multidisciplinary environment	Introduction of a brokering tier dedicated to mediation of access service interfaces and data formats harmonization in a transparent way for both users and data providers. Design and implementation of a brokering data model used to (i) harmonize and integrate the heterogeneous data formats brokered by GEOSS and (ii) expose the data formats well supported by GEOSS users. Extensible architecture of brokering to support new access service interfaces and data formats. Transformations facilitating reuse.
Velocity	*Discovery challenges* To manage the increasing rate at which metadata flows Fast metadata processing to satisfy users' needs	Operational data store that periodically extracts, integrates, and reorganizes brokered metadata records for operational inquire and ranking generation. Caches that provide instant access to the results of distributed queries while buffering data provider systems from additional load and performance degradation. Design of the DAB architecture that balances metadata latencies with GEOSS users' requirements, avoiding assuming that all data must be near real time. Incremental harvesting strategy. Live query distribution combined with caching of results. Load balancing to route incoming requests to machines with the lowest workload. Use of auto-scaling clusters to increase computing capacity in response of rapid workload growth.

(*Continued*)

TABLE 14.1 (*Continued*)

GEOSS Information Systems and GCI Strategies and Solutions to Address Big Data Challenges

Big Data Challenges		Solutions Adopted to Address the Challenges
	Access challenges To manage the increasing rate at which data flows Fast data processing to satisfy users' needs	Operational data store that periodically generates and stores preview tiled maps of brokered data for operational data preview. Caches that provide instant access to the results of previous access requests. Supplementing missing transformations allows limiting the local processing time. For extremely large processing requests, users are allowed to opt for an asynchronous version of the access functionality.
Veracity, value, and validity	*Challenges* Reduction of the "information noise" Retrieved data comparison Data trustiness for GEOSS decision makers Effective data reuse Data meaningfulness for user requests Data accuracy for intended use	The brokering data model includes a specific multidisciplinary quality extension. Implementation of a flexible ranking metrics including quality of service and metadata completeness as valuable indexes. The brokering metadata model supports a harmonized presentation of retrieved metadata facilitating their comparison. Use of GEOSS EVs as an additional parameter for improving the existing ranking metrics. The prototyped "fit-for-purpose" and users' feedback extensions aim to provide users with quality-aware results.
Visualization	*Challenges* Visualization speed Contextualized visualization	Support community portals and applications publishing DAB APIs for client development. Support the following visualization strategy: (1) provide an overview (trying to keep that simple and show important elements), (2) allow zoom and filter unnecessary clutter, and (3) provide more details if requested by users. Provide fast previews by generating preview tiles in batch.

Source: De Laurentis, D., Understanding transportation as a system of systems problem, in *System of Systems Engineering: Innovations for the 21st Century*, 2009, pp. 520–541.

14.5 Conclusions

Ecosystems are rapidly changing worldwide, most of which are a result of the multitude of interacting anthropogenic drivers. The changing ecosystems and the services they provide can seriously affect the benefits to humankind. For these reasons, it is necessary to quantitatively characterize the current state and ongoing changes of ecosystem functions, processes, and services both globally and locally, across the wide range of environments and landscapes that characterize our planet. The logical course of action then lies with developing user-friendly collection, archive, and dissemination of the large amount of heterogeneous EO data from diverse sources. However, a key remaining challenge is converting data to new understandings and knowledge that can be used for ecosystem conservation and management. The GEO ECO initiative, with its supporting activities (i.e., EU H2020 ECOPOTENTIAL and SWOS projects, and partner projects with USGS and ESRI), aims at contributing to this endeavor. In parallel to developing the observational infrastructure, GEO ECO plans to also collect a suite of data models able to incorporate/assimilate EO data to estimate future conditions (states) of selected ecosystems.

Taken in concert, the GEO data infrastructures will assist in developing the new vision of ecosystems as integral, coupled, complex geosphere–biosphere systems where new understandings will be derived to benefit society. All the results, information, and metadata will be made available through a system of portals that contribute to GEO/GEOSS and will provide the structure to derive new knowledge to advance science and improve conservation and management policies.

Acknowledgments

This work was partially funded by the H2020 project 641762 ECOPOTENTIAL: Improving Future Ecosystem Benefits through Earth Observations. We are grateful to C. Beierkuhnlein, R. Sayre, K. Weise, and to all participants in the H2020 ECOPOTENTIAL project.

Glossary

CBD: Convention for biological diversity
ConnectinGEO: http://www.connectingeo.net/

Copernicus: ESA, http://www.copernicus.eu/
ECOPOTENTIAL: www.ecopotential-project.eu
EFU: Ecological freshwater units
ELU: Ecological land units
EMU: Ecological marine units
EO: Earth observations
Esri: http://www.Esri.com/
FP7: European Union Framework Program 7
GEO: Group on Earth Observations
GEO ECO: GEO Global Ecosystem Initiative
GEOSS: Global Earth Observation System of Systems
GEOWOW: http://www.geowow.eu/
GWOS: Global Wetland Observation System
H2020: EU Horizon 2020 Programme
MAES: Mapping and assessment of ecosystems and their services, http://biodiversity.europa.eu/maes
SBA: Societal benefit area (in GEO)
SWOS: Satellite-Based Wetland Observation Service
USGS: United States Geological Survey

References

Convention on Biological Diversity Secretariat. *Earth Observation for Biodiversity Monitoring*. CBD Tech. Series 72, Montreal, Canada, 2014.

Daily, G.C. *Nature's Services*. Island Press, Washington, DC, 392pp., 1997.

De Laurentis, D. Understanding transportation as a system of systems problem. *In System of Systems Engineering: Innovations for the 21st Century*, Wiley, Hoboken, NJ, pp. 520–541, 2009.

European Commission. *Interoperability for Pan-European eGovernment Services*, COM, Brussels, Belgium, 2006.

GEO. *The Global Earth Observation System of Systems (GEOSS): 10-Year Implementation Plan*, Group on Earth Observation, 2005. https://www.earthobservations.org/documents/10-Year%20Implementation%20Plan.pdf, accessed April 10, 2016.

GEO. Strategic guidance for current and potential contributors to GEOSS, printed by JAXA on behalf of GEO. *Architecture and Data Committee*, 2007. https://www.earthobservations.org/documents/portal/25_Strategic%20Guidance%20Document, accessed April 10, 2016.

GEO. GEOSS strategic targets, 2009. https://www.earthobservations.org/documents/geo_vi/12_GEOSS%20Strategic%20Targets%20Rev1.pdf, accessed April 10, 2016.

GEO. The GEOSS common infrastructure (GCI), 2016. http://www.earthobservations.org/geoss.php, accessed April 10, 2016.

GEO. GEO Strategic Plan 2016–2025. Implementing GEOSS, 2016. http://www.earthobservations.org/documents/GEO_Strategic_Plan_2016_2025_Implementing_GEOSS.pdf, accessed April 10, 2016.

Heffernan, J.B. et al. Macrosystems ecology: Understanding ecological patterns and processes at continental scales. *Frontiers in Ecology and the Environment* 12 (2014): 5–14.

High Level Expert Group on Scientific Data of the European Union, Riding the wave: how Europe can gain form the raising tide of scientific data. Expert Group data and publication, 2010. http://ec.europa.eu/information_society/newsroom/cf/document.cfm?action=display&doc_id=707, accessed October 21, 2016.

Hobbs, R.J., E. Higgs, and J.A. Harris. Novel ecosystems: Implications for conservation and restoration. *Trends in Ecology and Evolution* 24 (2009): 599–605.

Jamshidi, M. System-of-systems engineering—A definition. In *IEEE SMC*, Big Island, HI, 2005.

Karcanias, N. and A. Hessami. Complexity and the notion of system of systems: General systems and complexity. *In 2010 World Automation Congress International Symposium on Intelligent Automation and Control* (*ISIAC*), Kobe, Japan, 2010.

Keith, D.A. et al. Scientific foundations for an IUCN red list of ecosystems. *PLoS ONE*, 8 (2013): e62111.

MAES Working Group. Mapping and assessment of ecosystems and their services—An analytical framework for ecosystem assessments under Action 5 of the EU Biodiversity Strategy to 2020. EU Technical Report – 2014-080, 2014.

Marris, E. The end of the wild. *Nature*, 469 (2011): 150–152.

Nativi, S., M. Craglia, and J. Pearlman, The brokering approach for multidisciplinary interoperability: A position paper. *International Journal of Spatial Data Infrastructures Research*, 7 (2012): 1–15.

Nativi, S., M. Craglia, and J. Pearlman. Earth science infrastructures interoperability: The brokering approach. *IEEE Journal of Selected Topics in Applied Earth Observation and Remote Sensing*, 6 (2013): 1118–1129.

Nativi, S., S.J. Khalsa, B. Domenico, M. Craglia, J. Pearlman, P. Mazzetti, and R. Rew. The brokering approach for earth science cyberinfrastructure. EarthCube white paper, US NSF, 2011. http://semanticommunity.info/@api/deki/files/13798/=010_Domenic, accessed April 10, 2016.

Nativi, S., P. Mazzetti, M. Santoro, F. Papeschi, M. Craglia, and O. Ochiai. Big data challenges in building the global earth observation system of systems. *Environmental Modelling and Software*, 68 (2015): 1–26.

Pereira, H.S. et al. Essential biodiversity variables. *Science*, 339 (2013): 277–278.

Potts, T. et al. Do marine protected areas deliver flows of ecosystem services to support human welfare? *Marine Policy*, 44 (2014): 139–148.

Santoro, M., E. Boldrini, F. Papeschi, A. Baldini, and S. Nativi. Specification of EUROGEOSS initial operating capacity. FP7 EuroGEOSS Deliverable D.2.2.2.a, Brussels, Belgium, 2010.

Soranno, P.A. et al. Cross-scale interactions: Quantifying multiscaled cause–effect relationships in macrosystems. *Frontiers in Ecology and the Environment*, 12 (2014): 65–73.

Soranno, P.A. and D.S. Schimel. Macrosystems ecology: Big data, big ecology. *Frontiers in Ecology and the Environment*, 12 (2014): 3.

Spalding, M., R. Brumbaugh, and I. Meliane. Mapping ocean wealth: Mainstreaming ecosystem services into decision making. *Global Partnership for Oceans*, draft working paper v2, 2013.

Stocker, M. Situation awareness in environmental monitoring. Doctoral Dissertation (PhD). Publications of the University of Eastern Finland. Dissertations in Forestry and Natural Sciences No. 192, University of Eastern Finland, Joensuu, Finland, 2015. ISBN: 978-952-61-1907-6. http://urn.fi/URN:ISBN:978-952-61-1908-3.

Tallis, H. et al. A global system for monitoring ecosystem service change. *Bioscience,* 62 (2012): 977–986.

Vaccari, L., M. Craglia, C. Fugazza, S. Nativi, and M. Santoro. Integrative research: The EuroGEOSS experience. Selected topics in applied earth observations and remote sensing, *IEEE Journal,* 5 (2012): 1603–1611.

Zins, C. Conceptual approaches for defining data, information, and knowledge. *Journal of the American Society for Information Science and Technology,* 58 (2007): 479–493. doi:10.1002/asi.20508.

15

Advancing the Software Systems of Environmental Knowledge Infrastructures

Markus Stocker

CONTENTS

15.1 Introduction

To acquire data about the environment is a core task of environmental infrastructures. Acquired data are for selected properties of certain elements of the environment, such as the temperature of air or the height of seedlings—where temperature and height are the properties of the elements air and seedlings, respectively. Data result in measurement, the "process of empirical, objective, assignment of numbers to properties" (Finkelstein, 1982). Measurement is repeated as the properties of elements are monitored in time and space (Meijers, 1986).

Environmental infrastructures often employ environmental sensor networks (Hart and Martinez, 2006) to automate data acquisition. Sensors automate monitoring, that is, automatically repeat measurement. A sensor generally monitors one property over time. A sensor system, with multiple sensors as its constituent parts, enables monitoring of multiple properties over time. Networked sensor systems, deployed at multiple locations, enable monitoring of properties over time and space.

Acquired data are processed to gain information about the environment, and information is transferred into knowledge. These tasks are typically performed manually by human agents. Environmental infrastructures thus consist of technical agents as hardware and software, for example, sensors and databases, and human agents, for example, technicians, engineers, and scientists. Environmental infrastructures are thus sociotechnical systems (Fox, 1995), consisting of technical and social subsystems. To increase human knowledge and understanding of the environment is arguably the primary aim of these sociotechnical systems. We thus speak of environmental *knowledge* infrastructures and of environmental knowledge *research* infrastructures if they serve primarily research. Edwards (2010) defined knowledge infrastructures as "robust networks of people, artifacts, and institutions that generate, share, and maintain specific knowledge about the human and natural worlds." Environmental knowledge infrastructures focus on generating, sharing, and maintaining specific knowledge primarily about *natural* worlds. Being "networks of people, artifacts, and institutions," Edwards also underscores the sociotechnical character of knowledge infrastructures.

Especially in large-scale environmental knowledge infrastructures (Kratz et al., 2006; Keller et al., 2008; Michener et al., 2011), data acquisition, curation, access, and processing are increasingly often left to technical subsystems as they enable automation, for example, data acquisition by means of sensors or data management by means of databases. The technical subsystems of environmental knowledge infrastructures are thus data-based systems. In contrast, data analysis and interpretation, that is, the acquisition of information from data and transfer of information into knowledge, are carried out by social subsystems, often with little support from technical subsystems.

We envision that the technical subsystems, in particular software systems, of future environmental knowledge infrastructures will advance from data-based systems to knowledge-based systems. The technical subsystems of future environmental knowledge infrastructures may thus more actively support human agents in information acquisition and support the curation of machine interpretable knowledge and automated knowledge processing. We thus envision future environmental knowledge infrastructures with knowledge-based technical subsystems.

In this chapter, we present the environmental knowledge infrastructures of two case studies and underscore the sociotechnical character of the infrastructures. We discuss the kinds of information the infrastructures acquire from data; the agents and methods involved in information acquisition and transfer of information into knowledge; and the kinds of resulting knowledge. We then present methods and technologies that enable technical subsystems of environmental knowledge infrastructures to more actively support human agents in—or altogether automate—information acquisition and support the curation of machine interpretable knowledge and thus automated knowledge processing. The discussed approaches are presented as possible building blocks toward knowledge-based technical subsystems.

Information and knowledge are *about the environment* monitored by the environmental knowledge infrastructure. Information is acquired from data about the monitored environment. Other information and knowledge types are obviously relevant to environmental knowledge infrastructures, for example, information about technical agents, workflows, or data. These are however not of concern here.

Other authors have discussed the idea of knowledge-driven sociotechnical systems that learn from scientific data. For instance, Peters et al. (2014) present the architecture of a Knowledge Learning and Analysis System that aims at being "a knowledge-driven, open access system that 'learns' and becomes more efficient and easier to use as streams of data, and the number and types of user interactions, increase." Peters et al. discuss the integration of hypothesis-driven and data-intensive machine learning scientific approaches. Ganguly et al. (2007) propose a framework for knowledge discovery on environmental data in scientific applications. Naturally, knowledge infrastructures—in particular also infrastructures that employ environmental sensor networks—are not limited to scientific applications. For instance, Parmiggiani and Monteiro (2016) discuss a knowledge infrastructure developed at a Norwegian oil and gas company and note that the infrastructure based on environmental monitoring attempts "to abstract the datasets into general representations of environmental risk that make sense for the oil and gas professionals." In contrast to these works for high-level architectural and system descriptions and analysis, this chapter discusses the application of concrete software methods and technologies that enable the development and implementation of knowledge-based technical subsystems in environmental infrastructures.

15.2 Case Studies

Environmental knowledge infrastructures are sociotechnical systems consisting of technical and social subsystems. Hardware and software are agents of technical subsystems, while humans and communities are agents of social subsystems. While data acquisition and the curation and processing of data are fundamental to environmental knowledge infrastructures, data are merely intermediate products from which infrastructures acquire information and transfer information into knowledge.

We present the environmental knowledge infrastructures of two case studies that underscore the sociotechnical character of the infrastructures and highlight how their overall aim is to acquire information, transfer information into knowledge, and curate and process knowledge. For each case study, we describe the relevant data, information, and knowledge as well as the technical and social agents involved in data processing, information acquisition, transfer of information into knowledge, and knowledge curation and processing. The case studies were originally developed in Stocker (2015).

15.2.1 Atmospheric New Particle Formation

The first case study is in aerosol science for the study of atmospheric new particle formation. The environmental knowledge infrastructure is thus a *research* infrastructure, and Stocker et al. (2014a) discuss the case study in more details.

Atmospheric new particle formation is an atmospheric phenomenon whereby new particles are formed and over time grow in size (Kulmala et al., 2004). The phenomenon has been documented in a wide variety of environments all over the world (Kulmala et al., 2004) and is studied because aerosol particles are known to scatter sunlight and influence quality of life, for instance, by affecting human health (Pope III et al., 2002). The scattering of radiation has a cooling effect on the climate (Solomon et al., 2007). The study of new particle formation is thus relevant to climate change research.

The environmental knowledge infrastructure involves the Finnish Station for Measuring Ecosystem–Atmosphere Relations (Hari and Kulmala, 2005, SMEAR), in particular the station located at the Puijo observation tower in Kuopio, Finland. This station is part of SMEAR IV, which is itself part of the wider SMEAR network with stations located in Eastern Lapland, Hyytiälä, Helsinki, and Kuopio.

The station consists of sensing devices for the monitoring of aerosols, weather, and atmospheric gases (Leskinen et al., 2009). Of interest here is the Differential Mobility Particle Sizer (DMPS) utilized to monitor the particle size distribution of polydisperse aerosols. A DMPS consists of a Differential Mobility Analyzer (DMA) and a Condensation Particle Counter (CPC). The particles of polydisperse aerosols are first classified according to diameter size by the DMA and then counted by the CPC (Kulkarni et al., 2011). The instrument measures the particle number concentration (cm^{-3}) for 40 discrete diameter sizes in the range of 7–800 nm, on average five times per hour.

In studying atmospheric new particle formation, a core task for the environmental knowledge infrastructure is to identify and classify individual events, that is, instances of the atmospheric phenomenon, as they occur in time and space. Different classification schemes have been proposed to characterize individual events (Dal Maso et al., 2005; Hamed et al., 2007; Vana et al., 2008). The identification and classification of events occur on processed DMPS data and are performed by human agents, in particular aerosol scientists.

Data acquisition in this environmental knowledge infrastructure, in particular measurement by the DMPS and collection over the network, is largely automated by the hardware and software agents of the technical subsystem. In contrast, the extraction of information about new particle formation events, the transfer of such information into knowledge, and the curation and processing of knowledge are performed by aerosol scientists and are—with the exception of software for statistical computing and reporting, that is, MATLAB® and Excel—hardly supported by the technical subsystem of the infrastructure.

Acquired data undergo an inversion from sensor data in (V) to particle number concentration in (cm^{-3}) (Wiedensohler et al., 2012). Such data processing is implemented in MATLAB. The resulting data are curated as text files consisting of an $m \times n$ data matrix, where m is the number of measurements over 24 h (1 day) for 40 particle diameter sizes and $n = 41$ (includes the timestamp). The daily text files are stored on a file system accessible to researchers. The technical subsystem of the environmental knowledge infrastructure largely automates these steps.

Researchers access processed data and create data products that support them in the visual assessment of new particle formation events on a particular day and location (Hamed et al., 2007). MATLAB is the software agent used to create figures for visual assessment. The figures display time (24 h) on the x-axis and particle number concentration for the 40 measured particle diameter sizes of monitored polydisperse aerosol on the y-axis. A color gradient is used to represent low-to-high concentration. On a day during which a clearly visible event occurred, the figure displays a characteristic so-called banana shape, reflecting the high concentration of very small particles that grow in diameter size over time.

Having identified a new particle formation event, aerosol scientists characterize the event. Among the extracted features, scientists may classify the event, for instance, based on its visual clarity, obtain an estimate for event start and end times, and compute formation and growth rates (Hamed et al., 2007). Acquired information about events is recorded and Excel is the software agent used for the curation of information. At a minimum, information includes the day at which an event occurs and the event class.

By recording information about a particular event of atmospheric new particle formation in the columns of an Excel row, the scientist creates a knowledge object about the event. The knowledge object integrates contextual information about the event and is curated in Excel. In this environmental knowledge infrastructure, we may further specialize the object as a *situational* knowledge object. Barwise and Perry (1980) and Devlin (1991) suggested that a situation is a structured part of reality that an agent manages to individuate. New particle formation events are objects in parts of reality, that is, connected regions of space–time (Barwise and Perry, 1981). They are thus objects in situations. The environmental knowledge infrastructure individuates structured parts of reality and is thus the agent that individuates situations.

Scientists utilize the recorded knowledge objects in further analysis, for example, to compute the monthly frequency of event classes, seasonal differences in hourly mean total particle concentration between event and nonevent days, or monthly mean event duration, formation, and growth rates (Hamed et al., 2007). Knowledge objects are thus processed. The results of such analysis are presented as figures and tables and are discussed in the natural language text of scientific journal articles. More abstractly, the results of such analysis are new information integrated into existing knowledge structures (Aamodt and Nygård, 1995).

Clearly, the infrastructure is an environmental *knowledge* infrastructure. Beyond acquiring, curating, and processing data about particle size distribution of polydisperse aerosols, the environmental infrastructure extracts, curates, and processes information and knowledge about events of atmospheric new particle formation. The infrastructure's aim is to increase human knowledge and understanding of atmospheric new particle formation.

The environmental knowledge infrastructure is furthermore a sociotechnical system as technical and social subsystems collaborate to attain the infrastructure's aim to increase human knowledge and understanding of atmospheric new particle formation. The technical subsystem consists of hardware and software agents. Hardware agents include the DMPS, communication links, and computers. Software agents include MATLAB, custom MATLAB scripts, and Excel. While hardware and software agents certainly do serve toward data processing and analysis, their role is primarily in data acquisition, curation, and access. The social subsystem consists of aerosol scientists and technicians. Human agents are involved in data acquisition, curation, and access, but the role of human agents, in particular scientists, is in data analysis, information acquisition, transfer of information into knowledge, and knowledge curation and processing. The technical subsystem of the environmental knowledge infrastructure is a data-based system and as such primarily concerned with tasks required prior to data analysis. The social subsystem builds on the data-based system and extends it with functionality for data analysis and interpretation. The social subsystem thus turns the infrastructure into a knowledge-based system.

15.2.2 Plant Disease Outbreaks

The second case study is in precision agriculture, for the assessment of (acute) disease outbreaks in plants. The environmental knowledge infrastructure serves agricultural advisors to farmers, and Stocker et al. (2016) discuss the case study in more details.

Plant disease is a threat to plant growth, quality, harvest, and thus economic return. Hence, farmers need to monitor disease progress to determine the right time when plants need to be protected, for example, by spraying chemical agents. Various factors other than disease progress influence decisions to protect plants, for example, regulations, utilized protective agent, or protection history. Decision-making thus depends on knowledge, that is, integrated information. Indeed, modern precision agriculture is "intrinsically information intensive" (Fountas et al., 2006). Farmers are guided by agricultural advisors, and together they are part of the social subsystem of an environmental knowledge infrastructure designed to support decision-making.

Agricultural advisors rely on computer models and systems that support them in information acquisition. In this case study, agricultural advisors utilize a mechanistic model for estimating disease pressure. Disease pressure is computed as the cumulative value $AR_t = AR_{t-1} + DR$, where AR_t is the accumulated disease pressure value on day t and DR is the change on a given day. DR is constructed from a base risk value modified by daily modifiers. The base risk depends on the susceptibility of the selected crop and farming history. The daily modifiers are computed from data for the weather on the given day, specifically average temperature, humidity, wind, and the amount of rainfall. Diseases included in the model are *Pyrenophora teres*, *Pyrenophora tritici-repentis*, and *Stagonospora nodorum* and follow this general disease pressure model. How the base risk or the daily modifiers are used, however, depends on the disease, as all diseases react to changes in the environmental variables in a unique manner. The model updates disease pressure once per day.

Agricultural advisors operate weather stations as part of the environmental knowledge infrastructure to monitor a range of environmental properties, including temperature, relative humidity, wind speed, and cumulative precipitation. The weather stations are part of the SoilWeather Wireless Sensor Network (WSN) (Kotamäki et al., 2009). Each weather station is a sensor system and consists of several sensing devices. Observation data can be accessed via a Web service. Observation data are complemented with seasonal data for the agricultural parcels in the region observed by the infrastructure. Such data include the preceding crop, current crop, current crop susceptibility, tillage method, and seeding date.

The environmental knowledge infrastructure utilizes data to compute disease pressure for the region observed by the infrastructure. The results are (daily) maps that display disease pressure as color-coded spatial features. Given such maps, agricultural advisors can monitor the progress of disease pressure in the region and obtain information about disease pressure that exceeds the threshold for which outbreaks are expected. Knowledge about possible outbreaks is obviously of interest to farmers.

The infrastructure is clearly an environmental knowledge infrastructure. As in the previous case study, beyond acquiring, curating, and processing data about weather parameters and agricultural parcels, the infrastructure acquires, curates, and processes knowledge about disease outbreaks. The infrastructure's aim is to inform decision-making.

The environmental knowledge infrastructure is again a sociotechnical system as technical and social subsystem collaborate to attain the infrastructure's aim. Among other devices and systems, the SoilWeather WSN is an important hardware component of the technical subsystem. The technical subsystem is primarily tasked with data acquisition, curation, and access; it is therefore a data-based system. The social subsystem consists, primarily, of agricultural advisors and farmers. Their primary role is, as in the previous case study, in information acquisition and transfer of information into knowledge, as well as decision-making. Their involvement in lower-level tasks such as data acquisition, curation, and access is minor. While data-based functionality is largely automated—in particular, the continuous acquisition of data via SoilWeather, the management of data by the database, and the access to data by the Web service—knowledge-based functionality is "implemented" by human agents. Most importantly, the technical subsystem of the infrastructure is not involved in knowledge curation. Indeed, knowledge about outbreaks is only implicit in the color-coded spatial features of maps for disease pressure. The technical subsystem does not have explicit representations of knowledge about outbreaks. Knowledge is thus not curated by the technical subsystem.

In the following section, we present software methods and technologies that can advance the data-based technical subsystems of state-of-the-art environmental knowledge infrastructures into knowledge-based systems. As a result, the technical subsystems of future environmental knowledge infrastructures will more actively support information acquisition, transfer of information into knowledge, as well as knowledge curation, access, and processing.

15.3 Approaches

We have presented the environmental knowledge infrastructures of two case studies to highlight how infrastructures acquire information about a monitored environment, transfer information into knowledge, and curate and process knowledge. We argued that the infrastructures are knowledge-based sociotechnical systems because social and technical subsystems collaborate to further human knowledge and understanding about the environment. We highlighted that in state-of-the-art environmental knowledge infrastructures it is because of social subsystems that the infrastructures are knowledge-based systems. Technical subsystems are prevalently data-based systems and provide the social subsystems with little support for information acquisition from data, transfer of information into knowledge, and knowledge curation and processing.

Our claim is that the technical subsystems of future environmental knowledge infrastructures will advance from data-based systems to knowledge-based systems. In other words, the technical subsystems will more actively support, and possibly largely automate, the execution of higher-level tasks currently mostly carried out by social subsystems.

As the data volumes acquired and processed by environmental knowledge infrastructures steadily increase—and as infrastructures become more interoperable, thus facilitating data fusion—this advancement is arguably a necessity. Daily visual extraction of situational knowledge about new particle formation events by an aerosol scientist is feasible for a single location. However, the task becomes increasingly expensive as new particle formation is to be identified at more locations. Furthermore, as the identification and characterization of new particle formation events are relatively straightforward and repetitive tasks, it makes good sense to automate them.

In this section, we present how information acquisition, transfer of information into knowledge, and knowledge curation and processing may be implemented by the technical subsystems of environmental knowledge infrastructures. We present relevant software methods and technologies.

We distinguish data, information, and knowledge and do so following the Data–Information–Knowledge model proposed by Aamodt and Nygård (1995). According to the model, data are syntactic entities. The syntactic entities resulting in the process of measurement, for example, sensor data for particle number concentration, are a kind of data relevant here. Data are input to an interpretation process. Information, according to the model, is interpreted data, that is, data with meaning or *semantic entities*, and is the output of an interpretation process. The semantic entities resulting in the process of information acquisition, for example, identified atmospheric new particle formation, are a kind of information relevant here. Finally, knowledge is learned information, that is, information incorporated into an existing body of knowledge. A situational knowledge object that integrates information about a new particle formation event (situation) is a kind of knowledge relevant here. Knowledge is itself a semantic entity, one that relates semantic entities.

To execute information acquisition, the technical subsystems of environmental knowledge infrastructures require one or more software agents designed to extract information from data. Technical subsystems must be able to control and execute the agents. Agents implement an interpretation process. Data are input to agents and information is the output.

Software agent design follows a model. Two model types are of particular interest: data driven and physically based. Data-driven models, also known as empirical models, may be *supervised*, that is, trained by labeled examples. Preconceived knowledge about the modeled phenomenon does not influence model development. In contrast, physically based models, also known as mechanistic models, are developed to include some degree of understanding about the processes underlying the modeled phenomenon. Mulligan and Wainwright (2004) provide an overview of (environmental) models and modeling and discuss in more depth the characteristics of various model types.

For the aerosol science case study, the technical subsystem of the environmental knowledge infrastructure could employ a supervised data-driven software agent trained with labeled examples for processed daily particle number concentration data and corresponding information for whether or not new particle formation occurred during the day. Given a data-driven agent trained with such labeled examples, the technical subsystem can then automate the classification of new input data to output information for identified new particle formation. In other words, equipped with a trained data-driven agent, the technical subsystem of the environmental knowledge infrastructure can automate the task otherwise carried out visually by a scientist. The degree of confidence in the accuracy of automated classification can be estimated empirically and can hint at how well the technical subsystem will perform—or how carefully the automated assessment ought to be curated by the scientist.

For the case study in agriculture, the technical subsystem of the environmental knowledge infrastructure could employ a software agent that implements the presented mechanistic model. The agent uses preconceived expert knowledge about plant disease infection development, current observation data for weather parameters, and seasonal data about the crop, pathogen, and agricultural parcel in an equation that estimates daily disease pressure. The technical subsystem can execute the agent to generate information about disease outbreaks, that is, situations in which disease pressure exceeds defined thresholds.

In both environmental knowledge infrastructures, data-driven and physically based agents have the same purpose: to automate data interpretation. Software agents enable the technical subsystems to automate information extraction in environmental knowledge infrastructures. The automated assessment by technical subsystems is accurate to a certain degree. Extracted information thus needs to be curated by the social subsystem. Of particular interest is quality control of extracted information.

A description for what an environmental knowledge infrastructure observes, for example, a description for an event of new particle formation, generally involves different kinds of information. What infrastructures observe is located in space and time. Descriptions thus involve information for temporal and spatial locations, for example, timestamp, latitude, and longitude. A symbolic identifier for the observed phenomenon, for example, a character string for an instance of new particle formation, is information that enables reference to the observed phenomenon in a description. Descriptions also characterize the observed phenomenon, for example, describe the class of new particle formation and the duration of the event. Characterization results in additional information.

Descriptions for what an environmental knowledge infrastructure observes are thus structures that relate information. We call such structures knowledge objects. Knowledge objects integrate information in a body of knowledge. A particular type of knowledge object is the *situational* knowledge object. It is a description of a situation. Other knowledge object types, for example, for descriptions of processes, are of interest as well.

The integration of information in a knowledge object follows a pattern. Technical subsystems may implement such patterns to automate the integration of information in knowledge objects. Situation theory (Devlin, 1991) provides a pattern for integrating information about a situation in a situational knowledge object.

In addition to models for information extraction and patterns for information integration into knowledge objects, technical subsystems of environmental knowledge infrastructures also require a framework for the representation of knowledge objects. The framework provided by the Semantic Web (Berners-Lee et al., 2001) and its technologies is one approach to equip technical subsystems of environmental knowledge infrastructures with functionality for knowledge representation.

The Web Ontology Language (W3C OWL Working Group, 2012) is a core technology of the Semantic Web. In information science, ontology is classically defined by Gruber (1993) as "an explicit specification of a conceptualization." Guarino et al. (2009) provide a succinct analysis of Gruber's definition. Some authors have extended Gruber's definition (Borst, 1997; Studer et al., 1998) while others have provided alternatives (Neches et al., 1991; Swartout et al., 1996; Hendler, 2001). For the purpose here, an ontology is a document that specifies the concepts and relations of some domain so that the semantics of specified terms are interpretable by both software and human agents. To specify concepts and relations we need a language with formal semantics. Today, the Web Ontology Language (OWL) is arguably the *de facto* standard ontology language. It is also the language adopted here for knowledge representation in environmental knowledge infrastructures.

OWL language constructs support the formal specification of the semantics of concepts and relations as class axioms and property axioms, respectively. For instance, the language enables us to state that `C` and `D` are classes and that they are equivalent or that `Q` is an inverse property of the property `P`. The language also supports the specification of individuals. Concept assertions and role assertions specify the class membership and property values of individuals, respectively. For instance, we can state that `a` and `b` are individuals. The concept assertion `C(a)` states that the individual `a` is a member (instance) of the class `C`; `D(b)` thus states that `b` is a member of `D`. The role assertion `P(a,b)` states that the individuals `a` and `b` are related by property `P`. Given that `Q` is inverse of `P` it holds that `Q(b,a)`.

Ontologies are a means for the social subsystems of environmental knowledge infrastructures to convey the semantics of relevant concepts and relations to technical subsystems. As a result, the two subsystems share term semantics. For software agents that implement the language, `C(a)` is not merely a string. If the social subsystem states that `C` is a subclass of `B` (formally $C \sqsubseteq B$), then software agents automatically conclude `B(a)`. Ontologies are also a means for the technical subsystems of environmental knowledge infrastructures to convey social subsystems information objects automatically acquired in data interpretation and knowledge objects resulting from automated information integration. Ontologies are thus a key component for the representation of knowledge communicated between the technical and social subsystems of environmental knowledge infrastructures.

Information objects extracted from data by technical subsystems are semantic entities and, specifically, entities of an OWL ontology. Given the class axiom NewParticleFormation $\sqsubseteq$ AtmosphericPhenomenon, the technical subsystem of our environmental knowledge infrastructure represents the symbolic identifier f for observed new particle formation as the class assertion NewParticleFormation(f). The class assertion is an information object and a semantic entity of an OWL ontology. Information objects for locations in time and space are represented similarly as members of a class and with property values. For the representation of time and space, there exist ontologies that both the technical and the social subsystems can adopt. A candidate ontology for time is OWL-Time (Hobbs and Pan, 2006), which, among other terms, provides definitions for Instant and Interval. For space, an infrastructure may adopt GeoSPARQL (Perry and Herring, 2012), which provides definitions for spatial Feature and Geometry.

Knowledge objects resulting from information objects automatically integrated by technical subsystems following determined integration patterns are also ontological semantic entities. For the particular case of situational knowledge objects, describing situations observed by an environmental knowledge infrastructure, we may adopt the Situation Theory Ontology (STO) (Kokar et al., 2009). An event s for NewParticleFormation(f) may thus be represented as the assertion Situation(s). The definition of the class Situation in the STO follows the Situation Theory developed by Barwise and Perry (1981) and Devlin (1991), which specifies how information about an observed situation is integrated in a situational knowledge object.

In the Semantic Web, the Resource Description Framework (RDF) (Cyganiak et al., 2014) is the data model utilized to encode the axioms and assertions of OWL ontologies. Originally conceived as a model of data about Web resources (Lassila and Swick, 1999), RDF can be utilized as a model of data about any resource, including physical objects, abstract concepts, or any entity that can be named by a Uniform Resource Identifier (URI) (Berners-Lee et al., 2005). The RDF *statement* is at the core of the framework and is a triple consisting of a resource, a property, and the value for the property of the resource. These are the subject, the predicate, and the object of the statement, respectively. According to the framework, the concept assertion C(a) is encoded as the triple <a,type,C> and the role assertion R(a,c) as the triple <a,R,c>, whereby a, type, C, and R are URIs and c may be a URI or a literal value, such as a string.

The subjects, predicates, and objects of two or more statements can share the same URI. Such statements *join* over the shared identifier. Common joins are subject–subject and object–subject. The former are statements about the same subject resource. It can be easily seen that a set of statements form a directed labeled graph of subject and object nodes related by predicates, which act as vertexes directed from the subject to the object. Adding a new statement for an existing resource is akin to expanding the graph with an additional vertex. Two resources can be related by simply adding a new vertex to the graph. The flexibility of the graph data structure is arguably one of the interesting features of RDF.

Figure 15.1 is an example situational knowledge object for a new particle formation event, represented using the discussed Semantic Web technologies. Clearly visible is the directed labeled graph structure of the knowledge object. Object semantics are interpretable by the technical subsystem of environmental knowledge infrastructures. To facilitate machine readability of knowledge objects, infrastructures can adopt one of several syntaxes for RDF, such as RDF/XML (Gandon and Schreiber, 2014).

There exist several RDF database systems, which generally also implement the SPARQL Protocol and RDF Query Language (SPARQL; Harris and Seaborne, 2013) to support querying, updating, or deleting RDF statements managed by the database. For popular programming languages, libraries are available to support reading, processing, and writing RDF data (e.g., Beckett, 2002; Broekstra et al., 2002; Carroll et al., 2003). Libraries designed for programmatic interaction with OWL ontologies, specifically, are available for some programming languages (e.g., Horridge and Bechhofer, 2009). Some software packages for statistical computing also support loading RDF data.

SPARQL supports formulating queries with complex graph patterns. The language thus enables us to formulate queries for knowledge objects meeting certain criteria. For instance, an agent may interrogate an RDF database for events (situations) of strong new particle formation that occurred in 2015 in a particular region of Finland, with perimeter determined by the coordinates of a polygon geometry. As the subsystems commit to a shared ontology, agents of two or more subsystems understand how information about situations is represented; that "strong" is a category of the classification by Hamed et al. (2007); that "2015" is a date–time interval; and that we are constraining our search for situations that occurred within the given polygon geometry. In other words, agents involved in knowledge acquisition represent acquired knowledge objects according to the same ontology used by agents that access knowledge objects. The semantics of relevant terms are specified externally to agents.

Together, these technologies enable environmental knowledge infrastructures with technical subsystems that (1) acquire information from data using data-driven or physically based models, (2) integrate information into knowledge objects according to patterns, and (3) formally and explicitly represent knowledge objects. The technologies thus facilitate the curation, access, and processing of knowledge objects in technical subsystems of environmental knowledge infrastructures. Such technical subsystems are knowledge-based systems.

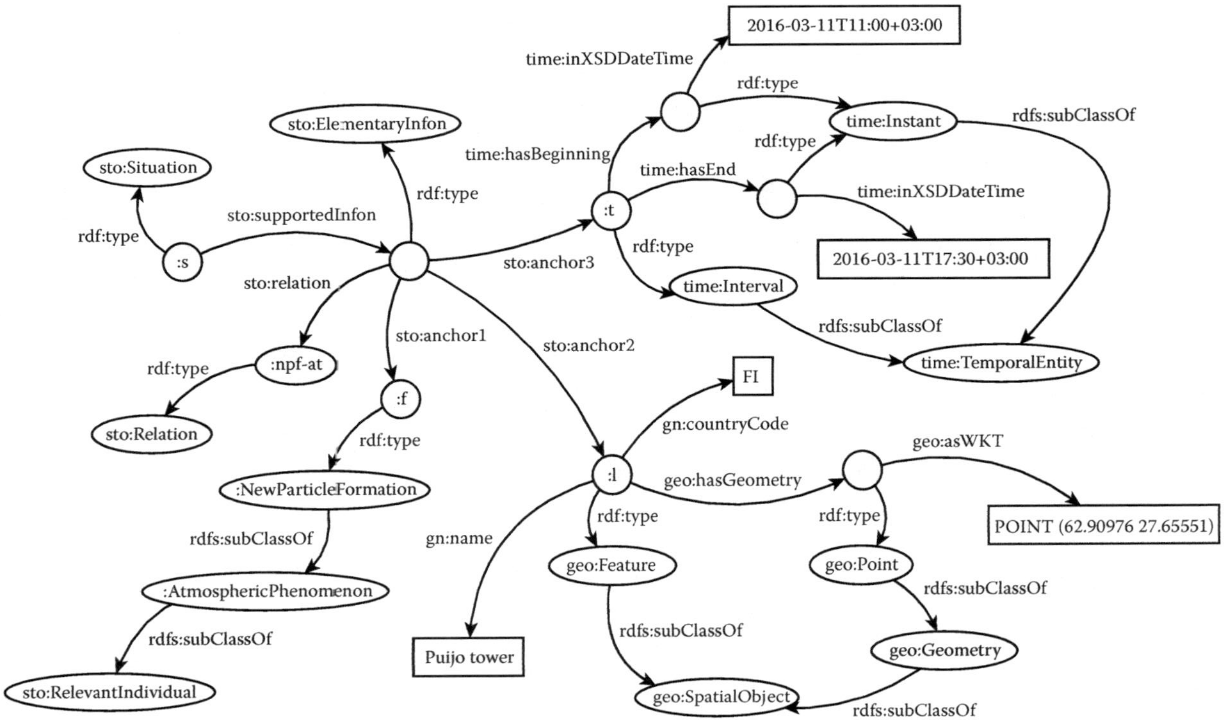

FIGURE 15.1
Example situational knowledge object `:s` represented in RDF for an event of new particle formation `:f` at spatial location `:l` and temporal location `:t`. In Situation Theory, a situation supports one or more elementary infons, that is, information about the situation, which are tuples consisting of a relation (`:npf-at`), a set of anchored objects, and a polarity (omitted here). The prefixes of terms, for example, `sto:`, reflect the ontologies.

15.4 Challenges

Earth and environmental science research communities have recently started to systematically study environmental research infrastructures. Based on an analysis of six infrastructures of the European Strategy Forum on Research Infrastructures (ESFRI)—that is, ICOS, EURO-Argo, EISCAT-3D, LifeWatch, EPOS, and EMSO*—Chen et al. (2013b) present a reference model for environmental research infrastructures. The reference model, called ENVRI-RM, structures identified and shared functionality of environmental research infrastructures into subsystems—namely, data acquisition, data curation, data access, data processing, and community support—and captures the requirements of the "archetypical" environmental research infrastructure from three viewpoints: science, information, and computational. The science viewpoint describes the requirements for "the people who perform their tasks and achieve their goals as mediated by the infrastructure" (Chen et al., 2013a). The information viewpoint describes the requirements for information handled by the infrastructure. The computational viewpoint describes the requirements for expected computational objects and the interfaces by which they interact.

The ENVRI-RM makes evident that state-of-the-art environmental research infrastructures are data-based systems and are thus modeled as such. Indeed, following the ENVRI-RM, technical subsystems are expected to support data acquisition, in particular with sensors; data curation, in databases and on storage systems; data access, for example, via Web portals; and data processing. This is despite that, beyond data, information and knowledge are arguably more important products of environmental research infrastructures. While data analysis and data mining are functionality of the ENVRI-RM data processing subsystem, it is unclear how the reference model and, thus, state-of-the-art environmental research infrastructures account for information and knowledge resulting from data analysis and mining. Accounting for knowledge objects, and their life cycle in environmental knowledge infrastructures, is important because the formal and explicit

*The Integrated Carbon Observation System (ICOS) aims at quantifying and understanding the greenhouse gas balance of Europe and neighboring regions (https://www.icos-ri.eu/); EURO-Argo is the European contribution to Argo, a global ocean observing system (http://www.euro-argo.eu/); EISCAT-3D primarily aims at investigating how the Earth's atmosphere is coupled to space (https://eiscat3d.se); LifeWatch aims at biodiversity and ecosystem research (http://www.lifewatch.eu/); the European Plate Observing System (EPOS) aims at developing a more holistic understanding of the processes underlying Earth's dynamics (https://www.eposip.org/); and the European Multidisciplinary Seafloor and water-column Observatory (EMSO) aims at long-term, high-resolution, (near) real-time monitoring of environmental processes including natural hazards, climate change, and marine ecosystems (http://www.emso-eu.org/).

representation of such objects enables automation in knowledge curation, access, and processing. Moreover, knowledge representation rests on methods and technologies different from those widely utilized for the representation of data objects, for example, tabular or relational data structures.

One of the expectations for the ENVRI-RM is to mature and serve as a blueprint for the implementation of environmental research infrastructures. The aim is harmonization and interoperability. The commitment by infrastructures toward the reference model is arguably a key requirement for its success. However, gaining such commitment is challenging. As experience shows, alone the commitment to a particular schema and format for data management and exchange is challenging to achieve (see, for instance, the brief account by Edwards et al. (2011) on the adoption of the Ecological Metadata Language in the Long-Term Ecological Research program).

Proposals for extending the ENVRI-RM with functionality for information acquisition, transfer of information into knowledge, knowledge representation, curation, access, and processing have recently been suggested in the literature (e.g., Stocker et al., 2015b). However, advancing the reference model and implementations to include such functionality presents major challenges. First, at the community level, a shift is needed toward understanding, modeling, and implementing environmental infrastructures as knowledge-based systems. Clearly, data are only an intermediary product in ICOS, EMSO, and other infrastructures—including those that serve purposes other than scientific research (e.g., Stocker et al., 2014b; Parmiggiani and Monteiro, 2016). For instance, ICOS acquires and processes gas flux observation data but is interested in information and knowledge about strong and weak carbon sinks and sources, for example, forests and cities. Using real-time data processing, EMSO is interested in early warning of tsunami (Best et al., 2014), a situated real-world phenomenon about which the technical subsystem of an early warning system ought to provide near real-time integrated information. As knowledge-based systems require first consolidated architectures and implementations for the lower-level data layers, it is expected that the advancement toward environmental knowledge infrastructures with knowledge-based technical subsystems will require time, significant resources, and commitment from interdisciplinary teams involving at least earth and environmental scientists and computer and information engineers and specialists.

Second, at the technical level, introducing new methods and technologies for knowledge-based systems further complicates already complex infrastructures. The successful design, implementation, and testing of software agents that implement data-driven or physically based models for information extraction are mostly a nontrivial task. The deployment of validated software agents into an environmental knowledge infrastructure typically comes with further technical challenges, such as near real-time execution of the agent, interface requirements, and performance issues. Ontology engineering and the implementation of knowledge-based systems with Semantic

Web technologies also require a specialized set of skills. For the case of research infrastructures, scientists familiar with these methods and technologies are arguably few.

Conversely, engineers capable of addressing technical challenges typically lack the science understanding. Such understanding is, however, required for information acquisition model development, ontology development, as well as for knowledge-based system application development, for example, applications for knowledge processing. Given an arbitrary environmental knowledge infrastructure, an important question engineers will typically have is what kind of information and knowledge are of interest to the infrastructure. The effort of building and maintaining environmental knowledge infrastructures is thus inherently an interdisciplinary endeavor where human agents in scientist and engineer roles need to collaborate. Unfortunately, interdisciplinary collaboration in science is plagued by what Edwards et al. (2011) call "science friction," "the difficulties encountered when two scientific disciplines working on related problems try to interoperate." Science friction "resists and impedes" and poses significant challenges to development.

Currently, a practical challenge is to build compelling case studies that demonstrate the architecture, implementation, and capabilities of environmental knowledge infrastructures with knowledge-based technical subsystems, in particular research infrastructures. Such case studies will highlight the significant opportunities in environmental knowledge infrastructures with advanced knowledge-based technical subsystems. We discuss some of the opportunities next.

15.5 Opportunities

We set forth the vision of future environmental knowledge infrastructures with knowledge-based technical subsystems that more actively support human agents in information acquisition, transfer of information into knowledge, as well as knowledge curation, access, and processing. We have also argued that, thanks to the employed technologies, technical subsystems may partially automate such tasks.

It is important to underscore that automation is *to some degree* and it is to support human agents in these tasks. The social subsystem of environmental knowledge infrastructure remains critical. Software agents for information acquisition need to be developed, for example, a supervised data-driven agent needs to be trained with labeled examples; acquired knowledge needs to be quality controlled; and knowledge serves toward decision-making. Labeling, quality control, and decision-making are generally performed, or at least supervised, by human agents. Hence, in environmental knowledge infrastructures, social and knowledge-based technical subsystems collaboratively learn from the wealth of data acquired and curated by the infrastructures. Collaborative learning in environmental knowledge infrastructures using the presented methods and technologies comes with several interesting opportunities.

One of the most interesting aspects of environmental knowledge infrastructures with knowledge-based technical subsystems is that acquired knowledge objects are machine readable and interpretable. Consider our two case studies. Knowledge about new particle formation events is recorded in Excel, and the results of statistical analysis are described in scientific articles, using tables, figures, and natural language text. In our second case study, knowledge about plant disease outbreaks is equally implicit, in the images for regional maps and the color scheme used to inform human agents about the acuteness of outbreaks in the region. Knowledge encoded in these forms is hardly machine processable. It is implicit in higher-level data products. Presented with an intuitively designed map, a human agent can effortlessly extract information and knowledge conveyed by the image. Unfortunately, the same cannot be said for technical agents. As a consequence of encoding information and knowledge implicitly in higher-level data products, human agents need to manually extract information from articles, for example, to perform a meta-analysis. Another practice is to attempt to algorithmically extract the characteristics of spatial features by processing image pixels (e.g., Epitropou et al., 2015; Stocker et al., 2015a).

As noted earlier, to delegate information acquisition and the transfer of information into knowledge to the technical subsystems of environmental knowledge infrastructures is particularly useful when such processes are well defined and repeated. The more often they are repeated in space–time, the greater is arguably the benefit of automation as it frees human agents from carrying out the processes manually. Automation can also eliminate subjective bias by individual human agents in manual assessment.

The distinction between data, information, and knowledge is not clear-cut. The knowledge objects curated by an environmental knowledge infrastructure may arguably be data to other systems. More concretely, knowledge objects generally relate data of primitive types, for example, numbers, which an agent may want to access and process further. Hence, data related in knowledge objects may become elements of a dataset in another system. However, from a technical perspective, the distinction can be more obviously made based on data structure. Data in an environmental knowledge infrastructure are often structured as dataset, where the columns most commonly represent variables and rows are observations with values for the variables. In contrast, graphs are more suited to structure information in knowledge objects as graphs integrate by linking objects. Any node in a graph can be flexibly expanded with further vertexes to nodes. If new information is available for a particular knowledge object, it can be integrated by accordingly expanding the corresponding graph. If the class of new particle formation has not been assessed for a particular event, the graph corresponding to the knowledge object simply has fewer vertexes compared to the one corresponding to a new particle formation event for which the class was assessed. Knowledge objects, in particular also those of the same type, can thus have varying types and counts of relations to objects without resulting into structures filled with `Null` values. New vertexes between existing nodes can also be added or removed flexibly. Hence, when an environmental knowledge infrastructure uncovers a new relation, the corresponding objects are simply linked by a new relationship.

RDF is a suitable data model to represent graphs and is thus arguably an interesting framework for the representation of knowledge objects curated by knowledge-based technical subsystems of environmental knowledge infrastructures. RDF addresses the syntactic interoperability of knowledge objects, while OWL addresses their semantic interoperability by formally restricting the meaning of terms to the one intended. Being a Web technology, RDF has further interesting aspects. Curated knowledge objects, as well as the information objects they integrate, are referred to by URI. They are thus globally identifiable and can be linked across distributed environmental knowledge infrastructures. Another potentially interesting aspect is the association of URIs with persistent identifiers (Hakala, 2010), such as Digital Object Identifiers, to enable location-independent reference to knowledge objects. Doing so could facilitate the citation of knowledge about environmental phenomena, for example, a hurricane, described by environmental research infrastructures. The human agents of the social subsystem responsible for the acquisition and curation of cited knowledge could be credited for their work.

Curated knowledge objects can be processed in various ways. An important type of processing is visualization. Environmental phenomena observed by environmental knowledge infrastructures are generally located in space–time. Knowledge about observed phenomena can thus be visualized along these two dimensions. For instance, situational knowledge for disease outbreaks in agriculture can be visualized for outbreak development over time and space. Given the commitment of situational knowledge objects to ontologies and underlying theories, such as Situation Theory, consumer applications that visualize situational knowledge in space–time can trivially extract spatial and temporal information from situational knowledge objects and utilize the information in processing for visualization. As a result, the environmental knowledge infrastructure visualizes knowledge for situations of disease outbreaks rather than data underlying situational knowledge acquisition, such as data for current weather, from which human agents have to draw knowledge about outbreaks manually. Furthermore, as knowledge is explicitly represented by knowledge-based technical subsystems, environmental knowledge infrastructures avoid having information and knowledge only implicit in higher-level data products. Explicitly represented knowledge can be reused for purposes other than those originally intended.

Knowledge processing forms other than visualization are also of interest. Consumers can fuse knowledge independently acquired and curated by two or more environmental knowledge infrastructures (e.g., Stocker et al., 2015a). The retrieval of knowledge from multiple infrastructures practically amounts to executing a federated SPARQL query (Prud'hommeaux and Buil-Aranda, 2013) over the distributed SPARQL services provided by the infrastructures. Automated reasoning, including rule-based reasoning, is a further possibility in knowledge processing. Interesting to evaluate is also the potential of curated knowledge for the empirical parameterization of high-level models, such as agent-based or Bayesian models.

15.6 Conclusion

For environmental infrastructures, in particular research infrastructures, we have highlighted that, beyond the acquisition and processing of data, they generate information and transfer information into knowledge. We thus argued that environmental (research) infrastructures are environmental (research) *knowledge* infrastructures. The emphasis on knowledge is important as it underscores that data are merely intermediate products in environmental infrastructures and that state-of-the-art architectural models, such as the ENVRI-RM, may want to reflect this aspect in order to represent environmental (research) infrastructures more holistically.

Discussing the environmental knowledge infrastructures of two case studies, we highlighted how the infrastructures are sociotechnical systems, that is, systems composed of technical and social subsystems. We highlighted how the technical subsystems are predominantly involved in lower-level functionality of the infrastructure, in particular data acquisition and curation, and how technical subsystems often to a great extent automate such functionality, for example, by means of environmental sensor networks and database systems. In contrast, the social subsystems are more actively involved in higher-level functionality of the infrastructure, that is, information acquisition, transfer of information into knowledge, and knowledge curation and processing. Functionality executed by social subsystems is not as automated as functionality executed by technical subsystems.

Of particular interest to well-defined information acquisition processes that are executed frequently over space–time, we discussed methods and technologies that could advance the technical subsystems of state-of-the-art environmental (research) infrastructures from data-based to knowledge-based systems. We argued that knowledge-based technical subsystems can better support the social subsystems of infrastructures in information acquisition, transfer of information into knowledge, and knowledge curation and processing. In some cases, knowledge-based technical subsystems may largely automate such functionality.

The most important aspect of knowledge-based technical subsystems is their ability to represent acquired knowledge objects formally and explicitly. Knowledge about the environment observed by infrastructures is hence accessible and interpretable not just for social subsystems but for technical subsystems as well. This is in stark contrast to how knowledge is curated, accessed, and shared via higher-level data products such as digital maps, figures, tables, or natural language text typically generated by state-of-the-art environmental infrastructures. Knowledge conveyed via such higher-level data products is implicit and hardly accessible for technical subsystems.

We discussed some of the challenges and opportunities that lay on the path toward environmental (research) infrastructures with knowledge-based technical subsystems. An important challenge is the additional complexity of knowledge-based methods and technologies, added to already complex infrastructures. One of the most interesting opportunities for the research community may be the possibility of associating persistent identifiers to knowledge descriptions for discovered environmental phenomena—thus making such knowledge objects citable.

Acknowledgment

This work was supported by funding from the European Union's Horizon 2020 research and innovation programme under grant agreement No 654182.

References

Aamodt, A. and Nygård, M. (1995). Different roles and mutual dependencies of data, information, and knowledge—An AI perspective on their integration. *Data & Knowledge Engineering*, 16(3):191–222.

Barwise, J. and Perry, J. (1980). The situation underground. In Barwise, J. and Sag, I., editors, *Stanford Working Papers in Semantics*, Vol. 1, pp. 1–55. Stanford Cognitive Science Group, Palo Alto, CA.

Barwise, J. and Perry, J. (1981). Situations and attitudes. *The Journal of Philosophy*, 78(11):668–691.

Beckett, D. (2002). The design and implementation of the Redland RDF application framework. *Computer Networks*, 39(5):577–588.

Berners-Lee, T., Fielding, R., and Masinter, L. (2005). Uniform Resource Identifier (URI): Generic syntax. Request for Comments 3986, IETF.

Berners-Lee, T., Hendler, J., and Lassila, O. (2001). The semantic web. *Scientific American*, 284(5):29–37.

Best, M., Beranzoli, L., Chierici, F., Delaney, J. R., Embriaco, D., Galbraith, N., Huber, R., Orcutt, J. A., and Weller, R. A. (2014). CoopEUS EMSO-OOI case study: Tsunami modelling and early warning systems for near source areas (Mediterranean, Juan de Fuca). In *AGU Fall Meeting Abstracts*. San Francisco, CA. https://agu.confex.com/agu/fm14/webprogram/Paper18074.html.

Borst, W. N. (1997). Construction of engineering ontologies for knowledge sharing and reuse. PhD thesis, Centre for Telematics and Information Technology, University of Twente, Enschede, the Netherlands.

Broekstra, J., Kampman, A., and van Harmelen, F. (2002). Sesame: A generic architecture for storing and querying RDF and RDF schema. In Horrocks, I. and Hendler, J., editors, *The Semantic Web—ISWC 2002, Lecture Notes in Computer Science*, Vol. 2342, pp. 54–68. Springer, Berlin, Germany.

Carroll, J. J., Dickinson, I., Dollin, C., Reynolds, D., Seaborne, A., and Wilkinson, K. (2003). Jena: Implementing the semantic web recommendations. Technical Report HPL-2003-146, HP Laboratories, Bristol, U.K.

Chen, Y., Hardisty, A., Preece, A., Martin, P., Atkinson, M., Zhao, Z., Magagna, B., Schentz, H., and Legré, Y. (2013a). Analysis of common requirements for environmental science research infrastructures. In *Proceedings of the International Symposium on Grids and Clouds (ISGC)*, Academia Sinica, Taipei, Taiwan. Proceedings of Science (SISSA).

Chen, Y., Martin, P., Magagna, B., Schentz, H., Zhao, Z., Hardisty, A., Preece, A., Atkinson, M., Huber, R., and Legré, Y. (2013b). A common reference model for environmental science research infrastructures. In Page, B., Fleischer, A. G., Göbel, J., and Wohlgemuth, V., editors, *Proceedings of the 27th International Conference on Environmental Informatics for Environmental Protection, Sustainable Development and Risk Management*, pp. 665–673, Hamburg, Germany.

Cyganiak, R., Wood, D., and Lanthaler, M. (2014). RDF 1.1 concepts and abstract syntax. Recommendation, W3C.

Dal Maso, M., Kulmala, M., Riipinen, I., Wagner, R., Hussein, T., Aalto, P., and Lehtinen, K. (2005). Formation and growth of fresh atmospheric aerosols: Eight years of aerosol size distribution data from SMEAR II, Hyytiälä, Finland. *Boreal Environment Research*, 10(5):323–336.

Devlin, K. (1991). *Logic and Information*. Cambridge University Press, Cambridge, UK.

Edwards, P. N. (2010). *A Vast Machine: Computer Models, Climate Data, and the Politics of Global Warming*. MIT Press, Cambridge, MA.

Edwards, P. N., Mayernik, M. S., Batcheller, A. L., Bowker, G. C., and Borgman, C. L. (2011). Science friction: Data, metadata, and collaboration. *Social Studies of Science*, 41(5):667–690.

Epitropou, V., Bassoukos, T., Karatzas, K., Karppinen, A., Wanner, L., Vrochidis, S., Kompatsiaris, I., and Kukkonen, J. (2015). Environmental data extraction from heatmaps using the airmerge system. *Multimedia Tools and Applications*, 75(3):1589–1613.

Finkelstein, L. (1982). Theory and philosophy of measurement. In Sydenham, P. H., editor, *Handbook of Measurement Science, Theoretical Fundamentals*, Vol. 1, pp. 1–30. John Wiley & Sons, Hoboken, NJ.

Fountas, S., Wulfsohn, D., Blackmore, B., Jacobsen, H., and Pedersen, S. (2006). A model of decision-making and information flows for information-intensive agriculture. *Agricultural Systems*, 87(2):192–210.

Fox, W. M. (1995). Sociotechnical system principles and guidelines: Past and present. *The Journal of Applied Behavioral Science*, 31(1):91–105.

Gandon, F. and Schreiber, G. (2014). RDF 1.1 XML syntax. Recommendation, W3C.

Ganguly, A. R., Omitaomu, O. A., Fang, Y., Khan, S., and Bhaduri, B. (2007). Knowledge discovery from sensor data for scientific applications. In Gama, J. and Gaber, M. M., editors, *Learning from Data Streams*, pp. 205–229. Springer, Berlin, Germany.

Gruber, T. R. (1993). A translation approach to portable ontology specifications. *Knowledge Acquisition*, 5(2):199–220.

Guarino, N., Oberle, D., and Staab, S. (2009). What is an ontology? In Staab, S. and Studer, R., editors, *Handbook on Ontologies, International Handbooks on Information Systems*, pp. 1–17. Springer, Berlin, Germany.

Hakala, J. (2010). Persistent identifiers—An overview. Technical report, The National Library of Finland, Helsinki, Finland.

Hamed, A., Joutsensaari, J., Mikkonen, S., Sogacheva, L., Dal Maso, M., Kulmala, M., Cavalli, F. et al. (2007). Nucleation and growth of new particles in Po Valley, Italy. *Atmospheric Chemistry and Physics*, 7(2):355–376.

Hari, P. and Kulmala, M. (2005). Station for measuring ecosystem-atmosphere relations (SMEAR II). *Boreal Environment Research*, 10(5):315–322.

Harris, S. and Seaborne, A. (2013). SPARQL 1.1 Query Language. Recommendation, W3C.

Hart, J. K. and Martinez, K. (2006). Environmental sensor networks: A revolution in the earth system science? *Earth-Science Reviews*, 78(3–4):177–191.

Hendler, J. (2001). Agents and the semantic web. *Intelligent Systems, IEEE*, 16(2):30–37.

Hobbs, J. R. and Pan, F. (2006). Time ontology in OWL. Working draft, W3C.

Horridge, M. and Bechhofer, S. (2009). The OWL API: A Java API for working with OWL 2 ontologies. In Hoekstra, R. and Patel-Schneider, P.F., editors, *Proceedings of the 6th International Workshop on OWL: Experiences and Directions (OWLED 2009)*, Chantilly, VA, Vol. 529, pp. 11–21. CEUR Workshop Proceedings. ISSN 1613-0073.

Keller, M., Schimel, D. S., Hargrove, W. W., and Hoffman, F. M. (2008). A continental strategy for the National Ecological Observatory Network. *Frontiers in Ecology and the Environment*, 6(5):282–284.

Kokar, M. M., Matheus, C. J., and Baclawski, K. (2009). Ontology-based situation awareness. *Information Fusion*, 10(1):83–98. Special Issue on High-Level Information Fusion and Situation Awareness.

Kotamäki, N., Thessler, S., Koskiaho, J., Hannukkala, A. O., Huitu, H., Huttula, T., Havento, J., and Järvenpää, M. (2009). Wireless in-situ sensor network for agriculture and water monitoring on a river basin scale in Southern Finland: Evaluation from a data user's perspective. *Sensors*, 9(4):2862–2883.

Kratz, T. K., Arzberger, P., Benson, B. J., Chiu, C.-Y., Chiu, K., Ding, L., Fountain, T. et al. (2006). Toward a global lake ecological observatory network. *Publications of the Karelian Institute*, 145:51–63.

Kulkarni, P., Baron, P. A., and Willeke, K. (2011). *Aerosol Measurement: Principles, Techniques, and Applications*. John Wiley & Sons, Hoboken, NJ.

Kulmala, M., Vehkamäki, H., Petäjä, T., Dal Maso, M., Lauri, A., Kerminen, V., Birmili, W., and McMurry, P. (2004). Formation and growth rates of ultrafine atmospheric particles: A review of observations. *Journal of Aerosol Science*, 35(2):143–176.

Lassila, O. and Swick, R. R. (1999). Resource Description Framework (RDF) model and syntax specification. *Recommendation*, W3C.

Leskinen, A., Portin, H., Komppula, M., Miettinen, P., Arola, A., Lihavainen, H., Hatakka, J., Laaksonen, A., and Lehtinen, K. E. J. (2009). Overview of the research activities and results at Puijo semi-urban measurement station. *Boreal Environment Research*, 14(4):576–590.

Meijers, E. (1986). Defining confusions—Confusing definitions. *Environmental Monitoring and Assessment*, 7(2):157–159.

Michener, W. K., Porter, J., Servilla, M., and Vanderbilt, K. (2011). Long term ecological research and information management. *Ecological Informatics*, 6(1):13–24.

Mulligan, M. and Wainwright, J. (2004). Modelling and model building. In Wainwright, J. and Mulligan, M., editors, *Environmental Modelling: Finding Simplicity in Complexity*, pp. 7–73. John Wiley & Sons, Hoboken, NJ.

Neches, R., Fikes, R., Finin, T., Gruber, T., Patil, R., Senator, T., and Swartout, W. R. (1991). Enabling technology for knowledge sharing. *AI Magazine*, 12(3):36–56.

Parmiggiani, E. and Monteiro, E. (2016). A measure of "environmental happiness": Infrastructuring environmental risk in oil and gas off shore operations. *Science & Technology Studies*, 29(1):30–51.

Perry, M. and Herring, J. (2012). OGC GeoSPARQL—A geographic query language for RDF data. *Technical Report OGC 11-052r4*, Open Geospatial Consortium Inc, Wayland, MA.

Peters, D. P. C., Havstad, K. M., Cushing, J., Tweedie, C., Fuentes, O., and Villanueva-Rosales, N. (2014). Harnessing the power of big data: Infusing the scientific method with machine learning to transform ecology. *Ecosphere*, 5(6):1–15.

Pope III, C. A., Burnett, R. T., Thun, M. J., Calle, E. E., Krewski, D., Ito, K., and Thurston, G. D. (2002). Lung cancer, cardiopulmonary mortality, and long-term exposure to fine particulate air pollution. *The Journal of the American Medical Association*, 287(9):1132–1141.

Prud'hommeaux, E. and Buil-Aranda, C. (2013). SPARQL 1.1 federated query. Recommendation, W3C.

Solomon, S., Qin, D., Manning, M., Chen, Z., Marquis, M., Averyt, K., Tignor, M., and Miller, H. L. (2007). *Climate Change 2007: The Physical Science Basis. Contribution of Working Group I to the Fourth Assessment Report of the Intergovernmental Panel on Climate Change*. Cambridge University Press, Cambridge, U.K.

Stocker, M. (2015). Situation awareness in environmental monitoring. PhD thesis. Publications of the University of Eastern Finland. Dissertations in Forestry and Natural Sciences No 192, University of Eastern Finland. ISBN: 978-952-61-1907-6.

Stocker, M., Baranizadeh, E., Portin, H., Komppula, M., Rönkkö, M., Hamed, A., Virtanen, A., Lehtinen, K., Laaksonen, A., and Kolehmainen, M. (2014a). Representing situational knowledge acquired from sensor data for atmospheric phenomena. *Environmental Modelling & Software*, 58:27–47.

Stocker, M., Kauhanen, O., Hiirsalmi, M., Saarela, J., Rossi, P., Rönkkö, M., Hytönen, H., Kotovirta, V., and Kolehmainen, M. (2015a). A software system for the discovery of situations involving drivers in storms. In Denzer, R., Argent, R. M., Schimak, G., and Hřebíček, J., editors, *Environmental Software Systems. Infrastructures, Services and Applications*, volume 448 of IFIP Advances in Information and Communication Technology, pp. 226–234. Springer International Publishing, Switzerland.

Stocker, M., Nikander, J., Huitu, H., Jalli, M., Koistinen, M., Rönkkö, M., and Kolehmainen, M. (2016). Representing situational knowledge for disease outbreaks in agriculture. *Journal of Agricultural Informatics*, 7(2):29–39.

Stocker, M., Rönkkö, M., and Kolehmainen, M. (2014b). Situational knowledge representation for traffic observed by a pavement vibration sensor network. *IEEE Transactions on Intelligent Transportation Systems*, 15(4):1441–1450.

Stocker, M., Rönkkö, M., and Kolehmainen, M. (2015b). Knowledge-based environmental research infrastructure: Moving beyond data. *Earth Science Informatics*, 9(1):47–65.

Studer, R., Benjamins, V., and Fensel, D. (1998). Knowledge engineering: Principles and methods. *Data & Knowledge Engineering*, 25(1–2):161–197.

Swartout, B., Patil, R., Knight, K., and Russ, T. (1996). Toward distributed use of large-scale ontologies. In *Proceedings of the 10th Knowledge Acquisition for Knowledge-Based Systems Wowrkshop*, Banff, Alberta, Canada.

Vana, M., Ehn, M., Petäjä, T., Vuollekoski, H., Aalto, P., de Leeuw, G., Ceburnis, D., O'Dowd, C. D., and Kulmala, M. (2008). Characteristic features of air ions at Mace Head on the west coast of Ireland. *Atmospheric Research*, 90(2–4):278–286.

W3C OWL Working Group (2012). OWL 2 web ontology language document overview, 2nd edn. Recommendation, W3C.

Wiedensohler, A., Birmili, W., Nowak, A., Sonntag, A., Weinhold, K., Merkel, M., Wehner, B. et al. (2012). Mobility particle size spectrometers: Harmonization of technical standards and data structure to facilitate high quality long-term observations of atmospheric particle number size distributions. *Atmospheric Measurement Techniques*, 5(3):657–685.

Section V

Infrastructure Integration and Perspectives

16

Australia's Terrestrial Ecosystem Research Network: A Network of Networks Approach to Building and Maintaining Continental Ecosystem Research Infrastructures

Nicole Thurgate, Andrew J. Lowe, and Timothy F. Clancy

CONTENTS

Abstract

The challenges of the Anthropocene require large-scale infrastructure investments to underpin long time scales, multiple spatial scales, and transdisciplinary ecology and ecosystem science. In Australia, the Terrestrial Ecosystem Research Network (TERN) was established as a first step to address this need for the terrestrial domain. The approach taken was one that built on existing networks and capabilities and strategically invested in supporting enhanced integration of facilities utilizing a range of methods (including remote sensing, ecological plots, and biogeochemical flux analysis). Some early successes of the approach included TERN being a catalyst for improved data storage and access, linking of science to policymakers through dedicated synthesis products, and increasing access to global science initiatives such as Fluxnet. However, the approach was not without significant shortcomings, most notably the level of investment relative to the aspirations of the networks, the short-term funding cycles and overall uncertainty, the lack of explicit treatment of operational costs in addition to the build investment, and the resistance of sectors of the ecosystem science community to more collaborative, open, and integrative approaches. Key learnings from the project have driven a need for a greater focus on a common set of long-term objectives and a narrower scope at the expense of maintaining all activities that individual components have been undertaking, as well as an explicit focus on data services and data usability. In the future, it is hoped TERN will be able to operate an integrated set of core capabilities that are well linked with global initiatives to drive high-level science outcomes to improve the health and sustainability of use of our natural and managed ecosystems.

Keywords: Research infrastructure, Informatics, Ecosystem change, Collaboration, Network, Capex, Opex

16.1 The Need for Large-Scale Scientific Infrastructure in Ecology and Ecosystem Science

The development and use of large-scale scientific research infrastructure is a relatively new concept for many areas in the natural sciences, especially ecology and ecosystem science (see Schimel and Keller 2015). The policy and operation of these infrastructure networks are still being developed globally, and there is no single template that will satisfy every need. However, there is a basic scientific imperative for such infrastructure, primarily because ecological and ecosystem processes occur across local, regional, national, and global scales (Peters et al. 2008). In addition, temporal depth is required to understand the complexities of many systems and that reaches beyond the time frames of individual researchers, relatively short-term monitoring projects, or many experiments (Lindenmayer et al. 2012). The unprecedented changes to Earth's ecosystems since the industrial revolution (Crutzen and Steffen 2003) have triggered a need that goes beyond knowledge of a system and its properties, into the realm of forecasting future changes and describing system mitigation and adaptation that may be needed (Carpenter et al. 2006). A deep understanding is necessary to ensure the continuance of essential ecosystem services and the persistence of a healthy planet and this can best be addressed through collaborative infrastructure.

The use of collaborative research infrastructure is not a novel concept in scientific inquiry. Astronomers and physicists are frequently highlighted as groups who have successfully argued for the need for common infrastructure, such as telescopes or particle accelerators. Many areas of ecology and ecosystem science have not traditionally relied on shared infrastructure as the study of biotic and abiotic interactions at certain scales can largely be satisfied at an individual researcher or laboratory level. Although certain areas of ecosystem science, such as biogeochemical fluxes and land cover estimates, have supported and used large infrastructure investments (e.g., flux towers or remotely sensed data derived from satellites) and marine and polar research has required investment in significant infrastructure (e.g., research vessels) to be feasible, this is still not generally the case in terrestrial ecology. Indeed, the concept of research infrastructure is relatively novel to ecological scientists and has not traditionally formed a nucleus of the discipline.

Those tasked with the design and development of such large-scale infrastructure in the terrestrial space have found themselves at the frontier of a new paradigm within their discipline with little precedent to guide them. The creation and maintenance of these programs needs collaboration, leadership, negotiation, creativity, flexibility, risk-taking, trust, and determination, not all of which are inherent in the scientific pursuit. As the concept of shared infrastructure has been embraced globally and more research infrastructures are being developed, it is timely to reflect on what has been learned from the recent past.

16.2 Designing Research Infrastructure

The development of large-scale scientific research infrastructure can take three main forms. The first is a "top-down" approach where the design, hypotheses, personnel needs, etc., of a program are considered before any research infrastructure is built or developed. The U.S. NEON (http://www.neonscience.org/) provides an admirable example of such infrastructure. This model has its own inherent challenges and opportunities, which have been discussed elsewhere (e.g., Loescher et al., this volume; Peters et al. 2014), and their journey provides valuable insights for new infrastructure programs.

Alternatively, there is a "grassroots" or "bottom-up" approach where a particular section of the research community sees the benefit of collaborative infrastructure and self-organizes into a larger community to maximize impact and value. Nutnet (http://www.nutnet.umn.edu/) provides an excellent example of such a group and highlights the benefits that can be realized through such a cooperative process. The development of this community has enhanced global knowledge of the deleterious effects to biodiversity of increased fertilization in grasslands (Hautier et al. 2014), something that would be beyond the scope of a single laboratory. Naturally there are also challenges facing bottom-up initiatives, such as when seeking coordinated funding for what are perceived as discrete, preexisting enterprises. This can become particularly problematic when one organization or government agency is asked to shoulder the fiscal and administrative burdens of centralized infrastructure that supports projects that appear well outside their purview.

A third approach involves the creation of novel infrastructure to address scientific and/or management gaps, which is integrated with existing suitable networks and projects into a single infrastructure program. This unified style allows for maximum flexibility in infrastructure design that maximizes new partnerships while benefitting from and including those that already exist. Peters et al. (2014) refer to this as an über network and wisely point out the limitations that need to be overcome to realize this model. It can be hampered by the need for a fundamental cultural change in the research community that cannot be mandated but must be negotiated. As the Terrestrial Ecosystem Research Network (TERN) infrastructure developed, we somewhat naively adopted such a model and this chapter presents the development, challenges, and learnings of such a system.

16.3 Creation of TERN

TERN is fundamentally a network of networks, some of which were wholly created using TERN funding and others that already existed in some form and were augmented by TERN. It should be emphasized that in the original conception of TERN, which was initiated after a consultation phase between the Australian government and the ecosystem science community, there was no single clear mandate on the systems' underlying design; rather, the network of networks was both a necessary response to the separate capabilities and their approaches and recognition that there were compelling reasons for improved integration.

The original concept of TERN talks about "putting in place the necessary building blocks—cooperative and data management frameworks—to initiate the development of TERN" and "trial the necessary organisational arrangements, networks and methodologies for systematic coordinated national data collection." That is, TERN was always envisaged to be an adaptive approach that could deal with the requirements of a national approach without having a specific endpoint in mind.

The current TERN infrastructure includes 11 facilities (Table 16.1) and a centralized coordination, management, and engagement facility. The TERN is distributed among the host institutions of each of these facilities, which spreads the infrastructure across Australia. The design intent of TERN was to connect ecosystem scientists and allow them to share, store, contribute, synthesize, collaborate, and integrate data and information from a range of sources. This in turn would increase Australia's capacity to manage important environmental resources in a more efficient, effective, and collaborative manner. The breadth of the capabilities represented across TERN reflects the multifaceted need for observation of Earth's systems. No one capability was considered adequate to address design needs that encompass an entire continent.

The development of TERN began with the National Collaborative Research Infrastructure Strategy (NCRIS) investment of AUD $20.375 million, awarded by the Commonwealth Department of Innovation, Industry, Science and Research in 2009. This represented the first major investment in ecosystem science research infrastructure in Australia and created a novel process for its implementation. The original funding was leveraged to attract major cash and in-kind contributions from a number of organizations including the Queensland government, the South Australian government, the University of Queensland, the University of Adelaide, Macquarie University, and CSIRO.

With these investments came a number of mandated objectives for TERN that were required to be fulfilled to meet the expectations of the funding organizations and the ecosystem science community more generally. These included

- Creating a framework to foster scientific interactions to establish a national terrestrial site and observing network
- Coordinating national observational networks to share experiences and outcomes
- Facilitating improved access to quality-assured ecosystem science data
- Improving links between policy and science
- Involving all ecosystem science stakeholders to strengthen design and outcomes (Clancy et al. 2013)

TABLE 16.1

Description of the Distributed Terrestrial Ecosystem Research Network Facilities and Their Current Status

Facility	Purpose	Host Institution	Status 2016
TERN central cffice	The TERN office, based at the University of Queensland, is providing a single point of entry to discover data services underpinning a national terrestrial ecosystem research and a coordinating framework to support intradisciplinary collaboration and data sharing.	University of Queensland	Active
ACEF	The Australian Coastal Ecosystem Facility (ACEF) is a leading national repository for the collection, distribution, and visualization of key coastal data sets used in policy and management decisions for the protection and use of Australia's coastal assets. It enables data collection from fine scale to satellite and airborne collections of flora, fauna, and biophysical properties of coastal ecosystems.	CSIRO	Minimum operation
AusCover	AusCover is delivering improved access to remote sensing–derived products (satellite and airborne imagery), associated with land-surface dynamics and biophysical variables. Also providing better access to large national and international technical expert networks (e.g., NEON, GOFC-GOLD), field methodology protocols, and in situ validation data, for use in ecosystem science and natural resources management.	CSIRO	Active
e-MAST	The Ecosystem Modelling and Scaling Infrastructure (e-MAST) synthesizes, develops, benchmarks, and applies ecosystem models by integrating data from all TERN facilities and international partners. Collaborative, data-model fusion is the only means of developing a robust scientific understanding of the consequences of environmental and land management impacts on the functioning, carbon and water exchanges of land ecosystems across a range of scales.	University of Sydney	Minimum operation

(Continued)

TABLE 16.1 (*Continued*)

Description of the Distributed Terrestrial Ecosystem Research Network Facilities and Their Current Status

Facility	Purpose	Host Institution	Status 2016
OzFlux	OzFlux is a network of 30 flux towers across Australia that continuously measure the water use, net CO_2 uptake/release and energy balance in a range of ecosystems. OzFlux provides necessary standardized core measurements required to understand ecosystem function and biogeochemical cycles for use in ecosystem and land surface models.	CSIRO	Active
AusPlots	AusPlots Forests is a plot-based network to improve the understanding of tree growth, productivity, and carbon dynamics in Australia's tall eucalypt forests. AusPlots Rangelands is a permanent plot-based network in the rangeland bioregion, undertaking baseline surveys of vegetation and soils.	University of Adelaide	Active
Australian Transect Network	The Australian Transect Network (ATN) comprises 4 subcontinental transects that span biomes and significant environmental and land-use gradients. They provide a capability to develop, calibrate, and validate ecological models and remote sensing products for an enhanced understanding of ecosystem structure and function and to monitor key ecological transitions in the context of global climate.	University of Adelaide	Active
LTERN	The Long-Term Ecological Research Network integrates leading Australian long-term field research sites across Australia to tackle critical questions associated with the impacts of disturbance on Australian ecosystems. The 12 contributing plot networks have been actively monitored for many years and in most cases for several decades and span a diverse range of ecosystems, land tenures, and land uses.	Australian National University	Active

(*Continued*)

TABLE 16.1 (*Continued*)

Description of the Distributed Terrestrial Ecosystem Research Network Facilities and Their Current Status

Facility	Purpose	Host Institution	Status 2016
SuperSites	SuperSites is a national network of multidisciplinary ecosystem observatories that improve our understanding of ecosystem response to environmental change. The 10 SuperSites each represent a significant Australian biome and support personnel and sensor-based ecosystem measurements providing detailed data sets on flora, fauna, and biophysical processes. Each SuperSite hosts an OzFlux tower.	James Cook University	Active
Eco-informatics	Eco-informatics brings rich ecological "plot" data together and provides access and discovery of these data in a contextual and highly detailed and flexible format via ÆKOS. This capability enables vast new opportunities for repurposing and informed reuse, leveraging the investment and data to address national challenges.	University of Adelaide	Active
ACEAS	The Australian Centre for Ecological Analysis and Synthesis (ACEAS) was a facility within the TERN designed to link ecosystem scientists and environmental managers to improve our understanding and management of Australian ecosystems. The ACEAS activities supported multidisciplinary integration, synthesis, and modelling of ecosystem data.	University of Queensland	Inactive

The original components of the TERN were deemed successful in meeting these early objectives and the network expanded through subsequent commonwealth investments under the Education Investment Fund (EIF) Super Science Initiative, bringing in a range of new partners, networks, and coinvestors. TERN was further developed with the direct involvement of 17 university partners; CSIRO and more than 25 other research organizations; Australian, state and territory government agencies; nongovernment organizations; and others and through further investment via the Collaborative Research Infrastructure Strategy and further NCRIS investment (2013, 2015). The initial phase of the TERN's build and implementation was from 2009 to 2013, after which funding levels contracted to operational funding (Clancy et al. 2013). This mandated some restructuring of the TERN and a review of its basic infrastructure and scientific frameworks, an overview of which is described in Lowe et al. (submitted). To meet this challenge, TERN created a science enabling plan (Phinn et al. 2013) that established the key TERN science questions and outlined infrastructure development and use to 2025. Subsequent to this, there has been a far-reaching consultation across the research community and stakeholder groups, which to a large extent has validated the overall approach, albeit recognizing the need for greater resourcing (Andersen et al. 2014; McVay et al. 2014).

16.4 Benefits of the TERN Approach to Research Infrastructure

Prior to the creation of the TERN, there was little that brought the ecosystem research community together in an integrative fashion. Networks that revolved around specific disciplines existed and are still a key driving force for research today. The disconnected nature of the ecosystem research communities meant that any investment that leads to greater collaboration was going to provide a range of benefits.

The TERN's role was catalytic in drawing together these key communities studying ecosystem science and allowing them to benefit from each other. This function was made easier by harnessing networks that already existed—the network of networks concept. As an example, remote sensing had been conducted by the CSIRO in Australia for many years providing a range of products that provide broad-scale information about changes across the landscape. Under the TERN, this infrastructure was harnessed to become the AusCover facility (Figure 16.1). Linking this capability with the TERN's ground-based plot networks allowed for greater refinement in calibration and validation of remotely sensed products, which in turn provided richer information for ecosystem science and management. Australia is of course a large and sparsely populated continent, and there were many regions where little or no on-ground information existed. The TERN infrastructure, in consultation with key ecosystem science stakeholders, created new capabilities (e.g., AusPlots) or enhanced existing facilities (e.g., the Australian Transect Network [ATN]) to fill knowledge gaps in a stratified and repeatable fashion. Therefore, old and new networks benefitted from coalescing under TERN.

Another benefit of integrating old and new networks was the avoidance of duplication (and just as important in the context of limited resources, avoiding the appearance of duplication). Where there was already an established network collecting suitable data, the TERN infrastructure could provide a conduit that not only increased interactions within this network but broadened the reach of that network and allowed for new members to join rather than creating competing solutions. This inclusiveness was crucial in a space where managers and policymakers already found harnessing the best scientific data and outcomes problematic. Programs that appeared duplicative were and still are viewed as poor investments and often languished when

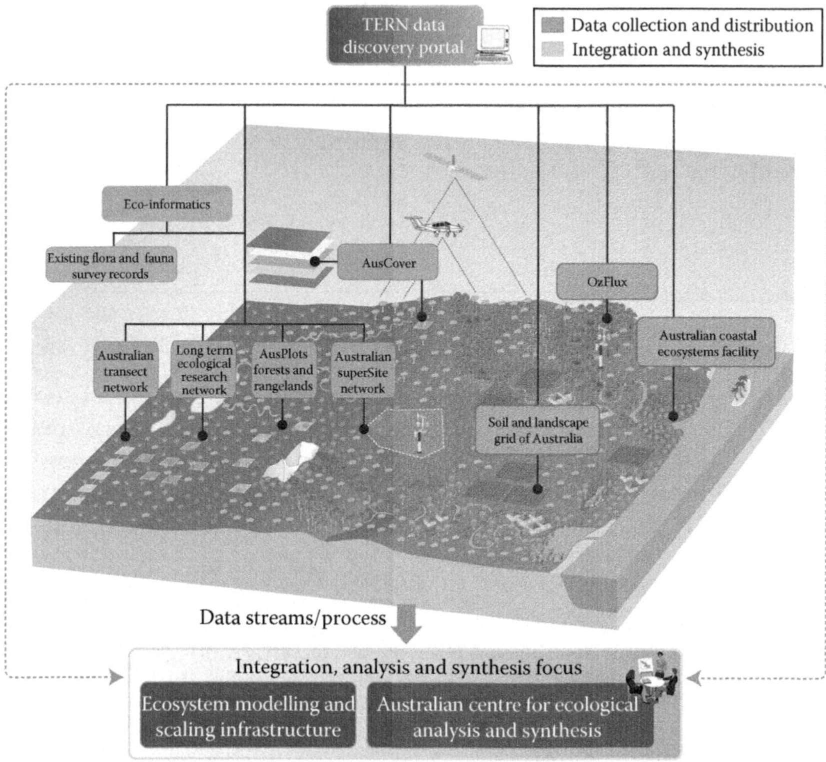

FIGURE 16.1
(See color insert.) TERN's infrastructure for ecosystem science: How TERN fits together.

users of ecosystem science infrastructure could not easily find the best information for their purposes. Inclusiveness also helped alleviate the problem of researchers feeling that they could not be part of the TERN infrastructure, as avenues existed to allow them to join networks and contribute to overall organizational goals.

Having a broad range of facilities with different capabilities also created a unique space for tailoring bespoke solutions to a range of emerging environmental issues. Where previously, a government agency or ecosystem manager would have to navigate their way through multiple, and often competing, research networks to find answers to emerging questions, the creation of national infrastructure provided a place where such information could be gathered from multiple parties with greater ease. So, a manager with interest in ecosystem structure, function, and composition could harness detailed information from plot-based TERN as well as remotely sensed data, models, and other information as they needed. While a range of environmental issues such as climate change, biodiversity loss, and land-use change are unlikely to stop being research priorities, emerging issues such as the effects of excessive carbon use can easily be incorporated into TERN functions. This provides an enduring need for TERN's research infrastructure beyond its current remit.

The existence of the TERN infrastructure provides key benefits for ecosystem science that are of more value than merely financial. The centralization of data management activities through the TERN data discovery portal (http://portal.tern.org.au/) allows the ecosystem science community to share, manage, and harness data in a novel way. Where previously data were inaccessible, being held by individual researchers and institutions, data infrastructure now allows information to be rapidly shared. TERN has commenced the facilitation of a cultural shift in the way Australian ecosystem scientists and managers deal with data and advance environmental knowledge (Clancy et al. 2013). This is done by delivering the national ecosystem data infrastructure that facilitates collaboration and new ways of sharing data. TERN offers a wide-ranging solution that gives consistent data collection, storage, discovery, publishing, licensing, and citation, all of which are freely available to scientists and managers anywhere.

This capacity to move efficiently through the ecosystem data management lifecycle helps to reduce replication and increases the opportunity for collaboration. In turn, this helps TERN to deliver greater returns on public investment in Australian environmental science and management. It also puts Australian scientists and managers in a position to rapidly and securely realize the benefits of the global movement toward open data and data publishing in the context of complex local, regional, and international environmental problems (Clancy et al. 2013). Some practical examples of success from the network include the following:

- Knowledge of Australia's diverse soil and landscape attributes is vital to enable the informed use and management of our soils and landscapes. Detailed digital maps of soil and landscape attributes for the Australian continent and islands have been completed using the TERN infrastructure (http://www.clw.csiro.au/aclep/soilandlandscapegrid/). The soils and landscape grid is a leap forward in both knowledge about soils and in technology and will be used to support food production and environmental conservation, among other things. The grid is a global first.
- TERN scientists developed the CSIRO publishing book—*Biodiversity and Environmental Change: Monitoring, Challenges, and Direction,* which involved 84 environmental professionals (Lindenmayer et al. 2014). It demonstrates the value of monitoring environmental change and biodiversity in the long term and draws out trends in ecosystem change. Its synthesis of ecological studies gives insights into systems and consequences of management

interventions that were previously poorly understood. Decision-makers need this knowledge to develop sustainable management responses. A companion policy handbook, which is free, makes it easier for policymakers to absorb the implications of the science when they consider options for managing different ecosystems (http://www.tern.org.au/Long-Term-Ecological-Research-Network-LTERN-pg17872.html#Policy%20Handbook). Data stories and downloads have also been recently made available from the book for use by educators and students (http://www.ltern.org.au/data/biodiversity-and-environmental-change-data).

- The TERN Australian Centre for Ecological Analysis and Synthesis (ACEAS) working group "Indigenous Biocultural Knowledge" produced a website to display the wealth of projects, research, and management plans about Indigenous Biocultural Knowledge (http://aibk.info/). This website is a platform for the community, both aboriginal and non-aboriginal, to better appreciate the information from indigenous knowledge that can be utilized for decision-making and research. The group won a 2014 Banksia sustainability award in recognition of this significant achievement.
- TERN Ozflux towers are now fully integrated as a national community and as an active part of an international network, with important implications for understanding and predicting local ecosystem productivity. This capability is fundamental to a range of practical outcomes including carbon budgets, improved water use efficiencies in plants and agricultural yields.
- TERN Ausplots has developed a series of agreed national surveillance monitoring protocols for the full range of Australia's unique environments (forests, woodlands, and arid and semiarid systems). This will allow measuring the state and trajectory of soil, vegetation, and faunal assemblages using consistent methods across the whole of Australia and provide critical information relevant to agriculture, water retention, and major ecological questions. The user base of this program includes state and federal government agricultural and environmental agencies, the agriculture and mining sectors, environmental impact consultancies, and community conservation and restoration organizations, and the methods and principles developed are a blueprint being adopted by other international programs.
- The ATN has brought together existing and new research infrastructure to understand the current and future impacts of climate change on Australian ecosystems. By examining a series of environmental gradients, we are able to understand which ecosystems are the most sensitive to these environmental changes, but also which systems are more resilient. This research also allows testing and trialling of key adaption strategies, for example, identification and

enlargement of refugia areas and increasing gene flow connectivity across landscapes. This research is supporting state and government environmental agencies to make key management decisions, particularly in relation to the National Biodiversity Corridors program and the National Climate Change Adaption Research Facility, and is supporting a Citizen Science program to identify the change of distribution of climate-sensitive indicator species.

After a very short operational life, TERN's greatest success has been in bringing together people. Defining this as a case for continued infrastructure investment is difficult but the enormous shift within the ecosystem science community is pivotal to TERN's success. This includes the many extraordinary individuals that have contributed to the design and implementation of TERN as well as external stakeholders who have helped guide and nurture infrastructure growth. The best large-scale scientific research infrastructure programs cannot overlook the imperative of a healthy community of practice. The engagement, enthusiasm, and hard work that have gone into creating TERN have exponentially added to the value of any built infrastructure.

16.5 Challenges of the TERN Infrastructure Approach

The original premise of TERN was broad, optimistic and far reaching. As previously stated, there was no infrastructure that unified terrestrial ecosystem science and ecology in Australia, and in 2009, there was no real role model to help with the design of TERN. This led to the original designers of TERN gravitating toward what Peters et al. (2014) later called the über network design. While this design is recommended by Peters et al. (2014) as the way of the future in network development, with respect to TERN, these features were as much a pragmatic response to the environment presented at the time rather than a deliberative design approach per se. Instead, TERN used an adaptive and integrative approach which while ultimately successful has uncovered many challenges and some solutions to building a large science infrastructure program around the concept of "a network of networks."

16.5.1 Disconnected Community and Culture

A challenge within the terrestrial ecosystem science community is the number of contributing disciplines and the geographic (working in different parts of the country and at different scales) and thematic (working on different problems from dissimilar perspectives) disconnect among the researcher communities involved. As stated previously, infrastructure has not featured in the way terrestrial field research programs are developed and this has stymied inter- and transdisciplinary collaboration. These disconnects have both been a contributing factor to the lack of a cohesive national ecosystem research agenda and have been exacerbated by the absence of a national strategy. However, to answer the environmental questions that need to be managed immediately, the connection within the community needs to happen rapidly and through targeted research infrastructure.

The culture of ecology and ecosystem science has also needed a massive shift away from a philosophy that rewards individuals for behavior that is not collaborative. Data have primarily been kept away from public scrutiny for fear of being misused or used before a researcher has published the information. Research grants are fiercely competitive and rely on an individual's history of publication and successfully attaining previous grants. There is no reward for working collaboratively within the Australian university system; instead, many of the most successful researchers are rewarded for quite the opposite. The sharing of data is likewise unrewarded as data citation is not currently a metric of academic achievement. Therefore, the development of a network of networks such as TERN relied on individual goodwill in contributing intellectual property and data for the greater good. After individuals agree to work together, each network within the TERN infrastructure also had to allow access to resources, provide agreement to follow the TERN vision, and manifest a willingness to trust the process of infrastructure creation. This remains challenging as it asks researchers and networks to choose between multiple competing influences; what is good for one may not be in the best interests of the whole.

16.5.2 Chronic Underinvestment

The Australian government and independent reviewers have recognized the sheer size of the task confronting any organization or initiative attempting to improve the quality and amount of data collected and integration of data and information systems, relative to the scale of the terrestrial ecosystem science issues and their complexity. This is a direct result of inadequate previous investment, particularly in the national and collaborative space. The resources available to TERN under NCRIS (and through the subsequent funding programs) were only enough to make a meaningful beginning to the creation of a tangible whole. These inadequate levels of investment have had two main flow-on effects:

1. Large sectors of the community viewed the investment as a way to address shortfalls in funding terrestrial ecosystem research per se, leading to misunderstandings about the primary aim of the funding and sometimes a disconnect with TERN goals. There was also a perception that funding for TERN was occurring at the expense of other research support programs, which led to resentment on behalf of some of the communities TERN was trying to integrate.

2. Difficulty in identifying the clear "highest-value" initial priorities to be addressed, both within and among the suite of capabilities to support Australia's broader ecosystem science agenda. This led to uncertainty about which networks should be included and which should be created to achieve a goal that was not clearly articulated at the outset.

These issues caused ongoing problems of competition within TERN for resources, and a tendency to promise to do too much with too little (see Section 16.5.3, below). Major progress has been made in addressing the issue of a collaborative versus a competitive approach, but it was a factor in the initial pace of infrastructure rollout and is still an issue that requires careful management into the future.

16.5.3 Short Funding Cycles

The science questions of interest to terrestrial ecosystem research community require a long-term commitment to data acquisition. The standard funding cycle of 1–3 years is extremely short when dealing with natural and even managed systems, where processes operate across temporal scales up to multiple decades due to the way intrinsic and extrinsic drivers affect the system being considered. For example, in wet sclerophyll forests, there can be complex multilayered ecosystem processes of fire, logging regimes, pathogen outbreaks, rainfall, and nutrient-driven impacts on productivity, nested within even longer-term impacts of climatic variability, CO_2 increases, rainforest patch expansion and contraction, etc., all operating on different spatial and temporal scales. While arid ecosystem processes may be somewhat simpler (at least at face value), the time scales can be significantly longer, driven by inherently more variable and extreme climate regimes. The infrastructure required to support this science needs to have two key features: planning for a long-term operational lifespan and a capacity to respond where necessary to the more infrequent and unpredictable (in both space and time) drivers that have an impact on systems. There is a trade-off in dealing with both requirements exacerbated by the chronic underinvestment/fragmented investment discussed earlier; however, the absence of genuinely long-term thinking stymies the effective establishment and operation of critical infrastructure components. This is in strong contrast with the situation in the United States (e.g., NEON that have planned for a 30-year operational lifespan), Japan (Monitoring Sites 1000 Project launched by the Ministry of the Environment in 2003 aimed at monitoring 1000 sites for 100 years, http://www.cbd.int/countries/profile/default.shtml?country=jp), European Union (e.g., Analysis and Experimentation on Ecosystems [AnaEE] program planned for 20-year operation after a 5-year build phase, http://www.anaee.com/images/pdf/esfri-strategy_report_and_roadmap.pdf), and New Zealand (12-year outcome-based investment model, http://www.biosecurity.govt.nz/biosec/sys/strategy/science).

Short and uncertain funding cycles make it difficult to ensure ongoing data collection and delivery systems can be maintained and are an impediment to the application of seamless methodologies, the absence of which causes significant spatial and temporal fragmentation in data and information. For example, the response TERN facilities invariably first receive when discussing issues of method standardization rollout and/or secure data ingestion into national collaborative supporting systems is "why should we engage in this process when it may not be present in the future?"

Short and uncertain funding cycles also have significant impacts on the staff and management of TERN and provide an unwelcome distraction for the true work of the infrastructure. For many years, even core individuals have operated on 1–2-year contracts and have no certainty of future employment. This leads to an almost annual sense of discomfort as individuals often do not have a new contract offered until 4–8 weeks before the old contract runs out. Many excellent employees have been lost from the network as they secure more sustained employment, taking organizational knowledge with them. To create the best possible infrastructure, TERN needs to ensure that staff are not lost and that the growth and development of the network can be sustained by those who understand it best.

16.5.4 No Clarity for Ongoing Operational Costs

Any view that programs funding large-scale ecosystem science infrastructure only need to provide resources to "create" infrastructure is specious. In fact, the original investment program for TERN really was only directed at creating the asset (Capex costs). That this was a mistake is now relatively well recognized; nevertheless, the lack of explicit treatment of ongoing operating costs (Opex) both in assessing their dimension and attribution of responsibility for them is a major structural shortcoming. From a TERN perspective, this shortcoming was somewhat ameliorated by the overlapping nature of some investments and more recent sustenance funding, but this is still not a sustainable long-term solution. Unlike roads, rail and other more classical ideas of infrastructure, ecosystem science infrastructure requires a significant investment in ongoing costs as there is no endpoint in building such a network.

16.5.5 Reliance on Partner Institution Goodwill

A consequence of the level of investment available, the short-term nature of the infrastructure program, the lack of response to ongoing operational requirements, and the geographically distributed nature of the infrastructure is that successful operation is dependent as much on goodwill of participating organizations as the legalities of direct subcontractual arrangements. Well over half of the establishment costs of the infrastructure were contributed by partners, and these partners expect return on investment. This funding approach is clearly a strength when considering the overall value from the support that was received. But this creates additional challenges in managing the trade-offs from all the competing interests of partners, while staying true to the core purpose of the ecosystem science infrastructure.

A flow-on effect from this, in the context of demonstration of the value of the investment, is that it is easy for TERN's contribution to be either lost in the representation of outcomes or, alternatively, for partners to potentially feel TERN is getting credit at the expense of the particular contributing institution. There is no simple solution to this and it is another area of ongoing management concern within the network itself.

16.5.6 No Direct Reporting Responsibilities for Subcontracts

Related to the earlier text is the situation that much of the core leadership and management responsibilities of TERN facilities are effectively delivered as an in-kind rather than a funded component of the infrastructure. Even when the management was a funded component, the relationship between TERN central office and component facilities is often very indirect. This meant that TERN-specific priorities are always competing against a range of other priorities. To a very large extent, this has been managed well but there were some consequences regarding milestone delivery slippage (more in timeliness as opposed to quality). Efforts have been made to address this through but it still is not completely resolved.

Likewise, there is an expectation of participation in planning meetings and communication stakeholder activities that are predominantly borne by institutional and individuals' discretion but there are limits to this. Currently, there are frustrations around this and potential for growing disengagement by some key people, particularly where they are already time-challenged. Again, much of TERN must rely on the goodwill of participants, as with no direct line management processes there is little that can be done if parts of the network or individuals within it behave in a suboptimal manner. It would be self-defeating to merely defund an individual or a network component for perceived wrongdoing as this leads to the potential of destabilizing the whole of TERN.

16.5.7 Overly Ambitious Scope

One of the TERN's biggest strengths in establishment (that of an extremely ambitious scope—the űber network) was also perhaps its greatest shortcoming when it came to implementation as, with the benefit of hindsight, too much was attempted relative to the resources available. This manifested in three negative consequences:

1. Significant achievements in the area may tend to be "marked down" as they are measured against essentially unrealistic objectives and/or sectors of the community feel their own particular infrastructure

needs were not addressed. There is also a great deal of expectation management to be done as there is a disconnect between expectations and reality. For example, some stakeholders and even parts of the network expect fully operational infrastructure within a relatively short time frame, leading to disappointment with good results as they are not the final product that was expected. Building good infrastructure at a continental level is a long-term investment and it may be years or even decades to fulfill early objectives.

2. Tendency for key contributing players to reduce their existing activities due to TERN being seen as now covering this area of responsibility (e.g., while not solely a factor, ongoing reduction in environmental monitoring by Australian state agencies may be exacerbated by TERN investment). This is turn leads to disengagement of key stakeholders and resentment at TERN's activities while at the same time TERN is asking the same stakeholders to contribute coinvestment. The maintenance and expansion of domestic partners has been a challenging but ultimately rewarding experience.
3. Confusion of and/or unrealistic ongoing expectations of the suite of proof of concept activities as being fully fledged and sustainable as ongoing capabilities. For example, there was never enough money invested to sustain all of the TERN facilities at an operating capacity. Smaller-scale parts of the network that produced excellent results as a proof of concept were not later invested in. This has led to the discontinuation of certain activities within TERN, for example, the ACEAS that provided excellent products but had to be sacrificed to ensure data collection could be maintained.

16.5.8 Ability to Distribute Funds for Strategic Priorities

The limits in available funds, the need to subcontract and allocate virtually all funds up front as well as the inherent complexities in amending or establishing new funding agreements were limitations because high-value, cross-facility infrastructure activities were difficult to fund as the program was rolled out. There was no money to invest in new and exciting cross-facility interactions within TERN. Where projects were externally funded, they were often not recognized as TERN products at all, even where TERN had been the catalyst in the development of a project. A great many excellent initiatives remain unrealized within TERN, which causes angst among network members who feel their ideas are being ignored, when in reality they are languishing for lack of funding.

16.5.9 Governance

The governance structure of TERN reflects its early status as a pilot project and the historical competitiveness between universities for limited research grants. At this time, this remains a huge challenge for TERN, which has moved beyond pilot status and needs refreshment in governance. An independent review noted that the board operates only in an advisory capacity and is supported by an advisory committee comprising facility leaders, some of whom are torn between representing the needs of their institutions and those of TERN. There is no clear distinction between science and operational matters.

A governance model that supports strategic decision-making and a stronger client focus is being implemented and will help TERN meet Commonwealth funding needs and its own objectives with greater effectiveness and efficiency. New governance arrangements could include a decision-making board that is accountable to funders and a streamlined implementation committee of facility leaders. A focused operational agenda and a separate science committee to handle science matters would help TERN retain its focus on science outcomes. Funders and the TERN network could consider making TERN a separate legal entity, which would bring about a new suite of challenges but certainly alleviate many of those addressed here.

16.5.10 Politics of Science

As with all large ecosystem science infrastructure projects, the heart of TERN is dependent on ongoing government investment. Governments operate on short-term cycles and are looking for infrastructure programs that lead to fast, tangible and saleable results to satisfy the voters whose money ultimately funds these schemes. As governments change, their departmental structures change and areas such as research infrastructure may be moved from one portfolio to another, under the guidance of a series of ministers and officials. Government priorities also change and there is a constant need for the infrastructure programs to show their relevance to a succession of government parties. Therefore, there can be a direct conflict between the inherent necessity of long-term infrastructure funding and short-term political goals.

Inevitably, funding expands and contracts under new governments and TERN has had to manage this process. This has led to some excellent facilities being "moth-balled" (i.e., ACEAS) and other areas being decommissioned to a "pilot light" level (i.e., eMAST, soils and coasts) (Table 16.1). However difficult it is, the design of TERN has had to manage this contraction for the greater good of the network. Facilities that have been negatively impacted await an increase in future funding that will allow the network to expand once again. This is another process that must be carefully managed to retain partner goodwill and ensure the network survives as a whole rather than becoming a shell of its original design intent.

16.5.11 Skills

In many instances, where researchers are the deliverers of infrastructure, such as the one described here with TERN—there was no formal understanding or training of people who suddenly found themselves managers of large, complex multidisciplinary research infrastructure schemes. One modulating factor against this has been that TERN is funded as part of a larger NCRIS program and there has been some cross-project collaboration in this space, especially in the e-infrastructure domain. Nevertheless, in future explicit training and development of staff should be part of focus. This issue is especially exacerbated by the fact that most participants are only partially funded by TERN and have a significant component of their time and duties focused on non-TERN activities.

16.6 Conclusion: A Test of the Network of Networks Approach (the Űber Network)

Peters et al. (2014) suggest there are a number of main components to building the best ecosystem science research infrastructure. These include

1. Data infrastructure that integrates data from multiple sites and across multiple disciplines allowing scaling
2. A network of networks that sensibly scales and prioritizes areas for new infrastructure and gap filling while allowing for synthesis across these networks
3. Interoperability across components of the network
4. A cultural shift within ecosystem science that embraces open data and collaboration and reaches across institutional and organizational boundaries (summarized from Peters et al. 2014)

TERN began without a clear scope or definition of what the infrastructure would ultimately mature into. We are unapologetic for this lack of program design as situational constraints necessitated harnessing available resources and networks and trying to harmonize them into a coherent whole. We have created world-leading data infrastructure (see Chapter 13) that allows data from across our network of networks to be accessed and analyzed in a completely novel way. We have established synthesis and modelling capabilities that allow these data to provide completely novel environmental outcomes. We have harnessed existing networks and created new ones to fill knowledge and geographic gaps. As resources and urgency permit TERN is ready to expand the network into these key regions and already has a plan to do so. There has been a shift toward network interoperability but this remains an ongoing process as many of the preexisting networks are reluctant to change protocols and procedures. In some instances, it would be nonsensical to expect them to do so, particularly for sites that have been operating for decades or more in a consistent fashion. Where feasible, this interoperability was built into new networks, which will allow greater use of this data in the future. Finally, we have challenged and overcome institutional and bureaucratic boundaries in an attempt to create a meaningful national ecosystem science infrastructure.

The journey of creation and development within TERN has not been simple and many challenges have arisen. The learnings from these challenges have strengthened the network, even when they were extremely difficult. We agree with Peters et al. (2014) that the network of networks approach remains feasible and optimal for the future of large-scale ecosystem research infrastructure, and we are sure that impediments will become smaller over time as the success of such programs is realized internationally. The perfect infrastructure will be global, recognizing the myriad of connections that influence the world's environment. There are significant commonalities of approaches evolving at a global level (e.g., Fu et al. 2010; O'Connor 2010; Schimel 2011 and numerous chapters in this book). And there are advantages of progressing with the network of network approaches at both the national and international levels. However, realism must temper optimism; this process will not happen rapidly but rather be a continuing evolution of ecology and ecosystem science and of the component networks. Realizing this goal will however provide the greatest benefit and fulfill the vision of an integrated ecosystem science network that provides solutions and outcomes as our environment continues to change.

Acknowledgments

TERN is funded by the Australian government through the NCRIS. Thanks to all past and current TERN facility Directors and staff, particularly S. Phinn, S. Long, B. Christiansen, and A. Gibson, who contributed to many components of this chapter.

References

Andersen, A., Beringer, J., Bull, C.M., Byrne, M., Cleugh, H., Christensen, R., French, K. et al. 2014. Foundations for the future: A long-term plan for Australian ecosystem science. *Austral Ecology*, 39: 739–748.

Carpenter, S.R., DeFries, R., Dietz, T., Mooney, H.A., Polasky, S., Reid, W.V. et al. 2006. Millennium ecosystem assessment: Research needs. *Science*, 314, 257–258.

Clancy, T.F., Long, S., Christensen, B., and Gibson, A. 2013. Transforming Australian ecosystem science. Final NCRIS report. www.tern.org.au/rs/7/sites/998/user_uploads/File/TERN%20Science%20Enabling%20Plan_v1_WEB.pdf (accessed Feburary, 2016).

Crutzen, P.J. and Steffen, W. 2003. How long have we been in the anthropocene era? *Climatic Change*, 61: 251–257.

Fu, B., Li, S., Yu, X., Yang, P., Yu, G., Feng, R., and Zhuang, X. 2010. Chinese ecosystem research network: Progress and perspectives. *Ecological Complexity*, 7: 225–233.

Hautier, Y., Seabloom, E.W., Borer, E.T., Adler, P.B., Harpole, W.S., Hillebrand, H., Lind, E.M. et al. 2014. Eutrophication weakens stabilizing effects of diversity in natural grasslands. *Nature*, 508: 521–525. doi:10.1038/nature13014.

Lindenmayer, D., Burns, E., Thurgate, N., and Lowe, A. 2014. *Biodiversity and Environmental Change: Monitoring, Challenges and Directions*. CSIRO Publishing, Melbourne, Victoria, Australia, 610pp.

Lindenmayer, D.B., Likens, G.E., Andersen, A., Bowman, D., Bull, C.M., Burns, E., Dickman, C.R. et al. 2012. Value of long-term ecological studies. *Austral Ecology*, 37(7): 745–747.

McVay, P., Sheridan, S., Shone, H., and Spencer, C. 2014. Review of the terrestrial ecosystem research network (TERN). Internal report to TERN, The University of Queensland, Brisbane, Queensland, Australia. Vista Advisory.

O'Connor, T.G. 2010. *Understanding Environmental Change in Complex Systems: SAEON Core Science Framework*. SAEON, Pretoria, South Africa.

Peters, D.P.C., Groffman, P.M., Nadelhoffer, K.J., Grimm, N.B., Collins, S.L., Michener, W.K., and Huston, M.A. 2008. Living in an increasingly connected world: A framework for continental-scale environmental science. *Frontiers in Ecology and the Environment* 6(5), 229–237. doi:10.1890/070098.

Peters, D.P.C., Loescher, H.W., SanClements, M.D., and Havstad, K.M. 2014. Taking the pulse of a continent:expanding site-based research infrastructure for regional-to continental-scale ecology. *Ecosphere*, 5(3), Article 29, 1–23. doi:10.1890/ES13-00295.1.

Phinn, S., Christensen, R., Lowe, A., Clancy, T., Deed, J., Guru, S., Long, S., and all TERN Facility Directors with input from the TERN Advisory Board, P Tennant (ANU) and A Hoffman (U Melbourne). 2013. TERN Science Enabling Plan 2013–2025. Transforming Australian ecosystem science. http://www.tern.org.au/rs/7/sites/998/user_uploads/File/TERN%20Science%20Enabling%20Plan_v1_WEB.pdf (accessed February, 2016).

Schimel, D. 2011. The era of continental-scale ecology. *Frontiers in Ecology and the Environment*, 9: 311–311.

Schimel, D. and Keller, M. 2015. Big questions, big science: Meeting the challenges of global ecology. *Oecologia*, 177(4): 925–934. doi:10.1007/s00442-015-3236-3.

17

Community-Driven Efforts for Joint Development of Environmental Research Infrastructures

Ari Asmi, Magdalena Brus, and Sanna Sorvari

CONTENTS

17.1 Introduction

17.1.1 Challenges in the Earth System

The earth is a highly complex system, formed by a large variety of subsystems, such as the biosphere, atmosphere, hydrosphere, and lithosphere, interacting by the exchange of energy and mass. All of these domains are highly interlinked, with many concurrent processes operating on a wide range of temporal and spatial scales. Many natural processes are self-organizing, giving rise to a high degree of variety and complexity in the system (Rial et al., 2004). At the same time, the earth system and human society are facing global environmental and societal challenges, many of which are not particular to any one of the domains (Figure 17.1). We can only understand, predict, and possibly solve them if we understand the systems where they originate. This is

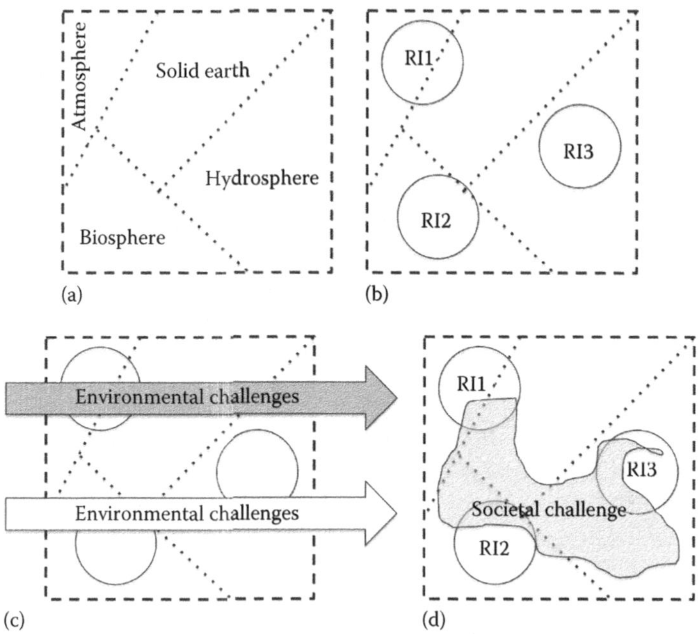

FIGURE 17.1
Cross-disciplinary challenges of environmental research. (a) Earth system is often considered (and researched) from the viewpoint of Earth system domains; (b) infrastructures do not capture the full field of potential research and often do not cover only one domain; (c) environmental challenges do not follow the boundaries or either domains of the research infrastructures; (d) societal challenges create yet another layer of complexity to the system, increasing the need to use different research infrastructures in unison. No single research infrastructure will be able to answer to the environmental or societal challenges—interoperability of the research infrastructures is crucial to get any answers to these challenges.

not possible without environmental research and its supporting global-scale data delivery with analyses from observations, experiments, and modeling.

For the earth system, as for any system, our capacity to understand is predicated on our capability to observe, analyze, and model the earth's subsystems and their interactions. Understanding our environment is not possible by simply extrapolating a given process from the single components of which it is composed (Donner et al., 2009). A different approach is needed—a *systems approach* (Steffen et al., 2001). The efforts to minimize the negative impacts of human activities on environment will only be successful if it is based on holistic knowledge and adequate predictive capabilities. By carefully exploring and studying the basic environmental processes in relevant time and spatial scales and synthesizing our observations into an overall understanding, new scientific breakthroughs can be achieved and environmental challenges can be tackled. However, acquiring such information requires long-term, standardized, openly available, well-documented, and trustworthy observations of the earth system.

17.1.2 Environmental Research Infrastructures

There is a clear evidence of the need to build a global integrated observing system for environmental sciences covering all the domains of the earth system (Houghton et al., 2012). Environmental research infrastructures with their service provision, especially on the delivery of observational data, represent an irreplaceable part of this system (OECD, 2010). They have a significant global role as knowledge providers, bringing together a wide variety of stakeholders to search for solutions to their own specific scientific problems. Indeed, most of the current environmental infrastructures were initially established separately from each other to serve a specific science community and therefore did mainly focus on limited aspects or phenomena of the earth system. The interlinked nature of the earth system however requires better organization of the research infrastructure landscape and scientific communities to transcend the well-established and familiar boundaries of disciplines and domains and work toward common holistic understanding of the environment as one system. Significant efforts toward the interoperability and harmonization of research infrastructures' operations must be taken to address today's environmental challenges.

This and the following sections are concentrated on the European research infrastructure development based on the authors' expertise and experiences. Even though the projects and initiatives are mostly European funded, the overall challenges seem to be largely usable to the other areas of the world as well, as demonstrated in the EU–U.S. research infrastructure projects (COOPEUS and COOP+; see Sections 17.2.2 and 17.4.2).

Most of the pan-European environmental research infrastructures are currently being built, and some are more matured than others. It is crucial to bring them together, to share the knowledge and to develop common solutions at all stages of their development—in their planning, design, construction, and operation to ensure their interoperability and to avoid unnecessary costs and duplication of efforts.

17.1.3 Environmental Research Infrastructure Development in Europe

There is no single accepted definition of "research infrastructure," as the term can mean different things in the different fields of science and in different regions even inside the same field. It is indeed a challenge to provide a comprehensive definition including all the elements of the research infrastructures and at the same time distinguish them enough from the other existing research facilities and organizations.

The most used definition in Europe is the one of European Strategy Forum on Research Infrastructures' (ESFRI), defining research infrastructures as "facilities, resources or services of a unique nature that have been identified by research communities to conduct top level activities in their fields. They may be single sited, distributed or virtual."

In terms of tools, they provide unique instruments or libraries enabling researchers to perform an excellent science. This can be, for example, telescopes when talking about astronomy, colliders in physics, observatories, and databases in the context of environmental sciences or biobanks in the field of life sciences.

The research infrastructures are changing the way in which the research is conducted by providing access to escalating amounts of data, by methodological innovation and novel approaches with respect to how data are gathered and used. Besides the key tools and services they provide to its scientific community, they also play an increasingly significant role in the dissemination of knowledge, scientific information, and know-how by way of training and network building. They propel cooperation across the scientific fields and national borders. By this, they are structuring the scientific community and play a key role in the construction of an efficient research and innovation environment in Europe and beyond.

Research infrastructures differ from other research facilities (Table 17.1) by their focus on providing the services, ability to gather the critical mass of people, knowledge, and funds. Long-term funding, together with a sustainable governance model and legal framework, ensures the long-term stability of the infrastructure and better possibility for strategic development and collaboration with other infrastructure facilities (e.g., on e-infrastructure side).

Research infrastructures are not a new approach to "big science." Many other scientific disciplines cannot conduct research otherwise. Such as the researchers in astronomy (e.g., European Southern Observatory, founded in 1962) or high-energy physics (e.g., European Organization for Nuclear Research, CERN, f. 1954) have, for example, required major international infrastructures also for practical reasons: the large investments needed to build and operate a research infrastructure are too large for any single

TABLE 17.1

Simplified Typical Features of Different Public Research Components

	University Research	Research Institutions	Research Infrastructure
Typical funding model	Projects	Projects and institutional funding	Governmental strategic funding
Typical time period of activity	Short (1–4 years)	Medium to long term	Long term
Main contribution to scientific process	Education of researchers, cutting-edge research	Medium- to long-term continuity in research, operational research, applied research	Maintaining the research capabilities and prerequisites (instruments, standards, policies)

Source: Adapted from Asmi, A. et al., ERIS—Environmental Research Infrastructures Strategy for 2030, ENVRI project publication, DOI:10.6084/m9.figshare.2067537, 2014.

European country to support it alone. There are also philosophical and political reasons for such cooperation, envisioning European states working together and in a coherent manner.

However, the mechanisms to build such international major collaborations needed many ad hoc legal and contractual solutions, making the overall institutional development a long and arduous process (Hermann et al., 1987). A more collective and uniform process for common research infrastructure development was needed.

In 2002, the Commission and the EU member states created ESFRI to support a "coherent and strategy-led approach to policy-making on research infrastructures in Europe." This process has created a specified pathway to create, develop, and evaluate pan-European research infrastructures, with ESFRI roadmap being a key instrument.

17.1.3.1 ESFRI Roadmap

The ESFRI roadmap identifies projects for implementation of new research infrastructures (or major upgrade on existing ones) based on their scientific excellence, pan-European relevance, and, the newly considered, socio-economic impact. A growing number of countries have prepared national roadmaps in order to prioritize the national and pan-European research infrastructures, using the ESFRI roadmap as a reference. This helps the countries to define their national budgets, to facilitate the political support, and to allow a long-term financial commitment.

The ESFRI roadmap is an ongoing process. Nine research infrastructures from the environmental field were included in the last update of the roadmap in 2010 and some of the ecology-related infrastructures, for example, research infrastructure for experimental manipulation of managed and unmanaged terrestrial and aquatic ecosystems (ANAEE, www.anaee.com)

and European Marine Biological Resource Center (www.embrc.eu) were also included from the biological and medical science field. The next update of the roadmap will be launched in 2016. It should be noted that even though the ESFRI organizes the strategic development and prioritization, it is not a funding agency. The main funding responsibility in the construction and operational phase lies in the member countries of the research infrastructure (EU member states). Regional structural funds from the EU and European Investment Bank funding instruments may provide additional sources of funding in the construction phase of the research infrastructures. In addition, the European Commission (EC) provides support for the pan-European coordination of the research infrastructure implementation, for example, by providing project funding for research infrastructure planning and implementation (design study, preparatory phase project, implementation, and cluster-level funding support), through the Horizon 2020 funding program.

It is worth mentioning that inclusion in the ESFRI roadmap does not guarantee that research infrastructure will be actually built. Before the construction and/or implementation itself, many decisions need to be taken with respect to issues such as the identification of funders, the financial plan for sustainability, the governance by involved stakeholders, the site and legal form of the managing organization, the architecture, and the service policies.

As the main funding comes from the member states, many national roadmap processes and funding cycles and policies and decision-making processes need to be run and streamlined in order to establish a European-level budget for the research infrastructures. This requires good coordination and communication practices among the member states and between the stakeholders and science communities in charge of the research infrastructure implementation. With the national-level and EC support funding, the research infrastructure can be implemented; however, the main challenge is to receive the right type of support in the right moment of the process. Thus, it is a challenge of "valleys of death" if the right type of support is not available either at the national or European level when needed. This can cause unnecessary delays and loss of expertise.

17.1.4 Common Challenges in the European Research Infrastructure Field

Most infrastructures serve (or intend to serve) a broad community of researchers working in a particular field of research by providing core data and derived data products, and physical, virtual, and/or data access and computational services for the users that have not been previously available to that community or were operating on a limited project basis.

The design of each research infrastructure service provision was originally based on research infrastructure's main science communities; however, each research infrastructure is also providing its data and services to wider user communities and thus contributing to the wider, trans- and interdisciplinary science questions and grand environmental challenges regardless of its particular field of interest. Because of this broad accessibility across disciplines and user groups, many research infrastructures early and/or key discoveries were not necessarily the ones for which the original plans and design were intended.

There are many common challenges that most of the research infrastructures share, for example, the collection, preservation, archival, quality control, integration and availability of data, governance, policies, performance indicators, training of staff, user community interaction, and providing the computational capability to perform the analyses of interest to researchers. Moreover, while each research infrastructure is separately concerned with the integration of data within its own domain of interest, it is also imperative to find efficient means to integrate data and computation across research infrastructures and domains to serve an increasingly multidisciplinary scientific community. These kinds of common challenges, together with the scientific requirements to answer the earth system challenges, lead to joint collaborative projects between the research infrastructures.

17.2 The Beginnings of Collaboration

17.2.1 ENVRI Project (2011–2014)

After the publication of ESFRI roadmap 2008, it was evident that all the selected environmental research infrastructures will face similar challenges in their implementation and as all the environmental research infrastructures are contributing to the earth system science, the idea of closer collaboration among them was identified. In 2010, the EC launched the call for enhancing the research infrastructure collaboration within the certain science domain (cluster), and the target actions toward collaborative work were initiated. As an outcome of the collaboration that began in 2011, the EC funded a project to cluster the major environmental research infrastructures in the ESFRI roadmap. The main goal of ENVRI "Common Operations of Environmental Research Infrastructures" (Chen and Hardisty, 2014) was to implement harmonized solutions and draw up guidelines for the common needs of the environmental ESFRI projects, with a special focus on issues as architectures, metadata frameworks, data discovery in scattered repositories, visualization, and data curation.

The target of ENVRI project was on developing common capabilities including software and services of the environmental and e-infrastructure communities to enable multidisciplinary scientists to access, study, and correlate data from multiple domains for system-level research. The common standards, deployable services, and tools developed were to be adopted by each infrastructure as it progresses through its construction phase. Due to differences in maturity level among the different research infrastructures, the adoption and implementation of ENVRI solutions within the European environmental research infrastructures is in some cases still underway (see Section 17.4 for current projects).

The project developed a common reference model (ENVRI RM) created by capturing the semantic resources of each participating research infrastructure. The model established taxonomy of terms, concepts, and definitions, which provides a common language for communication to unify understanding. Fundamentally, the model serves to provide a universal reference framework for discussing many common technical challenges facing all the research infrastructures collaborating in the ENVRI project. By drawing analogies between the reference components of the model and the actual elements of the infrastructures (or their proposed designs) as they exist now, various gaps and points of overlap could be identified (Chen et al., 2013). ENVRI RM also helped research infrastructures to get a clearer and more transparent description of their complex data streams and data processing steps, thus clarifying the respective responsibilities in the research infrastructure's construction phase.

However, many long-term benefits stem from more "soft" side of the project: ENVRI has created the nucleus of collaboration and trust among the individual environmental research infrastructures, and common understanding of the importance of their integration and codevelopment. One of the main outcomes of ENVRI was the agreement on the joint vision and strategy for the future development of the (in situ) European environmental research infrastructures (see Section 17.3).

17.2.2 COOPEUS (2012–2015)

The infrastructure collaboration in the European environmental sector was boosted by yet another project funded by the EC. COOPEUS "Strengthening the cooperation between the US and the EU in the field of environmental research infrastructures" (Koop-Jakobsen et al., 2014) brought together European research infrastructures with their U.S. counterparts in order to stimulate the creation of a truly global integration of existing infrastructures.

The main activities in COOPEUS were more oriented toward direct collaboration between the same (or similar) field research infrastructures, especially on the e-infrastructure development. An interesting challenge of the project was that it was separately funded from the EU and the US sides, with different funding models and periods.

The COOPEUS did create a similar sort of communality through an inter-Atlantic collaboration. The project included some interdisciplinary work, especially in the form of a common roadmap for environmental research infrastructure collaboration in EU–U.S. axis. Strictly speaking, as of now (2016), the COOPEUS is not a "past project" in the U.S. side due to the different lengths of the projects in EU and the United States.

17.3 Strategy of the European Environmental Research Infrastructures

In the ENVRI project, it became clear that a common plan is needed to ensure a concerted development of the European research infrastructures in the future. The European environmental research infrastructures strategy (ERIS) was thus a major product of the project and led to more strategic collaboration between the participants. The following sections will summarize the strategy.

17.3.1 Vision of ERIS Document

The ENVRI community defined in the ERIS strategy (Asmi et al., 2014) common aims and goals for the future:

The vision for environmental research infrastructures for 2030 is aiming toward universal understanding of our planet and its behavior. This should result in the evolution of a seamless holistic understanding of the earth system, an environmental system metamodel*, a framework of all interaction processes within the earth system, from solid earth to near space. Scientists that within their own science contribute with data, models, instruments, algorithms, and discoveries should feel that this serves a greater good, namely, a contribution to this understanding.

This holistic understanding will make it possible to approach the entire earth system from different perspectives and choose the portions of the whole conceptual understanding, which are relevant to the problem to be solved. This approach makes it possible to do new and flexible services and answer environmental and societal challenges. Most importantly, this approach is also aiming to be complete: any emerging issues can be tacked on the framework of this understanding, enabling tuning and improvement of the understanding and building the connections to other scientific fields such as social sciences.

Environmental sciences are rapidly moving to become one system-level science, mainly because modern science, engineering, and society are

* A term model here is not to mean a *computational model*, even though that can be one realization. Here the term is meant to describe a self-contained and consistent contextual model, which describes our understanding of the system, its linkages, and feedbacks.

increasingly faced with complex problems that can only be understood in the context of the full overall system they belong to. There are several reasons and enablers for this shift:

- *Technology push.* Technology innovations on, for instance, detectors and sensors with ever-increasing resolution allow deep observations of scientific phenomena important for the better understanding of the whole system. Connection between these new observations and the whole earth system requires ways to integrate between the domains. In addition, information technology innovations, such as digitalization of collections, also unlock resources at a systems level.
- *Demand pull.* The questions scientists are faced with nowadays (not only arising from curiosity but also from policies, like IPCC, IPBES, and GEO/GEOSS and societal needs) can simply not be solved using the traditional sources of information. Without access to information from adjacent disciplines, the answers scientists can give will increasingly be partial and incomplete and therefore less groundbreaking or even useful.
- *Globalization.* Like in economics and society, science is experiencing an upscaling due to globalization. Establishing and managing big data and information repositories often demand an international effort. This can also be observed from the ever-increasing aggregation of research funding such as ESFRI.
- *Resource integration.* Never before did researchers of so many domains have such a wealth of resources at their disposal. The integration of these worldwide available resources has further fueled system-level research. An important contribution of e-science as a system-level science is its potential for integration of information.
- *Science integration.* These developments in science, in general, approach the Earth as an integrated system and offer an excellent opportunity to include other related sciences, such as social sciences and life sciences.

17.3.2 Actions and Suggestions

The vision requires understanding a conceptual model that is capable of providing a definitive answer, which is reliable and credible. However, building such an integrated view is not straightforward. The development of this view requires resources, and we identified three interdependent sources of resource capital, which need to be improved to achieve this vision:

1. *Technological capital:* Capacity to measure, observe, compute, and store. This requires material, technologies, sensors, satellites, floats; software to integrate and to do analysis and modeling and processing;

building observational, computational, and storage platforms and networks.

2. *Cultural capital*: Open access to data, services, etc., from other research infrastructures. This requires rules, licenses, citation agreements, IPR agreements, technologies for machine–machine interaction, workflows, metadata, data annotations, etc. The goal should be to contribute always to the standard understanding of the systems approach; research infrastructures work together as a community on the policy level.
3. *Human capital*: Specialists to make it all work. This requires both data scientists and discipline scientists. The need of disciplinary specialists is not going away, but answering the societal and environmental challenges needs experts capable of truly inter- and transdisciplinary science, working between the typical scientific boundaries. This also calls for "generalists" who oversee more than just their own discipline. These specialists need their own curriculum and training. Additional human capital need comes from the challenges to include and train citizen scientists.

17.3.2.1 Technological Capital

The environmental science research community has much to gain from the rapid advance of technology in many areas. The research infrastructures in the field are mostly focused on collecting and acquiring observational data, which can be used to gain a comprehensive understanding of the earth system. The technology used for acquiring, storing, and processing environmental research data includes instrumentation such as sensors, floats, radars, and integrated measurement stations, as well as e-infrastructures (in their wide meaning), including networks, computing servers, and storage, together with software components enabling processing of these data. ERIS strategy listed the following key parts of the technology development, with examples of concrete actions:

- *Improved observation systems (instruments and platforms)*: Enlarging and developing observation coverage, supporting cross-use of experimental research platforms and vessels, scientific gap analysis, station integration between research infrastructures and new observation technologies.
- *Improvements of e-infrastructure*: Establishing fast network connections and data reduction on-site, improving data storage capabilities, developing software solutions for environmental big data applications, common interface development, common API development, and competence center collaboration.

17.3.2.2 Human Capital

To fulfill the need for a systems approach, a complete new type of data specialists is needed, concentrating on producing and maintaining the data products of the research infrastructures. Earth science data specialists have a strong information technology background, but also are well versed in the scientific questions and methodologies of the earth system sciences. They are capable of handling large data sets, understanding the limitations and benefits of different earth system observations and experiments, and are also responsible for the development of e-infrastructure of the research infrastructures. As they are concentrating on the data production and data assimilation, their career paths include far more data publications than other types of publication, and thus their positions are best developed using the data citation services. The career paths of these specialists require completely new positions recognizing their unique services to the science community. Additionally, the need for additional observations and data analysis has brought out the need for citizen science activities. These additional human observers, data scientists, science communicators, and innovators can be a great boon to the research. However, there is much to do to find out how to actually effectively use these resources. Policies on, for example, data integrity, research credit, funding, and trustworthiness are major challenges on using citizen scientists in research infrastructures, and much research and training will be needed in the future if they are to be integrated in the operational research.

The following key developments were identified, with partial associated actions:

- *Curriculum development*: Establishing the career paths for data scientists in the earth system sciences and establishing education and training requirements for data scientists
- *Multidisciplinary training*: Establishing curriculum and programs for cross-disciplinary studies and establishing cross-disciplinary mobility programs
- *Citizen science integration*: Establishing the methodology and practices for citizen scientist interaction with the research infrastructures

17.3.2.3 Cultural Capital

The cultural capital development describes the advances in nonmaterial aspects of the research infrastructures and research landscape. The research infrastructures do not only consist of physical instruments, produced data, or people handling them. A significant—for some cases the main—part of the research infrastructure is the development of methods, policies, and concepts needed for successful research endeavors. Without such developments, the goal of having integrated and complete understanding of the earth system cannot be realistically achieved. The following are the key developments and actions we suggested:

- *Disseminating the idea of a common goal in the earth system sciences*: Publication of the strategy and goals of the European environmental research infrastructures
- *Building the culture of open research*: Requirement of open data access as the standard clause in any public science funding, common and widely used data citation mechanisms and citation indices, taking data publication and citation as a key parameter for merit determination in research infrastructures
- *Standardize the language and terminology*: Harmonization of terminology within the earth system sciences, establishing standards for scientific data including contextualization, and developing and implementation of machine-readable documentation and licensing standards for the earth system data
- *Enlarging the view*: Integration efforts between the earth system communities, and toward including human activities on the earth system analyses and databases
- *Organizational framework*: Establishing coordination structure of environmental research infrastructures, dissemination of the standard model of the planet idea to scientists, funders, and end-user groups.

17.4 Current Activities (as of 2016)

The success and results of the previous ENVRI and COOPEUS (EU) projects motivated the community to continue its collaboration. In the Horizon 2020 program of the EC, two collaboration projects were proposed and successfully funded: ENVRIplus continuing the work of ENVRI and COOP+ similarly building on the COOPEUS activities.

17.4.1 ENVRIplus (2015–2019)

The ENVRIplus (www.envriplus.eu) follows in the footsteps of its predecessor with an aim to enhance the collaboration among the environmental research infrastructures in Europe and to build common synergistic solutions for pressing issues in research infrastructure construction and implementation. However, the ENVRIplus does not go just one, but many steps further. Whereas the main focus of ENVRI was on data and improved software solutions, ENVRIplus has six main objectives, called themes, spreading from the development of common technical solutions, through work on policies for multidisciplinary access to research infrastructures to transfer of knowledge through training activities, etc.

As said before, the overall course of this 4-year project is directly built on the earlier ENVRI experience, but due to the larger infrastructure participation (20 research infrastructures, projects, and networks) and budget (15 million euros), the project is far more ambitious and extensive.

17.4.1.1 ENVRIplus Themes

The ENVRIplus activities are divided to 19 work packages, organized in 6 overreaching activity groups, called "themes." Observant reader will notice many similarities between the ERIS strategy (Section 17.3) and the activities of the ENVRIplus. The themes of the ENVRIplus are as follows:

1. *Technical innovation*, working toward improved research infrastructures' abilities to observe the earth system, particularly through developing and testing new sensor technologies, harmonizing observation methodologies, and developing methods to overcome common problems associated with distributed remote observation networks.
2. *Data for science* will generate common solutions for shared information technology and data-related challenges of the environmental research infrastructures. Especially for challenges related to data and service discovery and use, workflow documentation, data citations methodologies, service virtualization, and user characterization and interaction.
3. *Access to research infrastructures* aims at developing harmonized policies for access (physical and virtual) for the environmental research infrastructures, including access services for multidisciplinary users.
4. *Societal relevance and understanding* will investigate the interactions between research infrastructures and society: find common approaches and methodologies to assess the research infrastructures' ability to answer economical and societal challenges, develop ethics guidelines for research infrastructures, and investigate the possibility to enhance the use of citizen science approaches in research infrastructure products and services.
5. *Knowledge transfer* will ensure the cross-fertilization and knowledge transfer of new technologies, best practices, approaches, and policies of the research infrastructures by generating training material for their personnel to use the new observational, technological, and computational tools and facilitate inter-research infrastructure knowledge transfer via a staff exchange program.

6. *Communication and dissemination* work toward creating research infrastructure communication and cooperation framework to coordinate activities of the research infrastructures toward common strategic development, improved user interaction, and interdisciplinary products and services.

17.4.1.2 Participation Concept

The environmental research infrastructures participating in ENVRIplus are illustrated in Figure 17.2. They represent a major investment from the member states. Even if counting only ENVRIplus research infrastructures in the ESFRI 2010 roadmap (the inner circle of the Figure 17.2), the construction costs are in excess of 1600 million euros, with annual running cost exceeding 300 million euros. The ENVRIplus collaboration and synergies between these research infrastructures thus have a tremendous potential for advantages in terms of added value, return for investment and support for science, and successively also economy and society.

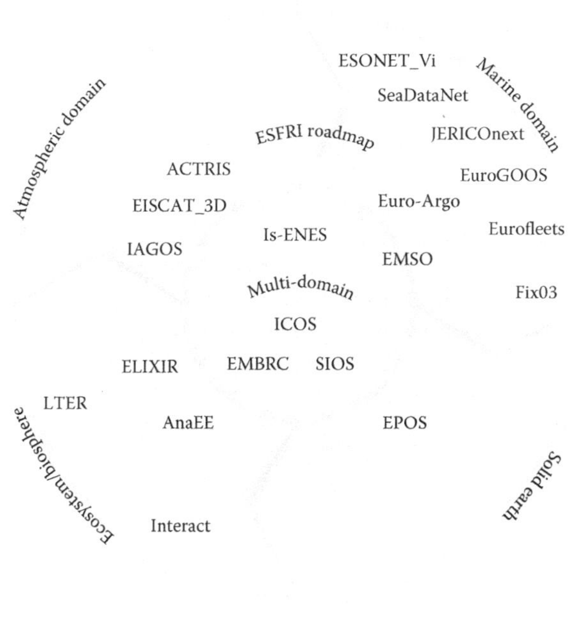

FIGURE 17.2
Research infrastructures from different earth system domains participating in ENVRI and ENVRIplus projects (indicated by symbols). More information about the collaborating infrastructures can be obtained at www.envriplus.eu/wp-content/uploads/2015/08/Booklet_EnvRIs-in-ENVRIplus.pdf.

The ENVRIplus research infrastructures are diverse in institutional aspects and maturity scales. Some of the research infrastructures have already built their facilities and/or are organized as legal entities (e.g., European Research Infrastructure Consortia or associations), some are still developing their logistical and administrative forms and can only be represented by key partners of preparation projects. This heterogeneity in organization creates further challenges for interoperability and is the key background of ENVRIplus: its approach is to integrate as much as possible over this diversity and organize cross-fertilization among research infrastructures. For that reason, ENVRIplus comprises in its partnership not only research infrastructures in the ESFRI roadmap but also research infrastructure projects that have reached sufficient maturity level (outside circle of the Figure 17.2).

The ENVRIplus concept is based on joining the efforts of the research infrastructures in the four environmental domains (atmospheric, marine, ecosystem/biodiversity, and solid Earth domains) to answer the challenges they share. Solutions developed within one domain may also be deployed by research infrastructures in other domains. Developing trans-boundary research requires exploring scientific questions at the intersection of different fields, conducting joint research projects, and developing concepts, devices, and methods that can be used to integrate knowledge. Establishing a cluster of environmental infrastructures is a starting point for reshaping the strategy to conduct research across disciplines.

17.4.1.3 Expected Impact

Impact of ENVRIplus products and solutions is expected on several different levels such as

- *Impact at single research infrastructure level*: The European environmental landscape is fast evolving and new research infrastructure communities are starting their operations. ENVRIplus will support these new research infrastructures and networks to develop and build the necessary structures and services for their users. They can directly adapt and use ENVRIplus services, tools, and products developed in the different work packages. Common modules at research infrastructure level also directly benefits wider communities of multidisciplinary users.
- *Impact on the EU research infrastructure structuration*: One of the main expected impacts of ENVRIplus will be on broadening, supporting, and sustaining the European environmental research infrastructure landscape while reducing the costs of the individual research infrastructure planning, construction, and operations. For example, the take-up of the ENVRI RM at the individual research

infrastructure level will help research infrastructures to adopt common approaches to the whole data management life cycle. This will lead to better interoperability among research infrastructures. Moreover, better interoperability will result in better economies of scale and improved, user-friendly access with greater use of the data and other research infrastructure services.

- *Impact on quality, quantity, and diversity of services offered by research infrastructures*: ENVRIplus improves the innovation capacity of the individual research infrastructures and domains by enabling cross-fertilization of the new knowledge to new solutions and services. An example for that is the implementation of a common data citation system. A single research infrastructure could have a small impact in such general challenge, but the ENVRIplus consortium as a whole may play an important role in cooperation with other projects and initiatives. Limiting duplicated actions and promoting cost-effective solutions for data management, implementing joint guidelines for access, training of staff, etc., will also improve efficiency of research infrastructures, which may revise internally their allocation for services turning them into benefits for users. The level of training in ENVRIplus, together with special outreach to secondary school students will raise educational levels across the European Research Area (ERA).
- *Use of data and data products*: The main barriers for the use (or reuse) of the cross-disciplinary data are not only technology or data issues per se. Several cultural and human-related issues may hinder the use of the cross-disciplinary data. Increased trust and user experience on the research infrastructure data and services is thus another important expected impact of ENVRIplus. Availability of multidimensional data and data products from environmental research infrastructures increases environmental information and brings innovative use of existing information.
- *Building new scientific knowledge*: Multidisciplinary/interdisciplinary research is an important avenue to new scientific breakthroughs. The complex challenges posed by the changing environment will require capacity to address new research concept across different disciplines.

The produced solutions, services, systems, and other project results will be made available to all environmental research infrastructure initiatives in Europe and other places through its Virtual Community Platform (ENVRI VCP, www.envri.eu; to be launched in May 2016). ENVRI Community Platform will be the long-term repository, communication and collaboration tool, and meeting point of all the environmental research infrastructures, projects, and networks. The platform ensures that the ENVRIplus is serving the entire environmental research infrastructure community regardless of their direct participation in the project.

Another key part of the ENVRIplus project, particularly important for the overall infrastructure development in Europe, is the establishment of the Board of European Environmental Research Infrastructures. This crucial component funded by the project is the first concrete step toward creating a strong long-term collaboration, decision-making, and information sharing platform. This board consists of directors or top managers of participating research infrastructures and thus has a strong voice as the representative of the whole environmental research infrastructure community and can act as an intermediary between the research infrastructure user communities (especially, researchers) and the funding agencies as well as international initiatives.

17.4.2 COOP+ (2016–2018)

COOP+ is a partial continuation of the COOPEUS project in the European side. Due to the different context of the Horizon 2020 calls, the scope of the COOP+ is partially smaller (in the context of the earth system domains covered) but in some sense larger (in the context of covered fields and interdisciplinary actions and geographical scope). The general goal of COOP+ is to strengthen the links and coordination of the ESFRI Research Infrastructures related to marine science, Arctic research and biodiversity with international counterparts and to leverage international scientific cooperation and data exchange with non-EU countries (the United States, Canada, Australia, and Brazil) aiming at creating a common ground for the development of a global network of research infrastructures. The project has an ambition to become the central hub for worldwide collaboration of the research infrastructures involved, coordinating all their common activities and fostering international agreements. In particular, the project will

1. Analyze how to address cross-disciplinary global challenges by joining resources including international research infrastructures in the same thematic area. Explore the complementarity of top-down and bottom-up approaches and understand how to combine geographical and temporally explicit scales.
2. Define the basis for an open coordination framework for global cooperation for research infrastructures in the field of environmental science by providing support to new agreements on reciprocal use or access to research infrastructures, openness, joint development of new resources, including cofinancing.
3. Develop and implement a plan for oriented synergistic dissemination and exploitation for research infrastructures at global level.

Identify the different relevant stakeholders, resources, results, and structures in the complex map of environmental research to assure the impact of research infrastructures' products and results. Identify new opportunities for further involvement of research infrastructures and, if possible, of European small- and medium-sized enterprises.

4. Assure the exchange of information and training on best practices and know-how, both for research infrastructure managers and technical personnel and for their user communities, exploiting their complementarity and different experiences.
5. Promote the use of global integrated platforms, in particular, those related to Group on Earth Observation, as a way to accelerate global access to data and address interoperability.

17.5 Conclusions and Outlook

Our planet faces many challenges. To address them for the earth and the society, we need ways to apply science across the traditional national, disciplinary, and earth system domain boundaries. Research infrastructures, as long-term strategic initiatives, play a strong role in generating such integrated and cross-disciplinary science. Current and past activities in Europe have already shown how powerful a tool such collaborations can be.

It is crucial that cooperation will continue in the future. Working just from one (sub)continent is not enough for understanding the global challenges. The collaboration within Europe (ENVRI and ENVRIplus activities) and with major research infrastructures outside Europe (COOPEUS and COOP+ activities) represents a key element in the environmental research development for the next decades.

References

Asmi A., Konijn J. and Pursula A. *ERIS – Environmental Research Infrastructures Strategy for 2030,* ENVRI project publication, available at http://project.envri.eu/, DOI:10.6084/m9.figshare.2067537, 2014.

Chen Y. and Hardisty A. A common reference model for environmental research infrastructures. In the *iLEAPS Newsletter: Special Issue on Environmental Research Infrastructures,* 2014. http://www.neonscience.org/sites/default/files/biblio-files/iLEAPS_Research_Infrastructures_2014.pdf (accessed March 14, 2016).

Chen Y., Martin P. and Magagna B. A common reference model for environmental science research infrastructures. In *the Proceedings of the 27th Conference on Environmental Informatics 2013*, Shaker Verlag, pp. 665–673, 2013.

Donner R., Barbosa S., Kuths, J. et al. Understanding the earth as a complex system—Recent advances in data analysis and modelling in Earth sciences. *European Physics Journal-Special Topics*, 174, 1–9, 2009.

Hermann A., Krige J., Mersits U. and Pestre D. (eds.). *History of CERN: Launching the European Organization for Nuclear Research*, Vol. 1. Amsterdam, the Netherlands: North Holland, 1987.

Houghton J., Townshend J., Dawson K. et al. The GCOS at 20 years: The origin, achievement and future development of the Global Climate Observing System. *Weather*, 67(9), 227–235, 2012.

Koop-Jakobsen K., Loescher H., Powers L., Huber R. and Waldmann Ch. COOPEUS – Building the framework for information exchange between the US and EU Environmental Research Infrastructures. In the *iLEAPS Newsletter: Special Issue on Environmental Research Infrastructures*, 2014. http://www.neonscience.org/sites/default/files/biblio-files/iLEAPS_Research_Infrastructures_2014.pdf (accessed March 14, 2016).

OECD. Global science forum: Establishing large international research infrastructures: Issues and options, 2010. http://www.oecd.org/science/sci-tech/47027330.pdf (accessed March 15, 2016).

Rial J.A., Pielke S.R.A. and Beniston M. Nonlinearities, feedbacks and critical thresholds within the earth's climate system. *Climate Change*, 65, 11–38, 2004.

Steffen W., Tyson P. and Jäger J. Earth system science: An integrated approach. *Environment*, 43(8), 21–27, 2001.

18

Synthesis Centres: Their Relevance to and Importance in the Anthropocene

Alison Specht

CONTENTS

Abstract

In a world where there is increasing demand for solutions to urgent and complex ecological questions requiring multidisciplinary participation and transdisciplinary insight, but where the abundance of information has become overwhelming, the analysis and synthesis centre has emerged recently as an effective and arguably indispensible tool. Using staff with multidimensional skills and considerable experience, the synthesis centre provides a supportive environment for groups of people with diverse disciplinary, organisational, cultural and geographical origins to work together using existing data to address big science questions. It fills a niche in time-pressed and discipline-centred lives, enabling wisdom to be drawn from a heterogeneous mass of data.

The successful performance of synthesis centres depends on a unique marriage of knowledge of learning, collaboration and teamwork behaviours, data management expertise (including legal aspects) and scientific understanding, coupled with the knowledge of national and international scientific and organisational landscapes.

The synthesis centre approach, described in this chapter, has proved to be remarkably successful, with high publication output and citation rates, and the delivery of many data packages for future use. Participation in a synthesis centre provides users with a uniquely stimulating experience and enables them to forge new and strong professional relationships. Without such centres, transdisciplinary insights will continue to be limited to existing partnerships and short-term goals and will lack the probity that original work on data and information can provide.

Keywords: Analysis and synthesis centre, Data deluge, Collaboration, Working group, Data reuse, Data delivery

18.1 Introduction

The technological changes experienced in our daily lives over the last 50 years as a result of the explosion in computational power are unprecedented (Marx 2013). They have led us into the era of 'big data' and to the emergence of fields such as genomics and spatial and sensor sciences, but they have also presented us with many challenges. Never before have we had the ability to compile and access such a wide range of data in order to solve complex problems, but as expressed by Wilson (1998), 'We are drowning in information while starving for wisdom. The world henceforth will be run by synthesizers, people able to put together the right information at the right time, think critically about it, and make important choices wisely'. This highlights the need for specialised expertise, dedicated space and time to deal with the challenging aspects of the 'data deluge' (Science Staff 2011, Thessen and Patterson 2011). Lack of sufficient and supported time for appropriate analysis and reflection has been blamed for a decline in confidence in the scientific enterprise and in its inability to handle the superabundance of information (Siebert et al. 2015).

The explosion in computational power and capacity has also been accompanied by, or indeed resulted in, an unparalleled rise in awareness of a range of complex, interconnected global social and environmental problems that need urgent attention. Understanding complex systems is itself challenging (Hmelo-Silver and Azevedo 2006), and solving today's complex problems rests not only on the availability of suitable data and our technical ability to deal with it but also in our ability to collaborate effectively across the necessary disciplinary and organisational boundaries (Crowston et al. 2015). Despite increased connectivity through social media, the discovery and development of working relationships with colleagues, especially those from other disciplines and across sectoral and geographical boundaries, requires directed effort (Cheruvelil et al. 2014, Specht et al. 2015a). Several challenges present themselves: (1) how to find the time to gather together the necessary information and to analyse and synthesise it in order to gain new understanding and insight; (2) how best to make sense of this information when we need to converse with each other across disciplinary, organisational and geographical boundaries; and (3) how to respond when our training systems are slow to react and careers are governed by disciplinary rewards.

The great explorations of the sixteenth and seventeenth centuries presented similar challenges. Explorers returning to the 'old world' with a deluge of information from the 'new world' challenged understanding of the meaning of civilisation. These discoveries, driven in large part by the acquisition of territory and wealth, drove the hunger of the community for exotica, the development of science and the systematisation of knowledge. As the bourgeoisie grew, with their aspirations, there was a demand for wider participation in discourse, something previously limited to the aristocracy (Chartier 1991, Kale 2002). The creation of egalitarian, often constructed, fora such as the Salons of France and greater Europe in the eighteenth century (Konczacki 1986, Vincent 2014), which continued until the early twentieth century (Kale 2002), was a societal response to create opportunities to assimilate, analyse and make sense of the plethora of new information and the accompanying cultural and social changes. Significantly for the purposes of this chapter, salons were described, *inter alia*, as 'civil working spaces of the project of the Enlightenment*' (Goodman 1994, p. 53) lying 'outside rule-bound institutions and established cultural bodies' (Chartier 1991 cited in Goodman 1994, p. 69). Their continuation beyond the Enlightenment attested to their popular utility, providing a protected space in which matters of concern of the day could be debated and discussed (Kale 2002). These 'salons' and their ilk were somewhat passive, however, compared with the demands and goals of the ecological/environmentally based *analysis and synthesis centre*, which has emerged arguably as a response to two pressures: the challenges of dealing with the 'information age' in an effective manner and the increasing pressure to answer global problems requiring critical analysis and integration of transdisciplinary information.

18.2 Analysis and Synthesis Centres

The well-established research workflow follows a classical pathway, from review of the literature (with particular focus on gap analysis), hypothesis setting, experimentation or observation, analysis of the results and

* The Enlightenment, *Siècle des lumières*, and *Aufklärung inter alia* are retrospective definitions of a phase of European history in the eighteenth century in which there was a growth in advocacy of rationalist thought (Lough 1985).

communication of the findings through the literature. Historically, this is largely an individual endeavour, certainly so for PhD students, the fundamental training for all scientists. Rarely do scientists 'synthesise material from linking facts and fact-based theory across disciplines to create a common groundwork for explanation' (Wilson 1998). Scientists by themselves generally do not have the time, collaboration skills, networking knowledge, resources or professional incentives to bring together the components required to build a broader understanding of ecosystems or to manage human interactions with them more effectively (Costanza 2003).

Scientific synthesis generally relates to an inductive process of integrating disparate elements (i.e. concepts, data, methods, analytical results) from one or more disciplines to develop a novel integrative insight or model as a primary outcome (Sidlauskas et al. 2010). Synthesis can be systematic and tied to particular methodologies that are quantitative or qualitative (Cooper and Hedges 2009). In its simplest form, 'synthesis' is a creative activity in which the aim is to produce new insights or outcomes that are greater and more meaningful than the constituent parts.

The National Center for Ecological Analysis and Synthesis (NCEAS) was established in the United States in 1995 in response to the recognition in the ecological community that the activity of synthesis was both essential and vastly under-supported (Costanza 2003), although 'synthesis centres' servicing data repositories had emerged previously (e.g. the synthesis centre attached to the Chinese Ecosystem Research Network [CERN] in 1988). NCEAS was designed to facilitate data analysis and synthesis around topics of concern (Rodrigo et al. 2013). NCEAS has been enormously successful in facilitating the production of many very highly cited papers, identifying and fostering many new frontiers in ecology, establishing the value of analysis and synthesis of existing data, initiating the development of the data management workflow and stimulating the development of many similar centres around the world (Table 18.1).

Each of these centres has interpreted the analysis and synthesis centre approach differently, dependent on their organisational structures and circumstances of how they come about. The Australian Centre for Ecological Analysis and Synthesis (ACEAS, established in 2010), for example, was designed to be inclusive of scientists, managers and policymakers. The intention was to fast-track the assembly of information around big science questions, enabling new insights and knowledge to guide long-term research and monitoring: the challenge of Lindenmayer and Likens (2010, p. 108). As stated in the funding application for ACEAS, it would '... support the integration and synthesis of ecosystem data and information across the many relevant disciplines and institutions in Australia... thereby enhancing and accelerating our knowledge and understanding of science particularly to improve our policy and management decision-making. ACEAS [will] achieve this by the sponsorship of integrative activity, mainly through working groups and workshops, in which groups of scientists and managers amalgamate

TABLE 18.1

Vital Statistics of Ecological (in the Broad Sense) Synthesis Centres Arranged from the Oldest Centre to the Newest Centre

Centre (Acronym)	Dates	Location	Linkage with Data Infrastructure
Chinese Ecosystem Research Network (CERN)	1988	China	Integral component of the CERN—observatory network
National Center for Ecological Analysis and Synthesis (NCEAS)	1995	USA	Created the Knowledge Network for Biodiversity, a major partner in DataONE
National Evolutionary Synthesis Center (NESCent)	2004–2015	USA	Supported the development of Dryad, a partner in DataONE
Biodiversity Synthesis Center (BioSync)	2007–2013	USA	No direct link
National Institute for Mathematical and Biological Synthesis (NIMBioS)	2008	USA	No direct link
Canadian Institute of Ecology and Evolution (CIEE/ICEE)	2008	Canada	No direct link
John Wesley Powell Center for Analysis and Synthesis	2009	USA	Associated with the USGS data repositories
Australian Centre for Ecological Analysis and Synthesis (ACEAS)	2010–2014	Australia	A component of the Terrestrial Ecosystem Research Network
Centre for the Synthesis and Analysis of Biodiversity (CESAB)	2010	France	A component of the Foundation for Research on Biodiversity (FRB), which has a fledgling data repository, ECOSCOPE
National Socio-Environmental Synthesis Center (SESYNC)	2011	USA	No direct link
Environmental 'Omics Synthesis Centre (EOS)	2012	UK	Linked with the Centre for Ecology and Hydrology and their infrastructure
Synthesis Centre for Biodiversity Sciences (sDiv)	2012	Germany	Linked with iDiv and through that TRY and other initiatives
Tansley Working Groups	2013	UK	No direct link

Note: More information and links to centre web sites can be found on www.synthesis-consortium.org.

unpublished and published data to develop a new understanding of an ecosystem problem'.

The focus of ACEAS was largely on transdisciplinary* integration of biophysical science and linked social science through environmental policy

* Transdisciplinary research is defined to mean research involving multiple scientific disciplines in collaboration with policy and management (not solely citizen or community engagement) (Lynch et al. 2015). This contrasts with interdisciplinary research which is defined here as research between academic disciplines in a nonadditive or non-transformational way, and multidisciplinary research as research between academic disciplines in an additive manner.

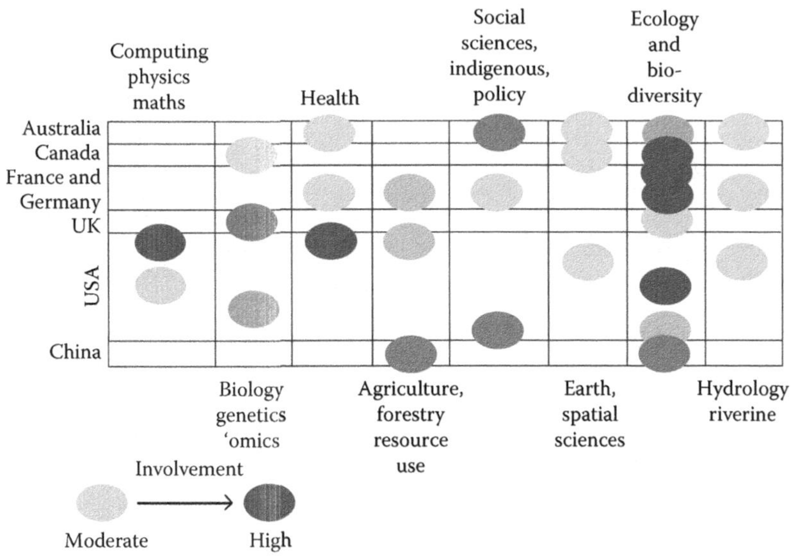

FIGURE 18.1
Diagram to illustrate disciplinary coverage of ecological and biodiversity synthesis centres.

and management. While NCEAS continued to have a broad non-engineered platform, centres in Europe (sDiv and CESAB) have focused on biodiversity, NESCent and EOS on evolutionary genomics and SESYNC (established in 2011) on the interface between ecology and society (Figure 18.1). Most have followed the NCEAS model of operation to a greater or lesser extent.

18.2.1 Function of Analysis and Synthesis Centres

The salons of the 1700s and 1800s brought together discussants, who would normally not meet, within an egalitarian social structure, which was revolutionary at the time. Synthesis centres do the same and have some other common programmatic elements. All centres provide funding for groups of people (working groups) to address complex questions using existing data. The groups are usually self-identified and submit proposals through annual funding rounds. Some centres provide merit-based or professional development funding for postdoctoral students, either separately (EOS) or in association with working groups, either directly linked (e.g. CESAB and sDiv) or independent (e.g. NCEAS and NESCent). There is no funding provided to gather new data, the centre proposition being that there are plenty of data in the world unused or unrealised, and analysis of these is the aim.

Together with the analysis of existing data, all centres have as central aims the analysis and synthesis of complex information (data) across normally unbridged disciplinary, organisational and geographical boundaries. Most follow a collaborative crowd-sourced working group model, where groups of people from different sectors who have some input to bear on the solution of a question propose that they can provide a new insight into an important problem if given the opportunity to work together. These 'sectors' are generally disciplinary but may also be organisational and geographical, as required by the question at hand.

The element of 'work' is critical to the function of the synthesis centre: these groups of people have to demonstrate the existence of data, qualitative or quantitative, with which to work. Synthesis centre participation is not a process of presentation (active) and critique (passive), or debate and conversation, although at various times these might occur. Active collaboration to analyse and synthesise real data causes many participants to work with people whom they may not have met before the working group was established (in ACEAS only half the group members knew each other personally before joining the group, and early analyses of CESAB participants indicate the same proportion) and with people from other disciplines and organisations. This can present communication challenges, as gulfs between disciplinary and organisational paradigms and norms have to be discovered and bridged. It is through the work process that trust is established between parties, critical to the success of the working group (Specht et al. 2015b).

Most centres advocate a flat organisational structure where, however senior or eminent an individual member may be, all participants are regarded as of equal importance for the achievement of the goal, that is, the product of a working group. Deliberations within the group are expected to range freely but confidentially; the synthesis centre provides a safe place to test ideas and interpretations, trial models, and argue semantics, without fear or favour. Some centres have ethics policies to ensure participants are reminded of, and commit themselves to, standards of collaborative behaviour.

Emphasis is usually placed on ensuring that group composition is appropriate, both for the successful functioning of the groups and for the wider benefit of the activity. Disciplinary expertise is usually fairly effectively determined by proponents, although assistance is often required to find individuals with appropriate expertise. Consideration of sociological elements is encouraged, such as ensuring wide representation across ages, gender, geographies, personality types and organisations (Figure 18.2). This advice is informed by educational and managerial team theory, as creation of a functional team from a mere assemblage of people is fundamental to success (Bammer 2013, Belbin 1993, Carpenter et al. 2009, Cooperstein and Kocevar-Weidinger 2004, Mutch 1998). Many of these considerations (age ranges, organisational and cultural diversity, gender diversity) not only promote effective team function but result in outcomes that are for the common good ('The range of skills/expertise/disciplines was very stimulating' KEIWS205*; 'we benefitted from the expertise of members using different approaches (modelling,

* Comments from ACEAS and CESAB feedback surveys.

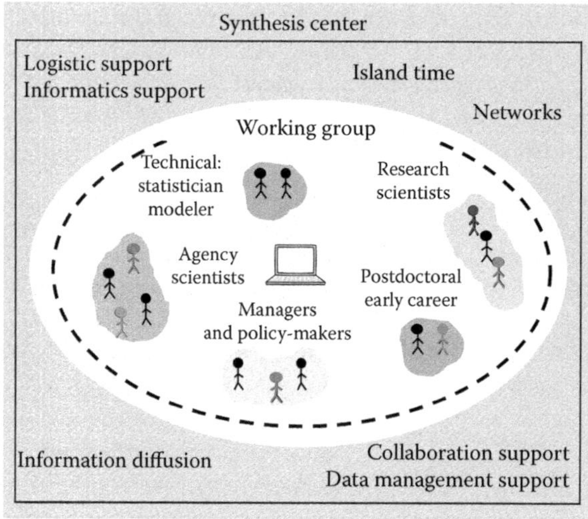

FIGURE 18.2
Synthesis centre structure showing elements of group composition and the main components of scaffolding provided by the centre. The indicated group composition shows different broad types of participants, different intensities indicating different aspects of relevant represented diversity (e.g. age, gender, discipline and geographical origin).

fieldwork, experimental virology) in different contexts which led to fruitful discussions and collaborative work'. BIO604). The benefit to a group participant has been shown to not only be an improved publication record (a well-known outcome of synthesis centre activity) but the effect of participation on subsequent practices in his or her workplace. This largely results from new techniques learnt, different insights and approaches experienced, and new relationships developed through the working group experience ('Great collaboration and learning as well as sharing knowledge. It was also a very good networking opportunity' SPEWS111).

For these reasons and more, it is recognised by all centres that sponsor working groups that the process of collaboration around the analysis and synthesis of existing data is not short term. The activity itself is part time for most participants, additional to their normal workplace commitments. Time is therefore often one of the most critical factors in the success of the process: limitations in time and a need for time. Time is required to develop communication languages, to develop trust, to obtain and clean data, and to analyse and test the result, to be in the 'zone'. Each interaction takes time, and there need to be rewards for participating.

Face-to-face meetings are fundamental to the development of relationships and trust, and depending on the question and the nature of the group, centres usually sponsor several such meetings. Face-to-face meetings have been found to be important for team productivity (Hall et al. 2012, Stokols et al. 2010). These meetings usually last from 3 to 8 days (although some are longer), and centres place importance on using meeting venues that separate participants from their day-to-day distractions. Several key criteria for suitable locations have been identified, which balance isolation and connectivity ('The perfect environment for stimulating productive teamwork'. DAVWS102).

When the following conditions are in place—collaborative intellectual stimulation and the combination of location and 'stolen' time (often called 'island' time; Figure 18.2)—a highly creative environment ensues, which aids the development of team identity, and also intellectual production (the 'hot moments' described by Parker and Hackett 2012).

Face-to-face meetings are by no means the only component of synthesis centre working group activity. Work must continue between formalised meeting times at the very least to prepare for the next. The between-meeting phases can be more or less effective, depending on the technological competence and commitment of the group and the support of the centre. Centres that have dedicated postdoctoral scholars (or staff) for each group depend greatly on their skills in maintaining the continuity of the group's work. Without this, much time can be lost. Regardless of the presence or otherwise of postdoctoral scholars, most centres engage in several strategies to maintain continuity, including (from Crowston et al. 2015)

- Making working group members aware that discontinuities are likely to arise but that they can be addressed with focused attention and a willingness to address them by creating continuities
- Establishing shared communication practices that facilitate the creation of continuities
- Allowing sufficient time for continuities to develop
- Ensuring the active participation of bridge builders such as librarians who know how to ask questions about disciplines other than their own

As may be clear from these strategies, centres provide a scaffolding* for their working groups, consistent with the type of support required for effective problem-based learning in higher education (Cooperstein and Kocevar-Weidinger 2004, Dickey 2006, Hmelo-Silver and Azevedo 2006) and other fields (Smith et al. 2007). This scaffolding (Figure 18.2) ranges from the clearly evident, such as logistic support—the aim of most centres is to make travel, accommodation and most meal arrangements effortless

* Scaffolding is support tailored to the needs of the client with the intention of helping them achieve his/her goals (paraphrase of Sawyer 2006).

for the participants—but also support for effective face-to-face and remote collaboration (e.g. through the provision of private wiki spaces), support of connectivity through networks (e.g. is there someone else in the world working on this problem, and can they help us?), informatics support (making sure that the Internet works, computing power is suitable and digital storage is available and secure), data management support and advice (diversification of data sources and assured access) and extension (information diffusion) support (including data delivery and often visualisation, support for scientific publications, conference presentations, web and social media presence).

18.3 Data Management

For an individual scientist, it can be a challenge to manage one's own data, let alone to discover, manipulate and manage the data produced by others. The disciplinary experts at a synthesis centre are primarily focused on the analysis phase of the research data life cycle (Specht et al. 2015b) and regard the management of data as a marginal activity in which they are little experienced (Volk et al. 2014). Scientists generally lack the competence or skills to publish their data for future reuse (Costello 2009) and lack confidence in using the data of others. Debate has emerged recently in an attempt to 'strike a balance' between the growing requirement to share data and restrictive factors relating to the willingness of owners to share, conditions on sharing (e.g. open data policies), protection of key datasets (e.g. propriety periods) and acknowledgement of data (e.g. Mauthner and Parry 2013, Mills et al. 2015, 2016, Tenopir et al. 2015, Whitlock et al. 2016). Few publications discuss the actual experience of data reuse and data management. The synthesis centre is arguably a microcosm of the wider community, as the entire data workflow from acquisition to publication of data is experienced (Figure 18.3), so it provides an invaluable insight for the participant and for the data management community (Specht et al. 2015b).

Data management specialists are generally obligatory members of the synthesis centre working groups (Figure 18.2). They can assist, but their competence varies widely and indeed their understanding of the scientific workflow is often limited being specialists themselves, thus making communication and workflows doubly difficult. The synthesis centre staff with their accumulated experience in all aspects of the data management and workflow with the disciplinary experts play an important role in facilitating, supporting and educating both the scientist and the data specialist in the practice of scientific, inter- and transdisciplinary data management.

Four components of the data workflow are relevant to the work of a synthesis centre. These are (Figure 18.3)

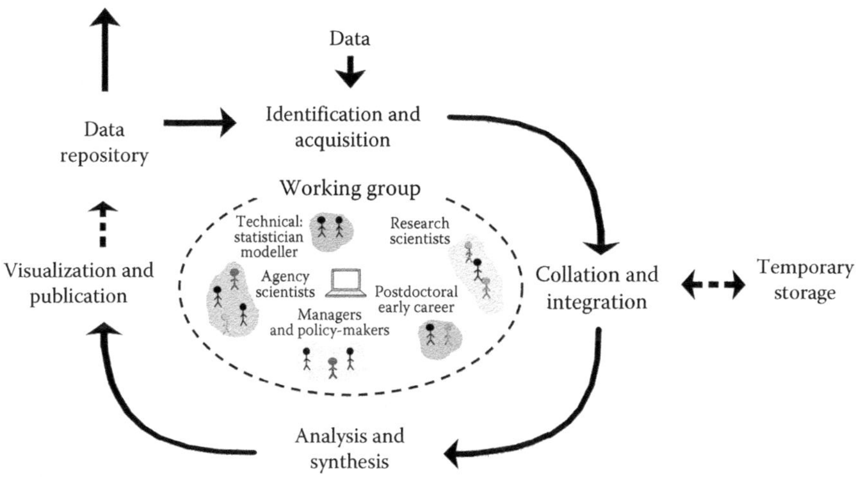

FIGURE 18.3
The modified synthesis centre data life cycle. (Adapted from Specht, A. et al., *Sci. Total Environ.*, 534, 144, 2015b.)

1. Identification and acquisition (discover)
2. Collation and integration
3. Analysis and synthesis
4. Publication and visualisation (deposition)

One common attribute of synthetic research is to compile data points that have been reported at various places in the literature into a single dataset that is amenable to comparative analysis or meta-analysis (Vision and Cranston 2015). Although most data will be identified prior to funding (a condition of funding for most centres), working groups usually have to physically obtain the data soon after they start and usually discover more in the process. Lack of metadata, and hence inability to assess the suitability of available data, is a common problem for data reuse experienced by working groups (Specht et al. 2015b). Once data are discovered, permission may be required to obtain them; then further time and skill is expended to ascertain their quality. The synthesis centre staff typically assist working group members in this phase by exploring data sources unknown to the participants, writing letters of support, talking to data providers and determining the conditions that would allow the working group's access and reviewing the data suitability with the working group participants. It is common that

- Researchers holding suitable data are reluctant to share their data due to concerns about reuse
- Reuse of data is challenging due to the question-specific context of the original studies

- Metadata is inadequate in many cases to be confident in reuse and the authors are often unable to provide information if contacted
- Difficulties occur in data discovery in open access repositories due to lack of awareness of repositories and lack of knowledge about how to interrogate the repositories

Once working groups identify and acquire their data, it has to be assembled and transformed into a suitable format so that it is fit for purpose and analyses can be conducted. Integrating datasets to allow for analysis can be very challenging. The main barriers are

- Blending data of dissimilar scales to the full, desired, geographic or temporal representation
- Time-consuming formatting required to correct data mismatches
- Correcting for different research purposes in data collection
- Adjusting to account for different data collection methods

Synthesis centres commonly provide support at this stage by assisting in the reorganisation of disparate datasets so they can be combined together for analysis. The magnitude of this task and the discipline required for organised data assembly is usually underestimated, but when done well, with good metadata and provenance description, the next steps are easier.

When significant challenges of the first two steps of the data workflow are overcome, working groups can begin to analyse their collated data. These collaborative activities often break new ground due to the amount of data used for analyses (supra-individual) and the high degree of multi-disciplinarity involved. However, establishing a consistent, standardised analysis technique is a major hurdle, and the group may find they do not have the suitable technical skills to accomplish their analyses, for example, write the code required, utilisation of cloud-based resources, interfaces with other large-scale relational databases or their associated applications. Frequently, groups develop their own approaches to deal with the data workflow challenges in partnership with synthesis centre staff.

Challenges also extend to the publication and visualisation of synthesised data, the least familiar of the steps in the data workflow for participants. There can be strong reluctance to publically release primary data from a technological and cultural viewpoint, for example (Specht et al. 2015b, Swauger and Vision 2015):

- Many journals won't publish articles if the data have already been released for open access.
- Some journals require *a priori* open access of all data and associated provenance and attribution.
- Some journals require *a priori* data management plans.

- Empowering a competitor or commercial partner prior to publication of articles based on the data.
- Insecurity about the quality of data (should it be released at all).
- Lack of knowledge of and trust in data repositories.

The importance of the establishment of data ownership and usage conditions in the acquisition phase often emerge at this stage, as, even though the data have been modified and transformed, the provenance needs to be acknowledged appropriately. Publication can be prevented if a data contributor cannot be contacted to obtain permission for data release; there is unknown ownership of a contributing dataset; there is uncertainty about institutional data publication policies; there are data licensing restrictions on some of the data (but perhaps not all); or there is confusion with respect to the correct attribution of the original data providers. All these matters can hinder data delivery, even of synthesised (and hence transformed) data. To solve them requires anticipation and planning and experience and knowledge of legal requirements and data licensing policies. The synthesis centre staff are vital at this stage enabling, quite often, rare and valuable 'long-tailed' data (*sensu* Heidorn 2008) to be delivered, 'cleaned' and enhanced by blending with other datasets.

18.3.1 Pioneering Tools for Data Management and Open Science

All centres, in partnership with their scientific communities, continue to create and promote standards and best practices for making data discoverable and accessible. Although many synthesis centre working groups develop new tools and models (e.g. release of r-scripts, controlled vocabularies, the SPEDDEXes group of ACEAS; www.emast.org.au/models/speddexes/), the close interaction between the informatics specialists in the centres and the energetic, highly qualified, innovative scientists has stimulated the production of new tools by centre staff which have wide applicability.

NCEAS changed the state of the art for data management in the ecological sciences. They recognised that synthesis has much to do with the development of semi-automated sharing, developed management tools (such as Morpho* and Metacat†) and analysis tools (Kepler‡), and built the web data repository Knowledge Network for Biocomplexity (KNB§), now

* Morpho Data Management Software allows the creation and management of data and enables sharing.

† Metacat is an open-source metadata catalogue and data repository that targets scientific data, particularly from ecology and environmental science.

‡ Kepler (https://kepler-project.org) is an open-source scientific workflow application designed to help scientists, analysts and computer programmers create, execute and share models and analyses across a broad range of scientific and engineering disciplines.

§ KNB is a data repository intended to facilitate ecological and environmental research: https://knb.ecoinformatics.org/.

a core partner of DataONE, a distributed discovery network for environmental data (Jones et al. 2006, Michener et al. 2012). The NCEAS staff also developed, with the Ecological Society of America, the Ecological Metadata Language (EML*), providing a mechanism for describing ecological information. This has become one of the standards for metadata description of ecological data.

Younger centres have responded similarly to the stimulating, problem-solving, environment of the synthesis centre. NESCent supported the development of the Dryad data repository (datadryad.org), while several centres have used geospatial visualisation methods for deposited data (www.aceas.org.au: 'products and outcomes' and www.betsi.cesab.org). The visualisations developed for ACEAS working groups fell into two categories: (1) those linked to metadata only and (2) those that provided a gateway to the raw data. Two ACEAS working groups felt that spatial delivery of their bibliographic collations was vital for understanding, prioritisation of effort and future conservation planning and worthy of spatial visualisation (Campbell et al. 2015, Pert et al. 2015). Other groups determined spatially coherent subsets of their synthesised data (from sub- to supra-continental scale) linking each zone to the raw data to provide evidence for their determinations (e.g. Bryan et al. 2013, McAlpine et al. 2015, Murphy et al. 2013, Thomson et al. 2015).

The wide variety of terminologies used in the ecological sciences is an impediment to effective communication and our ability to benefit from available data and information (Herrando-Perez et al. 2014, Jones et al. 2006). This is highlighted in the synthesis centre where combining data from different sources, across difference scales and from different domains is core business. The NCEAS staff and postdoctorates have been closely involved in the development of ontologies and thesauri (e.g. Bowers et al. 2010, Laporte et al. 2012, Madin et al. 2008) and the CESAB staff are participating with several groups in Europe, including the National Centre for Scientific Research (CNRS; www.cnrs.fr) to develop thesauri for reliable and explicit description of objects and measurements (e.g. t-sita.cesab.org; Pey et al. 2014a,b).

Organisation of the data from acquisition through to delivery also requires planning, and alongside encouraging the use of data management plans, CESAB is developing a decision support tool to assist biologists and ecologists create their databases. The National Institute for Mathematical and Biological Synthesis (NIMBioS) has developed evaluation approaches to quantify the value and benefit of publication (Bishop et al. 2014), and the

* EML is a metadata specification to document ecological data and a standard under ISO-19115-2. It is used by the Atlas of Living Australia (ALA; www.ala.org.au), DataONE (www.dataone.org), ECOSCOPE (www.fondationbiodiversite.fr/en/research/frb-programmes/ecoscope.html), Global Biodiversity Information Facility (GBIF; www.gbif.org) and the Terrestrial Ecosystem Research Network (TERN; www.tern.org.au), among others.

SESYNC is applying innovative approaches to enable effective collaboration across socioecological fields that will inevitably have a profound effect on the way science is practised.

18.4 Discussion

At the beginning of this chapter, some challenges were presented that we face today as we struggle to create the necessary wisdom to advance ecological thought to guide us into an ecologically sustainable future, despite being deluged with a superabundance of information. I have considered the first two challenges: (1) how to find the time to gather together the necessary information and to analyse and synthesise it in order to gain new understanding and insight and (2) how best to make sense of this information when we need to converse with each other across disciplinary, organisational, geographical and geopolitical boundaries. The collaborative working environment of the synthesis centre approach is demonstrably a very effective way for time-poor specialists to access valuable time for analysis and synthesis of a complex wealth of available information, in a manner that allows them to converse successfully across disciplinary, organisational and geographical boundaries.

It remains to consider the third challenge, 'how to respond when our training systems are slow to react and careers are governed by disciplinary rewards'. Through the collaborative and trusting environment and the support provided by the synthesis centre, participants not only analyse and synthesise data to answer their questions, but they learn new techniques and develop strong relationships that enable them to perform better in their own workplaces (Lynch et al. 2015). In other words, the synthesis centre experience for working group participants is a learning exercise and should be regarded as vital professional development. It provides many with an enriching experience that they have not felt since they were postgraduates: competing to be the last person to leave the laboratory, testing ideas, sharing coffee, successes and failures, learning about a new program/technique/model, helping on field trips, monitoring experiments, sorting that programming question and kicking that proverbial ecological can down the road. Learning from peers at that time was important, and the synthesis centre provides an opportunity for that to occur again, overcoming the professional isolation felt by many when they are embedded in their workplaces ('sharing thinking on a topic always helps me to distill and organise my own thoughts by drawing on other peoples' perspectives' BALWS112).

The results of the synthesis centres feed into established media and performance metrics. Publication rates from synthesis centres are high, and impact factors are also high (Baron et al. subm., Hampton and Parker 2011). A high proportion of all publications resulting from the activities of synthesis centres are published in leading multidisciplinary journals and are cited highly compared with other research institutions (NESCent 2011). Similar success factors are evident for data deposition, with the KNB data repository now reaching, at the time of writing, 21,174 data packages deposited as a result of NCEAS activity (knb.ecoinformatics.org/#data/page/0).

The typical workflow processes occurring in a synthesis centre shown are based on the operation of two synthesis centres (ACEAS and CESAB; Figure 18.4), and establishing the International Synthesis Consortium (www.synthesis-consortium.org), and close interaction with DataONE, the Terrestrial Ecosystem Research Network, and the Atlas of Living Australia. Figure 18.4 illustrates the attributes of the people, data and processes characteristic of the synthesis centre. The operation of the centre is rarely exposed beyond being a place where 'things happen' (the 'condensor' model of the diagram). As mentioned, the synthesis centre provides intelligent support (scaffolding) for the participants to enable them to achieve their goals. The people involved in providing this support are usually multitaskers and polymaths, and their role in the synthesis centre enhances those attributes. The expertise developed by the staff, through their exposure to the vast array of topics and people that pass through the centres, biased according to the focus of the

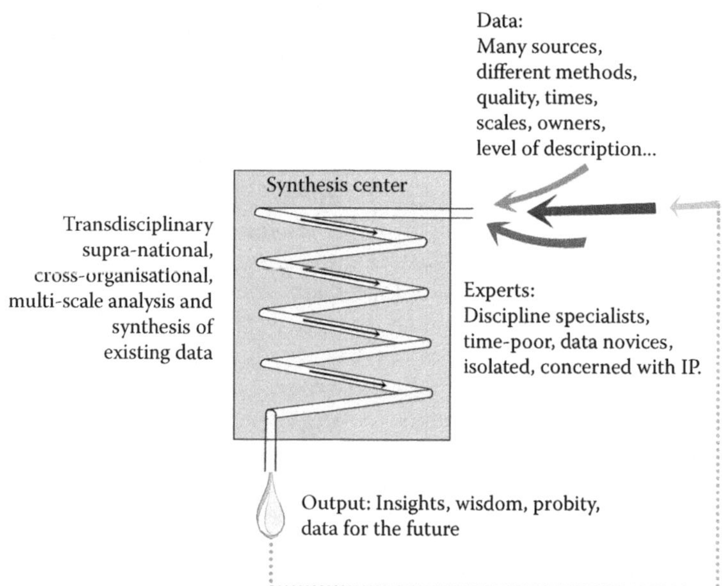

FIGURE 18.4
The function of the synthesis centre in facilitating the 'getting of wisdom' by scientists, policymakers and managers, beginning with the reuse of multiple, heterogeneous data and producing new insights and cleaned and synthesised data which can be made available for reuse.

centres (Figure 18.1), becomes a resource in its own right. This has resulted, for example, in the development of the tools mentioned in this chapter but also in a deep understanding of the breadth of topics, the interrelationships between disciplines and technologies and the characteristics of the people in the respective communities of practice. The idea of the synthesis consortium capitalises on this, as together, through sharing the particular skills and knowledge acquired in each centre, we can enhance the effectiveness of each and potentially create a new 'super' centre well suited to tackling global environmental questions beyond geographical and geopolitical boundaries.

Pivotal to the synthesis centre, of course, is the analysis and synthesis of existing data and information. They are places where existing data are given new life, legacy data are brought into the light and existing data are quality assured and value added in new ways. Centres are a place to test new ideas and analysis techniques outside normal institutional and funding structures and are thus in the fortunate position of being at the cutting edge of intellectual experimentation. This benefits not only the participants but also the centre staff and data providers. The synthesis centre concept is commended to scientists, policymakers and managers who are faced with complex ecosystem science questions and who wish to make the most of the emergent 'big data' world.

Acknowledgements

The author thanks the staff and participants of both ACEAS and CESAB and the members of the International Synthesis Consortium for many challenging discussions, support and collaboration. The author also thanks Drs. M. Schildhauer and M. Jones for his initiation into the world of data analysis in synthesis centres.

References

Bammer, G. 2013. *Disciplining Interdisciplinarity: Integration and Implementation Sciences for Researching Complex Real-World Problems*. ANU EPress, The Australian National University, Canberra, Australian Capital Territory, Australia.

Belbin, M. 1993. *Team Roles at Work*. Butterworth-Heinemann, Oxford, U.K.

Bishop, P., Huck, S.W., Ownley, B.O. et al. 2014. Impacts of an interdisciplinary research center on participant publication and collaboration patterns: A case study of the National Institute for Mathematical and Biological Synthesis. *Research Evaluation* 23: 327–340.

Bowers, S., Madin, J.S., and Schildhauer, M.P. 2010. Owlifier: Creating OWL-DL ontologies from simple spreadsheet-based knowledge. *Ecological Informatics* 5(1): 19–25.

Bryan, B.A., Meyer, W.S., Campbell, C.A. et al. 2013. The second industrial transformation of Australian landscapes. *Current Opinion in Environmental Sustainability* 5: 1–10.

Campbell, H.A., Beyer, H.L., Dennis, T.E. et al. 2015. Finding our way: On the sharing and reuse of animal telemetry data in Australasia. *Science of the Total Environment* 534: 79–84.

Carpenter, S.R., Armbrust, E.V., Arzberger, P.W. et al. 2009. Accelerate synthesis in ecology and environmental sciences. *BioScience* 59(8): 699–701.

Chartier, R. 1991. *The Cultural Origins of the French Revolution*. Translated by Lydia G. Cochrane. Princeton University Press, Princeton, NJ.

Cheruvelil, K.S., Soranno, P.A., Weathers, K.C. et al. 2014. Creating and maintaining high-performing collaborative research teams: The importance of diversity and interpersonal skills. *Frontiers in Ecology and Environment* 12: 31–38.

Cooper, H. and Hedges, L.V. 2009. Research synthesis as a scientific process. In *The Handbook of Research Synthesis and Meta-Analysis*, 2nd edn., eds. H.M. Cooper, L.V. Hedges, and J.C. Valentine. Russell Sage Foundation, New York, pp 4–16.

Cooperstein, S.E. and Kocevar-Weidinger, E. 2004. Beyond active learning: A constructivist approach to learning. *Reference Services Review* 32: 141–148

Costanza, R. 2003. A vision of the future of science: Reintegrating the study of humans and the rest of nature. *Futures* 35: 651–671.

Costello, M.J. 2009. Motivating online publication of data. *Bioscience* 59: 418–427.

Crowston, K., Specht, A., Hoover, C. et al. 2015. Perceived discontinuities and continuities in transdisciplinary scientific working groups. *Science of the Total Environment* 534: 159–172.

Dickey, M.D. 2006. Game design narrative for learning: Appropriating adventure game design narrative devices and techniques for the design of interactive learning environments educational technology. *Research and Development* 54: 245–263.

Goodman, D. 1994. *The Republic of Letters: A Cultural History of the French Enlightenment*. Cornell University Press, Ithaca, New York.

Hall, K.L., Vogel, A.L., Stipelman, B.A. et al. 2012. A four-phase model of transdisciplinary team-based research: Goals, team processes, and strategies. *Translational Behavioral Medicine* 2: 415–430.

Hampton, S.E. and Parker, J.N. 2011. Collaboration and productivity in scientific synthesis. *BioScience* 61: 900–910.

Heidorn, F.B. 2008. Shedding light on the dark data in the long tail of science. *Library Trends* 57: 280–299.

Herrando-Perez, S., Brook, B.W., and Bradshaw, C.J.A. 2014. Ecology needs a convention of nomenclature. *BioScience* 64: 311–321.

Hmelo-Silver, C.E. and Azevedo, R. 2006. Understanding complex systems: Some core challenges. *Journal of the Learning Sciences* 15: 53–61.

Jones, M.B., Schildhauer, M.P., Reichman, O.J. et al. 2006. The new bioinformatics: Integrating ecological data from the gene to the biosphere. *Annual Review of Ecology, Evolution, and Systematics* 37: 519–544.

Kale, S.D. 2002. Women, the public sphere, and the persistence of salons. *French Historical Studies* 25(1): 115–148.

Konczacki, J.M. 1986. Stanislaw August Poniatowski's "Thursday Dinners" and cultural change in late eighteenth century Poland. *Canadian Journal of History/ Annales Canadiennes d'Histoire* 21: 25–36.

Laporte, A., Mougenot, I., and Garnier, E. 2012. ThesauForm—Traits: A web based collaborative tool to develop a thesaurus for plant functional diversity research. *Ecological Informatics* 11: 34–44.

Lindenmayer, D. and Likens, G. 2010. *Effective Ecological Monitoring.* CSIRO Press, Melbourne, Victoria, Australia.

Lough J. 1985. Reflections on enlightenment and lumières. *Journal for Eighteenth Century Studies* 8(1): 1–15.

Lynch, A.J.J., Thackway, R., Specht, A. et al. 2015. Transdisciplinary synthesis for ecosystem science, policy and management: The Australian experience. *Science of the Total Environment* 534: 173–184.

Madin, J.S., Bowers, S., Schildhauer, M.P. et al. 2008. Advancing ecological research with ontologies. *Trends in Ecology and Evolution* 23: 159–168.

Marx, V. 2013. Biology: The big challenges of big data. *Nature* 498(7453): 255–260.

Mauthner, N.S. and Parry, O. 2013. Open access digital data sharing: Principle, policies and practices. *Social Epistemology* 27: 47–67.

McAlpine, C., Lunney, D., Melzer, A. et al. 2015. Conserving koalas: A review of the contrasting regional trends, outlooks and policy challenges. *Biological Conservation* 192: 226–236.

Michener, W.K. 2015. Ecological data sharing. *Ecological Informatics* 20: 33–44.

Michener, W.K., Allard, S., Budden, A. et al. 2012. Participatory design of DataONE—Enabling cyberinfrastructure for the biological and environmental sciences. *Ecological Informatics* 11: 5–15.

Mills, J.A., Teplitsky, C., Arroyo, B. et al. 2015. Archiving primary data: Solutions for long-term studies. *Trends in Ecology and Evolution* 30: 581–589.

Mills, J.A., Teplitsky, C., Arroyo, B. et al. 2016. Solutions for archiving data in long-term studies: A reply to Whitlock et al. *Trends in Ecology and Evolution* 31(2): 85–87. doi:10.1016/j.tree.2015.12.004.

Murphy, B.P., Bradstock, R.A., Boer, M.M. et al. 2013. Fire regimes of Australia, a pyrogeographic model system. *Journal of Biogeography* 40(6): 1048–1058.

Mutch, A. 1998. Employability or learning? Groupwork in higher education. *Education and Training* 40(2): 50–56.

NESCent. 2011. NESCent Assessment Report 2011. https://zenodo.org/record/17466/files/NESCentAssessmentReport.pdf, accessed November 3, 2016.

Parker, J.N. and Hackett, E.J. 2012. Hot spots and hot moments in scientific collaborations and social movements. *American Sociological Review* 77:21–44.

Pert, P.L., Ens, E.E., Locke, J. et al. 2015. An online spatial database of Australian Indigenous Biocultural Knowledge for contemporary natural and cultural resource management. *Science of the Total Environment* 534: 110–121.

Pey, B., Laporte, M.-A., Nahmani, J. et al. 2014a. A thesaurus for soil invertebrate trait-based approaches. *PLoS ONE* 9(10): e108985.

Pey, B., Nahmani, J., Auclerc, A. et al. 2014b. Current use of and future needs for soil invertebrate functional traits in community ecology. *Basic and Applied Ecology* 15: 194–206.

Rodrigo, A., Alberts, S., Cranston, K. et al. 2013. Science incubators: Synthesis centers and their role in the research ecosystem. *PLoS Biology* 11(1): 1–3.

Sawyer, R.K. 2006. *The Cambridge Handbook of the Learning Sciences.* Cambridge University Press, New York.

Science Staff. 2011. Challenges and opportunities. *Science* 331: 692–693.

Sidlauskas, B., Ganeshkumar, G., Hazkani-Covo, E. et al. 2010. Linking big: The continuing promise of evolutionary synthesis. *Evolution* 64: 871–880.

Siebert, S., Machesky, L.M., and Insall, R.H. 2015. Overflow in science and its implications for trust. *eLife* 4:e10825.

Smith, C., Felderhof, L., and Bosch, O.J.H. 2007. Adaptive management: Making it happen through participatory systems analysis. *Systems Research and Behavioural Science* 24: 567–587.

Specht, A. Gordon, I.J., Groves, H. et al. 2015a. Catalysing transdisciplinary synthesis in ecosystem science and management. *Science of the Total Environment* 534: 1–3.

Specht, A., Guru, S.M., Houghton, L. et al. 2015b. Data management challenges in analysis and synthesis in the ecosystem sciences. *Science of the Total Environment* 534: 144–158.

Stokols, D., Hall, K.L., Moser, R.P. et al. 2010. Cross-disciplinary team science initiatives: Research, training and translation. In J.T. Klein and C. Mitcham, Eds., *Oxford Handbook on Interdisciplinarity*. pp. 471–493. Oxford University Press, Oxford, UK.

Swauger, S. and Vision, T.J. 2015. What factors influence where researchers deposit their data? A survey of researcher submissions to data repositories. *International Journal of Digital Curation* 10: 68–81.

Tenopir, C., Dalton, E., Allard, S. et al. 2015. Changes in data sharing and data reuse practices and perceptions among scientists worldwide. *PLoS ONE* 10(8): e0134826.

Thessen, A.E. and Patterson, D.J. 2011. Data issues in the life sciences. *ZooKeys* 150: 15–51.

Thomson, J.R., Maron, M., and Grey, M.J. 2015. Avifaunal disarray: Quantifying models of the occurrence and ecological effects of a despotic bird species. *Diversity and Distributions* 21: 451–464.

Vincent, P. 2014. Readers, writers, salonnières: Female networks in Europe, 1700–1900; the life of Madame Necker: sin, redemption and the Parisian salon; British literary salons of the late eighteenth and early nineteenth centuries. *European Romantic Review* 25(1): 86–94.

Vision, T. and Cranston, K. 2015. Open data for evolutionary synthesis: An introduction to the NESCent collection. *Scientific Data* 1: 140030.

Volk, C.J., Lucero, Y., and Barnas, K. 2014. Why is data sharing in collaborative natural resource efforts so hard and what can we do to improve it? *Environmental Management* 53: 883–893.

Whitlock, M.C., Bronstein, J.L., and Bruna, E.M. 2016. A balanced data archiving policy for long-term studies. *Trends in Ecology and Evolution* 31(2): 84–85. doi:10.1016/j.tree.2015.12.001.

Wilson, E.O. 1998. *Consilience: The Unity of Knowledge*. Island Press, Alfred A. Knopf, New York.

19

Role of Long-Term Experiments in Understanding Ecosystem Response to Global Change

H. Henry Janzen and Benjamin H. Ellert

CONTENTS

19.1 Introduction

For many decades, scientists have sought to understand how ecosystems respond to stresses, especially to those imposed by humans. Already a century-and-a-half ago, for example, Marsh (1864) documented mounting evidence of how human activities were degrading the biosphere and looked for ways of reducing those damages. Now, understanding change has assumed even greater urgency because human stresses on ecosystems, locally and globally, have intensified, driven by growing population and expanding demands (Carpenter et al. 2009; Chapin et al. 2009; Collins and Childers 2014; Gunderson and Folke 2011). In the past, most human effects on ecosystems occurred locally; now, they often extend globally, notably through effects on atmospheric CO_2, which is rapidly increasing, affecting climates around the world (Le Quéré et al. 2015).

Ecosystem responses to human disturbance, however, can be properly understood only by studying them for periods of several decades or longer (Carpenter 2002; Knapp et al. 2012; Kümmerer et al. 2010; White 2013). Short-term studies, of several months or years, are indispensable in elucidating underlying processes and mechanisms, but they cannot reveal the eventual, final response. Many ecosystem properties change only slowly and the full response is seen only after time scales of a century; for example, responses of soil carbon to management change may unfold over many decades and may not even be discernible until 5 or 10 years have passed (Ellert et al. 2000). Furthermore, some ecosystem responses may be episodic, affected by fluctuating weather or other variables, so studies lasting only a few years may miss them entirely (Franklin 1989). Other responses may be delayed for several decades until a critical "tipping point" is reached (Villa et al. 2014); for example, after cultivating grassland, the cultivated land may initially be highly productive, but yields may eventually decline as native fertility is depleted (Janzen 2001). For these and other reasons, long-term experimental sites have been invaluable to ecologists in understanding ecosystem change and will likely remain so in the coming decades (Mooney et al. 2009; Southwood 1994). Indeed, given the intensity of projected climate and other global changes, some of which remain unpredictable (Boyd 2012), such long-term ecological sites (LTES) may become even more important in the future (Richter et al. 2007).

In light of the intensifying stresses on ecosystems and the long response time of ecosystems to these pressures, this brief review will address the following objectives: (1) to define "long-term ecological site or sites," (2) to propose some attributes that enhance the usefulness and longevity of such studies, and (3) to proffer examples of research questions that may merit more attention in long-term experiments.

19.2 Definition of Long-Term Ecological Sites

An LTES, as the phrase is used here, is a designated area of land maintained for long durations to measure eventual responses to human and other related stresses. By Tansley's definition, an ecosystem is any assemblage of biota and their myriad interactions with each other and their physical habitat (Tansley 1935). Strictly speaking, it can be of almost any size (Grove et al. 2013; Tansley 1935; Willis 1997), depending on questions being asked, so LTES may range in scale from a series of bottles buried in soil by Beal in 1879 (Greenfieldboyce 2012; Telewski and Zeevaart 2002) to a watershed (e.g., Peters et al. 2013) to the entire planet, as in monitoring sites for atmospheric CO_2 (Keeling 2008). Typically, however, LTES have areas measured in units of m^2 or ha, large enough to reflect the influences of a farmer, forester, or other ecosystem manager, but small enough to allow representative sampling. If the area is too small, it may be too susceptible to outside interferences (e.g., border effects) or inadequately represent the larger ecosystem; if the area is too large, sampling of ecosystem properties becomes prohibitively demanding because of spatial heterogeneity.

Often, LTES have more than one "treatment," each reflecting a distinct ecosystem or land use, thus allowing comparisons among ecosystems as well as measuring of changes over time. For example, many agricultural LTES have multiple plots of land, assigned management regimes varying in tillage, nutrient amendment, or crop rotation (Paul et al. 1997). Ideally, such LTES include a baseline treatment, against which prospective or extreme management regimes can be evaluated.

The fundamental purpose of an LTES is to monitor and understand the persistence of one or more ecosystems in delivering expected services. In the past, researchers, especially in studies of managed ecosystems, often focused on a system's continuing capacity to produce yield of food, timber, or other economic commodity. Increasingly, however, ecologists are interested also in other functions, such as maintaining biodiversity, filtering air and water, recycling wastes, sequestering atmospheric carbon, promoting aesthetic appeal, preserving recreational or cultural value, and sustaining soil health (Adhikari and Hartemink 2016; MEA 2005). By continuously or repeatedly measuring a system's capacity to furnish these services, either directly or using proxy indicators, researchers can determine whether an ecosystem's performance is improving, holding steady, or declining. In short, the purpose of LTES is to measure and understand the systems' resilience (Dudley 2011; Folke 2006; Folke et al. 2003; Holling 1973), although that term has now been used so widely, in so many ways, that its definition is becoming diluted (Downes et al. 2013; Hodgson et al. 2015).

The inherent value of LTES, of course, lies in their longevity. The most critical question, then, in defining an LTES is, what constitutes "long term"? How old does a study need to be before being designated an LTES? One way of resolving this question is to define an LTES prospectively, not retrospectively; to designate a site as an LTES not by time already elapsed but by lifespan intended. From this perspective, as proposed here, an LTES is a study site conceived and maintained deliberately to outlive the tenure of its founders, a site explicitly envisioned to be passed along, as a bequest to future ecologists (Haughland et al. 2010). By this definition, an ecological site established 1 year ago in such a way that it might furnish useful findings to later generations would already be deemed an LTES; but a complex agronomic experiment that has survived 10 years is not, if its continuation is threatened by stifling resource demands. This multigenerational perspective reorients the design and management of the LTES to focus less on answering currently topical questions and more on the challenges confronting future scientists. Just as scientists now derive enormous insight from studies established by their predecessors, so too the primary benefit of LTES established today may reside in what their successors may learn from them. Given the uncertainty over coming biospheric changes (Sutherland 2013; Villa et al. 2014; Weinberger 2012) and the challenges emanating from them, LTES begun today will likely have their greatest value in answering questions no one has yet foreseen.

19.3 Attributes of Enduring Long-Term Ecological Sites

LTES are almost universally prized and revered among ecologists, yet many face abandonment, threatened by accumulating costs and perceived obsolescence. How can LTES be designed to increase likelihood of their survival over multiple generations?

One important attribute is elegant simplicity—creative parsimony. A design too complex demands too many resources, increasing risk of termination during inevitable budgetary crises (Koehler and Melecis 2010; Leigh et al. 1994). In experiments at Lethbridge, for example, many plots with innovative cropping systems were established in 1911, but few survived; and those that did were not always the ecologically favored, but often just the simplest (Janzen 1995, 2001). Aside from its frugality, simplicity also confers flexibility, because it allows the individual plots to be of sufficient scale for management regimes gradually to be adapted or subdivided over time, avoiding obsolescence (Powlson et al. 2014).

A second attribute, following upon the first, is to select management regimes that best illuminate the functioning of the ecosystems under study. The purpose of LTES is not to identify or demonstrate the "best" system from among a range of conceivable options but rather to help researchers understand how ecosystems respond to stresses and management practices. Such understanding sometimes emerges most clearly from extreme management regimes, which may not always be commercially feasible or recommended (Palmer 1949; Powers and Van Cleve 1991). Saunders (1900), for example, in referring to a "long-conducted and extensive series of tests" observed that "these experiments were never intended to serve as model test plots such as farmers could copy to advantage in their general practice. On the contrary, it has been found necessary to use some fertilizers in extravagant quantities, and in other instances to more or less exhaust the soil…." A long-term experiment in Sweden, similarly, included a bare fallow treatment, clearly not viable economically, yet instrumental in developing a robust soil carbon model (Andrén and Kätterer 1997). In the past, LTES have sometimes been terminated because their management regimes were considered "obsolete"—no longer recommended for adoption by land managers. If the intent of LTES, as proposed here, is to understand unfolding ecosystem responses over time, then their value is not determined by how up to date their treatments are, but by how much insight they generate about ecosystem performance.

Third, the scientific value of any LTES, and hence the likelihood of its survival, is enhanced by melding it deliberately into multisite networks. From early on, researchers saw that many sites, seen together, give broader, more durable insights than the same sites studied alone. For example, agricultural scientists in Canada realized a century ago the importance of linking LTES across geographical regions (Anonymous 1924). Studies at many sites across the Canadian prairies were jointly established, with some common management regimes, and a consistent initial sampling of soils (Janzen 2001). This same motive led to the creation of LTES networks, perhaps most prominently in the U.S. LTER (Callahan 1984), but later extended also to Europe and beyond (LTER Europe 2016; Smith et al. 2002; Vanderbilt et al. 2015). Such networks have yielded bountiful understanding, but challenges remain; for example, scientists sometimes exhibit justifiable reluctance in sharing unpublished data (Hampton et al. 2013; Peters et al. 2013). Another challenge is resolving inevitable inconsistencies in research methods. For example, although soil carbon is widely measured in LTES, researchers do not always agree on uniform ways of sampling soils, of analyzing them, or even of expressing their carbon content (Jandl et al. 2014). Consequently, although progress in forging long-term research networks has been admirable (e.g., Richter et al. 2011), opportunity still remains to further enhance synergy among the LTES worldwide.

A further characteristic of enduring LTES is their integration with other research approaches. One reason that LTES sometimes falter is their slow delivery of findings, and the condescending perception that they amount to "mere monitoring," not innovative, hypothesis-driven research (Keeling 2008; Taylor 1989). Funding agencies are not always generous with studies promising answers decades away (Callahan 1984; Müller et al. 2010; Risser 1991). And researchers too may be reluctant to invest in LTES because such studies do not always produce many publications quickly (Anonymous 2007; Nisbet 2007). As articulated already a century ago (Allen 1922), this difficulty can be assuaged by linking LTES with process-level studies. In this way, the LTES become the birthplace of new hypotheses (and burial grounds for old ones), in an endless iterative cycle: the LTES spawn new enigmas, which are addressed in process-level studies, yielding hypotheses evaluated in the LTES, which invariably generate new questions (Figure 19.2) (O'Gorman and Woodward 2013). Long-term experiments at Lethbridge, for example, revealed an inexplicable imbalance in long-term nitrogen budgets, as seen also in other long-term experiments, prompting studies of N_2 fixation and N deposition, still ongoing (Bremer et al. 1995; McGinn et al. 2003). In the same way, LTES are also indispensable in the building and testing of models (Loveland et al. 2014; Nisbet et al. 2014; Pielou 1981). For example, a model might mimic nicely the changes in soil carbon but falter in tracing signals of radiocarbon measured in LTES (Ellert and Janzen 2006), thus keeping scientists properly humble and advancing further inquiry. LTES, then, can be seen as a connective, synthesizing venue of research, weaving together strands of science across dimensions of time and space, thereby enticing researchers (and maybe also funding agencies) (Figures 19.1 and 19.2). Acclaimed new findings in ecological science, often enough, are indebted to insights quietly gleaned from LTES without them always being recognized.

Another attribute of enduring LTES is the consistent archiving of records, data, and samples for use at a later time. The impressive archives at

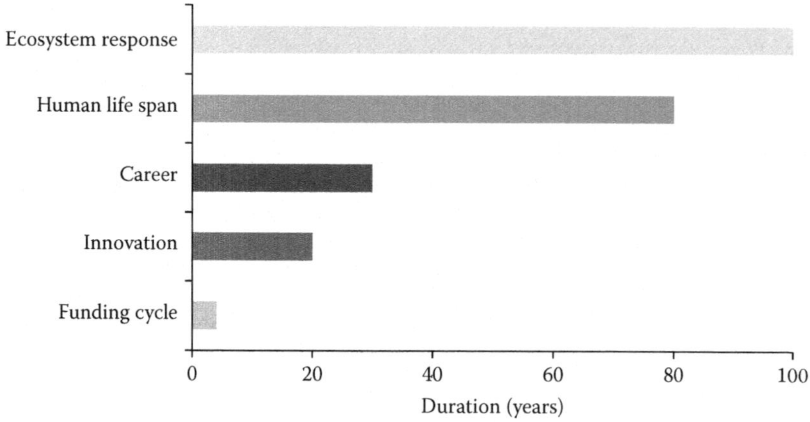

FIGURE 19.1
The duration of ecosystem response to management change or environmental upheaval, compared to that of a typical funding cycle, technological innovation, human career, and human lifespan. (From Kümmerer, K. et al., *J. Soil Water Conserv.*, 65, 141, 2010, based on Figure 1.)

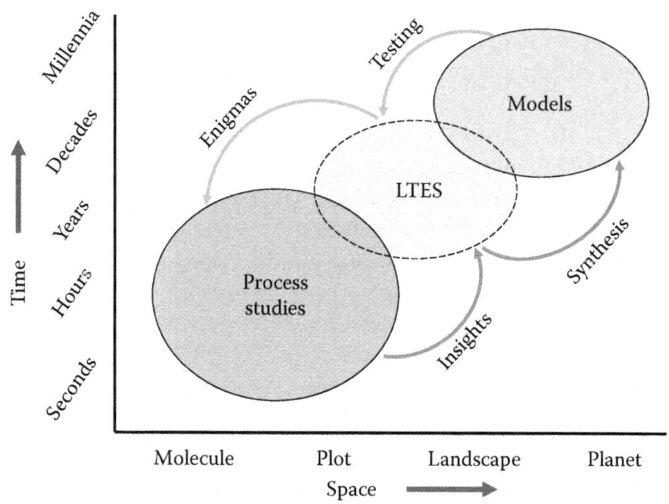

FIGURE 19.2
LTES as integrative elements in studies of ecosystem change. LTES generate questions (enigmas) that are addressed in process-level studies. Insights arising from such fundamental studies are evaluated over the long term in LTES, and findings are synthesized into ecosystem models. These models are evaluated over time in continuing LTES, invariably eliciting new enigmas, continuing the cycle. (Adapted from Janzen, H.H., *Glob. Chang. Biol.*, 15, 2770, 2009, Figure 3.)

Rothamsted, for example, are now as valuable as the experiments themselves (Leigh et al. 1994; Powlson et al. 2014). Long-term experiments maintained by INRA similarly have extensive archives of soils (van Oort 2013). Archived samples from these and other LTES serve as a repository for future analyses that those who collected the samples could not possibly have foreseen. For example, soil samples collected from Broadbalk, Rothamsted, in 1886 were used for innovative radiocarbon analysis a century later (Jenkinson et al. 1994). A challenge for LTES experimenters now is to decide what types of samples should be preserved for future scientists. For example, aside from collecting soil samples, experimenters may also want to consider storing samples of water, plant tissues, animal tissues, and DNA of plants, animals, and microbes. By far-sighted storage of samples from LTES, their value to science may continue even after the studies themselves have long expired.

Not all existing LTES should necessarily be continued indefinitely. Sometimes, the credibility of their findings is irretrievably compromised by flawed design, lapses in continuity, or disturbances such as urban encroachment. In other instances, the ecosystem under study is already adequately studied at other sites, resulting in unmerited duplication. Under such circumstances, terminating selected sites may help shift finite resources toward new or other ongoing studies. Always, however, LTES should be discontinued only after thorough, sober review; for example, scientists might have lost a treasured LTES if Lawes and Gilbert had not understood and vigorously defended the future value of Rothamsted experiments in the late nineteenth century (Johnston 1994).

19.4 New Directions

Existing LTES provide indispensable understanding of ecosystem response to human management and other drivers of global change. But the current slate of studies has gaps that may merit new or redirected LTES. Perhaps the most urgent is the establishment of more such studies in developing countries, where stresses on ecosystems are often most intense, lands are most vulnerable to global changes, and current LTES are most sparse (Fankhauser and McDermott 2014; Greenland 1994; Wheeler and von Braun 2013). Further, more LTES may be needed in urban lands, where most people now live and human stresses are often most pronounced (Pearson 2013; Vasenev et al. 2013). Although ecologists have sometimes shied away from studying such "novel" ecosystems, urban lands occupy increasing areas and can offer many services and potential opportunities for renewal (Ellis 2011; Lovell and Johnston 2009). More LTES are also required to monitor changes in polar ecosystems, where changes to climate may be more rapid and extreme than elsewhere (O'Gorman and Woodward 2013). Because these ecosystems may contain as much as one half of the global organic carbon stored belowground (Tarnocai et al. 2009), impending changes may induce substantial feedbacks to the global carbon cycles (Schaefer et al. 2011; Schuur et al. 2008; Tarnocai et al. 2009). Finally, more LTES may be justified on the vast areas of ecosystems already degraded by human activity (Smith et al. 2010). In the past, many ecological studies have focused, justifiably, on conserving ecosystems and avoiding degradation. But scientists are increasingly aware that ecosystems, once degraded, often demonstrate gradual recovery, if studied over long enough time periods (Jones and Schmitz 2009; Kareiva and Marvier 2012; Nelder 2013). For example, a soil stripped of its topsoil was shown in an LTES to slowly recover productivity and soil health over periods of decades, even without vigorous reclamation measures (Larney et al. 2016). The preceding are just a few examples of apparent gaps in the current suite of LTES. What is needed is a broad spectrum of such studies, distributed across the biomes, beyond the frequent focus on ecosystems with immediate economic returns.

A further challenge, aside from establishing LTES in new geographical locations, is devising new designs and approaches. A particular limitation of many current LTES is their small scale. Some important ecosystem services such as providing wildlife habitat, ensuring biodiversity, and filtering water can be properly studied only at landscape or larger scales (Francis et al. 2014; Godfray and Garnett 2014; Tscharntke et al. 2012; Wiens 2013), areas not easily encompassed in a typical LTES (Powlson et al. 2014). Possible approaches to resolve this difficulty include using "spatially nested" configurations (Syrbe et al. 2010) or establishing scaled-down ecosystems such as "farmlets" (Eisler 2014). With the growing scarcity of available land, every ecosystem will need to support multiple services concurrently (DeFries et al. 2012; Dosskey et al. 2012; Midgley 2012; Poppy et al. 2014). If, as proposed in this review, the intent of LTES is to study ecosystems' continuing capacity to furnish these services, then creative ingenuity will be required to envision new designs that broaden the range of services studied.

Another recent development, meriting stronger consideration in LTES, is the study of social-ecological systems (Cumming et al. 2012; Gragson 2013; Haberl et al. 2006; Mauz et al. 2012; Singh et al. 2013). Now, in the Anthropocene, humans are often the primary drivers of change in ecosystems (Barnosky et al. 2012; Crutzen and Stoermer 2000; Ellis 2011; Kareiva and Marvier 2012). At the same time, humans depend on land as much as ever and hence are vulnerable to changes in ecosystems over time. For example, soil degradation in ecosystems affects not only food supply but also human poverty and political stability (Lal 2008). One way of including human influences and dependencies may be the coupling of LTES with life-cycle analysis. Consider, for example, systems for producing livestock, now the largest anthropogenic user of land (Cassidy et al. 2013) and a source of both environmental harm and benefit (Eisler 2014; Franzluebbers et al. 2012; Janzen 2011; Pelletier et al. 2011; Smil 2013; Vinnari and Tapio 2012; Westhoek et al. 2014). Researchers have begun to use life-cycle analysis to evaluate the overall, system-wide effect of such systems on concerns such as greenhouse gas production (Beauchemin et al. 2010, 2011), but what is often missing is the contribution of gradual, slowly unfolding processes like soil carbon change (Herrero et al. 2013; Smith 2014; Soussana and Lemaire 2014). Melding long-term findings from LTES into life-cycle analysis thus enhances the understanding of social-ecological systems, but it represents just one example of potentially new ways of including human behavior more directly into LTES.

19.5 Conclusion

LTES remain an indispensable research tool for understanding ecosystem responses to impending global changes. Many ecosystem responses to these global stresses occur slowly, episodically, or in complex temporal patterns and can therefore be captured only by monitoring ecosystems for several decades or longer. The critical importance of this research tool and the changing nature of stresses imposed on ecosystems, however, may warrant some innovations in how and where such studies are conducted to ensure all vulnerable ecosystems are properly represented to broaden the measurements recorded and to devise innovative new approaches. A particular challenge will be including more explicitly the human dimension in social-ecological systems, now dominant on many lands.

Aside from improving and expanding LTES, scientists may also need to communicate more effectively the importance of such studies in ecological research. Many process-level studies and modelling efforts, at the forefront of research, are deeply dependent on findings and insights from LTES, but scientists have not always adequately highlighted this indebtedness. By communicating more explicitly the fundamental role of long-term studies to advancements in ecosystem research, today's scientists can help ensure that these priceless treasures survive and are bequeathed without compromise to future generations of scientists.

References

Adhikari, K. and A.E. Hartemink. 2016. Linking soils to ecosystem services—A global review. *Geoderma* 262:101–111.

Allen, E.W. 1922. The method of science in agriculture. *Scientific Agriculture* 2:181–185.

Andrén, O. and T. Kätterer. 1997. ICBM: The introductory carbon balance model for exploration of soil carbon balances. *Ecological Applications* 7:1226–1236.

Anonymous. 1924. *The Organization Achievements and Present Work of the Experimental Farms*. Ottawa, Ontario, Canada: Government Printing Bureau.

Anonymous. 2007. Patching together a world view. *Nature* 450:761.

Barnosky, A.D., E.A. Hadley, J. Bascompte et al. 2012. Approaching a state shift in Earth's biosphere. *Nature* 486:52–58.

Beauchemin, K.A., H.H. Janzen, S.M. Little, T.A. McAllister, and S.M. McGinn. 2010. Life cycle assessment of greenhouse gas emissions from beef production in western Canada: A case study. *Agricultural Systems* 103:371–379.

Beauchemin, K.A., H.H. Janzen, S.M. Little, T.A. McAllister, and S.M. McGinn. 2011. Mitigation of greenhouse gas emissions from beef production in western Canada—Evaluation using farm-based life cycle assessment. *Animal Feed Science and Technology* 166–167:663–677.

Boyd, I.L. 2012. The art of ecological modeling. *Science* 337:306–307.

Bremer, E., H.H. Janzen, and C. Gilbertson. 1995. Evidence against associative N_2 fixation as a significant N source in long-term wheat plots. *Plant and Soil* 175:13–19.

Callahan, J.T. 1984. Long-term ecological research. *Bioscience* 34 (1984):363–367.

Carpenter, S.R. 2002. Ecological futures: Building an ecology of the long now. *Ecology* 83:2069–2083.

Carpenter, S.R., H.A. Mooney, and J. Agard. 2009. Science for managing ecosystem services: Beyond the Millennium Ecosystem Assessment. *Proceedings of the National Academy of Sciences of the United States of America* 106:1305–1312.

Cassidy, E.S., P.C. West, J.S. Gerber, and J.A. Foley. 2013. Redefining agricultural yields: From tonnes to people nourished per hectare. *Environmental Research Letters* 8:034015.

Chapin, F.S., C. Folke, and G.P. Kofinas. 2009. A framework for understanding change. In *Principles of Ecosystem Stewardship. Resilience-Based Natural Resource Management in a Changing World*, eds. F.S. Chapin, G.P. Kofinas, and C. Folke, pp. 3–28. New York: Springer.

Collins, S.L. and D.L. Childers. 2014. Long-term ecological research and network-level science. *EOS, Transactions American Geophysical Union* 95:293–304.

Crutzen, P.J. and E.F. Stoermer. 2000. The anthropocene. *Global Change Newsletter* May 2000:41.

Cumming, G., P. Olsson, F. Chapin, and C. Holling. 2012. Resilience, experimentation, and scale mismatches in social-ecological landscapes. *Landscape Ecology*. 2012:1–12.

DeFries, R.S., E.C. Ellis, F.S. Chapin et al. 2012. Planetary opportunities: A social contract for global change science to contribute to a sustainable future. *Bioscience* 62:603–606.

Dosskey, M., G. Wells, G. Bentrup, and D. Wallace. 2012. Enhancing ecosystem services: Designing for multifunctionality. *Journal of Soil and Water Conservation* 67:37A–41A.

Downes, B., F. Miller, J. Barnett et al. 2013. How do we know about resilience? An analysis of empirical research on resilience, and implications for interdisciplinary praxis. *Environmental Research Letters* 8:1–8.

Dudley, N. 2011. *Authenticity in Nature. Making Choices about the Naturalness of Ecosystems*. New York: Earthscan.

Eisler, M.C. 2014. Agriculture: Steps to sustainable livestock. *Nature* 507:32–34.

Ellert, B. and H.H. Janzen. 2006. Long-term biochemical cycling in agroecosystems inferred from ^{13}C, ^{14}C, and ^{15}N. *Journal of Geochemical Exploration* 88:198–201.

Ellert, B.H., H.H. Janzen, and B.G. McConkey. 2000. Measuring and comparing soil carbon storage. In *Assessment Methods for Soil Carbon*, eds. J.M. Kimble, R.F. Follett, and B.A. Stewart, pp. 131–146. Boca Raton, FL: CRC Press.

Ellis, E.C. 2011. Anthropogenic transformation of the terrestrial biosphere. *Philosophical Transactions of the Royal Society A* 369:1010–1035.

Fankhauser, S. and T.K.J. McDermott. 2014. Understanding the adaptation deficit: Why are poor countries more vulnerable to climate events than rich countries? *Global Environmental Change* 27:9–18.

Folke, C. 2006. Resilience: The emergence of a perspective for social-ecological systems analyses. *Global Environmental Change* 16:253–267.

Folke, C., J. Colding, and B. Fikret. 2003. Synthesis: Building resilience and adaptive capacity in socio-ecological systems. In *Navigating Socio-Ecological Systems: Building Resilience for Complexity and Change*, eds. F. Berkes, J. Colding, and C. Folke, pp. 352–387. Cambridge, U.K.: Cambridge University Press.

Francis, C., J. Van Wart, and B. Johnson. 2014. How to regenerate rural community and ecoservices: Reversing the tragedy of the commons. *Agronomy Journal* 106:95–99.

Franklin, J.F. 1989. Importance and justification of long-term studies in ecology. In *Long-Term Studies in Ecology: Approaches and Alternatives*, ed. G.E. Likens, pp. 3–19. New York: Springer-Verlag.

Franzluebbers, A.J., L.K. Paine, J.R. Winsten et al. 2012. Well-managed grazing systems: A forgotten hero of conservation. *Journal of Soil and Water Conservation* 67:100A–104A.

Godfray, H.C.J. and T. Garnett. 2014. Food security and sustainable intensification. *Philosophical Transactions of the Royal Society B* 369:1–10.

Gragson, T.L. 2013. The contribution of anthropology to concepts guiding LTSER research. In *Long Term Socio-Ecological Research: Studies in Society-Nature Interactions Across Spatial and Temporal Scales*, eds. J.S. Singh, H. Haberl, M. Chertow, M. Mirtl, and M. Schmid, pp. 189–214. Dordrecht, the Netherlands: Springer.

Greenfieldboyce, N. 2012. Experiments that keep going and going and going. Long term science: When research outlives the researcher. http://www.npr.org/2012/11/23/165030844/experiments-that-keep-going-and-going-and-going (accessed March 2, 2016).

Greenland, D.J. 1994. Long-term cropping experiments in developing countries: The need, the history and the future. In *Long-Term Experiments in Agricultural and Ecological Sciences*, eds. R.A. Leigh and A.E. Johnston, pp. 187–209. Wallingford, CT: CAB International.

Grove, J.M., S.T.A. Pickett, A. Whitmer, and M.L. Cadenasso. 2013. Building an urban PTSER: The case of the Baltimore ecosystem study and the D.C./B/. ULTRA-Ex Project. In *Long Term Socio-Ecological Research: Studies in Society-Nature Interactions Across Spatial and Temporal Scales*, eds. J.S. Singh, H. Haberl, M. Chertow, M. Mirtl, and M. Schmid, pp. 369–408. Dordrecht, the Netherlands: Springer.

Gunderson, L. and C. Folke. 2011. Resilience 2011: Leading transformational change. *Ecology and Society* 16:30.

Haberl, H., V. Winiwarter, K. Andersson et al. 2006. From LTER to LTSER: Conceptualizing the socioeconomic dimension of long-term socioecological research. *Ecology and Society* 11:article 13.

Hampton, S.E., C.A. Strasser, J.J. Tewksbury et al. 2013. Big data and the future of ecology. *Frontiers in Ecology and the Environment* 11:156–162.

Haughland, D.L., J.-M. Hero, J. Schieck et al. 2010. Planning forwards: Biodiversity research and monitoring systems for better management. *Trends in Ecology and Evolution* 25:199–200.

Herrero, M. et al. 2013. Biomass use, production, feed efficiencies, and greenhouse gas emissions from global livestock systems. *Proceedings of the National Academy of Sciences of the United States of America* 110:20888–20893.

Hodgson, D., J.L. McDonald and D.J. Hosken. 2015. What do you mean, 'resilient'? *Trends in Ecology & Evolution*30:503–506.

Holling, C.S. 1973. Resilience and stability of ecological systems. *Annual Review of Ecology, Evolution and Systematics* 4:1–23.

ILTER. 2016. International long-term ecological research. http://www.ilternet.edu/ (accessed March 2, 2016) .

Jandl, R., M. Rodeghiero, C. Martinez et al. 2014. Current status, uncertainty and future needs in soil organic carbon monitoring. *Science of the Total Environment* 468–469:376–383.

Janzen, H.H. 1995. The role of long-term sites in agroecological research: A case study. *Canadian Journal of Soil Science* 75:123–133.

Janzen, H.H. 2001. Soil Science on the Canadian prairies—Peering into the future from a century ago. *Canadian Journal of Soil Science* 81:489–503.

Janzen, H.H. 2009. Long-term ecological sites: Musings on the future, as seen (dimly) from the past. *Global Change Biology* 15:2770–2778.

Janzen, H.H. 2011. What place for livestock on a re-greening earth? *Animal Feed Science and Technology* 166–167:783–796.

Jenkinson, D.S., N.J. Bradbury, and K. Coleman. 1994. How the Rothamsted classical experiments have been used to develop and test models for the turnover of carbon and nitrogen in soil. In *Long-Term Experiments in Agricultural and Ecological Sciences,* eds. R.A. Leigh and A.E. Johnston, pp. 117–138. Wallingford, CT: CAB International.

Johnston, A.E. 1994. The Rothamsted classical experiments. In *Long-Term Experiments in Agricultural and Ecological Sciences,* eds. R.A. Leigh and A.E. Johnston, pp. 9–37. Wallingford, CT: CAB International.

Jones, H.P. and O.J. Schmitz. 2009. Rapid recovery of damaged ecosystems. *PLoS ONE* 4:e5653.

Kareiva, P. and M. Marvier. 2012. What is conservation science? *Bioscience* 62:962–969.

Keeling, R.F. 2008. Recording earth's vital signs. *Science* 319:1771–1772.

Knapp, A.K., M.D. Smith, S.E. Hobbie et al. 2012. Past, present, and future roles of long-term experiments in the LTER network. *Bioscience* 62:377–389.

Koehler, H. and V. Melecis. 2010. Long-term observations of soil mesofauna. In *Long-Term Ecological Research: Between Theory and Application,* eds. F. Müller, C. Baessler, H. Schubert, and S. Klotz, pp. 203–220. Dordrecht, the Netherlands: Springer.

Kümmerer, K., M. Held, and D. Pimentel. 2010. Sustainable use of soils and time. *Journal of Soil and Water Conservation* 65:141–149.

Lal, R. 2008. Food insecurity's dirty secret. *Science* 322:673–674.

Larney, F.J., L. Li, H.H. Janzen, D.A. Angers, and B.M. Olson. 2016. Soil quality attributes, soil resilience and legacy effects following topsoil removal and one-time amendments. *Canadian Journal of Soil Science* 96:177–190.

Le Quéré, C., R. Moriarty, R.M. Andrew et al. 2015. Global carbon budget 2015. *Earth System Science Data* 7:349–396.

Leigh, R.A., R.D. Prew, and A.E. Johnston. 1994. The management of long-term agricultural field experiments: Procedures and policies evolved from the Rothamsted classical experiments. In *Long-Term Experiments in Agricultural and Ecological Sciences*, eds. R.A. Leigh and A.E. Johnston, pp. 253–268. Wallingford, CT: CAB International.

Loveland, P.J., F. Conen, and B. Van Wesemael. 2014. Batjes, N.H. 1996. Total carbon and nitrogen in the soils of the world. *European Journal of Soil Science* 47:151–163. Commentary on the impact of Batjes (1996). *European Journal of Soil Science* 65:4–21.

Lovell, S.T. and D.M. Johnston. 2009. Designing landscapes for performance based on emerging principles in landscape ecology. *Ecology Society* 14:article 44.

LTER Europe. 2016. Long-term ecosystem research in Europe http://www.lter-europe.net/ (accessed March 2, 2016).

Marsh, G.P. 1864. *Man and Nature; or, Physical Geography as Modified by Human Action*. New York: Charles Scribner.

Mauz, I., T. Peltola, C. Granjou et al. 2012. How scientific visions matter: Insights from three long-term socio-ecological research (LTSER) platforms under construction in Europe. *Environmental Science and Policy* 19–20:90–99.

McGinn, S.M., H.H. Janzen, and T. Coates. 2003. Atmospheric ammonia, volatile fatty acids, and other odorants near beef feedlots. *Journal of Environmental Quality* 32:1173–1182.

MEA (Millennium Ecosystem Assessment). 2005. *Ecosystems and Human Well-Being: Synthesis*. Washington, DC: Island Press.

Midgley, G.F. 2012. Biodiversity and ecosystem function. *Science* 335:174–175.

Mooney, H., A. Larigauderie, M. Cesario et al. 2009. Biodiversity, climate change, and ecosystem services. *Current Opinion in Environmental Sustainability* 1:46–54.

Müller, F., A. Gnauck, K.-O. Wenkel, H. Schubert, and M. Bredemeier. 2010. Theoretical demands for long-term ecological research and the management of long-term data sets. In *Long-Term Ecological Research: Between Theory and Application*, eds. F. Müller, C. Baessler, H. Schubert, and S. Klotz, pp. 11–25. Dordrecht, the Netherlands: Springer.

Nelder, C. 2013. Communication: Positive energy. *Nature* 498:293–295.

Nisbet, E.G. 2007. Earth monitoring: Cinderella science. *Nature* 450:789–790.

Nisbet, E.G., E.J. Dlugokencky, and P. Bousquet. 2014. Methane on the rise—Again. *Science* 343:493–495.

O'Gorman, E.J. and G. Woodward. 2013. Editorial commentary: Monitoring, manipulation and modelling of ecological responses to global change in multispecies systems. *Advances in Ecological Research* 48:xiii–xxii.

Palmer, A.E. 1949. *Progress Report 1936–1946 Dominion Experimental Station, Lethbridge, Alberta*. Ottawa, Ontario, Canada: Government of Canada.

Paul, E.A., K.H. Paustian, E.T. Elliott, and C.V. Cole. 1997. *Soil Organic Matter in Temperate Agroecosystems: Long Term Experiments in North America*. Boca Raton, FL: CRC Press.

Pearson, L.J. 2013. In search of resilient and sustainable cities: Prefatory remarks. *Ecological Economics* 86:222–223.

Pelletier, N., E. Audsley, S. Brodt et al. 2011. Energy intensity of agriculture and food systems. *Annual Review of Environmental Resources* 36:223–46.

Peters, D.P.C., C.M. Laney, A.E. Lugo et al. 2013. Long-term trends in ecological systems: An introduction to cross-site comparisons and relevance to global change studies. In *Long-Term Trends in Ecological Systems: A Basis for Understanding Responses to Global Change*, eds. D.P.C. Peters, C.M. Laney, A.E. Lugo et al., pp. 1–20. Las Cruces, NM: USDA Agricultural Research Service Department of Agriculture Technical Bulletin No. 1931.

Pielou, E.C. 1981. The usefulness of ecological models: A stock-taking. *Quarterly Review of Biology* 56:17–31.

Poppy, G.M., S. Chiotha, F. Eigenbrod et al. 2014. Food security in a perfect storm: Using the ecosystem services framework to increase understanding. *Philosophical Transactions of the Royal Society B* 369:1–13.

Powers, R.F. and K. Van Cleve. 1991. Long-term ecological research in temperate and boreal forest ecosystems. *Agronomy Journal* 83:11–24.

Powlson, D.S., A.J. MacDonald, and P.R. Poulton. 2014. The continuing value of long-term field experiments: Insights for achieving food security and environmental integrity. In *Soil as World Heritage*, ed. D. Dent, pp. 131–157. Dordrecht, the Netherlands: Springer.

Richter, D., A.R. Bacon, M. Mobley et al. 2011. Human-soil relations are changing rapidly: Proposals from SSSA's Cross-Divisional Soil Change Working Group. *Soil Science Society of America Journal* 75:2079–2084.

Richter, D., D. Holfmockel, M.A. Callaham, D.S. Powlson, and P. Smith. 2007. Long-term soil experiments: Keys to managing earth's rapidly changing ecosystems. *Soil Science Society of America Journal* 71:266–279.

Risser, P.G. 1991. Introduction. In *Long Term Ecological Research. An International Perspective*, ed. P.G. Risser, pp. 1–8. Chichester, U.K.: John Wiley & Sons Ltd.

Saunders, W. 1900. Annual report of the experimental farms. In *Appendix to the Report of the Minister of Agriculture*, ed. S.E. Dawson, pp. 5–66. Ottawa, Canada.

Schaefer, K., T. Zhang, L. Bruhwiler, and A.P. Barrett. 2011. Amount and timing of permafrost carbon release in response to climate warming. *Tellus B.* 63:165–180.

Schuur, E.A.G., J. Bockheim, J.G. Canadell, E. Euskirchen, C.B. Field, S.V. Goryachkin, et al. 2008. Vulnerability of permafrost carbon to climate change: Implications for the global carbon cycle. *BioScience* 58:701–714.

Singh, J.S., H. Haberl, M. Chertow, M. Mirtl, and M. Schmid. 2013. Introduction. In *Long Term Socio-Ecological Research. Studies in Society: Nature Interactions Across Spatial and Temporal Scales*, eds. J.S. Singh, H. Haberl, M. Chertow, M. Mirtl, and M. Schmid, pp. 1–26. Dordrecht, the Netherlands: Springer.

Smil, V. 2013. *Should We Eat Meat? Evolution and Consequences of Modern Carnivory*. Chichester, U.K.: Wiley-Blackwell.

Smith, P. 2014. Do grasslands act as a perpetual sink for carbon? *Global Change Biology* 20:2708–2711.

Smith, P., P.D. Falloon, M. Korschens et al. 2002. EuroSOMNET—A European database of long-term experiments on soil organic matter: The WWW metadatabase. *Journal of Agricultural Science* 138:123–134.

Smith, P., P.J. Gregory, D. van Vuuren et al. 2010. Competition for land. *Philosophical Transactions of the Royal Society B* 365:2941–2957.

Soussana, J.-F. and G. Lemaire. 2014. Coupling carbon and nitrogen cycles for environmentally sustainable intensification of grasslands and crop-livestock systems. *Agriculture, Ecosystems and Environment* 190:9–17.

Southwood, T.R.E. 1994. The importance of long-term experimentation. In *Long-Term Experiments in Agricultural and Ecological Sciences*, eds. R.A. Leigh and A.E. Johnston, pp. 3–8. Wallingford, CT: CAB International.

Sutherland, W.J. 2013. Policy: Twenty tips for interpreting scientific claims. *Nature* 503:335–337.

Syrbe, R.-U., W. Hierold, O. Bastian, and M. Roder. 2010. A conceptual framework for integrated functional landscape monitoring in the wider countryside of central Europe. In *Long-Term Ecological Research: Between Theory and Application*, eds. F. Müller, C. Baessler, H. Schubert, and S. Klotz, pp. 263–278. Dordrecht, the Netherlands: Springer.

Tansley, A.G. 1935. The use and abuse of vegetational concepts and terms. *Ecology* 16:284–307.

Tarnocai, C., J.G. Canadell, E.A.G. Schuur, P. Kuhry, G. Mazhitova, and S. Zimov. 2009. Soil organic carbon pools in the northern circumpolar permafrost region. *Global Biogeochemical Cycles* 23:GB2023.

Taylor, R.L. 1989. Objective and experiment in long term research. In *Long-Term Studies in Ecology: Approaches and Alternatives*, ed. G.E. Likens, pp. 20–70. New York: Springer-Verlag.

Telewski, F.W. and J.A.D. Zeevaart. 2002. The 120-yr period for Dr. Beal's seed viability experiment. *American Journal of Botany* 89:1285–1288.

Tingley, M.W. 2013. Ecosystems: Climate change must not blow conservation off course. *Nature* 500:271–272.

Tscharntke, T., Y. Clough, T.C. Wanger et al. 2012. Global food security, biodiversity conservation and the future of agricultural intensification. *Biology Conservation* 151:53–59.

van Oort, F. 2013. INRA's 42 plots design (Versailles, France) since 1928: A world's unique long-term bare fallow experiment. Grignon, Savoie, France: INRA, Centre de Versailles.

Vanderbilt, K.L., C.-C. Lin, S.-S. Lu et al. 2015. Fostering ecological data sharing: Collaborations in the International Long Term Ecological Research Network. *Ecosphere* 6: article 204.

Vasenev, V.I., J.J. Stoorvogel, and I.I. Vasenev. 2013. Urban soil organic carbon and its spatial heterogeneity in comparison with natural and agricultural areas in the Moscow region. *Catena* 107:96–102.

Villa, F., B. Voigt, and J.D. Erickson. 2014. New perspectives in ecosystem science to understand environmental securities. *Philosophical Transactions of the Royal Society B* 369:1–15.

Vinnari, M. and P. Tapio. 2012. Sustainability of diets: From concepts to governance. *Ecology Economics* 74:46–54.

Weinberger, D. 2012. To know, but not understand: David Weinberger on science and big data. *The Atlantic*, January 3, 2012. http://www.theatlantic.com/technology/archive/2012/01/to-know-but-not-understand-david-weinberger-on-science-and-big-data/250820/ (accessed October 24, 2016).

Westhoek, H., J.P. Lesschen, T. Rood et al. 2014. Food choices, health and environment: Effects of cutting Europe's meat and dairy intake. *Global Environmental Change* 26:196–205.

Wheeler, T. and J. von Braun. 2013. Climate change impacts on global food security. *Science* 341:508–513.

White, M.A. 2013. Sustainability: I know it when I see it. *Ecology Economics* 86:213–217.

Wiens, J. 2013. Is landscape sustainability a useful concept in a changing world? *Landscape Ecology* 28:1047–1052.

Willis, A.J. 1997. The ecosystem: An evolving concept viewed historically. *Functional Ecology* 11:268–271.

20

Integrated Carbon Observation System (ICOS): An Infrastructure to Monitor the European Greenhouse Gas Balance

Bert Gielen, Maarten Op de Beeck, Denis Loustau, Reinhart Ceulemans, Armin Jordan, and Dario Papale

CONTENTS

20.1 Introduction: Identify the Challenges

Climate change is one of the most challenging problems that humanity has to cope with in the coming decades. The Intergovernmental Panel on Climate Change (IPCC, 2013) has concluded that the observed rise in global temperature is very likely due to increasing greenhouse gases (GHGs) in the atmosphere, caused by anthropogenic emissions. These increased concentrations of CO_2 and CH_4, which exceed by far the natural range observed over the last 650,000 years, and its impact on the global biogeochemical cycles are a major driving force of current and future climate change. The current levels of CO_2 have increased by 40% from preindustrial times (Jackson et al., 2016). Moreover, the current atmospheric levels of CH_4 are nearly two and a half times the preindustrial value. The main sources of anthropogenic CH_4 emissions are fossil fuel combustion and modifications of global vegetation through land use change, in particular deforestation. Terrestrial vegetation and oceans absorb together about half of the yearly anthropogenic emissions (Le Quere et al., 2009). The question is whether these sinks will be persistent in the future, especially under changing climatic conditions and under increased human pressure. At the current atmospheric level of CH_4, the natural oxidizing power of the atmosphere removes almost all the CH_4 emitted by human activities and natural sources, but increased human activities are expected to increase emissions even further and result in elevated CH_4 concentrations in the atmosphere (Kirschke et al., 2013).

Enhanced understanding of the driving forces of climate change and its impact on the global biogeochemical cycles requires full quantification of the GHG sources and sinks and their evolution. Regional GHG flux patterns, tipping points, and vulnerabilities can be assessed by long-term, high-precision observations in the atmosphere and at the ocean and land surface. During the last decades, many European projects have been set up to study the terrestrial, oceanic, and atmospheric GHG budgets. It became, however, clear that the setup of an infrastructure with a long lifetime is essential to ensure high-precision measurements of the GHG balance of Europe and became inevitable to acquire more insight in the interaction between climate change and the geochemical cycles.

After an initial phase of preparation, the European Commission, in November 2015, recognized the Integrated Carbon Observation System European Research Infrastructure Consortium (ICOS ERIC). This legal framework had been established to manage the distributed ICOS research infrastructure (ICOS RI) (www.icos-ri.eu) with a minimum expected lifetime of 20 years (Figure 20.1).

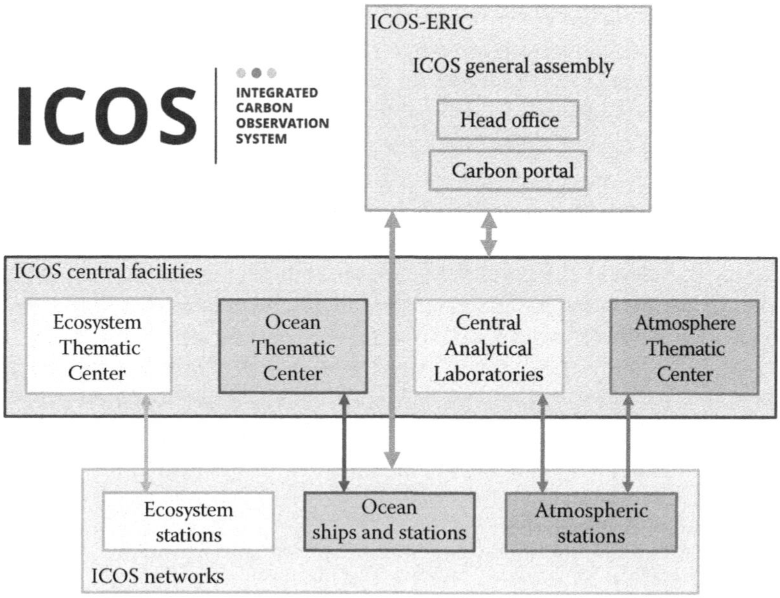

FIGURE 20.1
Schematic representation of the ICOS organizational structures with the four central facilities and the three station networks. ERIC, European Research Infrastructure Consortium.

ICOS aims

1. To provide long-term observations required to understand the present state and predict future behavior of the global carbon cycle and GHG emissions.
2. To monitor and assess the effectiveness of GHG mitigation activities on atmospheric composition levels, including attribution of sources and sinks by region and sector.
3. To serve the research community with high-quality data products to unravel the global geochemical cycles.

20.2 ICOS Structure: New Perspectives of Decentralized Research Infrastructure

To enable the operation of such a large distributed research infrastructure, a set of central facilities (CFs) has been set up to support the observations and do the coordination. ICOS comprises three large distributed observational networks: a network of atmospheric observation towers, a network of ecosystem observation towers, and a network of ocean observatories (ships and buoys). Each of these networks is coordinated by a thematic center that manages the establishment of standardized methodology and protocols across the network and performs centralized processing and quality control (QC) on the data. In addition, to guarantee the availability of highly accurate and internationally compatible calibration gases to the monitoring stations, a central analytical laboratory (CAL) was put into place. This laboratory also provides additional analyses of atmospheric components, which are not yet measured in situ at the stations, such as isotopes in CO_2. Furthermore, a data portal has been set up to distribute the high-level data sets to the end users. A head office has been installed to ensure the overall coordination of the entire research infrastructure. In the next section, each of the components is described in more detail.

20.2.1 Ecosystem Network

20.2.1.1 Ecosystem Stations

The ecosystem network of ICOS has the goal to monitor the terrestrial GHG balance of Europe and consists of a distributed network of eddy covariance towers placed over the main terrestrial ecosystems in Europe. The eddy covariance technique allows us to measure the net exchange of GHGs and energy between the ecosystem and the atmosphere with a temporal resolution of 30 min. In addition to these flux measurements, a whole suite of meteorological variables is measured as irradiation, precipitation, and temperature. While all these variables are recorded with a 30 min resolution, there is also a set of variables that are measured on a less frequent basis as biomass of the vegetation, leaf area index, litterfall, and carbon and nutrient contents of the soil and vegetation. The purpose of the ancillary measurements is to support process studies and to help to understand the physical and biotic factors controlling the GHG fluxes. All mandatory variables are listed in Table 20.1. The ICOS ecosystem network includes two levels of ecosystem stations (ES), referred to as Class 1 (complete) and Class 2 (basic) stations, which differ in costs of construction, running, and maintenance. This strategy enhances flexibility and ensures a high level of participation. Nonetheless, as a major characteristic of ICOS is standardization and data quality, all ES, either Class 1 or Class 2, are characterized by a strict standardization of instrumentation and procedure. This implies that costs can be reduced only by reducing the number of instrumentation/procedures supported and variables measured, not their quality. However, stations may be eligible for Class 1 ranking even if they do not support all of the mandatory measurements, provided it can be demonstrated (e.g., by the means of historical data or a short measurement campaign) that the measurements that are omitted are not relevant for the site.

TABLE 20.1

Mandatory Variables Measured at the Class 1 and Class 2 ICOS Ecosystem Stations for Six Ecosystem Types

Variable	Forest	Grassland	Cropland	Peatland	Marine	Lake
CO_2, H_2O, and sensible heat fluxes (eddy covariance)	1 and 2	1 and 2	1 and 2	1 and 2	1 and 2	1 and 2
Eddy covariance CH_4 and N_2O	1	1	1	1	1	1
Air CO_2 and H_2O vertical profile	1	1	1	1	Fac	Fac
Air CO_2 vertical profile	2	2	2	2	Fac	Fac
Air H_2O concentration	1	1	1	1	1	1
In, Out, and Net SW and LW radiation, surface temperature	1 and 2	1 and 2	1 and 2	1 and 2	1	1
SW incoming radiation high quality	Fac	Fac	Fac	Fac	Fac	Fac
PAR/PPFD incident	1 and 2	1 and 2	1 and 2	1 and 2	1 and 2	1 and 2
PAR/PPFD below canopy + ground reflected	Fac	Fac	Fac	N.R.	N.R.	N.R.
PAR/PPFD reflected	1 and 2	1 and 2	1 and 2	1 and 2	Fac	Fac
Diffuse PAR/PPFD radiation	1	1	1	1	Fac	Fac
Spectral reflectance	Fac	Fac	Fac	Fac	Fac	Fac
Soil heat flux	1 and 2	1 and 2	1 and 2	1 and 2	N.R.	N.R.
Air temperature and Rh profile	1 and 2	1 and 2	1 and 2	1 and 2	Fac	Fac
Main meteo vars (Ta, Rh, Swin, precipitation)	1 and 2	1 and 2	1 and 2	1 and 2	1 and 2	1 and 2
Total high accuracy precipitation	1	1	1	1	1	1
Rain precipitation	1 and 2	1 and 2	1 and 2	1 and 2	1 and 2	1 and 2
Snow precipitation	1	1	1	1	1	1
Snow height	1 and 2	1 and 2	1 and 2	1 and 2	Fac	Fac
Soil water content profile	1 and 2	1 and 2	1 and 2	1 and 2	N.R.	N.R.
Soil temperature profile	1 and 2	1 and 2	1 and 2	1 and 2	N.R.	N.R.

(Continued)

TABLE 20.1 (*Continued*)

Mandatory Variables Measured at the Class 1 and Class 2 ICOS Ecosystem Stations for Six Ecosystem Types

Variable	Forest	Grassland	Cropland	Peatland	Marine	Lake
Air pressure	1 and 2	1 and 2	1 and 2	1 and 2	1 and 2	1 and 2
Trunk and branches temperature	Fac	N.R.	N.R.	N.R.	N.R.	N.R.
Groundwater level	1 and 2	1 and 2	1 and 2	1 and 2	N.R.	N.R.
Trees diameter	1	N.R.	N.R.	N.R.	N.R.	N.R.
Phenology camera	1	1	1	1	N.R.	N.R.
Soil CO_2 automatic chambers	1	1	1	1	1	1
CH_4 and N_2O fluxes by automatic chambers	1	1	1	1	1	1
Wind speed and wind direction (additional)	1	1	1	1	1	1
LAI	1 and 2	1 and 2	1 and 2	1 and 2	N.R.	N.R.
Aboveground biomass	1 and 2	1 and 2	1 and 2	1 and 2	N.R.	N.R.
Soil carbon content	1 and 2	1 and 2	1 and 2	1 and 2	N.R.	N.R.
Litterfall	1	1	1	1	N.R.	N.R.
Leaf N content	1 and 2	1 and 2	1 and 2	1 and 2	N.R.	N.R.
Soil water N content	Fac	Fac	Fac	Fac	N.R.	N.R.
Dissolved organic carbon concentration	Fac	Fac	Fac	Fac	N.R.	N.R.
C and N import/export by management	1 and 2	1 and 2	1 and 2	1 and 2	N.R.	N.R.
Oxygen and pCO_2 surface concentration	N.R.	N.R.	N.R.	Fac	2	2
Oxygen, pCO_2, and pN_2O concentration profile	N.R.	N.R.	N.R.	Fac	1	1
Salinity	N.R.	N.R.	N.R.	N.R.	1 and 2	N.R.
Wave properties	N.R.	N.R.	N.R.	N.R.	Fac	Fac
Water temperature profile	N.R.	N.R.	N.R.	N.R.	1	1
Management and disturbances information	1 and 2	1 and 2	1 and 2	1 and 2	1 and 2	1 and 2

Note: N.R., Not Relevant.

20.2.1.2 Ecosystem Thematic Center

The ecosystem network is coordinated by a CF, the Ecosystem Thematic Center (ETC—www.icos-etc.eu). It consists of a consortium of three research groups located at the University of Tuscia (Italy), the University of Antwerp (Belgium), and the INRA in Bordeaux (France). The ETC is responsible to process all raw data that are transferred by the ES at a daily time scale. Wind speed and GHG concentrations are recorded at 20 Hz resolutions and sent to the ETC where they are processed into 30 min fluxes. Meteorological variables are measured at 1 Hz and sent. The main advantage of the centralized processing is the ability to apply a standardized quality check and QC starting from the raw data collected at the station. The ETC is also in charge of providing measurement protocols for all mandatory variables to ensure that all measurements are performed uniformly across the entire network. These protocols contain information of methodology, spatial sampling design, temporal sampling design, units, data formats, required accuracy of the instruments, and calibration. The choice of the instruments to be integrated into the ICOS-ES is made based on scientific literature review, user feedbacks, laboratory and field tests, and on the exchange with scientific partners. In addition, the ETC established a test unit that is in charge of following the newest technological development in the field of meteorological and flux measurements. The test unit has a laboratory facility to test and compare new instruments before they are considered to be ICOS compliant and be deployed in the ICOS network.

20.2.2 Atmospheric Network

20.2.2.1 Atmospheric Stations

Atmospheric stations (AS) are observatories continuously measuring GHG concentrations, focusing on the variability due to regional and global fluxes. The ICOS atmospheric network will rely on more than 40 sites distributed across Europe and adjacent regions. Species of interest include the main natural and anthropogenic GHGs CO_2, CH_4, and N_2O, as well as other atmospheric parameters that support the interpretation and analysis of the concentration measurements, such as isotope ratios and meteorological parameters. The full list of variables is shown in Table 20.2. As with ES, the ICOS AS modular character allows for different configurations. However, a range of mandatory parameters have been defined as key components that are measured in a harmonized way across the stations in Europe. Stations are further divided into two classes. Class 1 stations include a large range of mandatory measurements, whereas Class 2 stations operate only a subset of Class 1 stations' mandatory parameters (cf. Table 20.2).

TABLE 20.2

Overview of the Mandatory Gas Species, Meteorology, and Eddy Covariance Fluxes Measured at the Class 1 and Class 2 ICOS Atmospheric Stations

ICOS AS	Gas Species			
Category	**Continuous Sampling**	**Periodical Sampling**	**Meteorology**	**Eddy Fluxes**
Class 1 Mandatory parameters	CO_2, CH_4, CO: at each sampling height	–CO_2, CH_4, N_2O, SF_6, CO, H_2, CO_2 stable isotopes: weekly sampled –^{14}C (radiocarbon integrated samples): at highest sampling height	• Air temperature, relative humidity, wind direction, wind speed: at highest and lowest sampling height • Atmospheric pressure • Planetary boundary layer height	
Class 2 Mandatory parameters	CO_2, CH_4: at each sampling height		• Air temperature, relative humidity, wind direction, wind speed: at highest and lowest sampling height	
Recommended parameters	• Rn, N_2O, O_2/N_2 ratio • CO for Level 2 stations	CH_4 stable isotopes, O_2/N_2 ratio for Level 1 stations: weekly sampled at highest sampling height		CO_2 flux: at one sampling height

20.2.2.2 Atmospheric Thematic Center

The Atmospheric Thematic Center (ATC—www.icos-atc.lsce.ipsl.fr), located at LSCE in Paris (France), is the CF that coordinates the atmospheric network of ICOS. Its main target is to develop and operate the atmospheric data processing chains, ranging from data transmission from stations to the routine delivery of quality-checked data streams provided to the modelers. The data processing comprised three major steps. First, the raw data, ancillary data, and all parameters necessary to do the processing are transferred from the station to the ATC. Second, the air concentration measurements and the calibration information are extracted from the raw data. In this step, the GHG and other tracer concentrations are converted into geophysical units (e.g., ppm for CO_2). Once this step is concluded, higher-level data products are produced, which include trace gas concentrations, meteorological parameters, and some information about data quality. Online quality checks are performed during this processing stage. An automated quality check is applied based on the information given by each instrument (e.g., temperature, flow rates) and on a statistical algorithm to identify suspicious signals. Then, interactive tools are provided to the station principal investigators in order to flag the data, which correspond to malfunctions of the instruments or to local contaminations. To optimally serve the data user community, two distinct data streams are generated. The first one is the Near Real-Time (NRT) data stream. The maximum delay for NRT data after in situ measurement is 24 h. The second data set contains high-precision GHG time series. The precision of the latter product is increased as compared to NRT data, due to a posteriori revision of calibration information that is included in its computation. Also the final data set has gone through the inspection of the expertise of the station PI. Ultimately, higher-level data products (Level 2) are produced by the ATC. These data sets are built upon the Level 1 data. Using in situ meteorological data collected online at each station, Level 2 may incorporate information about air mass characteristics collected on an hourly basis, as well as on Lidar Boundary Layer Height measurements and in situ local CO_2 fluxes. Gap filling algorithms will be used in case of missing data intervals. Level 2 data incorporate information on data representativeness, which is useful for modelers. In addition, the ATC carries out regular measurement technology surveys and analyses and enables development of new sensors and their testing, for instance, through R&D programs. Furthermore, it provides services to the network by distributing spare instruments and organizing training sessions and high-level technical assistance.

20.2.3 Ocean Network

20.2.3.1 Ocean Stations

The ocean component of ICOS consists of an observational network of fixed stations like buoys–moorings, Voluntary Observation Ships (VOS), and repeat sections that are located and operational in the North Atlantic Ocean sector from the Arctic to the Southern Ocean also including marginal seas such as Nordic, the Baltic, and the Mediterranean Seas. VOS are commercial ships operating on regularly repeated routes, for example, ferry routes in European shelf and marginal seas, and cargo vessels on open ocean routes. The fixed time series are points in the ocean at which sustained time series observations are recorded by means of moorings and research ship monitoring. The VOS and fixed time series stations are equipped with a range of automated instrumentation to measure atmospheric and surface ocean partial CO_2 pressure (pCO_2), surface temperature, salinity, and related variables, and the VOS data will be the European contribution to the global database called SOCAT. The fixed stations are organized in a similar way. In addition, there is the repeat section in which the scientific community recently released the GLODAP2 database. A full list of variables can be found in Table 20.3. Measurements are repeated along similar transects at intervals

TABLE 20.3

Overview of the Mandatory Variable Measured at the ICOS Ocean Network Stations

Type	Variable	Frequency	Accuracy and Precision
Core	Atmospheric pCO_2	Continuous (30 min)	Essential to within 1 μatm Desirable to within 0.1 μatm
Core	Sea surface pCO_2	Continuous (30 min)	To within 1 μatm
Core	Barometric pressure	Continuous (30 min)	To within 0.5 mbar
Core	Sea surface temperature	Continuous (30 min)	To within 0.05°C
Core	Sea surface salinity	Continuous (30 min)	To within 0.1 units
Core	Nutrients ($NO_3 + NO_2$, PO_4, SiO_4)	Periodical, water sampling	To within 1 μM
Core	Dissolved inorganic carbon	Periodical, water sampling	Desirable to within 1 μmol/kg
Core	Total alkalinity	Periodical, water sampling	Desirable to within 1 μmol/kg
Additional	Atmospheric flask samples for CO_2, CH_4, N_2O, SF_6, CO, H_2, O_2/N_2, ^{13}C, ^{18}O, ^{14}C in CO_2	Depending on VOS route	As defined by WMO for atmospheric monitoring
Additional	Chlorophyll fluorescence		
Desirable	pH	Continuous (30 min)	To within 0.001 units
Desirable	Total dissolved gas pressure		
Desirable	Meteorological parameters	Continuous (30 min)	

Note: The frequency, accuracy, and precision of each parameter will also be provided.

of days to months. In addition to the core platforms used in marine ICOS, marine ICOS works with the wider community to develop new sensors for the carbon system for autonomous platforms such as Array for Real-time Geostrophic Oceanography (ARGO) floats and gliders. Furthermore, it collaborates to provide better coverage of more remote areas and work with the hydrographic community to provide full-depth carbon observations. Marine ICOS builds on expertise and results gained during EU-funded science projects (CAVASSOO, CARBOOCEAN [www.carboocean.org] and CARBOCHANGE [www.carbochange.b.uib.no]).

20.2.3.2 Ocean Thematic Center

The Ocean Thematic Center (OTC—www.icos-otc.eu), that is at the moment located in Bergen (Norway), has the task to coordinate the ocean station network of ICOS. The OTC will perform the ICOS labeling process of stations, offer technical support for the observations, perform QC, make sure that the data stream from the PI's or station owner to the Carbon Portal (CP) is running smooth, make sure of a version traceable, and secure archiving of the marine data, regular quality assessment of the network, processing and integration of data from ships and stations into products such as maps and regional budgets of CO_2 fluxes and transports, assessment of changes in pH ("acidification"), and interpretation and comparison with marine models and processes. In addition, the OTC will support new technological developments that can be deployed at the ICOS oceanic network. The United Kingdom might formally be a part of the ICOS OTC during the spring 2017 and Spain has also claimed their interest to cohost the OTC.

20.2.4 Central Analytical Laboratories

The mission of ICOS is to run a long-term monitoring network that produces compatible sets of high-precision observational data. These data should be of high density and high quality to allow for regular assessment of regional carbon fluxes from atmospheric observations using inversion models that aim at mapping the regional distribution of GHG fluxes with a grid size as low as 10 km. Thus, ICOS data might gauge the success of mitigation strategies in reducing GHG emissions, not just with inventories, but with measured data. Over the past few years, many research activities have aimed at high-resolution transport modeling. Most likely, further significant progress in this field will be made in the future. With decreasing model errors, the accuracy of observational data will emerge as the limiting factor, and even today calibration offsets between stations can possibly cause a bias in flux estimates. Information on changing emissions and associated carbon cycle responses will be reflected in long-term trends and changing continental gradients. This makes it particularly important to have highly consistent experimental records available over decades. A GHG monitoring system that aims at revealing carbon flux changes requires solid data, which are immune against any such uncertainties. The need for long-term data consistency is particularly challenging. The demand for highest accuracy and data continuity can best be met with a Central Analytical Laboratories (CAL—www.icos-cal.eu) as experience with similar institutions has proven elsewhere (e.g., in the United States, Canada, and Australia). A Flask and Calibration Laboratory (FCL) that conducts central flask analyses and provides calibration gases to the entire network of continuously monitoring stations will eliminate calibration biases, which will improve the emerging data sets considerably. A second part of the CAL, the Central Radiocarbon Laboratory (CRL), provides the necessary high-precision input data to separate the fossil component in the regional CO_2 offset measured at the European ICOS network. The CAL, located at the Max Planck Institute for Biogeochemistry (MPI-BGC) in Jena (FCL) and at Heidelberg University (UHEI-IUP) (CRL), serves the ICOS monitoring network by providing the following functions on the long term:

1. The FCL provides reference gases for calibration of in situ measurements performed at the continuous monitoring stations. This task comprises the preparation of reference standards and their calibration.
2. The FCL also ensures high-quality flask data. This function requires high-performance analytical procedures, a thorough QC strategy, and the routine test of flask material.
3. The CRL operates the ^{14}C monitoring network. Major responsibilities are the provision of the specific sampling equipment to the monitoring stations and the routine measurement of weekly integrated as well as regular diurnal cycle $^{14}CO_2$ air samples.
4. The CAL as a whole has an established data transmission routine to the ATC and other thematic centers.

20.2.5 Carbon Portal

The ICOS CP (www.icos-cp.eu), located at the University of Lund in Sweden, is envisioned as a virtual data center, that is, a place in which ICOS data can be discovered and accessed along with ancillary data and where users can post elaborated data products that are obtained from ICOS data. The CP also has the ability to address all the requirements stemming from these aspects. In order to facilitate the interaction with the users, the ICOS CP will be a user-friendly way to access ICOS data for users and the preferred access method across thematic centers. Furthermore, the implementation of a data set identifier system for ICOS data, such as a digital object identifier (DOI), important for tracking and referencing data resources (e.g., http://datacite.org/). Attaching a DOI to a data set is achieved by freezing the database at regular intervals (at least annually) and feeding data streams to the ICSU World Data System. The CP acts as an interface with the scientific community improving such products and generating novel approaches. It is vital for ICOS, as a research infrastructure, that as many research groups as possible in Europe and elsewhere use the data to generate Level 3 data products beyond operational GMES products and generate research publications as an output. In that sense, modelers are the main users of the ICOS Level 1 and Level 2 data, irrespective of the nature of the model used, going from simple conceptual to complex data assimilation models.

The ICOS CP organizes access to different Level 3 products derived from ICOS observational data and provides analytical tools for their detailed comparison. Therefore, it is important that users may post their elaborated data products to the ICOS CP and that these Level 3 products are made available in a user-friendly interface to the public. By its nature, the CP contributes directly and actively to the dissemination of ICOS results. ICOS will provide easily accessible and understandable science and education products through the CP. The CP produces data, graphics, and data interpretation tools needed for major ICOS outreach activities. It prepares and organizes outreach actions focused on the various ICOS data products at the European level.

20.2.6 Head Office

The Head Office (HO—www.icos-ri.eu), located in Helsinki (Finland), supports the ICOS RI governance in planning, integrating, developing, and coordinating the ICOS activities. The HO assists the planning and promoting of scientific and technical objectives. The HO has the responsibility to support the coordination of the infrastructure at the European level, its construction, commissioning, and transition to full operational mode. The HO coordinates and facilitates the network extension and supports ICOS RI's science and technological development. As the main guardian of ICOS legal entity, the HO strives toward an ICOS general identity, securing that ICOS as a distributed RI operates as a strong productive actor both at the European and global levels.

The HO is directed by the ICOS ERIC Director General (DG). The DG has the responsibility that General Assembly (GA) meetings are properly supported with preparation and follow-up actions.

20.3 Challenges, Solutions, and Lessons Learned

Building a large infrastructure, sparse in different countries and covering different domains such as ecosystems, atmosphere, and oceans, offers a number of challenges and difficulties. These require a strong coordination activity and a large democratic process in order to find the best solutions and strategies to address them. Building of the ICOS infrastructure required more than 10 years of continuous work, with hundreds of meetings at different levels, from the national coordination to European level scientific discussions and a long administrative and legal process.

20.3.1 Cross-Domain Dimension

The ICOS infrastructure provides high-quality measurements of GHG fluxes and concentrations and higher-level products that will be available for stakeholders and scientific users such as the modeling community in order to better understand, characterize, and monitor the global GHG budgets. In order to provide a comprehensive picture and the data needed to represent all the different components of GHG sources and sinks, the terrestrial, marine, and atmospheric domains need to be part of the infrastructure.

Although linked by a common development strategy and common targets (the GHG measurements), the ecosystem, atmospheric, and oceanic components of ICOS are often different in terms of methodology and techniques used, in their current development stage, background scientific community participating, and distribution of sites. This introduced a challenge in the integration of the different components in the same framework, essential to have a full development of ICOS and to ensure that all measurements needed for the ICOS users are provided with the same high quality level.

The integration of the three components has been obtained, thanks to different actions and organizational decisions:

1. The CP is a common data portal where all the ICOS measurements are gathered and distributed. This forces to find common data organization, definition, and quality and facilitate the user data access and interpretation.
2. Having CFs between the CP (and the Head Office) and the three component networks helped to organize discussion and strategic decisions. Also, the establishment of a RI Committee where representatives of the CFs and of the networks of sites are present is a good scientific management forum where common strategies and decisions are discussed and agreed under the HO.
3. An intense and continuous harmonization work among the components started already in the preparatory phase. The definition of quality standards, the agreement on a common strategy in the selection of the methods and sensors, the discussion, and then definition of common characteristics of the data products are examples of important cross-component synergies that helped to better integrate the three domains.

There are, however, still many aspects that need further cross-component activities and discussion to achieve full integration among components, in particular in relation to the data curation process, the selection of common variables, and the colocation of sites.

20.3.2 Network Design

The ICOS network has been established after more than 20 years of scientific activities and development, often with part of the funding coming from European research projects. The scientific activities also helped to develop a network of measurement sites that has been the basis of the ICOS network. For sure this helped to start from a strong background (in terms of knowledge, community, and infrastructures) but it limited the possibilities to build up a network following a "top-down" design increasing its representativeness or optimizing it in terms of site numbers and locations. Other networks (e.g., the American National Ecological Observation Network, United States) had the possibility to design the infrastructure following a predefined scheme, mainly because it started from scratch and with a centralized management and decision center.

Did the fact that a network design has not been applied affect the quality of the ICOS RI? This question is difficult to answer and probably the answers are different for the different components. A proper network design requires a clear definition of objective(s) and a clear selection of the method(s) that will use the measurements. Without this, it is impossible to design the best network because different objectives, different users or communities, and different methods and approaches require different solutions. In this context, the ICOS network, designed on the basis of multiple and different research questions, will probably not be optimal for a specific question/method, but an acceptable compromise to serve a broad plethora of users.

20.3.3 Agreement on Standards and Methods

One of the basic and fundamental characteristics of a research infrastructure is the definition of common and high-quality standards, methods, and protocols. Also in this activity, ICOS decided to follow a bottom-up approach where protocols have been discussed and defined by the community. In the ecosystem component, this has been done through the organization of working groups where ICOS participants but also external experts (also from other continents) discussed and prepared protocols defining all the details related to the data acquisition and processing.

The challenge in this case has been to find a common agreement among the experts contributing to the protocol preparation (about 100 participants per protocol; see http://www.icos-etc.eu/icos/working-groups) that often started from different backgrounds, experiences, and preferences. This clearly had an impact on the overall development of the documents and specifications because of the intense and long discussion but at the same time ensured a high quality level. Protocols developed by the community have the advantage to be at the same time accepted by the people that have to apply them and accepted by potential users of the data.

20.3.4 Legal and Financial Structure

ICOS established an ERIC in December 2015 with nine European countries participating. The process has been challenging, with a large number of difficulties, but this is out of the scope of this chapter. The complexity of the European Research Infrastructures, even after the ERIC establishment, is in the diffuse financial and legal responsibilities, where each Member State participating to the ERIC is also responsible for the support of the national part of the RI.

In addition to the risk that a key ERIC partner could step out (risk always present and in all kinds of infrastructures), there is an additional complexity due to different priorities, the timing and level of investments, across the participating countries. This creates heterogeneity in the readiness and actual measurement start, not helping a homogeneous network development. The solution to this problem can only come from the ICOS management and coordination bodies, primarily from the GA where stakeholders from all signing countries are represented.

References

IPCC, 2013, *Climate Change 2013: Fifth Assessment Report of the Intergovernmental Panel on Climate Change*, eds. Stocker, T.F., D. Qin, G.K. Plattner, M. Tignor, S.K. Allen, J. Bosc hung, A. Nauels, Y. Xia, V. Bex, and P.M. Midgley. Cambridge University Press, Cambridge, U.K., 1535pp.

Jackson, R.B., Canadell, J.G., Le Quere, C., Andrew, R., Korsbakken, J.I., Peters, G.P., Nakicenovic, N., 2016, Reaching peak emissions, *Nature Climate Change*, 6, 7–10.

Kirschke, S., Bousquet, P., Ciais, P., Saunois, M., Canadell, J.G., Dlugokencky, E.J., Bergamaschi, P. et al., 2013, Three decades of global methane sources and sinks, *Nature Geoscience*, 6, 813–823.

Le Quere, C., Raupach, M., Canadell, J., Marland, G., Bopp, L., Ciais, P., Conway, T. et al., 2009, Trends in the sources and sinks of carbon dioxide, *Nature Geoscience*, 2, 831–836.

Index

C

D

F

G

O

P

Q

R

S

T